Illustrated Guide to the National Electrical Code®

Charles R. Miller

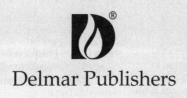

Delmar Publishers

an International Thomson Publishing company I(T)P®

Albany • Bonn • Boston • Cincinnati • Detroit • London • Madrid
Melbourne • Mexico City • New York • Pacific Grove • Paris • San Francisco
Singapore • Tokyo • Toronto • Washington

NOTICE TO THE READER

Cover design: Courtesy of Brucie Rosch

Delmar Staff:
Publisher: Alar Elken
Acquisitions Editor: Mark Huth
Developmental Editor: Jeanne Mesick
Project Editor: Megeen Mulholland
Production Coordinator: Toni Hansen
Art and Design Coordinator: Cheri Plasse
Editorial Assistant: Dawn Daugherty
Production Manager: Mary Ellen Black
Marketing Manager: Mona Caron

Copyright © 1999
By Delmar Publishers

an International Thomson Publishing Company I(T)P®

The ITP logo is a trademark under license
Printed in Canada

For more information contact:

Online Services

Delmar Online
To access a wide variety of Delmar products
and services on the World Wide Web, point your
browser to:
    http://www.delmar.com
    or email: info@delmar.com

A service of I(T)

Delmar Publishers
3 Columbia Circle, Box 15015
Albany, New York 12212-5015

International Thomson Publishing Europe
Berkshire House
168-173 High Holborn
London, WC1V7AA
United Kingdom

Nelson ITP, Australia
102 Dodds Street
South Melbourne,
Victoria, 3205 Australia

Nelson Canada
1120 Birchmont Road
Scarborough, Ontario
M1K 5G4, Canada

International Thomson Publishing France
Tour Maine-Montparnasse
33 Avenue du Maine
75755 Paris Cedex 15, France

International Thomson Editores
Seneca 53
Colonia Polanco
11560 Mexico D. F. Mexico

International Thomson Publishing GmbH
Königswinterer Straße 418
53227 Bonn
Germany

International Thomson Publishing Asia
60 Albert Street
#15-01 Albert Complex
Singapore 189969

International Thomson Publishing Japan
Hirakawa-cho Kyowa Building, 3F
2-2-1 Hirakawa-cho, Chiyoda-ku,
Tokyo 102, Japan

ITE Spain/Paraninfo
Calle Magallanes, 25
28015-Madrid, Espana

1 2 3 4 5 6 7 8 9 10 XXX 03 02 01 00 99

Library of Congress Cataloging-in-Publication Data
Miller, Charles R., 1955–
    Illustrated guide to the National Electrical Code / by Charles R. Miller.
      p. cm.
    Includes index.
    ISBN 0-7668-0529-8 (pbk. : alk. paper)
    1. Electric engineering—Insurance requirements—United States.
  2. Electric wiring—Insurance requirements—United States.
  3. National Fire Protection Association. National Electrical Code.
  I. National Fire Protection Association. National Electrical Code.
  II. Title.
TK260.M52   1999
621.319'24'021873—dc21
                             98-28423
                               CIP

Contents

Preface

Illustrated Guide to the NEC® offers an exciting new approach to understanding and applying the provisions of the *National Electrical Code®*. Unlike the *Code*, this text gathers and presents detailed information in a format, such as One-Family or Multi-Family Wellings, based upon "type of occupancy." Code specifications applicable to a given type of occupancy are logically organized in easy-to-read units and graphically enhanced by numerous technical illustrations. Going an extra step, the occupancy-specific material is sub-divided into specific rooms and areas. Information relevant to more than one type of occupancy is organized into independent units for easier reference. For instance, items such as raceways and conductors are covered in Unit 5 but are related to every type of occupancy.

Students who wish to acquire a comprehensive grasp of all electrical codes will want to study this text section by section and unit by unit. Practicing electricians who have specialized in one type of occupancy and who wish to understand an unfamiliar segment may want to focus on those new areas.

Example: An electrician who has been wiring commercial facilities for a number of years wants to wire a new house. Being unfamiliar with the codes concerning residential wiring, this individual can turn to Section 2, "One-Family Dwellings." Here, everything from receptacle placement to the placement of the service point is explained. Section 2 is made up of four units: Units 6 through 9. Unit 6, "General Provisions," contains general requirements for one-family dwellings, both interior and exterior. Unit 7, "Specific Provisions," addresses more complex issues, requiring additional provisions for specific areas such as kitchens, hallways, clothes closets,

bathrooms, garages, basements, etc. Unit 8, "Load Calculations," simplifies the standard as well as optional load calculation methods for one-family dwellings. Unit 9, "Services and Electrical Equipment," is divided into five sub-headings: Service-Entrance Wiring Methods, Service and Outside Wiring Clearances, Working Space Around Equipment, Service Equipment and Panelboards, and Grounding.

The "what," "when," "where" adoption of the provisions of the *NEC®* is under the discretionary control of state and local jurisdictions. State and local jurisdictions also have the liberty of appending additional codes, which in many cases may be more stringent than those outlined by the *NEC®*. The *Code* may be adopted in whole or in part. For example, while some local codes do not allow the use of non-metallic-sheathed cable for residential or commercial wiring, others allow its use in residential, but not in commercial, wiring applications. To ensure compliance, obtain a copy of any additional rules and regulations for your area.

This guide's objective is to provide the information needed to complete your project—without the necessity of learning the *NEC®* from cover to cover. *Illustrated Guide to the NEC®* will bring your project to life as quickly and as accurately as any text on the market today. In the electrical field, as in any career, the learning experience never ends. Whether you are an electrician's apprentice, a master electrician, or an electrical inspector, *Illustrated Guide to the National Eelectrical Code®* has something for you. We believe you will find it to be a valuable addition to your reference library. In fact, you may want to include it in your toolbox or briefcase!

ABOUT THE AUTHOR

For eighteen years, Charles R. Miller owned and operated a successful commercial electrical contracting company (Lighthouse Electric Co., Inc.) in Nashville, Tennessee. Throughout those years, he prided himself on solving problems abandoned by less-skilled or less-dedicated technicians. In 1988 he began operating a second company, dedicated to electrical-related training and known as Lighthouse Educational Services. Mr. Miller teaches custom-tailored classes and seminars covering various aspects of the electrical industry. Hundreds of students have taken advantage of his extensive experience in electrical contracting, regulatory exams (current electrical codes), and electrical-related business and law. Class and seminar attendees have included individuals employed by companies such as Ford, Textron, The Aerostructures Corporation, and Aladdin Industries; by academic institutions such as Tennessee State and Vanderbuilt Universities; and by governmental agencies, including NASA.

Charles Miller has dedicated over 5,000 hours to making *Illustrated Guide to the National Electrical Code®* a reality. His unsurpassed attention to detail is evident on every page. Since this book's conception, every day's waking hours have been consumed with careful planning and execution of content and design. His unwavering commitment to quality, from the first page in Unit 1 to the last page in Unit 19, has produced a technically superior, quintessentially user-friendly guide.

ACKNOWLEDGMENTS

First and foremost, I would like to express my sincere gratitude to Denise Sublett. Besides reading and, where needed, editing every sentence in the text, she looked at every illustration and offered valuable suggestions for improvement. I could not have finished this book without Denise's professional skills, support, and most of all, friendship.

I would like to say "thank you" to my children, Christin and Adam, for being patient and understanding during the extremely long hours and endless days working on this text. My mother, Evelyn Miller, gets a special "thank you" and "I love you" for a lifetime of support and encouragement. She called every day to check on me and quite often sent encouraging greeting cards which always came at just the right time.

Thank you to my Senior Editor at Delmar, Mark Huth, for the privilege of writing for such a professional publishing company. I also would like to thank the entire Delmar project team comprised of all of those listed under "Delmar Staff" at the front of this book.

Last, but not least, the author and Delmar would like to thank the following reviewers for their contributions:

Greg Fletcher
Kennebec Valley Technical College

David Gehlauf
Tri County Vocational School

Ralph Hathaway
Dean Institute of Technology

Fred Johnson
Champlain Valley Technical College

Lanny McMahill
City of Phoenix

Applicable tables and section references are reprinted with permission from NFPA 70-1999, *National Electrical Code®*, copyright © 1998, National Fire Protection Association, Quincy, MA 02269. This reprinted material is not the complete and official position of the NFPA on the referenced subject, which is represented only by the standard in its entirety.

DELMAR PUBLISHERS IS YOUR ELECTRICAL BOOK SOURCE!

In addition to *Illustrated Guide to the National Electrical Code®*, Delmar Publishers offers a comprehensive selection of electrically oriented best-sellers and all new titles designed to bring you technically accurate, leading-edge information. *Electrical Wiring—Residential*, *Electrical Wiring—Commercial*, and *Electrical Wiring—Industrial* may be of particular interest to you, as they complement the materials found in *Illustrated Guide to the National Electrical Code®*.

■ National Electrical Code

National Electrical Code® 1999/NFPA
Revised every three years, the *National Electrical Code®* is the basis of all U.S. electrical codes.
> Order # 0-8776-5432-8
> Loose-leaf version in binder
> Order # 0-8776-5433-6

National Electrical Code® Handbook 1999/NFPA
This essential resource pulls together all the extra facts, figures, and explanations you need to interpret the 1999 *NEC®*. It includes the entire text of the Code, plus expert commentary, real-world examples, diagrams, and illustrations that clarify requirements.
> Order # 0-8776-5437-9

Illustrated Changes in the 1999 National Electrical Code®/O'Riley
This book provides an abundantly-illustrated and easy-to-understand analysis of the changes made to the 1999 *NEC®*.
> Order # 0-7668-0763-0

Understanding the National Electrical Code®, 3E/Holt
This book gives users at every level the ability to understand what the *NEC®* requires, and simplifies this sometimes intimidating and confusing code.
> Order # 0-7668-0350-3

Illustrated Guide to the National Electrical Code®/Miller
Highly-detailed illustrations offer insight into Code requirements, and are further enhanced through clearly-written, concise blocks of text that can be read very quickly and understood with ease. Organized by classes of occupancy.
> Order # 0-7668-0529-8

Interpreting the National Electrical Code®, 5E/Surbrook
This updated resource provides a process for understanding and applying the *National Electrical Code®* to electrical contracting, plan development, and review.
> Order # 0-7668-0187-X

Electrical Grounding, 5E/O'Riley
Electrical Grounding is a highly illustrated, systematic approach for understanding grounding principles and their application to the 1999 *NEC®*.
> Order # 0-7668-0486-0

■ Electrical Wiring

Electrical Raceways and Other Wiring Methods, 3E/Loyd
The most authoritive resource on metallic and non-metallic raceways, provides users with a concise, easy-to-understand guide to the specific design criteria and wiring methods and materials required by the 1999 *NEC®*.
> Order # 0-7668-0266-3

Electrical Wiring Residential, 13E/Mullin
Now in full color! Users can learn all aspects of residential wiring and how to apply them to the wiring of a typical house from this, the most widely-used residential wiring book in the country.
> Softcover Order # 0-8273-8607-9
> Hardcover Order # 0-8273-8610-9

House Wiring with the NEC®/Mullin
The focus of this new book is the applications of the *NEC®* to house wiring.
> Order # 0-8273-8350-9

Electrical Wiring Commercial, 10E/Mullin and Smith
Users can learn commercial wiring in accordance with the *NEC®* from this comprehensive guide to applying the newly revised 1999 *NEC®*.
> Order # 0-7668-0179-9

Electrical Wiring Industrial, 10E/Smith and Herman
This practical resource has users work their way through an entire industrial building—wiring the branch-circuits, feeders, service entrances, and many of the electrical appliances and sub-systems found in commercial buildings.
> Order # 0-7668-0193-4

Cables and Wiring, 2E/AVO
This concise, easy-to-use book is your single-source guide to electrical cables—it's a "must-have" reference for journeyman electricians, contractors, inspectors, and designers.
> Order # 0-7668-0270-1

■ Electrical Machines and Controls

***Industrial Motor Control, 4E*/Herman and Alerich**
This newly-revised and expanded book, now in full color, provides easy-to-follow instructions and essential information for controlling industrial motors. Also available are a new lab manual and an interactive CD-ROM.
        Order # 0-8273-8640-0

***Electric Motor Control, 6E*/Alerich and Herman**
Fully updated in this new sixth edition, this book has been a long-standing leader in the area of electric motor controls.
        Order # 0-8273-8456-4

***Introduction to Programmable Logic Controllers*/Dunning**
This book offers an introduction to Programmable Logic Controllers.
        Order # 0-8273-7866-1

***Technician's Guide to Programmable Controllers, 3E*/Cox**
Uses a plain, easy-to-understand approach and covers the basics of programmable controllers.
        Order # 0-8273-6238-2

***Programmable Controller Circuits*/Bertrand**
This book is a project manual designed to provide practical laboratory experience for one studying industrial controls.
        Order # 0-8273-7066-0

***Electronic Variable Speed Drives*/Brumbach**
Aimed squarely at maintenance and troubleshooting, Electronic Variable Speed Drives is the only book devoted exclusively to this topic.
        Order # 0-8273-6937-9

***Electrical Controls for Machines, 5E*/Rexford**
State-of-the-art process and machine control devices, circuits, and systems for all types of industries are explained in detail in this comprehensive resource.
        Order # 0-8273-7644-8

***Electrical Transformers and Rotating Machines*/Herman**
This new book is an excellent resource for electrical students and professionals in the electrical trade.
        Order # 0-7668-0579-4

***Delmar's Standard Guide to Transformers*/Herman**
Delmar's Standard Guide to Transformers was developed from the best-seller Standard Textbook of Electricity with expanded transformer coverage not found in any other book.
        Order # 0-8273-7209-4

■ Data and Voice Communication Cabling and Fiber Optics

***Complete Guide to Fiber Optic Cable System Installation*/Pearson**
This book offers comprehensive, unbiased, state-of-the-art information and procedures for installing fiber optic cable systems.
        Order # 0-8273-7318-X

***Fiber Optics Technician's Manual*/Hayes**
Here's an indispensable tool for all technicians and electricians who need to learn about optimal fiber optic design and installation as well as the latest troubleshooting tips and techniques.
        Order # 0-8273-7426-7

***A Guide for Telecommunications Cable Splicing*/Highhouse**
A "how-to" guide for splicing all types of telecommunications cables.
        Order # 0-8273-8066-6

***Premises Cabling*/Sterling**
This reference is ideal for electricians, electrical contractors, and inspectors needing specific information on the principles of structured wiring systems.
        Order # 0-8273-7244-2

■ Electrical Theory

***Delmar's Standard Textbook of Electricity, 2E*/Herman**
This exciting full-color book is the most comprehensive book on DC/AC circuits and machines for those learning the electrical trades.
        Order # 0-8273-8550-1

***Industrial Electricity, 6E*/Nadon, Gelmine, and Brumbach**
This revised, illustrated book offers broad coverage of the basics of electrical theory and industrial applications. It is perfect for those who wish to be industrial maintenance technicians.
        Order # 0-7668-0101-2

■ Exam Preparation

***Electrician's Exam Preparation*/Holt**
This comprehensive exam prep guide includes all of the topics on journeyman and master electrician's competency exams.
        Order # 0-7668-0376-7

REFERENCE

■ Electrical Reference Series

This series of technical reference books is written by experts and designed to provide the electrician, electrical contractor, industrial maintenance technician, and other electrical workers with a source of reference information about virtually all of the electrical topics that they encounter.

Electrician's Technical Reference—Motor Controls/
Carpenter
Electrician's Technical Reference—Motor Controls is a source of comprehensive information on understanding the controls that start, stop, and regulate the speed of motors.
   Order # 0-8273-8514-5

Electrician's Technical Reference—Motors/
Carpenter
Electrician's Technical Reference—Motors builds an understanding of the operation, theory, and applications of motors.
   Order # 0-8273-8513-7

*Electrician's Technical Reference—Theory and Calculations/*Herman
Electrician's Technical Reference—Theory and Calculations provides detailed examples of problem-solving for different kinds of DC and AC circuits.
   Order # 0-8273-7885-8

Electrician's Technical Reference—Transformers/
Herman
Electrician's Technical Reference—Transformers focuses on the theoretical and practical aspects of single-phase and 3-phase transformers and transformer connections.
   Order # 0-8273-8496-3

*Electrician's Technical Reference—Hazardous Locations/*Loyd
Electrician's Technical Reference—Hazardous Locations covers electrical wiring methods and basic electrical design considerations for hazardous locations.
   Order # 0-8273-8380-0

*Electrician's Technical Reference—Wiring Methods/*Loyd
Electrician's Technical Reference—Wiring Methods covers electrical wiring methods and basic electrical design considerations for all locations, and shows how to provide efficient, safe, and economical applications of various types of available wiring methods.
   Order # 0-8273-8379-7

*Electrician's Technical Reference—Industrial Electronics/*Herman
Electrician's Technical Reference—Industrial Electronics covers components most used in heavy industry, such as silicon control rectifiers, triacs, and more. It also includes examples of common rectifiers and phase-shifting circuits.
   Order # 0-7668-0347-3

RELATED TITLES

*Common Sense Conduit Bending and Cable Tray Techniques/*Simpson
Now geared especially for students, this manual remains the only complete treatment of the topic in the electrical field.
   Order # 0-8273-7110-1

*Practical Problems in Mathematics for Electricians, 5E/*Herman
This book details the mathematics principles needed by electricians.
   Order # 0-8273-6708-2

*Electrical Estimating/*Holt
This book provides a comprehensive look at how to estimate electrical wiring for residential and commercial buildings with extensive discussion of manual versus computer-assisted estimating.
   Order # 0-8273-8100-X

*Electrical Studies for Trades/*Herman
Based on Delmar's Standard Textbook of Electricity, this new book provides non-electrical trades students with the basic information they need to understand electrical systems.
   Order # 0-8273-7845-9

SECTION 1
Foundational Provisions

Introduction to the National Electrical Code®

UNIT 1

Objectives

After studying this unit, the student will:

- be given a brief account of electricity in its infancy.
- be able to identify the catalyst which brought about the *National Electric Code (NEC)*®.
- understand how the *NEC*® began and its purpose.
- understand how changes to the *Code* come about.
- be familiar with the terminology, presentation and format of the *NEC*®.
- know what type of information is found in the *NEC*® (its layout).
- understand the *NEC*®'s concern with equipment and material standards.

- be able to recognize trademark logos that denote listed and labeled products.
- comprehend the role of Nationally Recognized Testing Laboratories (NRTL), as well as the National Electrical Manufacturers Association (NEMA), and the expanded role of the National Fire Protection Association (NFPA).
- be familiar with this book's layout, text conventions and illustration methods.
- be advised on how to study the *Illustrated Guide to the NEC*®.
- be aware that electrical requirements, in addition to the *NEC*®, may exist, and if so, that compliance is required.

THE *NATIONAL ELECTRICAL CODE*®

Just as an extensive education is required for doctors to perform the duties of their chosen field, a working knowledge of the *National Electrical Code (NEC)*® is a necessity for anyone practicing a profession in the electrical industry. The *NEC*® provides the standards by which all electrical installations are judged. Although other requirements, such as local ordinances and manufacturer instructions, must be applied, the *NEC*® is the foundation upon which successful installations are built. It is the most widely recognized and utilized compilation of technical rules for the installation and operation of electrical systems in the world today. Because of its widespread effect on the industry, it is important to understand the history of the *NEC*®.

1

■ The Beginning

In 1882, New York City was home to the first central-station electric generating plant developed by Thomas A. Edison. The Pearl Street Station began operation at 3:00 P.M. on Monday, September 4. Fifty-nine customers had reluctantly consented to have their houses wired on the promise of three free months of electric light. They were given the option of discarding the service if it proved to be unsatisfactory. But this new way of lighting was more than satisfactory . . . it was a sensation. The number of customers tripled in only four months. And, as they say, the rest is history. The new industry swept the nation: New construction included the installation of electricity, while property owners demanded that existing structures be updated, as well. New materials and equipment were developed and manufactured, and methods for installing and connecting these items to the electrical source were devised. For more than a decade, manufacturers, architects, engineers, inventors, electricians, and others worked independently to develop their contributions to the new technology. By 1895, there were as many as five different electric installation codes in use, and no single set of codes was accepted by all. To further complicate matters, there was an unexpected hazard darkening the prospects of this new industry.

■ Purpose and History of the *NEC*®

Electrically caused fires were becoming commonplace and, by 1897, the problem was reaching epidemic proportions. A diverse group of knowledgeable, concerned individuals assembled to address this critical issue. The need for standardization was apparent. The consensus of more than 1,200 individuals produced the first set of nationally-adopted rules to govern electrical installations and operations—the *National Electrical Code*®

The *NEC*® states its purpose as ". . . the practical safeguarding of persons and property from hazards arising from the use of electricity." This objective has remained constant throughout the *NEC*®'s existence, and the principles it contains continue to grow and change with the dynamic electrical industry.

■ Code Changes

The *NEC*® is regularly revised to reflect the evolution of products, materials, and installation techniques. Since 1911, the National Fire Protection Association (NFPA) of Quincy, Massachusetts, has been responsible for the maintenance and publication of the *NEC*®. The 1999 edition, which contains hundreds of reworded, as well as new, regulations, represents the diligent work of twenty code-making panels, composed strictly of volunteers from all professions within the electrical industry.

These panels are complemented by a host of private individuals who submit proposals (more than 4,200 for the 1999 edition), or comment on proposals already submitted, for changes to the *NEC*®. Anyone who wishes to participate can contact the National Fire Protection Association, 1 Batterymarch Park, Quincy, MA 02269, and request their free booklet, "The NFPA Standards-Making System." The current edition of the *NEC*® provides a form in the back of the book for submitting code change suggestions. (See the form which follows.)

Now that the background of the *NEC*® is an open book, let's take a closer look at what's inside it, and how we can go about understanding it.

FORM FOR PROPOSALS FOR 2002 *NATIONAL ELECTRICAL CODE*®

Mail to: Secretary, Standards Council
   National Fire Protection Association
   1 Batterymarch Park, P.O. Box 9101
   Quincy, Massachusetts 02269-9101
   Fax to: 617-770-3500

FOR OFFICE USE ONLY
Log # _____
Date Rec'd _____

Notes: 1. All proposals must be received by 5:00 p.m. EST on Friday, November 5, 1999.
    Proposals received after 5:00 p.m. EST, Friday, November 5, 1999, will be returned to the submitter.
  2. Type or print legibly in black ink. Limit each proposal to a SINGLE section. Use a separate copy for each proposal.
  3. If supplementary material (photographs, diagrams, reports, etc.) is included, you may be required to submit sufficient copies for all
   members and alternates of the technical committee.

Please indicate in which format you wish to receive your ROP/ROC: ❏ electronic or ❏ paper

Date _____ Name _____ Tel. No. _____

Company _____

Street Address _____

Please Indicate Organization Represented (if any) _____

1. Section/Paragraph _____

2. Proposal Recommends: (Check one) ❏ new text ❏ revised text ❏ deleted text

3. **Proposal** (include proposed new or revised wording, or identification of wording to be deleted): (Note: Proposed text should be in legislative format: i.e., use underscore to denote wording to be inserted (inserted wording) and strike-through to denote wording to be deleted (deleted wording).

4. **Statement of Problem and Substantiation for Proposal:** (Note: State the problem that will be resolved by your recommendation; give the specific reason for your proposal including copies of tests, research papers, fire experience, etc. If more than 200 words, it may be abstracted for publication.)

5. ❏ **This Proposal is original material.** (Note: Original material is considered to be the submitter's own idea based on or as a result of his/her own experience, thought, or research and, to the best of his/her knowledge, is not copied from another source.)
❏ **This Proposal is not original material;** its source (if known) is as follows: _____

If you need further information on the standards-making process, please contact the
Standards Administration Department at 617-984-7249.
For technical assistance, please call NFPA at 617-770-3000.

I hereby grant the NFPA the non-exclusive, royalty-free rights, including non-exclusive, royalty-free rights in copyright, in this proposal and I understand that I acquire no rights in any publication of NFPA in which this proposal, in this or another similar or analogous form, is used.

_____
Signature (Required)

PLEASE USE SEPARATE FORM FOR EACH PROPOSAL

■ *NEC*® Terminology, Presentation, and Format

• •

Tables present a requirement's multiple application possibilities.

Table 210-21(b)(3). Receptacle Ratings for Various Size Circuits

Circuit Rating (Amperes)	Receptacle Rating (Amperes)
15	Not over 15
20	15 or 20
30	30
40	40 or 50
50	50

Extracted text, denoted by a superscript "x", contains text taken from other NFPA documents (identified in Appendix A).

ˣ**517-31. Emergency System.** Those functions of patient care depending on lighting or appliances that are connected to the emergency system shall be divided into two mandatory branches: the life safety branch and the critical branch, described in Sections 517-32 and 517-33.

 The branches of the emergency system shall be installed and connected to the alternate power source so that all functions specified herein for the emergency system shall be automatically restored to operation within 10 seconds after interruption of the normal source.

ˣ**517-32. Life Safety Branch.** No function other than those listed in (a) through (f) shall be connected to the life safety branch. The life safety branch of the emergency system shall supply power for the following lighting, receptacles, and equipment.

Diagrams, or figures, are used to further clarify *NEC*® applications.

Figure 551-46(c) Configurations for grounding-type receptacles and attachment plug caps used for recreational vehicle supply cords and recreational vehicle lots.

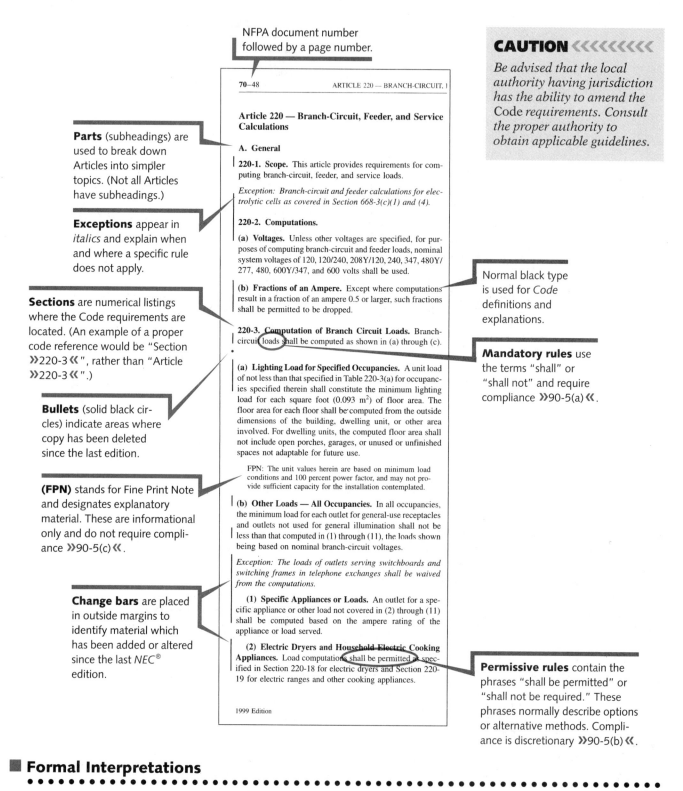

NFPA document number followed by a page number.

CAUTION «««««««

Be advised that the local authority having jurisdiction has the ability to amend the Code requirements. Consult the proper authority to obtain applicable guidelines.

70–48   ARTICLE 220 — BRANCH-CIRCUIT, 1

Article 220 — Branch-Circuit, Feeder, and Service Calculations

A. General

220-1. Scope. This article provides requirements for computing branch-circuit, feeder, and service loads.

Exception: Branch-circuit and feeder calculations for electrolytic cells as covered in Section 668-3(c)(1) and (4).

220-2. Computations.

(a) Voltages. Unless other voltages are specified, for purposes of computing branch-circuit and feeder loads, nominal system voltages of 120, 120/240, 208Y/120, 240, 347, 480Y/277, 480, 600Y/347, and 600 volts shall be used.

(b) Fractions of an Ampere. Except where computations result in a fraction of an ampere 0.5 or larger, such fractions shall be permitted to be dropped.

220-3. Computation of Branch Circuit Loads. Branch-circuit loads shall be computed as shown in (a) through (c).

(a) Lighting Load for Specified Occupancies. A unit load of not less than that specified in Table 220-3(a) for occupancies specified therein shall constitute the minimum lighting load for each square foot (0.093 m²) of floor area. The floor area for each floor shall be computed from the outside dimensions of the building, dwelling unit, or other area involved. For dwelling units, the computed floor area shall not include open porches, garages, or unused or unfinished spaces not adaptable for future use.

FPN: The unit values herein are based on minimum load conditions and 100 percent power factor, and may not provide sufficient capacity for the installation contemplated.

(b) Other Loads — All Occupancies. In all occupancies, the minimum load for each outlet for general-use receptacles and outlets not used for general illumination shall not be less than that computed in (1) through (11), the loads shown being based on nominal branch-circuit voltages.

Exception: The loads of outlets serving switchboards and switching frames in telephone exchanges shall be waived from the computations.

(1) Specific Appliances or Loads. An outlet for a specific appliance or other load not covered in (2) through (11) shall be computed based on the ampere rating of the appliance or load served.

(2) Electric Dryers and Household Electric Cooking Appliances. Load computations shall be permitted as specified in Section 220-18 for electric dryers and Section 220-19 for electric ranges and other cooking appliances.

1999 Edition

Parts (subheadings) are used to break down Articles into simpler topics. (Not all Articles have subheadings.)

Exceptions appear in *italics* and explain when and where a specific rule does not apply.

Sections are numerical listings where the Code requirements are located. (An example of a proper code reference would be "Section »220-3«", rather than "Article »220-3«".)

Bullets (solid black circles) indicate areas where copy has been deleted since the last edition.

(FPN) stands for Fine Print Note and designates explanatory material. These are informational only and do not require compliance »90-5(c)«.

Change bars are placed in outside margins to identify material which has been added or altered since the last *NEC*® edition.

Normal black type is used for *Code* definitions and explanations.

Mandatory rules use the terms "shall" or "shall not" and require compliance »90-5(a)«.

Permissive rules contain the phrases "shall be permitted" or "shall not be required." These phrases normally describe options or alternative methods. Compliance is discretionary »90-5(b)«.

■ Formal Interpretations

• •

Section 90-6 clearly states: *To promote uniformity of interpretation and application of the provisions of this* Code, *Formal Interpretation procedures have been established and are found in the NFPA Regulations Governing Committee Projects.*

■ The *NEC*® Layout

The table of contents in the *NEC*® gives a detailed breakdown of the information found in the book. Chapters 1 through 4 contain the most-often used Articles in the *Code*, because they include general, or basic, provisions. Chapter 1, while relatively brief, includes definitions essential to the proper application of the *NEC*®. It also includes an introduction and a variety of general requirements for electrical installations. More general requirements are found in Chapters 2, 3, and 4, addressing Wiring and Protection, Wiring Methods and Materials, and Equipment for General Use. Special issues are covered in Chapters 5 through 7. Chapter 5 contains information on Special Occupancies; Chapter 6, Special Equipment; Chapter 7, Special Conditions. The contents of these chapters are applied in addition to the general rules given in earlier chapters. Chapter 8 covers Communications Systems and is basically independent of other chapters, except where cross-references are given. The final chapter, Chapter 9, contains Tables and Examples. Each chapter contains one or more Articles, while each Article contains Sections. Sections may be further subdivided by the use of lettered or numbered paragraphs. The *Code* is completed by Appendices A, B, C, D, and E; an index; and a Proposal Form.

WIRING SYSTEM PRODUCT STANDARDS

In addition to installation rules, the *NEC*® is concerned with the type and quality of electrical wiring system materials. Two terms are synonymous with acceptability in this area: they are **Labeled** and **Listed**. These definitions, found in Article 100, are very similar. Similarities include: (1) an organization that is responsible for providing the listing or labeling; (2) these organizations must be acceptable to the authority having jurisdiction; (3) both are concerned with the evaluation of products; and (4) both maintain periodic inspection of the production (or manufacturing) of the equipment or materials which have been listed or labeled. A manufacturer of labeled equipment (or material) must continue to comply with the appropriate standards (or performance) under which the labeling was granted. Listed also means that the equipment, materials, and services meet identified standards or have been tested and found suitable for a specified purpose. This information is compiled and published by the organization. The Fine Print Note under **Listed** states that each organization may have different means for identifying listed equipment. In fact, some do not recognize equipment as listed unless it is also labeled. Listed or labeled equipment must be installed and used as instructed »110-3(b)«.

The organizations described below directly affect the *Code* as it relates to equipment and material acceptability, and plays a role in developing and maintaining the standards set forth in the *NEC*®.

■ Nationally Recognized Testing Laboratories (NRTLs)

Prior to 1989, there were only two organizations perceived as capable of providing safety certification of products that would be used nationwide. Because there were only two, innovative technology was slow to be tested and approved. When Congress created the Occupational Safety and Health Administration in the early 1970s, OSHA was directed to establish safety regulations for the workplace and for the monitoring of those regulations. OSHA adopted an explanation from the *National Electrical Code*® and included it in the Code of Federal Regulations. In part, it reads: "an installation or equipment is acceptable to the Assistant Secretary of Labor . . . if it is acceptable or certified, or listed, or labeled, or otherwise determined to be safe by a nationally recognized testing laboratory. . . ."

This was a start, but it did not identify the requirements for becoming a NRTL. Although OSHA introduced "Accreditation of Testing Laboratories" in 1973, the process through which a laboratory would receive accreditation was still missing. Vigorous cooperative efforts produced the OSHA regulation finalized in 1988 and called "OSHA Recognition Process for Nationally Recognized Testing Laboratories."

OSHA's NRTL program greatly benefits manufacturers by providing a system that certifies that a product meets national safety standards. Just as important, the door was opened for a greater number of laboratories to provide certification, and manufacturers are now better able to meet the demands of today's highly competitive market.

The aim of NRTLs is to ensure that electrical products properly safeguard against reasonable, identifiable risks. An extensive network of field personnel conduct unannounced inspections at manufacturing facilities which use the laboratory's "seal of approval." Some of the better-known trademarks of testing laboratories are shown below:

Some of the labels that appear on evaluated and certified electrical products, such as the ones below, will carry the trademarks of the testing laboratory, or the laboratory's standards being used for comparison.

■ MET Laboratories

MET Laboratories, working with the Department of Labor as well as other agencies, served as a working example for the accreditation process for independent testing laboratories. In fact, MET became the first United States laboratory to successfully complete the process (1989) and thus became the first NRTL licensed by OSHA.

MET field inspectors interact with local electrical inspectors throughout the country to ensure product acceptance by all federal and state regulatory officials. The MET label is accepted by all fifty states, the federal government, and major retailers.

■ Underwriters Laboratories, Inc.

Prior to the formalization of NRTLs in 1989, electrical product standards were primarily written, and certification testing performed by, Underwriters Laboratories, Inc. (UL). Standards written by UL are still widely used. The appearance of the UL logo on a label indicates that the product complies with the UL standard. It does not mean, necessarily, that UL did the product testing. Although one of many NRTLs, Underwriters Laboratories is perhaps the most widely-recognized and respected testing laboratory in operation today. Founded in 1894, UL is a not-for-profit corporation whose mission is bring safer products to the marketplace and to serve the public through rigorous product safety testing. This organization offers a wide range of services, which include, but are not limited to: product listing, classification, component recognition, field certification, field engineering, facility registration, inspection, fact-finding, and research. As one can see from this list, UL plays a major role in guiding the safety of the electrical industry.

■ Intertek Testing Services

Select laboratories of Intertek Testing Services (ITS) have passed OSHA's stringent NRTL accreditation procedures and thereby have earned the right to issue product approvals and list products using the familiar ETL Listed and CE Marks. ITS has been conducting performance and reliability tests to nearly 200 safety standards applicable to workplace-related products since 1896. Intertek's comprehensive program includes testing, listing, labeling and quarterly follow-up inspections. While recognized internationally by its many Listed Marks, the ETL Listed Mark is accepted throughout the United States, by all jurisdictions for electrical products, when denoting compliance with nationally recognized standards such as ANSI, IEC, UL, CSA and CGA.

■ National Electrical Manufacturers Association (NEMA)

Founded in 1926, this organization is comprised of companies who manufacture equipment for all facets of electrical application, from generation through utilization. Their expansive objectives include product quality maintenance and improvement, safety standards for product manufacture and usage, and a variety of product standards, such as ratings and performance. NEMA contributes to the development of the *National Electrical Safety Code* as well as the *NEC®*.

■ National Fire Protection Association (NFPA)

The NFPA, more than a century old, dedicates itself to safety standards, gathering of statistical data, research, providing crucial information on fire protection, prevention and suppression methods, and much more. Boasting an internationally diverse membership of more than 65,000, this leading nonprofit organization publishes over 270 widely recognized codes and standards, including the *NEC®*. In addition, the NFPA is involved in training and education. Its primary pursuit is to protect lives and property from the often catastrophic hazards of fire.

THIS BOOK

The *Illustrated Guide to the NEC®* is designed to teach through visualization. If a picture is truly worth a thousand words, this book should provide a more in-depth look at the *National Electrical Code®* than can be found in any other single publication. These highly detailed illustrations are complemented with concise, easy to understand written information. Not intended as a how-to book, the *Illustrated Guide to the NEC®* instead strives to translate difficult material into simpler, straightforward principles. Once the reader understands how the *Code* translates in a specific area, the same techniques can be applied throughout.

■ Its Layout

Not only is the presentation of material different in this text from others on the market, but the organization of information also offers an new approach. After covering the fundamental provisions in the balance of Unit 1, this text proceeds to address code requirements by type of occupancy. Comprehensive information is given for one-family dwellings, multi-family dwellings, commercial locations, and special occupancies. To accomplish this task, information has been gathered logically from throughout the *Code* book and concentrated in one section, under the appropriate occupancy. Each occupancy type is broken down into its finite components, and each component thoroughly discussed and illustrated. (See Table of Contents.)

■ Text Conventions

General text will be grouped in small areas surrounding an illustration. **Notes** provide additional information, considered relevant to the point being discussed. **Cautions** indicate that particular care is needed during application. **Warnings** indicate impending danger, and are intended to prevent misunderstanding of a given rule.

■ Studying This Text
• •

As the title implies, frequent references are made to the *National Electrical Code*®. Keep a copy of the latest edition close at hand. Any confusion about terminology not cleared up by the "Definitions" section of this text, may be explained by consulting the *Code's* Article 100: Definitions section. Whenever direct references are made to the *Code*, benefits will be gained by taking the time to read the suggested Article or Section. The *Illustrated Guide to the NEC*® is not intended, in any way, replace the *Code*. Each unit's "Competency Test" requires a thorough understanding of related *NEC*® subject matter. Use of this text alone is insufficient to successfully complete the test. It is, however, intended as an indispensable supplement to the *NEC*®.

ADDITIONAL ELECTRICAL REQUIREMENTS

■ Local Ordinances
• •

The importance of local (state, city, etc.) electrical codes cannot be over-emphasized. Local agencies can adopt the *NEC*® exactly as written or can amend the *Code* by incorporating more stringent regulations. While the *National Electrical Code*® represents the minimum standards for safety, some jurisdictions have additional restrictions. These restrictions may, for example, prohibit the installation of nonmetallic-sheathed cable. Obtain a copy of additional requirements (if any) for your area.

Engineers and/or architects who design electrical systems may also set requirements beyond the provisions of the *NEC*®. For example, an engineer might require the installation of 20 ampere circuits in areas where the *NEC*® allows 15 ampere circuits. Requirements from engineers and/or architects are found in additional documents, such as the following.

■ Plans and Specifications
• •

If plans and specifications are provided for a project by knowledgeable engineers and/or architects, this information must be considered, and if need be, compared to the requirements set forth by the *NEC*®. It is unlikely that the plans or specifications provided by competent professionals will conflict with or contradict the *Code*. Nonetheless, it is best to be diligent in applying the governing principals of the *NEC*®.

■ Manufacturer Instructions
• •

Some equipment or material may include instructions from the manufacturer. These instructions must be followed. For example, baseboard heaters generally include installation instructions. The *NEC*® does not prohibit the installation of receptacle outlets above baseboard heaters, but the manufacturer's instructions may prohibit the installation of their heater below receptacles.

SUMMARY

This introduction attempts to answer each question as it relates to the electrical industry in general, and to the *National Electrical Code*® in particular. It is not possible to do justice to the importance of the *Code* in a few short pages. With only a glimpse into its history and present-day supporting structure, this text moves on to the task of understanding the contents of the *Code*. The *Illustrated Guide to the NEC*® presents visually stimulating information in an occupancy organized, concise format. To begin the journey through the 1999 edition of the *National Electrical Code*®, simply turn the page, read, look, and understand.

Definitions

Objectives

• •

After studying this unit, the student will:

✔ understand the meaning of the term **accessible** (1) when applied to wiring methods, and (2) when applied to equipment.

✔ be able to identify accessible equipment which is not readily accessible.

✔ be able to accurately evaluate a location as accessible, readily accessible, or not readily accessible.

✔ be able to readily identify equipment classified as appliances.

✔ be familiar with the four categories of branch circuits and be able to list their differences.

✔ be able to distinguish the difference between enclosed and guarded.

✔ be able to determine whether a load is continuous or noncontinuous.

✔ know the difference between branch circuit conductors and feeder conductors.

✔ understand the terminology associated with grounded and grounding.

✔ know the maximum distance permitted for **within sight** situations.

✔ be able to give examples of damp, dry, and wet locations.

✔ comprehend the electrical vocabulary associated with the word **service**.

✔ be familiar with what constitutes a separately derived system.

Introduction

• •

Understanding the terminology in the *National Electrical Code* is of utmost importance. Knowing the correct definition of words and phrases as found in Article 100 is crucial to installing a hazard-free electrical system. Article 100 does not include commonly defined general or technical terms from related codes and standards. Normally, only terms used in two or more articles are defined in Article 100. Other terms are defined within the article in which they appear but may be referenced in Article 100. Part A of Article 100 contains definitions to be applied wherever the terms are used throughout the *NEC*. Part B contains definitions applicable only to the parts of articles specifically covering installations and equipment operating at over 600 volts, nominal.

What is the difference between accessible and readily accessible? Which is appropriate in a given application? When sizing a branch circuit or feeder, is the electrical load considered continuous or non-continuous? What is the difference between a damp location and a wet location?

These and many other questions can be accurately answered only through a thorough understanding of *NEC* terminology.

DEFINITIONS

■ Accessible (As Applied to Wiring Methods)

Wiring components are considered accessible when: (1) access can be gained without damaging the structure or finish of the building, or (2) they are not permanently closed in by the structure or finish of the building »Article 100 «.

Conductors in junction boxes behind lighting fixtures are considered accessible if, by removing the fixture, access to the conductors is available.

Conductors connected to switches and receptacles are accessible by removing the cover-plate and device.

Receptacles, behind furniture, are accessible because the furniture can be moved out of the way to allow access.

■ Accessible

Ready accessibility to wiring in lighting fixtures is not required. In most cases access can be gained through the use of a ladder, scaffolding, etc.

Conductors within junction boxes of recessed fixtures can be accessed by removing part of the fixture, such as, trim, lamp, internal shell, etc.

■ Accessible, Readily (Readily Accessible)

The service disconnecting means must be readily accessible. It may be located either outside or inside, near the entry point of the service conductors »230-70(a) «.

Readily accessible means capable of being reached quickly (for operation, renewal, or inspections) without having to climb over or remove obstacles, or resort to portable ladders, etc. »Article 100 «.

■ Accessible (As Applied to Equipment)

Accessible equipment is equipment not guarded by locked doors, elevators, or other effective means »Article 100 «. Equipment installed in locations requiring the use of portable means, such as a ladder, is considered accessible, but not *readily accessible*.

H.V.A.C. Equipment

Overcurrent devices do not have to be **readily accessible**, if located adjacent to the equipment, where access is achieved by the use of portable means »240-24(a)(4) «.

■ Appliance

Appliance designates utilization equipment (generally not industrial) of commonly built-in standardized types and sizes, installed as a unit to perform specific function(s), i.e., laundry, air conditioning, food preparation, etc. ≫ Article 100 ≪.

■ Not Readily Accessible

While receptacles installed in garages normally require ground-fault circuit-interrupter (GFCI) protection, receptacles that are not readily accessible do not ≫ 210-8(a)(2) *Exception No. 2* ≪. For example, a receptacle installed for a garage door opener is not readily accessible.

■ Attachment Plug (Plug Cap) (Plug)

An attachment plug (plug cap) (plug) is a device that, when inserted into a receptacle, establishes connection between the conductors of the attached flexible cord and the conductors permanently connected to the receptacle ≫ Article 100 ≪.

■ Bonding (Bonded)

Bonding is the permanent joining of metallic parts to form an electrically conductive path ensuring continuity and the capacity to conduct safely any current likely to be imposed ≫ Article 100≪.

Bonding shall be provided where necessary not only to ensure electrical continuity but also to provide ample capacity to conduct safely any likely fault current ≫ 250-90≪.

■ Bonding Jumper, Equipment

An equipment bonding jumper is the connection between two or more portions of the equipment grounding conductor ≫ Article 100≪.

■ Bonding Jumper and Bonding Jumper, Main

Meter base cabinet

Terminating point for grounded (neutral) circuit conductors

The main bonding jumper is the connection at the service between the grounded circuit conductor and the equipment grounding conductor ≫ Article 100≪.

Main service equipment

Electrical connected metal parts require a reliable conductor (referred to as a bonding jumper) to ensure electrical conductivity ≫ Article 100≪.

■ Branch Circuit, Appliance

An appliance branch circuit supplies energy to one or more outlets for the purpose of connecting appliance(s). These circuits exclude the connection of lighting fixtures unless they are part of the appliance being connected ≫ Article 100≪.

■ Branch Circuit

The circuit conductors found between a circuit's final overcurrent protective device (such as the last fuse or breaker) and the circuit's outlet(s) is called a branch circuit »Article 100 «. Branch circuits are divided into four categories: appliance, general purpose, individual, and multiwire.

■ Branch Circuit, Individual

An individual branch circuit supplies only one appliance »Article 100 «.

■ Branch Circuit, General Purpose

A general purpose branch circuit supplies a number of outlets for lighting and appliances »Article 100 «. Any number of, or combination of, receptacles, lights, and appliances can be fed by a general purpose branch circuit.

■ Branch Circuit, Multiwire

A voltmeter will not register a potential difference (voltage) when connected to the same ungrounded (hot) phase. Therefore, a multiwire circuit must not consist of conductors connected to the same phase.

A multiwire branch circuit consists of two or more ungrounded (hot) conductors having a potential difference *between* the conductors »Article 100 «.

240/120 volt, single-phase panelboard

The one grounded (neutral) conductor of a multiwire circuit must be connected to the neutral or grounded conductor of the system »Article 100 «.

All conductors of a multiwire branch circuit must originate from the same panelboard »210-4(a) «.

There must be only one grounded (neutral) conductor, and there must be an equal potential difference between the grounded conductor and each ungrounded conductor of the circuit »Article 100 «.

■ Concealed

Concealed means rendered inaccessible by the structure or finish of the building »Article 100 «.

Conductors in concealed raceways, even though they may become accessible by withdrawing them, are still considered concealed »Article 100 «.

Raceway containing branch circuit conductors in concrete slab.

■ Receptacles on Multiwire Branch Circuits

A multiwire receptacle circuit consists of one (or more) duplex receptacles, one (or more) multiple receptacles, two (or more) single receptacles, or combinations thereof.

A duplex receptacle, with two circuits on the same yoke, is also considered a multiwire circuit. Two circuits connected to one duplex receptacle are not considered separate, and dwelling units must have a means to simultaneously disconnect all ungrounded (hot) conductors ≫210-4(b)≪. This is accomplished through the use of either one double-pole breaker or two single-pole breakers with tie handles.

WARNING

In multiwire branch circuits, the continuity of a grounded conductor shall not be dependent upon the device ≫300-13(b)≪. If breaking the grounded conductor at the receptacle breaks the circuit down the line, then the grounded conductors must not be connected to the receptacle. Simply splice the grounded conductors and install a jumper wire to the receptacle.

As long as both circuits are not connected to the same receptacle, the two circuits can be thought of as separate circuits ≫210-4(a)≪. The means of disconnect can be two single-pole breakers.

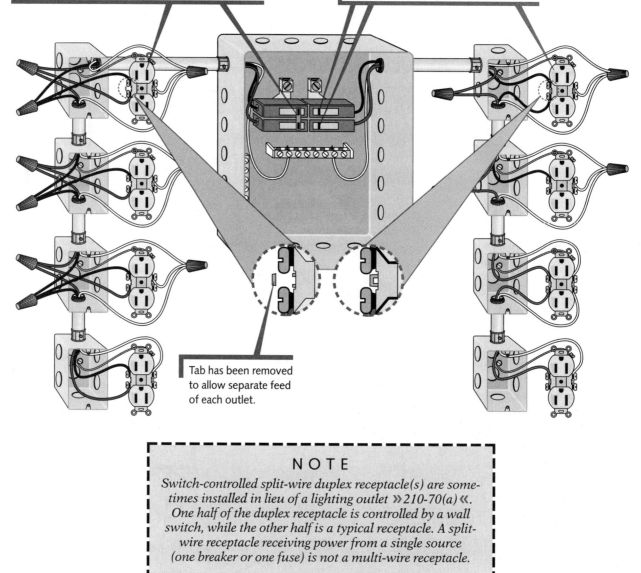

Tab has been removed to allow separate feed of each outlet.

> ## N O T E
> *Switch-controlled split-wire duplex receptacle(s) are sometimes installed in lieu of a lighting outlet ≫210-70(a)≪. One half of the duplex receptacle is controlled by a wall switch, while the other half is a typical receptacle. A split-wire receptacle receiving power from a single source (one breaker or one fuse) is not a multi-wire receptacle.*

■ Conduit Body (Condulet)

A conduit body is a separate portion of a conduit (or tubing) system providing access to the interior of the system through a removable cover(s) at a junction of multiple sections or at a terminal point of the system » Article 100 « .

FS, FD, and larger boxes (cast or sheet metal) are not classified as conduit bodies » Article 100 « .

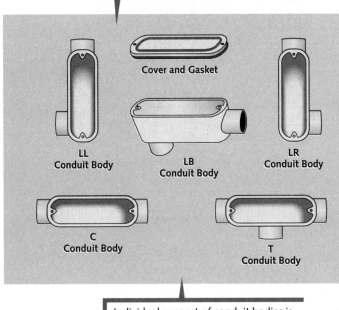

Cover and Gasket

LL Conduit Body

LB Conduit Body

LR Conduit Body

C Conduit Body

T Conduit Body

FS Box

FSS Box

A single conduit is not permitted as sole support for an FS-type or weather-proof junction box » 370-23(e) and (f) « .

Individual support of conduit bodies is not required when one of the exceptions of Section 370-23(e) or (f) is met.

■ Continuous Load

A load having the maximum level of current sustained for three hours or more is referred to as a continuous load » Article 100 « . Office lighting is an example of a continuous load.

■ Device

A unit of an electrical system, which carries but does not utilize electrical energy is known as a device » Article 100 « .

■ Enclosed

Equipment, conductors, etc., surrounded by a case, housing, fence or walls that prevent persons from accidentally contacting energized parts are referred to as enclosed ≫ Article 100 ≪.

Conductors are also considered enclosed when installed in panels, wireways, raceways, etc.

Electrical equipment installed within the perimeter of a fence, or similar area, qualifies as enclosed.

■ Enclosure

Any case, housing, apparatus, fence or walls surrounding an installation, designed to prevent physical damage to personnel or equipment, serves as an enclosure ≫ Article 100 ≪.

Enclosure for a motor starter.

START
STOP
Miller
RESET

■ Equipment

Devices

Materials

Equipment is a generic term describing material, fittings, devices, appliances, fixtures, apparatus, etc. used either directly or indirectly in an electrical installation ≫ Article 100 ≪.

Appliances

Fixtures

Apparatus

Fittings

Similar Items

▪ Feeder

● ●

A feeder consists of all circuit conductors located between the service equipment, the source of a separately derived system, or other power supply source and the final branch-circuit overcurrent device ≫ Article 100 ≪.

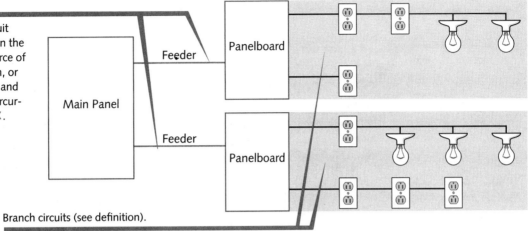

Branch circuits (see definition).

▪ Festoon Lighting

● ●

Festoon lighting is a string of outdoor lights suspended between two points ≫ Article 100 ≪.

Overhead conductors for festoon lighting shall not be smaller than No. 12 unless supported by messenger wires ≫ 225-6(b) ≪.

Messenger wire, together with strain insulators, are used to support conductors in all spans exceeding 40 feet in length. Conductors shall not be attached to any fire escape, downspout, or plumbing equipment ≫ 225-6(b) ≪.

▪ Fitting

● ●

A fitting is an accessory such as a locknut, bushing, etc. whose function is primarily mechanical, rather than electrical, in nature ≫ Article 100 ≪.

■ Grounded, Effectively Grounded, and Grounded Conductor

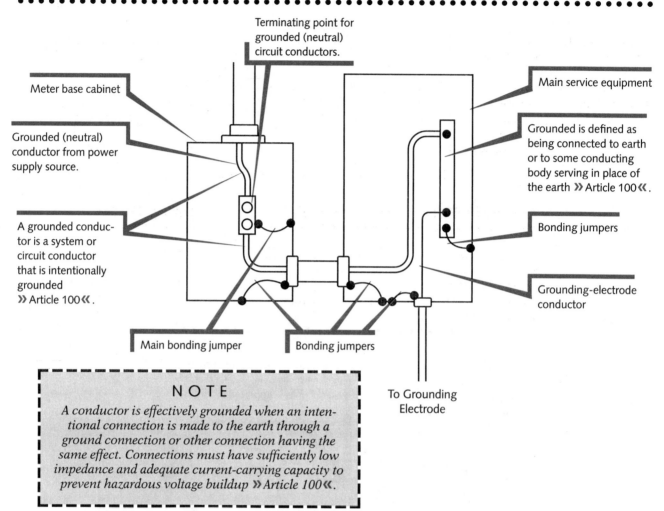

Terminating point for grounded (neutral) circuit conductors.

Meter base cabinet

Main service equipment

Grounded (neutral) conductor from power supply source.

Grounded is defined as being connected to earth or to some conducting body serving in place of the earth » Article 100 «.

A grounded conductor is a system or circuit conductor that is intentionally grounded » Article 100 «.

Bonding jumpers

Grounding-electrode conductor

Main bonding jumper

Bonding jumpers

To Grounding Electrode

> **NOTE**
>
> *A conductor is effectively grounded when an intentional connection is made to the earth through a ground connection or other connection having the same effect. Connections must have sufficiently low impedance and adequate current-carrying capacity to prevent hazardous voltage buildup » Article 100 «.*

■ Grounding Electrode Conductor and Equipment Grounding Conductor

Main service equipment

Branch Circuit for Equipment

The conductor which connects the non-current-carrying metal parts of equipment, raceways, etc. to the system grounded conductor, the grounding electrode conductor (or both) at the service equipment (or source of a separately derived system) is called an equipment grounding conductor Article 100.

Bonding jumpers

The grounding electrode conductor is the conductor connecting the grounding electrode and the equipment grounding conductor, the grounded conductor (or both) at the service equipment (or source of a separately derived system) » Article 100 «.

System grounded (neutral) conductor

Bonding jumpers

Grounding electrode

Guarded

Guarded implies that conductors are covered, shielded, fenced, enclosed, or otherwise protected by means of suitable covers, casings, barriers, rails, screens, mats, or platforms effectively removing the likelihood of approach or dangerous contact by persons or objects »Article 100 «.

When normally enclosed live parts are exposed for inspection or servicing, the working space, if in a passageway or general open space, shall be suitably guarded »110-26(b) «.

In Sight from (Within Sight from, Within Sight)

The NEC terms "in sight from," "within sight from," or "within sight," etc., applied to equipment indicates that the specified items of equipment are visible and are not more than 50 feet apart »Article 100 «.

A motor controller disconnecting means must be located in sight from the controller location as required by Section 430-102(a).

Maximum Distance 50 Feet

Combination Starter and Disconnect

Lighting Outlet

An outlet intended for the direct connection of a lampholder, a lighting fixture, or a pendant cord terminating in a lampholder is called a lighting outlet »Article 100 «.

Location, Damp

Damp locations are those subject to moderate degrees of moisture. Exterior partially protected locations under canopies, marquees, roofed (open) porches, and similar sites are considered damp locations as well as interior locations such as some basements, barns, and cold-storage warehouses »Article 100 «.

Location, Wet

Installations in any of the following categories are wet locations: underground, within concrete slabs, in masonry (directly contacting the earth), areas subject to saturation (water or other liquids), and locations unprotected from weather »Article 100 «.

■ Location, Dry

Dry locations are those not normally subject to moisture, except on a temporary basis such as a building under construction ≫ Article 100 ≪.

■ Multioutlet Assembly

A surface or flush raceway designed to hold conductors and receptacles (assembled in the field or at the factory) is called a multi-outlet assembly ≫ Article 100 ≪.

■ Outlet

A point in a wiring system from which current is taken to supply utilization equipment is known as an outlet ≫ Article 100 ≪.

■ Plenum

The space above a suspended ceiling used for environmental air-handling purposes is an example of **other space used for environmental air** as described in Section 300-22(c).

A compartment or chamber having one or more attached air ducts and forming part of the air distribution system is known as a plenum ≫ Article 100 ≪.

Receptacle

A contact device installed at an outlet for the connection of an attachment plug is a receptacle ❯❯ Article 100 ❮❮.

A single receptacle is a single contact device with no other contact device on the same yoke ❯❯ Article 100 ❮❮.

A multiple receptacle is a single device consisting of two or more receptacles ❯❯ Article 100 ❮❮.

Separately Derived System

Article 450 contains provisions for transformers.

Premises wiring deriving power to supply conductors originating in another system from systems having no direct electrical connection (such as generators, transformers, or converter windings), and systems having direct electrical connection (such as a battery, or solar photovoltaic system) is a separately derived system. Systems having no direct electrical connection include systems with a connected grounded circuit conductor ❯❯ Article 100 ❮❮.

Service

The conductors and equipment which deliver energy from the electricity supply system to the wiring system of the premises are called the service ❯❯ Article 100 ❮❮.

Service Conductors

The conductors from the service point to the service disconnecting means are known as service conductors ❯❯ Article 100 ❮❮.

■ Service Drop

The overhead conductors from the last pole (or other aerial support), including splices (if any) to the service-entrance conductors at the building (or other structure) is the service drop »Article 100 «.

■ Service Lateral

The underground service conductors between the main and the first point of connection to the service entrance conductors in a terminal box, meter base, etc. (inside or outside the building wall) is the service lateral. This includes any risers at a pole (or other structure) or from transformers »Article 100 «.

To utility company

■ Service Equipment

Service equipment is that equipment (usually circuit breaker(s), switch(es), fuse(s) and accessories) necessary to constitute the main control and cutoff which is connected to the load end of service conductors »Article 100 «.

■ Service Point

The service point is the point of connection between the facilities of the serving utility and the premises wiring »Article 100 «.

■ Special Permission

Special permission is the written consent of the authority having jurisdiction »Article 100«. In most cases, special permission grants exceptions to certain customary guidelines.

The authority having jurisdiction for enforcing the National Electrical Code may grant exception for the installation of conductors and equipment (not under the exclusive control of the electric utilities) used to connect the electric utility supply system to the service-entrance conductors of the premises served (provided such installations are outside a building or terminate immediately inside a building wall) »90-2(c)«.

Summary

- Conductors within junction boxes must be accessible without damaging the construction or finish of the building or structure.
- Certain equipment, such as the service disconnecting means, must be readily accessible.
- The term appliance denotes more than just kitchen equipment.
- Branch circuits are divided into four categories: appliance, general purpose, individual, and multiwire.
- General purpose branch circuits may feed lights and receptacles or any combination thereof.
- An individual branch circuit feeds only one piece of equipment.
- The terms bonded and grounded are not interchangeable.

- A dwelling multiwire circuit connected to one duplex receptacle must have a means to simultaneously disconnect all ungrounded (hot) conductors.
- A load having the maximum level of current sustained for three hours or more is a continuous load.
- Equipment is a general term encompassing a wide variety of items.
- A grounded conductor and a grounding conductor have different functions.
- One duplex receptacle is not defined as a single receptacle.
- Special permission is the written consent of the authority having jurisdiction.

Unit 2 Competency Test

• •

NEC® Reference   Answer

_____   _____   1. A(n) _____ branch circuit that supplies a number of outlets for lighting and appliances.

_____   _____   2. An electric circuit that controls another electric circuit through a relay is referred to as a(n) _____.

_____   _____   3. A(n) _____ branch circuit consists of two or more ungrounded conductors having a potential difference between them, and a grounded conductor having equal potential difference between it and each ungrounded conductor of the circuit and that is connected to the neutral or grounded conductor of the system.

_____   _____   4. An intermittent operation in which load conditions are regularly recurrent is the definition of _____.

_____   _____   5. The _____ is the connection between the grounded circuit conductor and the equipment grounding conductor at the service.

_____   _____   6. A(n) _____ conductor is used to connect equipment or the grounded circuit of a wiring system to a grounding electrode or electrodes.

_____   _____   7. Rainproof, raintight, or watertight equipment can fulfill the requirements for _____ where varying weather conditions other than wetness, such as snow, ice, dust, or temperature extremes, are not a factor.

_____   _____   8. The purpose of the *National Electrical Code*® is the practical safeguarding of _____ from hazards arising from the use of electricity.

_____   _____   9. A(n) _____ is a manually-operated device used in conjunction with a transfer switch to provide a means of directly connecting load conductors to a power source, and of disconnecting the transfer switch.

_____   _____   10. A type of surface or flush raceway, designed to hold conductors and receptacles, assembled in the field or at the factory is called a(n) _____.

_____   _____   11. A raceway encased in 4 inches of concrete on the fourth floor shall be considered a(n) _____ location.

_____   _____   12. A(n) _____ may consist of one or more sensing elements integral with the motor-compressor and an external control device.

_____   _____   13. An enclosure designed either for surface or flush mounting and provided with a frame, mat, or trim in which a swinging door or doors are or can be hung is called a(n) _____.

_____   _____   14. A(n) _____ is a shaftway, hatchway, well hole, or other vertical opening of space in which an elevator or dumbwaiter is designed to operate.

_____   _____   15. A multiwire branch circuit can supply:

       I. 120/240 volts to only one utilization equipment

      II. 120/240 volts where all ungrounded conductors are opened simultaneously

    a) I only   b) II only   c) either I or II   d) neither I nor II

_____   _____   16. An overcurrent protective device with a circuit opening fusible part that is heated and severed by the passage of overcurrent through it is the definition of a(n)_____.

_____   _____   17. _____ is a string of outdoor lights suspended between two points.

NEC® Reference   **Answer**

_____   _____   18. Name two items that must be present when defining an area as a bathroom.

_____   _____   19. A continuous load is where the _____ current is expected to continue for three hours or more.

   a) 80%   b) average   c) maximum   d) 125%

_____   _____   20. A device is _____ if it is intentionally connected to earth through a ground connection and has sufficient current-carrying capacity to prevent the buildup of voltages.

_____   _____   21. A(n) _____ is a point on the wiring system at which current is taken to supply utilization equipment.

_____   _____   22. Circuit conductors between the final overcurrent device protecting the circuit and the outlet(s) is called a(n) _____.

_____   _____   23. _____ enclosures are constructed or protected so that exposure to a beating rain will not result in the entrance of water under specified test conditions.

_____   _____   24. Continuous duty is an operation at a substantially constant load for _____.

   a) 1 hour or more   b) 1½ hours or more
   c) 3 hours or more   d) an indefinitely long time

_____   _____   25. Ampacity is defined as the current in amperes that a conductor can carry continuously under the conditions of use without exceeding _____.

_____   _____   26. When a disconnecting means must be located within sight from a motor controller, the disconnect must be visible and not more than _____ feet from the controller.

_____   _____   27. A building containing three dwelling units is called a(n) _____.

UNIT 3

Boxes and Enclosures

Objectives

After studying this unit, the student will

- ✔ accurately determine the cubic-inch capacity of boxes (metal and nonmetallic) when installing No. 6 and smaller conductors.
- ✔ know which items are counted when calculating box fill and which are not.
- ✔ be aware that two identical switches, mounted side by side in a two-gang device box, could each have different cubic-inch volume allowances.
- ✔ be able to determine the minimum box size (including plaster ring, extension ring, etc.) that is needed, if the number of conductors (No 6 and smaller) is known.

- ✔ be familiar with box requirements when using nonmetallic-sheathed cable.
- ✔ know the minimum conductor length to be left inside boxes.
- ✔ understand that boxes and conduit bodies must remain accessible after installation.
- ✔ be familiar with mounting and supporting provisions for boxes and conduit bodies.
- ✔ be able to determine the type of box needed for various applications.
- ✔ understand calculation procedures for junction boxes containing No. 4 and larger conductors.

Introduction

Choosing the right type and size of box (or enclosure) is very important to installing a system essentially free from hazard. Article 370 covers a variety of provisions concerning boxes (outlet, device, pull, and junction), conduit bodies, and fittings. Box selection must be based on requirements for a given location (such as dry, damp, wet, or hazardous). Boxes have particular requirements concerning the maximum number of conductors. Boxes containing No. 6 and smaller conductors are required to have a minimum cubic-inch capacity which is determined by the size and number of conductors. Boxes containing No. 4 and larger conductors are required to have a minimum height, width, and depth which is determined by the size and number of raceway entries. Article 370 contains provisions for installing, as well as supporting, boxes and conduit bodies. Boxes must be rigidly and securely fastened in place whether mounted *on* the surface, mounted to a framing member, or mounted *in* a finished surface. Under certain conditions, the only means of support a box needs is two threaded conduits. Access must be available to conductors and devices located within boxes and conduit bodies. Article 370 covers more information than the length of this book allows, such as, manholes and boxes (pull and junction) for systems over 600 volts, nominal.

BOX FILL CALCULATIONS

■ Metal Boxes

Cubic-inch capacities do not have to be marked on metal boxes listed in Table 370-16(a) 370-16(a)(2).

A 3-inch by 2-inch device box, 2½ inches deep, has a volume of 12.5 cubic inches »Table 370-16(a)«.

Metal box size: 4"h x 4"w x 1-1/2"d

Metal box size: 3"h x 2"w x 2-1/2"d

A 4-inch square box (commonly referred to as a 1900 box, or a 4-S box) with a depth of 1½ inches, has a volume of 21 cubic inches »Table 370-16(a)«.

The maximum number of conductors (for sizes No. 18 through No. 6) permitted in various standard size metal boxes are listed in Table 370-16(a). Minimum cubic-inch capacities are also shown.

■ Additional Cubic-Inch Capacity

Additional capacity can only be calculated when the device is clearly marked with a cubic-inch capacity or when it corresponds in size to a box listed in Table 370-16(a).

Cubic-inch capacity can be increased by using plaster (mud) rings, domed covers, extension rings, or similar devices »370-16(a)«.

7.5 CU. IN.

Box 4" x 4" x 1-1/2"

Box 4" x 4" x 1-1/2" Extension Ring 4" x 4" x 1-1/2"

The combined volume of this box and plaster (mud) ring is 28.5 (21 + 7.5) cubic inches »Table 370-16(a)«.

The combined volume of the box and extension ring is 42 (21 + 21) cubic inches »Table 370-16(a)«.

A 4-inch square extension ring, having a depth of 1½ inches, has a capacity of 21 cubic inches because it has the same dimensions as a box listed in Table 370-16(a).

■ Nonmetallic Boxes

Cubic-inch capacity shall be marked on all boxes with a volume of 100 cubic inches or less, except for nonmetallic boxes and boxes listed in Table 370-16(a) »370-16(a)(2)«.

18 CU. IN.

■ Equipment Grounding Conductor Fill

Equipment grounding conductor(s) or equipment bonding jumpers entering a box count as one conductor »370-16(b)(5)«. Only the largest equipment grounding conductor shall be counted if multiple grounding conductors (of different sizes) enter the box.

Pigtails are not counted.

NOTE
One or more additional equipment grounding conductors (isolated equipment grounding conductors) as permitted in Section 250-146(d) shall be counted as one conductor »370-16(b)(5)«. When multiple isolated equipment grounding conductors of different sizes enter the box, only the largest is counted.

■ Determining the Number of Conductors

Raceway fittings (connectors, hubs, etc.) are not counted.

Metal box 4" x 4" x 1-1/2"

A conductor which passes through the box without splice or termination (unbroken) counts as one conductor ≫ 370-16(b)(1) ≪. This black unbroken conductor is counted as one conductor.

Each conductor originating outside the box which is terminated or spliced inside the box counts as one conductor [370-16(b)(1)]. These red conductors are counted as two conductors.

Each conductor originating outside the box which is terminated or spliced inside the box counts as one conductor ≫ 370-16(b)(1) ≪. These white conductors are counted as two conductors.

Wirenuts are not counted.

This box, as pictured, contains five conductors. If all of the conductors are No. 12, the box could hold a maximum of nine ≫ Table 370-16(a) ≪. Four more conductors could be added.

A conductor that does not leave the box, such as equipment bonding jumpers and pigtails, are not counted ≫ 370-16(b)(1) ≪.

■ Cable Clamps and Connectors

External cable connectors are not counted.

Cable connector(s) with the clamping mechanism outside the box are not counted ≫ 370-16(b)(2) ≪.

Two internal cable clamps count as one conductor.

Internal cable clamp(s), whether factory or field supplied, shall be counted as one conductor ≫ 370-16(b)(2) ≪. Where more than one size conductor is present in the box, the clamp shall be counted as the largest conductor.

■ Fixture Studs and Hickeys

Fixture stud(s) shall be counted as one conductor ≫ 370-16(b)(3) ≪.

One or more hickeys shall be counted as one conductor ≫ 370-16(b)(3) ≪.

If conductors of different sizes are present in the box, the fixture stud (or hickey) shall be counted as the largest conductor ≫ 370-16(b)(3) ≪.

Devices or Equipment

No. 12

No. 14

This duplex receptacle counts as two No. 12 conductors.

This single-pole switch counts as two No. 14 conductors.

Each mounting yoke or strap counts as two conductors. A mounting yoke or strap can contain one or more devices, such as a single receptacle, a duplex receptacle, a single switch, a double switch, a triple switch, or any combination. The size of the two conductors (when calculating box fill) shall be equal in size to the largest conductor connected to the device ≫ 370-16(b)(4) ≪.

Cubic-Inch Volume Per Conductor

No. 18 = 1.50 cu. in.

No. 16 = 1.75 cu. in.

No. 14 = 2.00 cu. in.

No. 12 = 2.25 cu. in.

No. 10 = 2.5 cu. in.

No. 8 = 3.0 cu. in.

No. 6 = 5.0 cu. in.

> **NOTE**
> *Conductor insulation is not a factor when determining box fill calculations.*

Table 370-16(b) lists the cubic-inch volume for conductors, sizes No. 18 through No. 6.

Boxes, enclosures, and conduit bodies containing conductors, size No. 4 or larger, must also comply with Section 370-28 provisions ≫ 370-16 ≪.

Box Fill Calculation

These three conductors, although spliced, still count as *three* conductors.

All Conductors are No. 12

An unbroken conductor counts as *one* conductor.

Metal Box 4" x 4" x 1-1/2"

These three equipment grounding conductors count as *one* conductor.

A single unbroken conductor is counted as *one* conductor.

Two conductors which terminate in the box still count as *two* conductors.

These two conductors terminating in the box are counted as *two* conductors.

6 CU. IN.

One receptacle counts as *two* No. 12 conductors.

Wirenuts, pigtails, locknuts, bushings, raceway connectors, grounding screws, and equipment bonding jumpers are not factors when calculating box fill.

Total No. 12 conductors	12
Volume per No. 12	2.25 cu. in. (each)
Minimum cubic-inch volume for conductors	12 × 2.25 = 27
Volume for box (including raised cover)	21 + 6 = 27 cu. in.

This installation complies with Section 370-16 provisions.

■ Domed Covers and Canopies

Four or fewer fixture wires (smaller than No. 14) and an equipment grounding conductor can be omitted from box fill calculations where they enter a box from a domed fixture (or similar canopy) and terminate within that box ≫ 370-16(b)(1) *Exception* ≪.

The size of the domed cover or canopy is not a factor.

Two fixture wires and one equipment ground are not counted.

■ Splices Inside Conduit Bodies

Splices, taps, or devices are permitted inside conduit bodies, provided that the cubic inch capacity is durably and legibly marked on the conduit body by the manufacturer ≫ 370-16(c)(2) ≪.

A conduit body can be supported by rigid metal conduit, intermediate metal conduit, rigid nonmetallic conduit, or electrical metallic tubing ≫ 370-23(e) *Exception* ≪.

8 CU. IN.

LL conduit body

6 inches

All conductors are No. 14

At least 6 inches of free conductor (measured from the point in the conduit body where it emerges from its raceway or cable sheath) shall be left at each outlet, junction, and switch point ≫ 300-14 ≪.

The maximum number of conductors shall be calculated in accordance with Section 370-16(b). Four No. 14 conductors with a volume of 2.0 cubic inches each require a total volume of 8 cubic inches.

A raceway supporting a conduit body must be secured within 3 feet of the conduit body.

> **N O T E**
> *The maximum number of conductors permitted shall be the maximum number permitted by Table 1 of Chapter 9 for the conduit to which it is attached* ≫ 370-16(c)(1) ≪.

GENERAL INSTALLATION

■ Securing Cables to Metal Boxes

A connection must be made between the equipment grounding conductor(s) and a metal box by means of a listed grounding device or a grounding screw serving no other purpose ≫ 250-148(a) ≪. Grounding conductors *must not* be connected to enclosures with sheet metal screws ≫ 250-8 ≪.

12 inches

Nonmetallic-sheathed cable shall be secured within 12 inches of every cabinet, box, or fitting. The cable can be secured by using staples, cable ties, straps, or similar fittings which are designed and installed so that the cable remains undamaged ≫ 336-18 ≪.

Cables must be securely fastened to any metal box or conduit body that they enter ≫ 370-17-(b) ≪. The clamping mechanism can be internal or external to the box or conduit body.

Conductors must be protected from abrasion where entering boxes, conduit bodies, or fittings ≫ 370-17 ≪.

Unused cable or raceway openings in boxes and conduit bodies must be closed so that the protection provided is at least equal to that provided by the wall of the box or conduit body ≫ 370-18 ≪.

■ Single-Gang Nonmetallic Boxes

Nonmetallic-sheathed cable must be secured within 8 inches of a single-gang box unless the cable is secured to the box ≫370-17(c) *Exception*≪.

At least 6 inches of free conductor (measured from the point in the box where it emerges from its raceway or cable sheath) shall be left at each outlet, junction, and switch point ≫300-14≪.

The cable sheath shall extend into the box at least ¼ inch through the cable knockout or opening ≫370-17(c)≪.

> **N O T E**
> *All permitted wiring methods shall be secured to the box, unless the exception in Section 370-17(c) has been met.*

■ Securing Cables to Nonmetallic Boxes

Nonmetallic-sheathed cable shall be secured within 12 inches of every box ≫336-18≪.

Cables entering a nonmetallic box must be secured to the box, unless it is a single-gang nonmetallic box ≫370-17-(c)≪.

The cable sheath shall extend into the box at least ¼ inch through the cable knockout or opening ≫370-17(c)≪.

> **N O T E**
> *The exception pertaining to nonmetallic-sheathed cable entering a box which has no means of securing the cable to the box applies only to **single-gang** nonmetallic boxes ≫370-17 Exception≪.*

■ Boxes in Combustible and Noncombustible Materials

Boxes in walls or ceilings constructed of non-combustible material (concrete, tile, etc.) shall be installed so that the front edge of the box will be within ¼ inch of the finished surface ≫370-20≪.

Combustible Material — flush

Non-combustible Material — 1/4 inch maximum

Boxes in walls or ceilings constructed of combustible material, such as wood, shall be flush with or extend beyond the finished surface ≫370-20≪.

■ Gaps or Open Spaces

Damaged or incomplete plaster, dry wall, or plasterboard surfaces must be repaired so that no gap or open space greater than ⅛ inch surrounds the box or fitting ≫370-21≪.

■ Exposed Surface Extensions

Concealed boxes requiring surface extensions, must have such box (or extension ring) mounted and mechanically secured to the concealed box »370-22 «.

Equipment grounding, where required, must be in accordance with Article 250.

Extension Ring

Concealed Wall Box

■ Surface Extensions Made from Covers

The cover of a concealed box can provide a surface extension where the cover is designed so that it is unlikely to fall off, or be removed if its securing means becomes loose. The wiring method must be flexible and be arranged so that any required grounding continuity is independent of the connection between the box and the cover »370-22 *Exception* «.

■ Minimum Internal Depth

6 cu. in.

All metal boxes shall be grounded according to Article 250 »370-4 «.

All boxes must have an internal depth of at least ½ inch. Boxes intended to enclose flush devices must be at least 15⁄16 inch deep »370-24 «.

■ Covers and Canopies

To complete the installation, each box must have a cover, faceplate, or fixture canopy »370-25 «.

■ Metal Faceplates Covering Nonmetallic Boxes

Snap switches (including dimmer switches) must be effectively grounded and must also provide a means to ground faceplates, even if a metal faceplate is not installed »380-9(b) «.

Both metal and nonmetallic covers and plates shall be permitted. Metal covers or plates, when used, must comply with Section 250-110 grounding requirements »370-25(a) «.

NOTE

Always ground metal faceplates which cover nonmetallic boxes »410-56(d) «.

CAUTION <<<<<<<<<<<<<<<<<<<<

Isolated ground receptacles, in nonmetallic boxes, must be covered with either a nonmetallic faceplate, or with an effectively grounded metal faceplate »410-56(c) «.

Should the snap switch enclosure, or the wiring method used, not provide for an equipment ground, a snap switch without a grounding connection can be used for replacement purposes only. If a snap switch is installed under this exception, which is located within reach of a conducting surface (floor or otherwise), a faceplate of nonconducting, noncombustible material must be installed »380-9(b) *Exception* «.

There are two acceptable methods for grounding snap switches effectively: (1) The switch is mounted to a metal box using metal screws or to a nonmetallic box equipped with a means for grounding devices; (2) An equipment grounding conductor, or equipment bonding jumper, is connected to the equipment grounding termination on the snap switch »380-9(b) «.

■ Receptacles Mounted on Covers

Where receptacles are mounted, and supported by a cover, they must be secured by more than one screw, unless the box cover or device assembly is listed and identified as single-screw mounting »410-56(f)(3) «.

■ Floor Boxes

CAUTION ‹‹‹‹‹‹‹
Only a limited number of boxes are listed specifically for wood floor construction.

Boxes installed in floors must be listed for the specific application, such as, "floor box for wood floors" »370-27(b) «.

Boxes must have covers which are compatible with the construction of the box and suitable for the conditions of use »370-28(c) «.

N O T E
Boxes located in elevated floors (such as show windows) do not have to be listed specifically as floor boxes, if the authority having jurisdiction judges them free from likely exposure to physical damage, moisture, and dirt. Receptacles and covers shall be listed as an assembly for this type of location »370-27(b) Exception «.

■ Access to Outlet Boxes

A means for connecting an equipment grounding conductor must be provided for fixtures with exposed metal parts »410-20 «.

Fixtures and equipment are considered grounded when mechanically connected to an equipment grounding conductor as specified in Section 250-118, and sized in accordance with Section 250-122 »410-21 «.

Electric-discharge lighting fixtures (such as fluorescent) which are surface-mounted over concealed boxes (outlet, pull, or junction) shall have suitable openings in back of the fixture providing access to the box »410-14(b) «.

Branch-circuit conductors within 3 inches of a ballast (inside the ballast compartment) must have an insulation temperature rating not lower than 90° C »410-31 «.

N O T E
Supplementary overcurrent protective devices, when used within lighting fixtures, do not have to be readily accessible »240-10 «.

BOX AND FIXTURE SUPPORT

■ Box and Enclosure Supports

Nails are not acceptable fasteners unless they are used to attach brackets on the outside of the enclosure, or they extend through the interior within ¼ inch of the back or ends of the enclosure ≫370-23(b)(1)≪.

An enclosure, supported from a structural member of a building, shall be rigidly supported either directly, or by use of a brace (metal, polymeric, or wood) ≫370-23(b)≪.

Wood braces must have a cross section of at least 1 inch by 2 inches ≫370-23(b)(2)≪.

NOTE

An enclosure mounted on a building, or other surface, shall be rigidly and securely fastened in place. If the mounting surface does not provide rigid and secure support, additional support must be provided in accordance with Section 370-23 provisions ≫370-23(a)≪.

■ Mounting Enclosures in Finished Surfaces

NOTE

No support is required for unbroken lengths (without coupling) of EMT that is fished between access points, concealed in the finished surface, and where such securing of the raceway is impractical ≫348-13 Exception No. 2≪.

No support is required where the cable is fished between access points, concealed in the finished surface, and where such supporting of the cable is impractical. Armored cable support provisions are found in Section 333-7, for metal-clad cable in 334-10, and for nonmetallic-sheathed cable in 336-18.

An enclosure mounted in a finished surface must be rigidly secured to the surface by clamps, anchors, or fittings identified for the application ≫370-23(c)≪.

■ Enclosures in Suspended Ceilings

Raceways cannot be supported by ceiling grid support wires. Raceways can be secured to independent (additional) support wires that are secured at both ends ≫300-11(a)≪.

An enclosure mounted in a suspended ceiling system shall be fastened to framing members by mechanical means (bolts, screws, rivets, clips, etc.) identified for use with the enclosure(s) and ceiling framing member(s) employed. The framing members must be adequately supported and securely fastened to one another as well as to the building structure ≫370-23(d)(1)≪.

NOTE

Boxes can be secured to independent (additional) support wires that are attached at both ends ≫300-11(a)≪.

CAUTION ≪≪≪≪≪≪≪≪

See Section 300-11(a)(1) and (a)(2) for wiring located within the cavity of a fire-rated and nonfire-rated floor-ceiling (or roof-ceiling) assembly.

■ Enclosures (No Device or Fixture) Supported by Raceways

An enclosure which contains no devices, **or** which support no fixtures, can be supported by entering raceways when *all* of the following conditions are met: (1) the enclosure does not exceed 100 cubic inches in size; (2) the enclosure has threaded entries or hubs identified for the purpose; (3) the enclosure is supported by two or more conduits threaded wrenchtight into the enclosure or hubs; and (4) each conduit is secured within 36 inches of the enclosure, unless all entries are on the same side ≫ 370-23(e) ≪ .

Only rigid or intermediate metal conduit with threaded ends are permitted in Section 370-23(e).

■ Enclosures (with Devices or Fixtures) Supported by Raceways

An enclosure that contains devices, supports fixtures, or both can be supported by entering raceways when **all** of the following conditions are met: (1) the enclosure does not exceed 100 cubic inches in size; (2) the enclosure has threaded entries or have hubs identified for the purpose; (3) the enclosure is supported by two or more conduits threaded wrenchtight into the enclosure or hubs; and (4) each conduit is secured within 18 inches of the enclosure ≫ 370-23(f) ≪ .

An outlet box can support a light fixture weighing 50 pounds or less, unless the box is listed for a weight greater than the fixture ≫ 410-16(a) ≪ .

Section 370-23(f) provisions do not require that conduit length(s) be unbroken.

Rigid Coupling

Rigid Coupling

Rigid metal or intermediate metal conduit **only**

■ Enclosures with Conduit Entries on One Side

An enclosure can be supported by conduits entering on the same side when *all* of the following conditions are met: (1) the enclosure does not exceed 100 cubic inches in size; (2) the enclosure has threaded entries or hubs identified for the purpose; (3) the enclosure is supported by two, or more, conduits threaded wrenchtight into the enclosure or hubs; and (4) each conduit is secured within 18 inches of the enclosure ≫ 370-23(e) ≪ .

Rigid or intermediate metal conduit *only*

18 inches

Unused cable or raceway openings in boxes and conduit bodies must be closed so that the protection provided is at least equal to that provided by the wall of the box or conduit body ≫ 370-18 ≪ .

Boxes must be supported within 18 inches, whether or not they contain devices or support fixtures, if all the conduits enter on the same side, unless the requirements of Section 370-23(f) *Exception No. 2* are met.

> **N O T E**
>
> *Boxes, regardless of contents, cannot be supported by only one conduit. Boxes which support fixtures cannot be supported by one conduit unless the requirements of Section 370-23(f) Exception No. 2 are met.*

■ Support for Conduit Bodies (Condulets)

Conduit body support must be rigid and secure »370-16(c)(2)«.

A conduit body of any size (including one constructed with a single conduit entry) requires only one means of support. Conduit bodies, without devices or fixtures, can be supported by rigid or intermediate metal conduit, rigid nonmetallic conduit, or electrical metallic tubing (provided the conduit body is not larger than the largest trade size of the conduit or electrical tubing) »370-23(e) *Exception* «.

The conduit must be secured within 18 inches of the conduit body »370-23(e) and (f)«.

> ### N O T E
> *One rigid or intermediate metal conduit is all that is required to support a conduit body that contains devices or supports fixtures, or both. The conduit body must not be larger than the largest trade size of the conduit* »370-23(f) *Exception No. 1* «.

■ Supporting Fixtures Using Lengths of Conduit Longer than 18 Inches

Section 370-23(f) *Exception* permits unbroken length(s) of rigid or intermediate metal conduit to support a box used for fixture support, or to support a wiring enclosure within a fixture where all of the following conditions are met:

a. The length of conduit extending beyond the last point of securely fastened support does not exceed 3 feet.

b. The unbroken conduit before the last point of support is 12 inches or greater, and that portion of the conduit is securely fastened not less than 12 inches from its last point of support.

d. A fixture supported by a single conduit does not exceed 12 inches, in any direction, from the point of conduit entry.

e. The weight supported by any single conduit cannot exceed 20 pounds.

f. At the fixture end, each conduit (if more than one) is threaded wrench-tight into the box or wiring enclosure, or into hubs identified for the purpose.

c. Where accessible to unqualified persons, the fixture's lowest point is at least 8 feet above grade (or standing area) and at least 3 feet (measured horizontally to the 8 foot elevation) from windows, doors, porches, fire escapes, or similar locations.

■ Strain Relief for Flexible Cords

Multiconductor cord or cable, used to support a box, must be protected in an approved manner so that the conductors are not subjected to strain. A strain-relief connector threaded into a box with a hub would be acceptable **≫**370-23(h)(1) **≪**.

■ Fixture Hanger

Section 410-16(a) requires an independent support, such as a fixture hanger, for fixtures that weighs more than 50 pounds.

When raceway fittings are used to support light fixture(s), they must be capable of supporting the combined weight of the fixture assembly and lamp(s) **≫**410-16(f) **≪**.

Fixtures, lampholders, and receptacles must be supported securely **≫**410-16(a) **≪**.

■ Conduit Stems Supporting Fixtures

An outlet box can support a light fixture weighing no more than 50 pounds, unless the box is listed for a weight greater than the fixture. A separate means for supporting a light fixture (independent of the outlet box) is also permitted **≫**410-16(a) **≪**.

Stems longer than 18 inches shall be connected to the wiring system with flexible fittings suitable for the location. At the fixture end, the conduit(s) shall be threaded wrenchtight into the box or wiring enclosure, or into hubs identified for the purpose **≫**370-23(h)(2) **≪**.

A box supporting lampholders (or lighting fixtures) or wiring enclosures within fixtures used in lieu of boxes in compliance with Section 300-15(b), must be supported by rigid or intermediate metal conduit stems **≫**370-23(h)(2) **≪**.

18 inches

12 inches   12 inches

A fixture supported by a single conduit shall not exceed 12 inches, in any horizontal direction, from the point of conduit entry **≫**370-23(h)(2) **≪**.

CAUTION ≪≪≪≪≪≪≪≪≪≪≪≪≪≪≪≪≪≪≪≪≪≪≪≪≪≪

Any point of a fixture supported by a single conduit must be at least 8 feet above grade (or standing area) and at least 3 feet (measured horizontally to the 8 foot elevation) from windows, doors, porches, fire escapes, or similar locations, unless an effective means to prevent the threaded joint from loosening (such as a set screw) is used ≫370-23(h)(2) ≪.

■ Device Boxes Supporting Fixtures

Wall-mounted light fixtures weighing no more than 6 pounds *and* having no dimension greater than 16 inches, can be supported by boxes (such as device boxes) not specifically designed to support light fixtures, provided the fixture (or its supporting yoke) is secured to the box with at least two No. 6 or larger screws **≫**370-27(a) *Exception* **≪**.

Not over 6 pounds

Not over 16 inches in any dimension

Holes for two No. 6 (6/32) screws

■ Mounting Nonmetallic Boxes

Supporting screws for nonmetallic boxes must be mounted outside of the box, unless the box is constructed in a manner which prevents contact between the conductors in the box and the supporting screws **≫**373-43 **≪**.

■ Lighting Fixture Outlet Boxes

Boxes used at lighting fixture outlets must be designed for the purpose. Every box used exclusively for lighting, must be designed and installed so that a lighting fixture may be attached ≫ 370-27(a)≪ .

An outlet box can support a light fixture weighing no more than 50 pounds, unless the box is listed for a weight greater than the fixture ≫ 410-16(a)≪ .

Inspection of the connections between fixture conductors and circuit conductors must be possible without having to disconnect any part of the wiring (unless the fixtures are connected by attachment plugs and receptacles) ≫ 410-16(b)≪ .

N O T E
Typically, two No. 8 (8/32) screws are used to attach a light fixture (or its supporting yoke) to an outlet box.

■ Ceiling-Suspended (Paddle) Fan Boxes

An outlet box cannot be used as the sole means of support for a ceiling-suspended (paddle) fan, unless the box is listed for the application, such as, "ceiling fan outlet box" ≫ 370-27(c) and 370-27(c) *Exception*≪ .

Ceiling-suspended (paddle) fans with a combined weight including accessories of not more than 35 pounds can be supported by outlet boxes identified for such use and installed in accordance with Sections 370-23 and 370-27 ≫ 422-18(a)≪ .

N O T E
Ceiling-suspended (paddle) fans, including accessories, if any, exceeding 35 pounds, must be supported independently of the outlet box ≫ 422-18(b)≪ .

JUNCTION AND PULL BOX SIZING

■ Straight Pull—Two Raceways

Box calculations for No. 4 and larger conductors (under 600 volts) are performed based on the size and numbers of raceways ≫ 370-28≪ .

Boxes containing straight pulls are sized according to the largest raceway entering the box. The length must be at least eight times the trade diameter of the largest raceway ≫ 370-28(a)(1)≪ .

A No. 4 or larger grounding conductor (insulated or covered) can be permanently identified as an equipment grounding conductor. During installation, the grounding conductor must be identified at every accessible point ≫ 250-119(a)≪ .

The box width must be large enough to provide proper installation of the raceway (or cable), including locknuts and bushings.

The box depth shall be large enough to provide proper installation of the raceway (or cable), including locknuts and bushings.

Straight Pull

$$\frac{2}{\text{(trade diameter of largest raceway)}} \times 8 = \underline{\textbf{16}} \text{ inches}$$

Section 370-16 is used to determine box size requirements for No. 6 and smaller conductors. Calculations were based on the sizes and numbers of *conductors*. Section 370-28 is used to determine the box size requirements for No. 4 and larger conductors (under 600 volts). Calculations here are based on the sizes and numbers of *raceways*.

■ Straight Pull—Multiple Raceways

The length must be at least eight times the trade diameter of the largest single raceway. No extra space is required for additional raceways when calculating the minimum length of straight pulls. However, additional space is needed for the width (and depth) of additional raceways, including locknuts and bushings »370-28(a)(1)«.

Conduit bodies and boxes (junction, pull, and outlet) must be installed so that the wiring they contain can be made accessible without removing any part of the building »370-29«.

Straight Pull

$$\underset{\substack{\text{(trade diameter of}\\\text{largest raceway)}}}{\underline{\textbf{3}}} \times 8 = \underline{\textbf{24}} \text{ inches}$$

24 inches

CAUTION ≪≪≪≪≪≪≪≪≪

All metal boxes shall be grounded in accordance with Article 250 provisions »370-4«.

■ Angle Pull—Multiple Raceways

Where splices, or where angle or U pulls are made, the distance between each raceway entry into the box and the opposite wall must be at least six times the trade diameter of the largest raceway in a row. This distance is increased for additional raceway entries (in the same row on the same wall of the box) by the sum of the diameters of all other raceway entries »370-28(a)(2)«.

N O T E

All pull boxes, junction boxes, and conduit bodies must have covers which are compatible with the box or conduit body construction and must be suitable for the conditions of use »370-28(c)«.

Angle Pull for "Y" Dimension

$$\underset{\substack{\text{(trade diameter of}\\\text{largest raceway)}}}{\underline{\textbf{3½}}} \times 6 + \underset{\substack{\text{(all other raceways,}\\\text{in the same row,}\\\text{on the same wall)}}}{\underline{\textbf{2+2+2}}} = \underline{\textbf{27}} \text{ inches}$$

To calculate the dimension of a box with angle pulls, start with one wall where the raceways enter the box, and find the distance to the opposite wall of the box. The path of the conductors is irrelevant to this calculation.

First, pick one wall and multiply the largest raceway (trade diameter) by 6. Add to that number the trade diameter of all other raceway(s) in the same row, on the same side of the box.

27 inches — The vertical "Y" dimension

29 inches — The horizontal "X" dimension

Angle Pull for "X" Dimension

$$\underset{\substack{\text{(trade diameter of}\\\text{largest raceway)}}}{\underline{\textbf{4}}} \times 6 + \underset{\substack{\text{(all other raceways,}\\\text{in the same row,}\\\text{on the same wall)}}}{\underline{\textbf{2½+2½}}} = \underline{\textbf{29}} \text{ inches}$$

Angle Pull—Multiple Rows

Where multiple rows of raceways enter a box, calculate each row separately. Use the single row that provides the maximum distance ➤➤370-28(a)(2)◀◀.

Imaginary divider line separating each row

NOTE

In pull or junction boxes having any dimension over 6 feet, all conductors shall be cabled (or racked) in an approved manner ➤➤370-28(b)◀◀.

Angle Pull for "Y" Dimension (Back Row)

$$\underset{\substack{\text{(trade diameter of}\\\text{largest raceway)}}}{4} \times 6 + \underset{\substack{\text{(all other raceways,}\\\text{in the same row,}\\\text{on the same wall)}}}{3\frac{1}{2} + 3\frac{1}{2}} = \textbf{31} \text{ inches}$$

Angle Pull for "Y" Dimension (Front Row)

$$\underset{\substack{\text{(trade diameter of}\\\text{largest raceway)}}}{3\frac{1}{2}} \times 6 + \underset{\substack{\text{(all other raceways,}\\\text{in the same row,}\\\text{on the same wall)}}}{3 + 3 + 2\frac{1}{2} + 2\frac{1}{2}} = \textbf{32} \text{ inches}$$

one row for "Y" dimension

one row for "Y" dimension

Angle Pull for "X" Dimension (Back Row)

$$\underset{\substack{\text{(trade diameter of}\\\text{largest raceway)}}}{4} \times 6 + \underset{\substack{\text{(all other raceways,}\\\text{in the same row,}\\\text{on the same wall)}}}{4 + 4} = \textbf{32} \text{ inches}$$

Angle Pull for "X" Dimension (Front Row)

$$\underset{\substack{\text{(trade diameter of}\\\text{largest raceway)}}}{3\frac{1}{2}} \times 6 + \underset{\substack{\text{(all other raceways,}\\\text{in the same row,}\\\text{on the same wall)}}}{3 + 2 + 2 + 2} = \textbf{30} \text{ inches}$$

Perform each row calculation as if in a separate box.

one row for "X" dimension

one row for "X" dimension

Raceways Enclosing the Same Conductors

$$\underset{\substack{\text{(trade diameter of}\\\text{largest raceway)}}}{2} \times 6 + \underset{\substack{\text{(all other raceways,}\\\text{in the same row,}\\\text{on the same wall)}}}{0} = \textbf{12} \text{ inches}$$

Raceways containing the same conductors cannot be closer than six times the larger raceway, even if they are on different walls ➤➤370-28(a)(2)◀◀.

Since no other raceways enter on the same wall of the box, no additional raceway diameters are added.

Raceways Enclosing the Same Conductors

$$\underset{\substack{\text{(trade diameter of}\\\text{largest raceway)}}}{2} \times 6 = \textbf{12} \text{ inches}$$

$$\underset{\substack{\text{(trade diameter of}\\\text{largest raceway)}}}{2} \times 6 + \underset{\substack{\text{(all other raceways,}\\\text{in the same row,}\\\text{on the same wall)}}}{0} = \textbf{12} \text{ inches}$$

■ U Pull—Two Raceways

Use the angle pull method to calculate U pulls. A box with conduit entries only on one wall has a minimum distance to the opposite wall. Multiply the largest raceway by six and add the sum of the trade diameters of the other raceway(s) entering the same wall ≫370-28(a)(2) ≪.

U Pull for "Y" Dimension

$$\underset{\substack{\text{(trade diameter} \\ \text{of largest} \\ \text{raceway)}}}{2} \times 6 + \underset{\substack{\text{(all other} \\ \text{raceways in} \\ \text{the same row,} \\ \text{on the same wall)}}}{2} = \textbf{14} \text{ inches}$$

The distance between raceways enclosing the same conductor(s) must be at least 6 times the trade diameter of the largest raceway ≫370-28(a)(2) ≪.

The minimum box width must include:12 inches between raceways plus the thickness of the two raceways (including enough area to provide proper installation of locknuts and bushings).

Raceways Enclosing the Same Conductors

$$\underset{\substack{\text{(trade diameter of} \\ \text{largest raceway)}}}{2} \times 6 = \underline{\textbf{12}} \text{ inches}$$

NOTE

Ungrounded conductors (No. 4 and larger) entering a cabinet, box, enclosure, or raceway from another raceway must be fitted with a substantial, smoothly rounded insulating surface, except where the conductors are insulated from the fitting or raceway by a securely fastened-in-place material (such as a plastic bushing) ≫300-4(f) ≪.

■ Raceways Entering Opposite Removable Covers

Where angle pulls are made, the distance between each raceway entry into the box and the opposite wall of the box must be at least six times the trade diameter of the largest raceway ≫370-28(a)(2) ≪. Where no other raceways enter the same wall of the box, no additional raceway diameters are added.

Where a raceway or cable enters the wall of a box (or conduit body) opposite a removable cover, the distance from the entry wall to the cover can be determined by the distance requirements for one wire per terminal found in Table 373-6(a) ≫370-28(a)(2) *Exception* ≪.

The minimum distance between raceways, enclosing the same conductor(s), is six times the trade diameter of the largest raceway ≫370-28(a)(2) ≪.

Front View   Side View

Raceways or Cables Entering Boxes Opposite from Removable Covers	
Size AWG/ kcmil	Minimum Distance from Wall to Cover (in inches)
4–3	2
2	2½
1	3
1/0–2/0	3½
3/0–4/0	4
250	4½
300–350	5
400–500	6
600–700	8
750–900	8
1000–1250	10
1500–2000	12

■ Boxes and Conduit Bodies Not Meeting Section 370-28 and Chapter 9

WARNING

Conduit bodies do not have to be large enough to contain the same size and number of conductors permitted in the conduit which enters the conduit body. Some conduit bodies are marked with the maximum size and number of conductors.

Smaller-size conduit bodies can generally hold the same number and sizes of conductors (No. 6 and smaller) as the conduit entering the conduit body. Conduit bodies (and boxes) are manufactured which are smaller than those required by Section 370-28(a)(1) and (2). These may be used if approved and permanently marked showing the maximum number and size conductors allowed 》370-28(a)(3)《.

Three 250 kcmil conductors are the maximum number and size permitted in this conduit body.

3 - 250 kcmil max. size conductors

3"

3"

According to Chapter 9 and Table C8, four 500 kcmil THHN conductors can be installed in a 3 inch rigid metal conduit.

If a conduit body is durably and legibly marked by the manufacturer with the cubic-inch capacity, it can contain splices, taps, or devices 》370-16(c)(2)《.

160 cu. in. volume

Summary

- Limitations apply to the size and number of conductors permitted in boxes (junction, pull, etc.) and conduit bodies.
- One strap (yoke) counts as two of the largest conductors connected to the device.
- Nonmetallic boxes may list the maximum number of conductors permitted in the box, but that number must be reduced if switches, receptacles, etc., are installed in the box.
- Plaster rings, raised covers, extension rings, etc., can provide additional cubic-inch capacity to boxes.
- Conduit bodies can contain splices, taps, or devices, if the conduit body is durably and legibly marked by the manufacturer with the cubic-inch capacity.
- The minimum length of free conductor (6 inches) is measured from the point in the box where it emerges from its raceway or cable sheath.

- All metal boxes must be grounded in accordance with Article 250 provisions.
- Boxes and conduit bodies must remain accessible.
- In suspended ceilings, boxes can be secured by support wires that are installed in addition to the ceiling grid support wires.
- Where boxes enclose No. 4 and larger conductors, different calculation methods apply to straight pulls than to angle pulls.
- There is a minimum distance required between raceways enclosing the same conductors (No. 4 and larger).
- Conduit bodies may not allow the same size and number of conductors as is permitted in the conduit (or tubing) which enters the conduit body.

Unit 3 Competency Test

● ●

NEC® Reference   **Answer**

_____   _____ 1. In pull boxes or junction boxes having any dimension over _____ inches, all conductors shall be cabled or racked up in an approved manner.

 a) 6   b) 12   c) 36   d) 72

_____   _____ 2. A nonmetallic surface extension can be run in any direction from an existing outlet, but cannot be run on the floor or within ____ from the floor.

_____   _____ 3. What is the minimum dimension for "X" and "Y" in the drawing below?

_____   _____ 4. No box shall have an internal depth of less than ____, unless the box is intended to enclose flush devices.

_____   _____ 5. A 4¹¹⁄₁₆" square box that is 2⅛" deep already contains two receptacles, two internal cable clamps, four No. 12 THHN copper conductors (2 black and 2 white), two grounding conductors, three wirenuts, two pigtails, and one equipment bonding jumper. (A flat plaster ring was used to secure the receptacles to the box.) How many more No. 12 THHN copper conductors can be added to this box?

_____   _____ 6. In damp or wet locations, surface type enclosures (within the scope of Article 373) shall be so placed or equipped as to prevent moisture or water from entering and accumulating within the cabinet or cutout box, and shall be mounted so there is at least ____ inch(es) airspace between the enclosure and the wall or other supporting surface.

_____   _____ 7. The minimum thickness for a metal box measuring 6" × 4" × 3½" is ____ thick.

_____   _____ 8. Covers of outlet boxes and conduit bodies having holes through which flexible cord pendants may pass shall be provided with approved ____ or shall have smooth, well-rounded surfaces upon which the cord may bear.

_____   _____ 9. A weatherproof metal junction box measuring 4" h × 4" w × 6" d has been installed to support a light fixture. Two rigid metal conduits have been threaded wrenchtight into the threaded entries of the enclosure. The box has not yet been secured to the brick wall, but both conduits have been strapped to the wall 16 inches from the box. What additional support (from the list below) is required for this box?

 I. Two metal screws with plastic anchors

 II. Two toggle bolts

 III. Two drive pins

 a) none   b) I or II only   c) I or III only   d) I, II, or III

NEC® Reference	Answer

_____ _____ 10. Where permanent barriers are installed in a junction box, each section shall be considered as a(n) _____.

_____ _____ 11. What is the maximum number of No. 12 THW conductors permitted in a 4" octagon box that is 1½" deep? (The box contains a fixture stud and a hickey.)

_____ _____ 12. A means must be provided in each metal box for the connection of a(n) _____.

_____ _____ 13. In walls or ceiling of noncombustible material, boxes shall be installed so that the front edge of the box will not be set back of the finished surface more than _____.

_____ _____ 14. Nonmetallic boxes shall be suitable for the lowest _____ conductor entering the box.

_____ _____ 15. An electrician needs to install a surfacemounted, nonmetallic, weatherproof cabinet on the outside of a concrete block building. Since the cabinet must be mounted in a wet location, it must be mounted so there is at least _____ inch(es) airspace between the cabinet and the wall.

_____ _____ 16. How many No. 14 THW conductors are permitted in a 3" × 2" × 2½" device box?

_____ _____ 17. Nonmetallic surface extensions must be secured in place by approved means at intervals not exceeding _____.

_____ _____ 18. A 2½" rigid metal conduit enters the back of a pull box with a removable cover on the opposite side. The raceway encloses four 4/0 THHN copper conductors and one No. 4 equipment grounding conductor. The minimum depth (distance from the entry wall to the cover) for this pull box is _____.

_____ _____ 19. Where nonmetallic boxes are used with open wiring or concealed knob-and-tube wiring, the conductors must enter the box through _____.

_____ _____ 20. Light fixtures weighing no more than _____ and having no dimension greater than _____, can be supported by boxes (such as device boxes) not specifically designed to support light fixtures, provided the fixture (or its supporting yoke) is secured to the box with at least two No. 6 or larger screws.

_____ _____ 21. For straight pulls, the length of the box shall not be less than _____ the outside diameter, over sheath, of the largest shielded conductor or cable entering the box on systems over 600 volts.

Cables

Objectives

●●●

After studying this unit, the student will:

✔ know that cables must be installed at least 1¼ inches from the nearest edge of wood framing members, unless a steel plate (or bushing) has been installed.

✔ understand that nonmetallic-sheathed cable passing through metal framing must be protected by bushings (or grommets) covering all metal edges.

✔ be aware that openings around electrical penetrations through fire-resistant-rated construction must be sealed using approved methods to maintain the fire resistant rating.

✔ be able to determine what cables are permitted in spaces used for environmental air-handling purposes.

✔ know the support requirements for MC, AC, and nonmetallic-sheathed cable.

✔ be aware that a minimum bending radius must be maintained with cables.

✔ understand that cables must be protected from physical damage.

✔ be familiar with general, as well as specific, installation provisions for MC, AC, and nonmetallic-sheathed cable.

✔ understand conductor identification and the permissible re-identification of certain conductors.

✔ have a good grasp of underground installation provisions.

✔ be introduced to flat conductor, integrated gas spacer, mineral-insulated, and medium voltage cables.

Introduction

●●●

Unit 4 contains regulations for cable systems that are used often (such as metal-clad cable and nonmetallic-sheathed cable), as well as cable systems that are used rarely (such as integrated gas spacer cable and flat conductor cable. A key provision found in most of the cable articles (and some of the raceway articles) states that the cable (or raceway) must not be installed where subject to physical damage. Cable protection can be achieved by several methods, including: installing the cable a minimum distance from the outside edge of a framing member; installing a steel plate over the cable; or installing the cable in an approved raceway system. Cables installed in attics have different provisions dependant upon the type of attic entrance. Minimum support distances (from box or enclosure to the first support, or between supports) are included in this unit. Conductor identification (and permissible re-identification) for cables is also presented in this unit.

Most of the cable Articles, found in Chapter 3 of the *NEC*®, have sections titled **Uses Permitted** and **Uses Not Permitted**. These sections give locations (some specific) where the cable system can (or cannot) be installed. Since most of these provisions are clearly understandable, this book will not repeat much of this information. Also, be aware that state and local jurisdictions restrict the usage of certain cables in some (if not all) types of occupancies.

GENERAL INSTALLATION

■ Setback for Bored Holes

Holes must be bored so that the edge of the hole is at least 1¼ inch from the nearest edge of the wood member »300-4(a)(1)«.

The 1¼-inch setback applies to all locations (concealed and exposed).

The 1¼-inch setback applies to studs, joists, rafters, etc.

> **N O T E**
> *The largest hole that can be bored in a 3½-inch wood stud without using a steel plate (clip or sleeve), is one inch (provided the drill bit cuts straight through the center of the stud). The 1¼-inch setback must be maintained from both edges (front and back). 3½ inches minus 2½-inches (1¼ setback, two edges) leaves a remainder of one inch.*

■ Metal Framing Members

Nonmetallic-sheathed cable passing through holes (or slots) in metal framing must be protected by bushings (or grommets) covering all metal edges and securely fastened in the opening prior to cable installation »300-4(b)(1)«.

Section 300-4 applies to all locations (concealed and exposed).

Openings (slots or holes) can be punched (either factory or field), cut, or drilled into the metal members »300-4(b)(1)«.

> **N O T E**
> *Where nails (or screws) are likely to penetrate nonmetallic-sheathed cable (or electrical nonmetallic tubing), a steel plate (clip or sleeve) at least 1/16-inch thick shall be used to protect the cable or tubing »300-4(b)(2)«.*

■ Steel Plates

Where the bored hole is less than 1¼ inch from the edge, the cable (or raceway) shall be protected from penetration, by screws or nails, by a steel plate (or bushing) »300-4(a)(1)«.

The steel plate (or bushing) must be at least 1/16-inch thick, and of appropriate length and width to adequately cover the wiring »300-4(a)(1)«.

Cables, installed in a groove, covered by wallboard, paneling, carpeting, etc., shall be protected by 1/16-inch thick steel plate (sleeve or equivalent) or by at least 1¼ free space for the full length of the groove in which the cable is installed »300-4(e)«.

1¼" Free Space

The steel plate (with a minimum thickness of 1/16-inch) must be installed before the building finish is completed »300-4(a)(2)«.

Where there is no potential of weakening building structure, cables (or raceways) can be laid within notches in wood studs, joists, rafters, etc. »300-4(a)(2)«.

Requirements for notched wood members apply to all locations (concealed and exposed).

■ Parallel to Framing Members

Cables installed parallel to framing members (studs, joists, etc.,) shall be installed and supported so that the nearest outside surface of the cable is at least 1¼ inches from the nearest edge of the framing member where nails (or screws) are likely to penetrate »300-4(d)«.

If a 1¼-setback is not possible, the cable must be protected from penetration by nails (or screws) by a minimum ⅟₁₆-inch thick steel plate (sleeve, etc.) »300-4(d)«.

> **NOTE**
> *For concealed work in finished buildings or finished panels in prefabricated buildings, where such support is impracticable, cables can be fished between access points »300-4(d) Exception No. 2«.*

■ Maintaining the Integrity of Fire-Resistant-Rated Construction

CAUTION «««««««««««««««««
Some state and local jurisdictions may require that all penetrations (fire rated and non-fire rated) be sealed.

Electrical installation openings which penetrate fire-resistant-rated structures (walls, partitions, floors, ceilings, etc.) must be firestopped using approved methods that maintain the fire resistance rating »300-21«.

> **NOTE**
> *In hollow spaces, vertical shafts, and ventilation (or air-handling) ducts, electrical installations must not substantially increase the possible spread of fire or products of combustion »300-21«.*

■ Wiring in Ducts, Plenums, and Other Air-Handling Spaces

> **NOTE**
> *Metallic manufactured wiring systems (without nonmetallic sheath) having listed prefabricated cable assemblies is permitted in this type of installation »300-22(c)(1)«.*

CAUTION ««««««««««««««
Ducts used to transport dust, loose stock, or flammable vapors cannot contain wiring systems of any type. No wiring system (of any type) shall be installed within a duct or shaft containing only such ducts used for vapor removal or for ventilation of commercial-grade cooking equipment »300-22(a)«.

Only Type MI and MC cables, employing a smooth or corrugated impervious metal sheath without an overall nonmetallic covering, can be installed in ducts (or plenums) specifically fabricated to transport environmental air »300-22(b)«. Section 300-22(b) also lists other raceways permitted in ducts and plenums of this type.

Section 300-22(c) applies to spaces used for environmental air-handling purposes other than ducts and plenums specified in Section 300-22(a) and (b). Types AC, MI, MC (without an overall nonmetallic covering), or other factory-assembled multiconductor cables (control or power) listed specifically for the use must be installed in these spaces. All other types of cables and conductors must be installed in one of the raceways (or wireways) named in Section 300-22(c)(1).

■ Back-to-Back Boxes in Fire-Resistant-Rated Walls

Qualified testing laboratories publish electrical construction material directories listing installation restrictions which apply to maintaining fire resistive ratings of assemblies involving penetrations, or openings. (An example is the minimum 24-inch horizontal separation usually required between boxes on opposite sides of the wall). These directories and product listings offer assistance in Section 300-21 compliance »300-21 (FPN)«.

Fire-Resistant-Rated Walls

Fire-Resistant-Rated Walls

← 24″ →

Non-Fire-Resistant-Rated Walls

Permitted

■ Wiring Within Air-Handling Spaces in Dwelling Units

Section 300-22 does not apply to **dwelling unit** joist or stud spaces where wiring passes perpendicular to the long dimension of such spaces »300-22(c) *Exception*«.

Conductors must remain at least 1¼ inch from the nearest edge of a wood member »300-4(a)(1)«.

Joist or stud spaces, used for environmental air-handling purposes, cannot contain nonmetallic-sheathed cable which runs along the long dimension.

■ Securing Cables

Type MC cable must be supported and secured at intervals not exceeding 6 feet »334-10(a)«.

Type AC cable must be supported and secured at intervals not exceeding 4½ feet »333-7«.

Nonmetallic-sheathed cable must be supported and secured at intervals not exceeding 4½ feet »336-18«.

> **NOTE**
> *Cables fished between access points, where concealed in finished buildings (or structures) and where support is impracticable, can remain unsupported and unsecured »333-7(b), 334-10(b), and 336-18«.*

MC Cable

AC Cable

Nonmetallic-Sheathed Cable

Generally, all cables must be secured to the box, cabinet, etc. »370-17(b) and (c)«.

6ft

4½ft

4½ft

12in

12in

12in

8in

Nonmetallic-sheathed cable **not** secured to a single-gang box must be secured within 8 inches »370-17(c) *Exception*«.

Type MC cables containing no more than four conductors, sized No.10 AWG or smaller, must be secured within 12 inches of every box, cabinet, fitting, etc. »334-10(a)«.

Type AC cable must be secured within 12 inches of every box, cabinet, or fitting »333-7«.

Nonmetallic-sheathed cable (**not** secured to the box) must be secured within 12 inches of every box, cabinet, or fitting »336-18«.

Staples, cable ties, straps, etc. must be designed to secure cables and shall be installed so that cables are not damaged.

■ Cables Passing Through Framing Members

Cables running horizontally (or diagonally) are considered supported and secured when passing through a framing member (wood, metal, etc.), unless the support intervals exceed those listed for the specified cable. Cables installed in notches and protected by a ¹⁄₁₆-inch steel plate are also considered secured ≫ 333-7(a), 334-10(a)(1), and 336-18 ≪.

Cables must be secured within 12 inches of every box, cabinet, fitting, etc.

This cable is not "secured" because the support interval exceeds the maximum distance permitted.

■ Attics Without Permanent Stairs or Ladders

Attics and roof spaces not accessible by permanent stairs or ladders require protection only within 6 feet of the nearest edge of the scuttle hole or attic entrance ≫ 333-12(a) ≪.

Protection is required for cables located within 6 feet (measured both vertically and horizontally) of the attic entrance.

■ Bending Radius

Type AC cable must have a bending radius at least five times the diameter of the cable ≫ 333-8 ≪.

Type MC cable (interlocked-type or corrugated sheath) has a bending radius of seven times the external diameter of the metallic sheath ≫ 334-11(b) ≪. For Type MC cable with a smooth sheath, refer to Section 334-11(a). For Type MC cable with shielded conductors, refer to Section 334-11(c).

Nonmetallic-sheathed cable has a minimum bending radius of five times the cable diameter ≫ 336-16 ≪.

Radius

Strap

A curve radius is measured from the inner edge of the bend.

Example: If the diameter of Type MC cable (interlocked-type) is ½ (.5) inch, then the radius will need a measurement of 3½ inches (.5 × 7 = 3.5 inches).

NOTE
All bends shall be made so that the cable will not be damaged.

■ Attics With Permanent Stairs or Ladders

Cables more than 7 feet above the floor (or floor joist) can be run across the face of framing members »333-12(a)«.

Where run within 7 feet of the floor (or floor joist) across the face of rafters (or studding) in attics and roof spaces that are accessible, the cable shall be protected by substantial guard strips that are at least as high as the cable »333-12(a)«.

Cables running across the top of floor joists must be protected by substantial guard strips that are of sufficient height to adequately protect the cable »333-12(a)«.

Cable installed parallel to the sides of rafters, studs, or floor joists does not require guard strips or running boards; however, the installation must also comply with Section 300-4(d) »333-12(b)«.

Cables not installed on the face (or surface) do not require guard strips or running boards.

Cables installed in attics (and roof spaces) accessible by permanent stairs (or ladders) must meet Section 333-12(a) provisions.

> **N O T E**
> *Type MC cable and nonmetallic-sheathed cable must also comply with Section 333-12 »334-10(h) and 336-6(d)«.*

■ Insulating (Anti-Short) Bushings

All armor termination points of AC cable shall have a fitting to protect the wires from abrasion, unless the outlet boxes are designed to afford equivalent protection. In addition, an insulating bushing (or its equivalent protection) shall be provided between the conductors and the armor »333-9«.

Type AC cable shall have a flexible metal tape armor. The insulated conductors must comply Section 333-20 »333-19«.

Type AC cables shall have an internal bonding strip (copper or aluminum) in intimate contact with the armor for the entire length of the cable »333-19«.

The connector or clamp which fastens Type AC cable to boxes or cabinets must be designed so that the insulating bushing (or its equivalent) is visible for inspections »333-9«.

Some insulating (anti-short) bushings are manufactured so that part of the bushing extends past the connector or clamp (once installed), effectively acting as a flag. This flag increases the visibility of the bushing after installation.

> **N O T E**
> *Insulating bushings in Type MC cable are optional rather than required. Although Article 334 does not require an insulating bushing between the conductors and the armor, it is a good practice to use one.*

Exposed Runs of Type AC and MC Cables Under Joists

Exposed runs of Type AC cable installed on the underside of joists must be supported (and secured) at *every* joist and must be located so that physical damage is avoided ≫333-11 ≪.

Type MC cable must be supported (and secured) at intervals of 6 feet or less ≫334-10(a) ≪.

Exposed Nonmetallic-Sheathed Cable in Unfinished Basements

Exposed nonmetallic-sheathed cable (smaller than No. 8 if three-conductor, or No. 6 if two-conductor), running at angles with joists in unfinished basements, must either be run through bored holes in the joists or on running boards ≫336-6(c) ≪.

Three-conductor No. 8, or two-conductor No. 6, or larger, can be secured directly to the lower edges of the joists ≫336-6(c) ≪.

The nearest outside surface of installed and supported cable must be at least 1¼ inch from the nearest edge of the framing member where nails or screws are likely to penetrate ≫300-4(d) ≪.

Where the number of current carrying conductors in multiconductor cables are bundled (or stacked) longer than 24 inches (without maintaining spacing), reduce the allowable ampacity of each conductor as shown in Table 310-15(b)(2)(a) ≫310-15(b)(2)(a) ≪.

Exposed Nonmetallic-Sheathed Cable Passing Through a Floor

Nonmetallic-sheathed cable passing through a floor, must be enclosed in rigid (or intermediate) metal conduit, electrical metallic tubing, Schedule 80 PVC rigid nonmetallic conduit, listed surface metal (or nonmetallic) raceway, or other metal pipe that extends at least 6 inches above the floor ≫336-6(b) ≪.

N O T E

If subject to physical damage, conductors must be adequately protected ≫300-4 ≪. Nonmetallic-sheathed cable can be protected by conduit, electrical metallic tubing, Schedule 80 PVC rigid nonmetallic conduit, pipe, guard strips, listed surface metal (or nonmetallic) raceway, or other means ≫336-6(b) ≪.

N O T E

Raceway (or cable) openings in a fire-resistant-rated floor must be firestopped using approved methods to maintain the fire resistance rating ≫300-21 ≪.

6″

Nonmetallic-Sheathed Cable Enclosed in EMT

Nonmetallic-Sheathed Cable Exposed

Cables entering or exiting conduit (or tubing) used for support or protection against physical damage require fittings on the conduit (or tubing) ends to prevent cable abrasion ≫300-15(c) ≪.

■ Structures Exceeding Three Floors

Nonmetallic-Sheathed Cable
Permitted

4 floors
(or more)

One- or Two-family Dwelling

Nonmetallic-Sheathed Cable
Permitted

3 floors

Multifamily Dwelling

For the purpose of Article 336, the first floor of a building must have at least 50 percent of the exterior wall surface level with (or above) finished grade ≫ 336-5(a)(1) ≪.

> **NOTE**
>
> *Nonmetallic-sheathed cable (Type NM and NMC) can be used in any dwelling or structure exceeding three floors above grade when installed as branch circuit **temporary wiring** ≫ 305-4(c) ≪.*

"Romex" is a generic electrical term for non-metallic-sheathed cable. The name "Romex" is a registered trademark of General Cable Corporation.

Nonmetallic-Sheathed Cable
Permitted

3 floors

Multifamily Dwelling

Any first level not designed for human habitation, used only for vehicle parking, storage, or similar use, is not counted ≫ 336-5(a)(1) ≪.

Nonmetallic-Sheathed Cable
Not Permitted

4 floors

Multifamily Dwelling

Type NM, NMC, and NMS cables cannot be used in any dwelling or structure exceeding three floors above grade. ≫ 336-5(a)(1) ≪.

> **WARNING**
>
> Some state and city jurisdictions limit, or even prohibit, the installation of nonmetallic-sheathed cable.

CONDUCTOR IDENTIFICATION

■ General Conductor Identification Provisions

Generally, a conductor with a continuous white (or natural gray) covering shall be used only as a grounded (or neutral) conductor ≫ 200-7 ≪.

Metal-Clad Cable

Nonmetallic-Sheathed Cable

Type NMB

Equipment grounding conductors can be bare, covered, or insulated. Individually covered (or insulated) equipment grounding conductors must have a continuous outer finish that is either green or green with yellow stripe(s) ≫ 250-119 ≪.

Conductors used as ungrounded (hot) conductors, whether single conductors or in multiconductor cables, must be finished in a way that clearly distinguishes them from grounded and grounding conductors ≫ 310-12(c) ≪. Ungrounded (hot) conductors (except for a **high leg** conductor) can be any color other than white, natural gray, green, or green with yellow stripe(s).

■ Three-Phase Conductor Identification

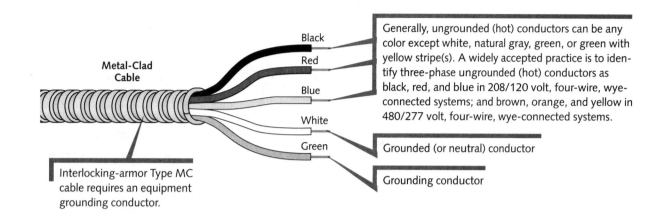

Metal-Clad
Cable

Black

Red

Blue

White

Green

Generally, ungrounded (hot) conductors can be any color except white, natural gray, green, or green with yellow stripe(s). A widely accepted practice is to identify three-phase ungrounded (hot) conductors as black, red, and blue in 208/120 volt, four-wire, wye-connected systems; and brown, orange, and yellow in 480/277 volt, four-wire, wye-connected systems.

Grounded (or neutral) conductor

Grounding conductor

Interlocking-armor Type MC cable requires an equipment grounding conductor.

■ White Conductor Used as Ungrounded Switch Leg (Loop)

A cable's white (or natural gray) conductor can be used as a switch loop (leg) in single-pole, 3-way, and 4-way switch installations, even though it is an ungrounded (hot) conductor ≫ 200-7(c)(2) ≪.

Black
Ungrounded
(Hot)
Conductor

White
Conductor

Octagon
Box

To supply

White
(Grounded)
Conductor

Black
(Return)
Conductor

The white (or natural gray) conductor's new use must be permanently re-identified by painting, or other effective means, at its terminations, and at each location where the conductor is visible and accessible ≫ 200-7(c)(1) ≪.

Type MC
Cable

Red tape has been wrapped around the white conductor, re-identifying it as an ungrounded (hot) conductor. White, natural gray, green, or green with yellow strip(s) tape (or paint) cannot be used.

White
Conductor

Device
Box

The white (or natural gray) conductor can be used to feed a switch or can be a traveler in 3-way (or 4-way) switch installations. The white (or natural gray) conductor cannot be used as a return conductor from the switch to the switched outlet ≫ 200-7(c)(2) ≪.

CAUTION ⟨⟨⟨⟨⟨⟨⟨⟨⟨⟨⟨⟨⟨

*A white (or natural gray) conductor shall **not** be used as a **return** conductor from the switch to the switched outlet (light, receptacle, etc.) ≫ 200-7(c)(2) ≪.*

■ White Conductor Used as an Ungrounded (Hot) Conductor

The white (or natural gray) conductor can be used as an ungrounded (hot) conductor if it is part of a cable assembly. It must be permanently re-identified as an ungrounded conductor, by painting (or other effective means) at its termination, and at each location where the conductor is visible and accessible ≫ 200-7(c)(1) ≪.

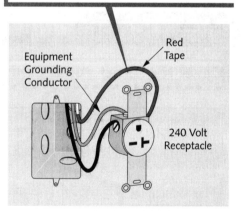

Equipment
Grounding
Conductor

Red
Tape

240 Volt
Receptacle

WARNING

A white (or natural gray) conductor in a conduit (or raceway) cannot be used (or re-identified) as an ungrounded conductor.

■ White Conductor Used as a Traveler

Although grounded circuit conductors are not required at switch locations, new product technology (such as occupancy sensors, home automation switches, programmable controllers, etc.) may require a grounded circuit conductor. For this reason, installing a grounded circuit conductor at switch locations is a design issue worth consideration.

The white (or natural gray) conductor can feed a switch or can be a traveler in 3-way (or 4-way) switch installations. The white (or natural gray) conductor must not be used as a return conductor from the switch to the switched outlet »200-7(c)(2)«.

NOTE
Switching in 3-way and 4-way configurations is done only in the ungrounded (hot) circuit conductor »380-2(a)«.

WARNING
A grounding conductor (green, green with yellow stripe(s) or bare) cannot be used (or reidentified) as an ungrounded (hot) or a grounded (white) conductor.

The white (or natural gray) conductor's new use must be permanently re-identified by painting (or other effective means), at its terminations, and at each location where the conductor is visible and accessible »200-7(c)(1)«.

Grounded circuit conductors are not required at switch locations »380-2(a) *Exception*«.

Different colors can be used for travelers in 3-way and 4-way switches.

UNDERGROUND INSTALLATIONS

■ Cover for Direct Burial Cables and Conductors

NOTE
If the cover depth requirements of Table 300-5 cannot be met due to the presence of solid rock, the wiring must be installed in a raceway (metal or nonmetallic) approved for direct burial. The raceways must then be covered by a minimum of 2 inches of concrete which extends down to the solid rock surface »Table 300-5, Note 5«.

Cover measurement commences at the top surface of any direct burial wiring method (conductor, cable, conduit or other raceway) and ends at the closest point of the finished grade's top surface (including concrete, etc.) »Table 300-5«.

NOTE
Raceways must enclose cables and conductors located under buildings »Table 300-5«.

Direct burial cables or conductors.
Table 300-5

All locations **NOT** specified in Table 300-5

Streets, highways, roads, alleys, driveways, and parking lots

Under minimum of 4-inch thick concrete exterior slab with no vehicular traffic (the slab must extend at least 6 inches beyond the underground installation)

In trench below 2 inch thick concrete or equivalent

One- and two-family dwelling driveways (and outdoor parking areas) used only for dwelling-related purposes **AND** airport runways, including adjacent areas where trespassing is prohibited

Protection of Conductors and Cables

Cable installed under a building must be in a raceway which extends beyond the building's outside walls »300-5(c) «.

Direct buried conductors and cables emerging from the ground must be protected by enclosures or raceways »300-5(d) «.

Protection must extend at least 8 feet above finished grade »300-5(d) «.

Conductors entering a building shall be protected to the point of entry »300-5(d) «.

Raceways and boxes shall be fastened in place securely »300-11(a) «.

Excavation backfill cannot contain large rock, paving materials, cinders, large or sharp substances, or corrosive materials where damage to raceways or cables may occur. Provide protection in the form of granular or selected material, suitable running boards, suitable sleeves, or other approved means to prevent physical damage to the raceway or cable »300-5(f) «.

Protection must extend below grade to a depth matching Table 300-5 cover requirements, up to a maximum depth of 18 inches »300-5(d) «.

Where the conductors or cables emerge as a direct burial wiring method, install a bushing or fitting to protect conductors from abrasion on the end of the conduit (or tubing) that terminates underground. A seal incorporating the same protective characteristics can be used in lieu of a bushing »300-5(h) and 300-15(c) «.

Section 300-5 recognizes "**S**" loops in underground direct burial to raceway transitions, expansion joints in raceway risers to fixed equipment, and the provision of flexible connections to equipment subject to settlement or frost heaves »300-5(j)(FPN) «.

Direct buried conductors, raceways, or cables subject to movement by settlement (or frost) must be arranged to prevent damage to both the enclosed conductors and the equipment connected to the raceways »300-5(j) «.

Cables Permitted Underground

Cable Types	Article
IGS Integrated Gas Spacer Cable	325
MV Medium Voltage Cable	326
MI Mineral-Insulated, Metal-Sheathed Cable	330
MC Metal-Clad Cable (where identified for such use)	334
USE Underground Service-Entrance Cable	338
UF Underground Feeder (and Branch Circuit) Cable	339

■ Cover for Residential Branch Circuits

● ●

> **N O T E**
> *Eighteen inches is the mini-mum cover for residential branch circuits rated no more than 120 volts, with GFCI protection, and having a max-imum overcurrent protection of 20 amperes, in or under airport runways (including adjacent areas where tres-passing is prohibited)* »*Table 300-5, Note 5* «.

> **N O T E**
> *Type UF cable installed as nonmetallic-sheathed cable must meet the requirements of Article 336 and must be a multiconductor cable* »*339-3(a)(4)* «.

Raceways must enclose cables and conductors located under buildings. These raceways must extend beyond the building's outside walls »300-5(c) and Table300-5 «.

> **N O T E**
> *Lesser depths are per-mitted where cables and conductors rise for terminations (or splices) or where access is otherwise required* »*Table 300-5, Note 3* «.

Residential branch circuits rated 120 volts (or less) with GFCI protection and maximum overcurrent protection of 20 amperes. Table 300-5

All locations **NOT** specified in Table 300-5

One- and two-family dwelling driveways (and outdoor parking areas) used only for dwelling-related purposes

Under minimum of 4-inch thick concrete exterior slab with no vehicular traffic (the slab must extend at least 6 inches beyond the underground installation)

In trench below 2-inch thick concrete or equivalent

Streets, highways, roads, alleys, driveways, and parking lots

■ Cover for Low Voltage Circuits

● ●

> **N O T E**
> *Where a wiring method listed in Table 300-5, columns 1–3, is used for a circuit type in columns 4 and 5, the shallower depth of burial is allowed* »*Table 300-5, Note 4* «.

> **N O T E**
> *Low-voltage conductors in a trench beneath 2-inch thick concrete re-quire a minimum cover of 6 inches.*

Circuits for control of irrigation and landscape lighting limited to not more than 30 volts and installed with Type UF or in other identified cable (or raceway). Table 300-5

All locations **NOT** specified in Table 300-5

Under minimum of 4-inch thick concrete exterior slab with no vehicular traffic (the slab must extend at least 6 inches beyond the underground installation)

Streets, highways, roads, alleys, driveways, and parking lots

One- and two-family dwelling driveways (and outdoor parking areas) used only for dwelling-related purposes **AND** airport runways, includ-ing adjacent areas where trespassing is prohibited

SPECIAL APPLICATION CABLES

■ Flat Conductor Cable (Under Carpet Wiring)

Power feed, grounding connection, and shield system connection between the FCC system and other wiring systems shall be accomplished through a transition assembly identified for this use ≫328-15 ≪.

Carpet squares are required to cover floor-mounted Type FCC cable, cable connectors, and insulating ends. They cannot be larger than 36 inches square (3 feet × 3 feet). Release-type adhesives must be used when adhering carpet squares to the floor ≫328-10 ≪.

All FCC system components must be securely anchored to the floor (or wall) using an adhesive or mechanical system identified for this use. Floors must be prepared to ensure attachment of the FCC system to the floor until the carpet squares are placed ≫328-16 ≪.

Type FCC cable can be installed on hard, solid, smooth, continuous floor surfaces made of concrete, ceramic, composition flooring, wood, or similar materials ≫328-4(b) ≪.

A metal top shield shall be installed over all floor-mounted Type FCC cable, connectors, and insulating ends. The metal top shield must completely cover all cable runs, corners, connectors, and ends (328-12(a)) and must be electrically continuous to the supplying branch circuit's equipment-grounding conductor ≫328-13 ≪.

Type FCC cable consists of three or more flat copper conductors placed edge-to-edge and separated by an enclosing insulation assembly ≫328-2 ≪. Type FCC cable can be rated either 20 or 30 amperes. While the maximum current rating for general-purpose and appliance branch circuits is 20 amperes, individual branch circuits can have a current rating up to 30 amperes ≫328-6(b) ≪.

Transition Assembly

36 inches

36 inches

36 inches

Carpet Square

Receptacle Pedestal

Hold-Down Tape

Metal Top Shield

Type FCC Cable

Bottom Shield

White
Green
Black
Red
Blue

All receptacles, receptacle housings, and self-contained devices used with the FCC system shall be identified for this use and shall be connected to the Type FCC cable as well as the metal shields. Connection from any grounding conductor of the Type FCC cable to the shield system must be made at *each* receptacle ≫328-14 ≪.

All Type FCC cable, connectors, and insulating ends require a bottom shield, either metallic or nonmetallic ≫328-12(b) and 328-35(a) ≪.

NOTE
FCC systems cannot be used (1) outdoors or in wet locations; (2) where subject to corrosive vapors; (3) in any hazardous (classified) location; or (4) in residential, school, and hospital buildings ≫328-5 ≪.

NOTE
Listed insulating ends (caps) must be installed on all bare cable ends to seal against dampness and liquid spillage ≫328-11 ≪.

Integrated Gas Spacer Cable (Type IGS)

The maximum bending radius is found in Table 325-11.

A run of Type IGS cable between pull boxes (or terminations) cannot contain more than the equivalent of four quarter (90°) bends (4 × 90 degrees = 360° total), including those bends located immediately adjacent to the pull box (or terminations) »325-12 «.

Terminations and splices must be identified as suitable for maintaining the gas pressure of 20 pounds per square inch gauge (psig) within the conduit. A valve and cap must be provided for each length of the cable and conduit for monitoring and maintenance of gas pressure in the conduit »325-13 «.

Each conductor contains ½-inch diameter solid aluminum rod(s). A conductor assembly contains from one to nineteen of these rods, which are laid parallel. Conductor size ranges from a 250-kcmil minimum to a 4,750-kcmil maximum »325-20 «.

Type IGS cable is a manufactured assembly of conductor(s), each individually insulated and enclosed in a loose-fit nonmetallic flexible conduit as an integrated gas spacer cable rated 0-600 volts »325-1 «. It is used in underground (direct burial) applications as service-entrance, feeder, or branch circuit conductors »325-3 «.

Continuous Kraft Paper Spacer

Polyethylene Conduit

Aluminum Shield Tape

SF_6 Gas (Sulfur Hexafluoride)

Aluminum Tape Strand Shield

Solid Aluminum Rod Conductor (250 kcmil)

```
┌─────────────────────────────────┐
│           N O T E               │
│  Table 325-14 provides the maxi- │
│  mum ampacity for single (or multi- │
│  conductor) Type IGS cable.      │
└─────────────────────────────────┘
```

Medium Voltage Cable (Type MV)

Black P.V.C. Jacket

Copper Shielding Tape

Extruded Insulation Shield

Ethylene-Propylene-Rubber Insulation (EPR)

Extruded Conductor Shield

Annealed Copper Conductor

Type MV is a solid dielectric insulated cable (single or multiconductor) rated 2,001 volts or higher »326-1 «.

Conductor ampacity is found in Section 310-15, unless the cable is installed in a cable tray, in which case Section 318-13 gives the ampacity »326-6 «.

Type MV cables can be used on power systems rated up to 35,000 volts, nominal »326-3 «.

Type MV cables must have copper, aluminum, or copper-clad aluminum conductors and be constructed in accordance with Article 310 »326-5 «.

■ Mineral-Insulated, Metal-Sheathed Cable (Type MI)

Type MI (mineral-insulated, metal-sheathed) cable is a manufactured assembly of conductor(s) insulated with a highly compressed refractory mineral insulation and enclosed in a liquidtight (and gastight) continuous copper (or alloy steel) sheath 》330-1《.

The cable must not be damaged when bent. The radius of the inner edge of any bend cannot be less than five times the external diameter of the metallic sheath for cable that is not more than ¾-inch in external diameter. Cable diameters from ¾-inch to 1-inch must have a bend radius not less than ten times the diameter 》330-13《.

When using single-conductor cables, all phase conductors and the neutral conductor (if any) must be grouped together to minimize induced voltages on the sheath 》330-16《.

Liquidtight and Gastight Continuous Copper Sheath

Single-Conductor Cable

Solid Copper Conductors

Two-Conductor Cable

Highly Compressed Refractor Mineral Insulation

Nonferrous Plate

Only fittings identified for this use can connect Type MI cable to boxes, cabinets, or other equipment 》330-14《.

The installation of Type MI cable through studs, joists, rafters, or similar wood members must comply with Section 300-4 》330-11《. This cable must be securely supported at intervals no greater than 6 feet 》330-12(1)《.

Where Type MI cable terminates, a seal shall be provided *immediately* after stripping to prevent the entrance of moisture into the insulation. Each conductor extending beyond the sheath must be provided with an insulating material 》330-15《.

If these single-conductor cables had entered a ferrous enclosure, compliance with Section 300-20 would be required to prevent heating from induction 》300-16《.

> **NOTE**
> *General installation provisions for Type MI cable are similar to Type MC cable.*

Summary

- Cables, when installed, must be protected from the possibility of physical damage.
- Conductors must be installed at least 1¼ inches from the nearest edge of the framing member, unless a steel plate (at least ¹⁄₁₆-inch thick) has been installed to protect the conductor(s).
- Conductors passing through metal studs (or framing members) may require protection in the form of bushings or steel plates.
- Fire-resistant-rated construction must remain intact once the electrical system has been installed.
- In dwellings, wiring can pass through the short dimension of a joist (or stud) space being used for environmental air-handling purposes.
- Cables must be secured within a certain distance from boxes (junction, device, etc.), unless the cable has been fished between access points.

- The maximum distance between supports depends upon the type of cable.
- The minimum bending radius must be observed when installing a cable.
- The installation of nonmetallic-sheathed cable is prohibited in structures exceeding three floors above grade.
- Certain sizes of nonmetallic-sheathed cable can be attached directly to the underside of exposed joists in unfinished basements.
- In a cable system, a white conductor used as a feed (or traveler) with switches (single-pole, 3-way, and 4-way) must be re-identified.
- Table 300-5 contains minimum cover requirements for direct buried cables and conductors.
- Chapter 3 contains provisions for certain specialized cables.

Unit 4 Competency Test

● ●

NEC® Reference   Answer

_____   _____   1. Nonmetallic-sheathed cable must be secured within ____ inches from a two-gang plastic device box mounted in a wall.

_____   _____   2. Floor-mounted Type FCC cable, cable connectors, and insulating ends shall be covered with carpet squares no larger than ____ square.

_____   _____   3. Type AC cable shall be secured by staples, cable ties, straps, hangers, or similar fittings at intervals not exceeding ____ feet.

_____   _____   4. The minimum bending radius for three-inch diameter integrated gas spacer cable is ____ inches.

_____   _____   5. ____ is a factory assembly of one or more insulated circuit conductors with or without optical-fiber members enclosed in an armor of interlocking metal tape, or a smooth or corrugated metallic sheath.

_____   _____   6. Where Type MI cable terminates, a seal shall be provided immediately after stripping to prevent the entrance of ____ into the insulation.

_____   _____   7. Where nonmetallic-sheathed cable is run at angles with joists in unfinished basements, it shall be permissible to secure cables not smaller than three No.____ conductors directly to the lower edges of the joists.

       a) 12   b) 10   c) 8   d) 6

_____   _____   8. ____ is a single or multiconductor solid dielectric insulated cable rated 2001 volts or higher.

_____   _____   9. The overall covering of Type UF cable shall be ____.

       I. flame retardant

       II. suitable for direct burial in the earth

       III. moisture, fungus, and corrosion resistant

       a) II only   b) I and II only   c) II and III only   d) I, II and III

_____   _____   10. Power feed, grounding connection, and shield system connection between the FCC system and other wiring systems shall be accomplished in a(n) ____ identified for this use.

_____   _____   11. Type AC cable shall be permitted to be unsupported where the cable is not more than ____ inches in length at terminals where flexibility is necessary.

       a) 2   b) 6   c) 24   d) 72

_____   _____   12. The temperature limitations on which the ampacities of mineral-insulated, metal-sheathed, cable are based shall be determined by the ____.

       a) insulation of the cable

       b) composition of the outer sheath

       c) insulating materials used in the end seal

       d) highest ambient temperature surrounding the cable

_____   _____   13. The minimum bending radius for metal-clad (corrugated sheath) cable is ____ times the external diameter of the metallic sheath.

_____   _____   14. In both exposed and concealed locations, where a cable or raceway-type wiring method is installed through bored holes in joists, rafters, or wood members, holes shall be bored so that the edge of the hole is not less than ____ from the nearest edge of the wood member.

_____ _____ 15. Where run across the top of floor joists, or within _____ of the floor or floor joists across the face of rafters or studding, in an attic that has permanent stairs, nonmetallic-sheathed cable must be protected by substantial guard strips that are at least as high as the cable.

_____ _____ 16. Type FCC cable shall be clearly and durably marked on both sides at intervals of not more than 24 inches with the information required by Section 310-11(a) and with what additional information?

      I. material of conductors and maximum ampacity

      II. rated frequency and number of phases

      III. maximum temperature rating

    a) I only   b) I and II only   c) I and III only   d) I, II, and III

_____ _____ 17. _____ is a factory assembly of one or more conductors insulated with a highly compressed refractory mineral insulation and enclosed in a liquidtight and gastight continuous copper or alloy steel sheath.

_____ _____ 18. Armored cable installed in thermal insulation shall have conductors rated at _____. The ampacity of the cable installed in these applications shall be that of 60°C conductors.

    a) 60°C   b) 75°C   c) 90°F   d) 75°F

_____ _____ 19. Type MC cable must be supported (and secured) at intervals not exceeding _____ inches.

    a) 4½   b) 6   c) 54   d) 72

_____ _____ 20. The maximum ampacity of a 500 kcmil, integrated gas spacer cable (Type IGS) is _____ amperes.

_____ _____ 21. Type UF cable, buried underground, feeding a 15-ampere, ground-fault circuit-interrupter in a one-family dwelling must have a minimum cover of _____ inches.

    a) 6   b) 12   c) 18   d) 24

_____ _____ 22. Splices, in knob-and-tube wiring, shall be _____ unless approved splicing devices are used.

_____ _____ 23. A nonmetallic-sheathed cable, passing through a bored hole less than 1¼ inches from the edge of a wood framing member, must be protected by a steel plate (or bushing) at least _____ thick.

_____ _____ 24. In FCC systems, general-purpose and appliance branch circuits have a maximum current rating of _____ amperes, while individual branch circuits have a maximum current rating of _____ amperes.

_____ _____ 25. Type MI cable shall be supported securely at intervals not exceeding _____ by straps, staples, hangers, or similar fittings so designed and installed so as not to damage the cable.

_____ _____ 26. Metal-clad cable, installed outside of buildings or as aerial cable, shall comply with what additional Article(s)?

_____ _____ 27. Type UF cable, buried under a one-family dwelling driveway, must have a minimum cover of _____ inches.

UNIT 5

Raceways and Conductors

Objectives

After studying this unit, the student will:

- ✔ know the maximum number of bends for raceways installed between pull points.
- ✔ have a good understanding of rigid and intermediate metal-conduit provisions.
- ✔ be able to determine the type of raceway best suited for each installation.
- ✔ be familiar with IMC, RMC, and EMT support requirements.
- ✔ understand the unique characteristics of the four types of rigid nonmetallic conduits.
- ✔ thoroughly understand rigid nonmetallic conduit and electrical nonmetallic tubing provisions, both general and specific.
- ✔ be familiar with the different types of flexible conduit, and be able to locate each related Article.

- ✔ be aware of raceway types other than conduit and tubing.
- ✔ successfully calculate the electrical trade-size conduit required for each project.
- ✔ have a basic grasp of conductor properties.
- ✔ be familiar with conductor temperature limitations.
- ✔ know and understand the provisions pertaining to conductors connected in parallel.
- ✔ understand ampacity correction factors, including continuous loads, and how to apply them.

Introduction

Raceways are the conduits, or similar housing, whose dual function is to facilitate the installation (or removal) of electrical conductors and to provide protection from mechanical damage. Article 100 defines raceway as an enclosed channel of metal (or nonmetallic) materials designed expressly for holding wires, cables, or busbars, with additional functions as permitted in the *Code*. The term **raceway** applies to more than just tubular conduit. Raceways include, but are not limited to, rigid metal conduit, rigid nonmetallic conduit, intermediate metal conduit, liquidtight flexible conduit, flexible metallic tubing, flexible metal conduit, electrical nonmetallic tubing, electrical metallic tubing, underfloor raceways, cellular concrete floor raceways, cellular metal floor raceways, surface raceways, wireways, and busways. Most of these raceways are explained in Unit 5.

This unit contains regulations for raceway systems that are frequently used (such as electrical metallic tubing and wireways), as well as raceway systems that are rarely used (such as, cellular concrete floor raceways.) In addition to defining different raceways, other important information is covered, such as minimum and maximum size, installation provisions, fittings, and support requirements. Instructions are presented on how to perform raceway fill calculations, a prerequisite in determining the size of raceway needed for particular circuit(s). A brief explanation of conductors, including ampacity correction factors and conductor temperature limitations, can be found near the conclusion of this unit. Understanding this unit will make finding the best raceway system for a particular project much easier.

Most raceway Articles have sections entitled **Uses Permitted** and **Uses Not Permitted** (*NEC*® Chapter 3). These sections give locations (some specific) where the raceway system can (or cannot) be installed. Because most of these provisions are clearly understandable, this book will not repeat much of this information. Although not addressed in this unit, another important area is underground raceway installation (Section 300-5). Underground installations were covered in Unit 4, which explained minimum cover requirements for cables. Use Table 300-5 (Minimum Cover Requirements) to find the minimum burial depth for different types of raceways installed in a variety of locations.

GENERAL PROVISIONS

■ Suspended (Drop) Ceilings

Raceways can be supported by independent (additional) support wires, secured at both ends. Support wires and associated fittings must provide adequate support >> 300-11(a)《 .

Boxes can be secured by attaching independent (additional) support wires at both ends. Secure support must be provided by the support wires and associated fittings >> 300-11(a) and 370-23(d)(2)《 .

EMT conduit must be fastened securely within 3 feet of all boxes (outlet, junction, device, etc.) >> 348-13《 .

Ceiling grid support wires cannot be used to support raceways.

EMT Conduit

Ceiling Grid Support Wire

Ceiling Grid Support Wire

Raceway Support Wire (Secured at Both Ends)

Junction Box Support Wire (Secured at Both Ends)

Certain boxes can also be mounted to suspended ceiling framing members >> 370-23(d)(1)《 .

Suspended ceiling system framing members, used to support fixtures, must be securely fastened to one another as well as to the building structure at appropriate intervals. Fixtures must be securely fastened to the ceiling framing member by mechanical means (bolts, screws, or rivets, etc.). Listed clips, identified for use with the type of ceiling framing member(s) and fixture(s), are also permitted >> 410-16(c)《 .

■ Maximum Bends in One Run

The equivalent of four quarter bends (360° total) is the maximum allowed between pull points, e.g., conduit bodies and boxes. Because the total bends in this conduit run is 340 degrees, this installation falls within *NEC*® specifications.

The bend maximum of 360° applies to the following raceways: IGS (325-12), ENT (331-10), NUC (343-11), IMC (345-11), RMC (346-11), RNC (347-14), EMT (348-12), FMC (350-16), LFMC (351-10), and LFNC (351-30).

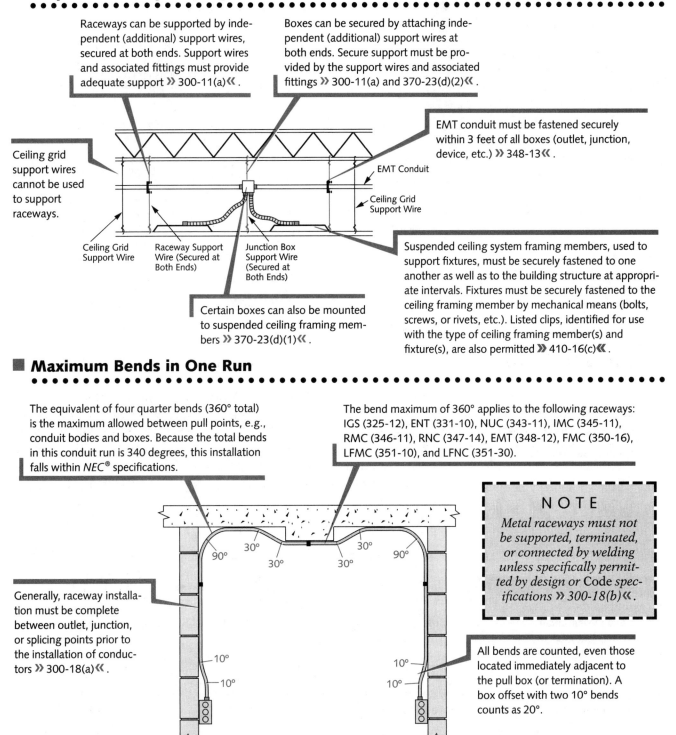

```
N O T E
Metal raceways must not
be supported, terminated,
or connected by welding
unless specifically permit-
ted by design or Code spec-
ifications >> 300-18(b)《 .
```

Generally, raceway installation must be complete between outlet, junction, or splicing points prior to the installation of conductors >> 300-18(a)《 .

All bends are counted, even those located immediately adjacent to the pull box (or termination). A box offset with two 10° bends counts as 20°.

■ Insulated Fittings

An insulated fitting is not required if a smoothly rounded (or flared) entry for conductors is provided by threaded hubs (or bosses) that are an integral part of a cabinet, box enclosure, or raceway ≫ 300-4(f) *Exception* ≪.

Insulated Fitting Not Required

No. 6 Conductors

Where raceways containing No. 4 or larger ungrounded (hot) conductors enter a cabinet (box, enclosure, or raceway), the conductors must be protected by a substantial fitting which provides a smoothly rounded insulating surface, unless the conductors are separated from the fitting (or raceway) by other securely attached substantial insulating material ≫ 300-4(f) ≪.

No. 4 Conductors

Plastic Bushings

No Locknut

Conduit bushings, constructed completely of insulating material, cannot be used to secure a fitting or raceway ≫ 300-4(f) ≪. An insulating (plastic, thermoplastic, etc.) bushing cannot replace a locknut.

■ Raceways Supporting Raceways

Conduits (not used as raceways) may support raceways, if fastened securely in place with approved fittings.

Raceways *cannot* be used to support other raceways, cables, conductors, or non-electric equipment, unless identified for that purpose ≫ 300-11(b) ≪.

This conduit contains power supply conductors for HVAC unit.

Class 2 Conductor

HVAC Unite

Raceways, containing power supply conductors for electrically controlled equipment, can support Class 2 circuit conductors (or cables) used solely for connection to the equipment control circuits ≫ 300-11(b)(2) ≪.

■ Conduit Thickness

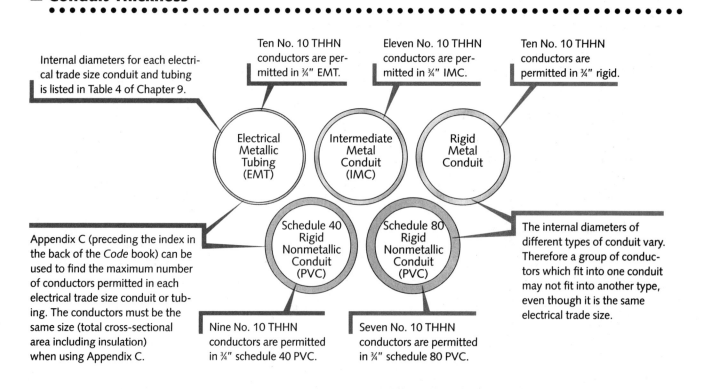

Internal diameters for each electrical trade size conduit and tubing is listed in Table 4 of Chapter 9.

Ten No. 10 THHN conductors are permitted in ¾" EMT.

Eleven No. 10 THHN conductors are permitted in ¾" IMC.

Ten No. 10 THHN conductors are permitted in ¾" rigid.

Electrical Metallic Tubing (EMT)

Intermediate Metal Conduit (IMC)

Rigid Metal Conduit

Schedule 40 Rigid Nonmetallic Conduit (PVC)

Schedule 80 Rigid Nonmetallic Conduit (PVC)

Appendix C (preceding the index in the back of the *Code* book) can be used to find the maximum number of conductors permitted in each electrical trade size conduit or tubing. The conductors must be the same size (total cross-sectional area including insulation) when using Appendix C.

Nine No. 10 THHN conductors are permitted in ¾" schedule 40 PVC.

Seven No. 10 THHN conductors are permitted in ¾" schedule 80 PVC.

The internal diameters of different types of conduit vary. Therefore a group of conductors which fit into one conduit may not fit into another type, even though it is the same electrical trade size.

Bending Radius

Bends in raceways must be so made that the raceway remains undamaged with its internal diameter basically undiminished. For any field bend, the radius of the curve's inner edge must not be less than indicated in Table 346-10. Raceways which must comply with this table include: electrical metallic tubing, intermediate metal conduit, rigid metal conduit, rigid nonmetallic conduit, electrical metallic tubing, flexible metal conduit, liquidtight flexible metal conduit, and liquidtight flexible nonmetallic conduit.

Table 346-10 Minimum Bending Radius (in inches)

Electrical Trade Size (in inches)	ENT, IMC, RMC RNC, EMT, FMC LFMC, LFNC
½	4
¾	5
1	6
1¼	8
1½	10
2	12
2½	15
3	18
3½	21
4	24
5	30
6	36

Bends made with specially designed single operation (one shot) bending machines must have a minimum radius not less than that indicated in Table 346-10, Exception. Intermediate metal conduit, rigid metal conduit, and electrical metallic tubing raceways can use this exception.

Table 346-10, *Exception.* Minimum Bending Radius (in inches)

Electrical Trade Size (in inches)	IMC, RMC EMT
½	4
¾	4½
1	5¾
1¼	7¼
1½	8¼
2	9½
2½	10½
3	13
3½	15
4	16
5	24
6	30

NON-FLEXIBLE CONDUIT (AND TUBING)

Rigid Metal Conduit

Rigid Metal Conduit (RMC) is a listed metal raceway of circular cross section (together with integral or associated couplings) approved for the installation of electrical conductors ≫ 346-1 ≪. RMC is manufactured in both ferrous (such as steel) and nonferrous (such as aluminum and brass) metal.

The minimum approved electrical trade size for RMC is ½ inch ≫ 346-6(a) ≪.

The maximum approved electrical trade size for RMC is 6 inches ≫ 346-6(b) ≪.

Rigid metal conduit can be used under all atmospheric conditions, within all types of occupancies ≫ 346-3(a) ≪.

RMC is the heaviest (thickest walled) classification of metal conduit.

Contact of dissimilar metals (except for the combination of aluminum and steel) should be avoided wherever possible to lessen the potential for galvanic action ≫ 346-3(a) ≪.

Threadless Set Screw Coupling

Threadless Compression Coupling

Threadless couplings and connectors used with conduit must be made tight. Those buried in masonry (or concrete), must be the concrete-tight type. Installed in wet locations, they must be the raintight type ≫ 346-9(a) ≪.

90° Elbow (Ell)

Coupling

RMC usually ships in standard lengths of 10 feet, including the coupling. Normally one coupling is furnished with each length ≫ 346-16(a) ≪.

Coupled with listed fittings, RMC provides electrical continuity; and therefore, can serve as an equipment grounding conductor ≫ 346-1 ≪.

Threadless Set Screw Connector

Threadless Compression Connector

> ### N O T E
> *Rigid metal conduit, elbows, couplings, and fittings can be installed in concrete, in direct contact with the earth, or in areas subject to severe corrosive influences where corrosion protection is provided and where judged suitable for the condition ≫ 346-3(b) ≪.*

> ### N O T E
> *An exception for enclosing motor leads, as permitted in Section 430-145(b), applies to the minimum size for most raceway types.*

■ Intermediate Metal Conduit

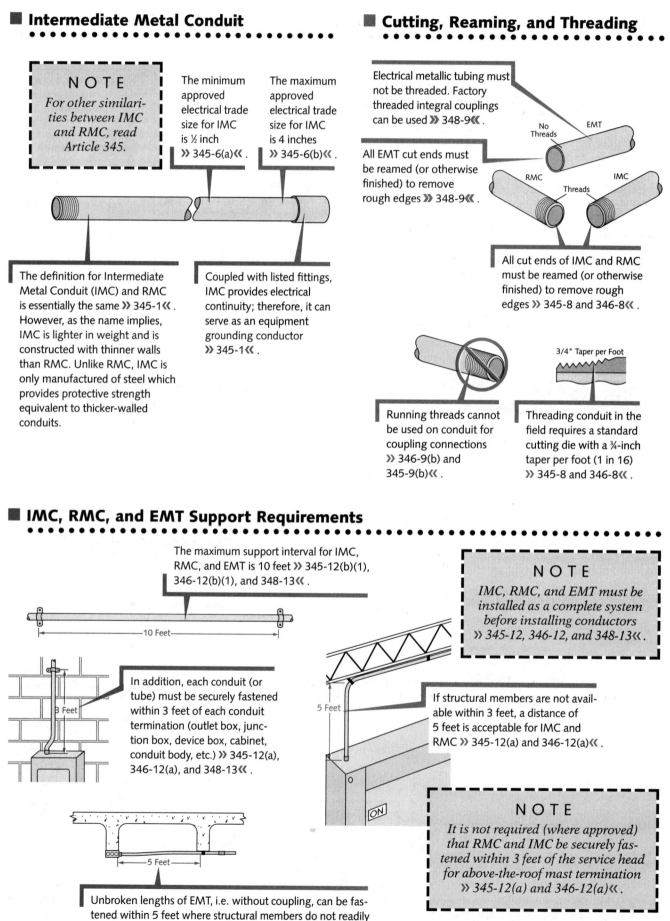

NOTE

For other similarities between IMC and RMC, read Article 345.

The minimum approved electrical trade size for IMC is ½ inch 》345-6(a)《.

The maximum approved electrical trade size for IMC is 4 inches 》345-6(b)《.

The definition for Intermediate Metal Conduit (IMC) and RMC is essentially the same 》345-1《. However, as the name implies, IMC is lighter in weight and is constructed with thinner walls than RMC. Unlike RMC, IMC is only manufactured of steel which provides protective strength equivalent to thicker-walled conduits.

Coupled with listed fittings, IMC provides electrical continuity; therefore, it can serve as an equipment grounding conductor 》345-1《.

■ Cutting, Reaming, and Threading

Electrical metallic tubing must not be threaded. Factory threaded integral couplings can be used 》348-9《.

No Threads · EMT · RMC · Threads · IMC

All EMT cut ends must be reamed (or otherwise finished) to remove rough edges 》348-9《.

All cut ends of IMC and RMC must be reamed (or otherwise finished) to remove rough edges 》345-8 and 346-8《.

3/4" Taper per Foot

Running threads cannot be used on conduit for coupling connections 》346-9(b) and 345-9(b)《.

Threading conduit in the field requires a standard cutting die with a ¾-inch taper per foot (1 in 16) 》345-8 and 346-8《.

■ IMC, RMC, and EMT Support Requirements

The maximum support interval for IMC, RMC, and EMT is 10 feet 》345-12(b)(1), 346-12(b)(1), and 348-13《.

—10 Feet—

NOTE

IMC, RMC, and EMT must be installed as a complete system before installing conductors 》345-12, 346-12, and 348-13《.

3 Feet

In addition, each conduit (or tube) must be securely fastened within 3 feet of each conduit termination (outlet box, junction box, device box, cabinet, conduit body, etc.) 》345-12(a), 346-12(a), and 348-13《.

5 Feet

If structural members are not available within 3 feet, a distance of 5 feet is acceptable for IMC and RMC 》345-12(a) and 346-12(a)《.

ON

NOTE

It is not required (where approved) that RMC and IMC be securely fastened within 3 feet of the service head for above-the-roof mast termination 》345-12(a) and 346-12(a)《.

—5 Feet—

Unbroken lengths of EMT, i.e. without coupling, can be fastened within 5 feet where structural members do not readily permit fastening within 3 feet 》348-13 *Exception No. 1*《.

IMC and RMC With Threaded Couplings

Straight runs of IMC and RMC, made up with threaded couplings, can be supported in accordance with Table 346-12(b)(2), provided such supports prevent transmission of stresses to termination where conduit is deflected between supports ≫346-12(b)(2)≪.

The distance between supports increases as the conduit size increases.

1/2" and 3/4" — 10 Feet
1"
12 Feet
1-1/4" and 1-1/2" — 14 Feet
2" and 2-1/2" — 16 Feet
3" and Larger — 20 Feet

Conduit Supported at the Top

20 Feet

Threaded Coupling

Conduit Support at the Bottom

The distance between supports can be increased to 20 feet for exposed vertical risers from industrial machinery, provided: (1) the conduit is made up with threaded couplings; (2) the conduit is firmly supported at the top and bottom of the riser; and (3) no other means of intermediate support is readily available ≫346-12(b)(3)≪.

Horizontal Runs Through Framing Members

Supported, but Not Fastened

3 Feet

10 Feet, or Less

Horizontal runs of IMC, RMC, and EMT which are supported by openings through framing members (at intervals not greater than 10 feet) and are fastened securely within 3 feet of termination points, are permitted ≫345-12(b)(4), 346-12(b)(4), and 348-13≪.

EMT Fished in Walls

For concealed work in finished buildings (or prefinished wall panels) where standard securing is impracticable, unbroken lengths (without coupling) of electrical metallic tubing can be fished ≫348-13 *Exception No. 2*≪.

No coupling in fished portion of wall.

Normally, EMT must be fastened within 3 feet of device boxes.

Finished wall provides no access to secure tubing.

EMT Used as an Equipment Grounding Conductor

EMT, can serve as an equipment grounding conductor as permitted in Section 250-122(a).

EMT

A raceway used as the equipment grounding conductor, as provided in Sections 250-118 and 250-134(a), must comply with Section 250-2(d) ≫250-122(a)≪.

Electrical metallic tubing must be supported according to Section 348-13 provisions.

Electrical Metallic Tubing

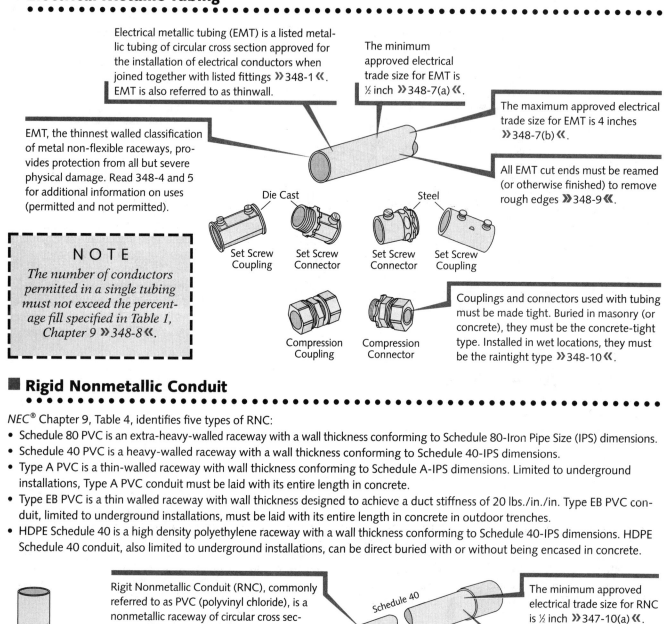

Electrical metallic tubing (EMT) is a listed metallic tubing of circular cross section approved for the installation of electrical conductors when joined together with listed fittings »348-1«. EMT is also referred to as thinwall.

The minimum approved electrical trade size for EMT is ½ inch »348-7(a)«.

The maximum approved electrical trade size for EMT is 4 inches »348-7(b)«.

EMT, the thinnest walled classification of metal non-flexible raceways, provides protection from all but severe physical damage. Read 348-4 and 5 for additional information on uses (permitted and not permitted).

All EMT cut ends must be reamed (or otherwise finished) to remove rough edges »348-9«.

Die Cast

Steel

Set Screw Coupling

Set Screw Connector

Set Screw Connector

Set Screw Coupling

NOTE

The number of conductors permitted in a single tubing must not exceed the percentage fill specified in Table 1, Chapter 9 »348-8«.

Compression Coupling

Compression Connector

Couplings and connectors used with tubing must be made tight. Buried in masonry (or concrete), they must be the concrete-tight type. Installed in wet locations, they must be the raintight type »348-10«.

Rigid Nonmetallic Conduit

NEC® Chapter 9, Table 4, identifies five types of RNC:
- Schedule 80 PVC is an extra-heavy-walled raceway with a wall thickness conforming to Schedule 80-Iron Pipe Size (IPS) dimensions.
- Schedule 40 PVC is a heavy-walled raceway with a wall thickness conforming to Schedule 40-IPS dimensions.
- Type A PVC is a thin-walled raceway with wall thickness conforming to Schedule A-IPS dimensions. Limited to underground installations, Type A PVC conduit must be laid with its entire length in concrete.
- Type EB PVC is a thin walled raceway with wall thickness designed to achieve a duct stiffness of 20 lbs./in./in. Type EB PVC conduit, limited to underground installations, must be laid with its entire length in concrete in outdoor trenches.
- HDPE Schedule 40 is a high density polyethylene raceway with a wall thickness conforming to Schedule 40-IPS dimensions. HDPE Schedule 40 conduit, also limited to underground installations, can be direct buried with or without being encased in concrete.

Coupling

Rigit Nonmetallic Conduit (RNC), commonly referred to as PVC (polyvinyl chloride), is a nonmetallic raceway of circular cross section, with or without integral couplings.

Schedule 40

Schedule 80

The minimum approved electrical trade size for RNC is ½ inch »347-10(a)«.

The maximum approved electrical trade size for RNC is 6 inches »347-10(b)«.

Female Adapter

All joints between lengths of conduit, and between conduit and couplings, fittings, and boxes, must be made by an approved method »347-6«.

All cut ends must be trimmed inside and out to remove rough edges »347-5«.

Expansion Fitting

Expansion and contraction problems generally do not arise in underground RNC applications.

Male Adapter

Only listed fittings can be used with rigit nonmetallic conduit »347-15«.

Expansion fittings must be provided for RNC to compensate for thermal expansion and contraction where the length change is expected to be 0.25 inch or greater, in accordance with »Table 347-9(A) and (B)«, in a straight run between securely mounted items such as boxes, cabinets, elbows, or other conduit terminations »347-9 and 300-7(b)«.

NOTE

Read Sections 347-2 and 3 for a description of uses (permitted and not permitted) of rigid nonmetallic conduit.

Bell End

Plug

■ Securing Rigid Nonmetallic Conduit

> ### NOTE
> *Horizontal runs of rigid nonmetallic conduit which are supported by openings through framing members (at intervals not greater than those in Table 347-8) and are securely fastened within 3 feet of termination points, are permitted »347-8(b)«.*

> ### NOTE
> *Rigid nonmetallic conduit, listed for securing at other than 3 feet and support at spacings other than as shown in Table 347-8, can be installed in accordance with the listing »347-8(a)«.*

Termination Point

1/2" through 1"
3 Feet

Each conduit must be fastened securely within 3 feet of all termination points »347-8(a)«.

1/2" through 1"
3 Feet

1-1/4" through 2"
5 Feet

Table 347-8 support provisions must be followed for RNC »347-8«.

2-1/2" through 3"
6 Feet

3-1/2" through 5"
7 Feet

RNC must be fastened so that movement from thermal expansion or contraction is permitted »347-8«.

6"
8 Feet

■ Bending Rigid Nonmetallic Conduit

Factory Elbows

90° Elbow

45° Elbow

30° Elbow

Rigid nonmetallic conduit bends must be made so that the conduit remains undamaged and the internal diameter of the conduit is not reduced »347-13«.

The conduit must not be damaged.

Bends must be made with equipment identified for the purpose.

Electric PVC Heater 1/2"-2"

Field bends must be made only with bending equipment identified for the purpose. The radius of the curve of the inner edge of such bends must not be less than shown in Table 346-10 »347-13«.

PVC Plugs

Installing PVC plugs in the ends of larger conduits (2 to 6 inches) can prevent conduit collapses or other deformities.

■ Securing ENT

The equivalent of four quarter bends (360° total) is the maximum between pull points, e.g., conduit bodies and boxes »331-10«.

Electrical Nonmetallic Tubing must be secured every 3 feet or less »331-11«.

Bends must be made so that the tubing is undamaged and the tubing's internal diameter is not reduced. While bends can be made manually (without auxiliary equipment) the radius of the curve of the inner edge of each bend must not be less than shown in Table 346-10 »331-9«.

ENT must be fastened securely within 3 feet of each outlet box, device box, junction box, cabinet, or other termination point »333-11«.

Connector

3 Feet

3 Feet

Electrical Nonmetallic Tubing

Electrical Nonmetallic Tubing (ENT) is a pliable corrugated raceway of circular cross section with integral (or associated) couplings, connectors, and fittings listed for the installation of electrical conductors. It is composed of a material that is flame-retardant as well as resistant to moisture and chemical atmospheres.

The minimum approved electrical trade size for ENT is ½ inch 》331-5(a)《.

The maximum approved electrical trade size for ENT is 2 inches 》331-5(b)《.

A pliable raceway can be bent by hand with a reasonable force, but without other assistance 》331-1《.

All cut ends must be trimmed inside and out to remove rough edges 》331-7《.

Outside diameters are such that standard rigid PVC conduit couplings and connectors can be used on ENT of PVC construction. ENT installation instructions outline the procedure to follow when installing PVC conduit fittings which are cemented in place. Specific cement requirements and application methods are provided.

Schedule 40 PVC Male Adapter

CAUTION 《《《《《《

ENT must not be stored or installed where exposed to direct sunlight, unless identified as "sunlight resistant" 》331-4(9)《.

Snap-in Connector

Male Adapter

Coupling

An approved method must be used for all joints between lengths of tubing and between tubing and couplings, fittings, and boxes 》331-8《.

NOTE

Electrical nonmetallic tubing must comply with Section 300-4 protection requirements

For the complete listing of ENT uses (permitted and not permitted), read Sections 331-3 and 4.

ENT Through Framing Members

If the bored hole is less than 1¼ inch from the edge of the framing member, the raceway must be protected from penetration (by screws, nails, etc.) by a steel plate (or bushing). The steel plate must be at least ¹⁄₁₆ inch thick, and of appropriate length and width to adequately cover the wiring 》300-4(a)(1)《.

Holes must be bored so that the edge of the hole is at least 1¼ inch from the nearest edge of the wood member 》300-4(a)(1)《.

ENT must be securely fastened in place within 3 feet of all termination points 》331-11《.

Horizontal runs of electrical nonmetallic tubing can be supported by openings through framing members at intervals not greater than 3 feet 》331-11《.

ENT Fixture Whips

ENT must be secured within 3 feet of all termination points 》331-11《.

6 Feet or Less

No Support Required

Securing is not required for electrical nonmetallic tubing in lengths not more than 6 feet from a fixture terminal connection (for tap connections to lightings fixtures) 》331-11 *Exception*《.

FLEXIBLE CONDUIT (AND TUBING)

■ Flexible Metal Conduit

The minimum electrical trade size for FMC is ½ inch, unless a Section 350-10(a) provision is met.

The maximum approved electrical trade size for FMC is 4 inches ≫ 350-10(b)≪ .

The equivalent of four quarter bends (360° total) is the maximum between pull points, e.g., conduit bodies and boxes ≫ 350-16≪ .

Flexible Metal Conduit (FMC) is a ferrous (steel) or nonferrous (aluminum) metal raceway of circular cross section made of helically wound, formed, interlocked metal strip ≫ 350-1≪ . FMC must be listed, and can be used in both exposed and concealed locations ≫ 350-4≪ .

Bends must be made so that the conduit is not damaged and the conduit's internal diameter is not effectively reduced. The radius of the curve of the inner edge of any field bend must not be less than shown in Table 346-10 ≫ 350-16≪ .

Flexible metal conduit is often referred to as *Greenfield*, or simply *Flex*.

All cut ends must be trimmed (or otherwise finished) to remove rough edges, except where fittings are used which thread into the convolutions ≫ 350-22≪ .

Fittings used with FMC must be listed ≫ 350-20≪ .

> **N O T E**
> *For areas where flexible metal conduit is not allowed, see Section 350-5.*

Straight Connector

Screw-in Connector

90° Angle Connector

Angle connectors cannot be used for concealed raceway installations ≫ 350-20≪ .

Flexible Metal Conduit to EMT Coupling

■ FMC Support

Flexible metal conduit must be fastened securely in place (by an approved means) at intervals of 4½ feet, or less ≫ 350-18≪ .

FMC must be secured within 12 inches of each box, cabinet, conduit body, or other conduit termination ≫ 350-18≪ .

4-1/2 Feet ←→ 1 Foot

1/2" Flex

3/8" Flex

Refer to Table 350-12 for the maximum number of conductors permitted in ⅜ inch flex.

If the length of FMC from a fixture terminal for tap connections, as permitted in Section 410-67(c), to light fixtures is 6 feet, or less, strapping is not required ≫ 350-18 *Exception No. 3*≪ .

> **N O T E**
> *Horizontal runs of flexible metal conduit can be supported by openings through framing members (at intervals not greater than 4½ feet), if securely fastened within 12 inches of each termination point ≫ 350-18≪ .*

Support is not required for fished FMC ≫ 350-18 *Exception No. 1*≪ .

FMC of ⅜ inch electrical trade size can be used: (1) for enclosing the leads of motors as permitted in Section 430-145(b); (2) in lengths not in excess of 6 feet as part of a listed assembly, or for tap connections to lighting fixtures as permitted in Section 410-67(c), or for utilization equipment; (3) for manufactured wiring systems as permitted in Section 604-6(a); (4) in hoistways, as permitted in Section 620-21(a)(1); (5) as part of a listed assembly to connect wired fixture sections as permitted in Section 410-77(c) ≫ 350-10(a)≪ .

■ FMC Installation

Grounding conductors are required for circuits over 20 amperes.

For lengths of 3 feet, or less, requiring flexibility, the 12-inch securing distance can be waived ≫ 350-18 *Exception No. 2* ≪.

30 Ampere Circuit

1 Foot

20 Ampere Circuit

Grounding Conductor Required

1 Foot

20 Ampere Circuit

If an equipment bonding jumper is required around FMC, it must be installed in accordance with Section 250-102 ≫ 350-14 ≪.

6 Feet, or Less . . . No Grounding Conductor Required

1 Foot

Listed flexible metal conduit can serve as a grounding means if: (1) the conduit is terminated in fittings listed for grounding; (2) the circuit conductors contained therein are protected by overcurrent devices rated at 20 amperes or less; (3) the total length in any ground return path is 6 feet or less; and (4) the conduit is not installed for flexibility ≫ 250-118(b) ≪.

FMC must be secured within 12 inches of each box, cabinet, conduit body, or other conduit termination ≫ 350-18 ≪.

CAUTION ≪≪≪≪≪≪≪≪
An equipment grounding conductor must be installed in FMC used to connect equipment where flexibility must be maintained ≫ 350-14 ≪.

■ Liquidtight Flexible Metal Conduit

The minimum electrical trade size for LFMC is ½ inch, except as provided in Section 350-10(a) ≫ 351-5(a) ≪. The number of conductors allowed in ⅜ inch conduit must not exceed the limit set in Table 350-12 ≫ 351-6(b) ≪.

The maximum approved electrical trade size for LFMC is 4 inches ≫ 351-5(b) ≪.

Liquidtight Flexible Metal Conduit (LFMC) is a listed raceway of circular cross section having an outer liquidtight, nonmetallic, sunlight-resistant jacket over an inner flexible metal core with associated couplings, connectors, and fittings and approved for the installation ≫ 351-2 ≪.

Metal

Must be Listed

Liquidtight flexible metal conduit is often referred to as *Sealtight* (a registered trademark.)

> N O T E
> *Section 351-4 contains both permitted and non-permitted LFMC uses.*

Use only listed terminal fittings with LFMC ≫ 351-7 ≪.

Angle connectors cannot be used for concealed raceway installations ≫ 351-7 ≪.

The equivalent of four quarter-bends (360° total) is the maximum allowed between pull points, e.g., conduit bodies and boxes ≫ 351-10 ≪.

Straight Connector

90° Angle Connector

Straight Connector with Exterior Bonding Lug

> N O T E
> *Support requirements for liquidtight flexible metal conduit are the same as those for flexible metal conduit ≫ 351-8 ≪. (See FMC support illustrations.)*

Grounding Liquidtight Flexible Metal Conduit

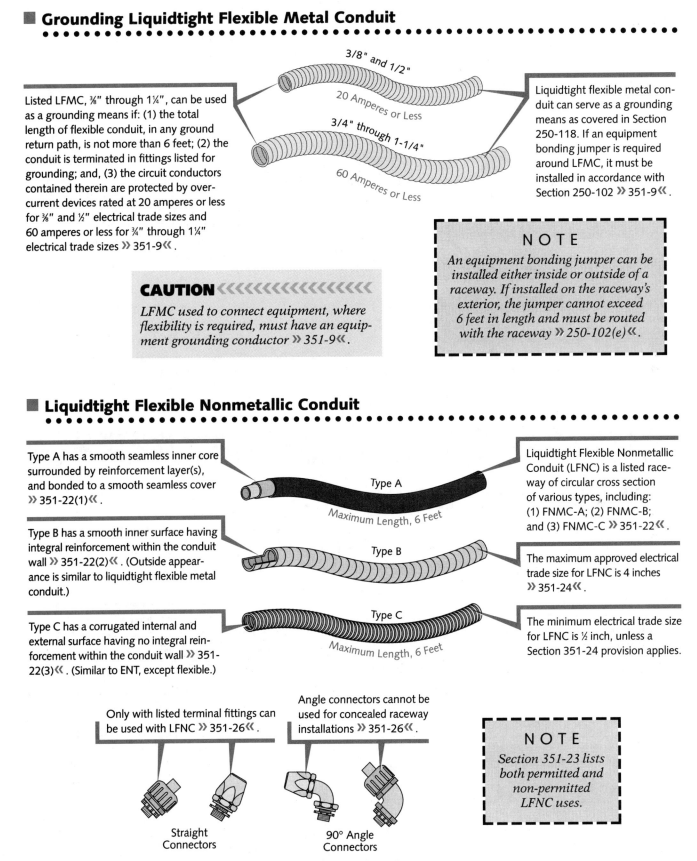

Listed LFMC, ⅜" through 1¼", can be used as a grounding means if: (1) the total length of flexible conduit, in any ground return path, is not more than 6 feet; (2) the conduit is terminated in fittings listed for grounding; and, (3) the circuit conductors contained therein are protected by over-current devices rated at 20 amperes or less for ⅜" and ½" electrical trade sizes and 60 amperes or less for ¾" through 1¼" electrical trade sizes 》351-9《.

3/8" and 1/2"
20 Amperes or Less

3/4" through 1-1/4"
60 Amperes or Less

Liquidtight flexible metal conduit can serve as a grounding means as covered in Section 250-118. If an equipment bonding jumper is required around LFMC, it must be installed in accordance with Section 250-102 》351-9《.

NOTE
An equipment bonding jumper can be installed either inside or outside of a raceway. If installed on the raceway's exterior, the jumper cannot exceed 6 feet in length and must be routed with the raceway 》250-102(e)《.

CAUTION «««««««««««««««««
LFMC used to connect equipment, where flexibility is required, must have an equipment grounding conductor 》351-9《.

Liquidtight Flexible Nonmetallic Conduit

Type A has a smooth seamless inner core surrounded by reinforcement layer(s), and bonded to a smooth seamless cover 》351-22(1)《.

Type A
Maximum Length, 6 Feet

Liquidtight Flexible Nonmetallic Conduit (LFNC) is a listed raceway of circular cross section of various types, including: (1) FNMC-A; (2) FNMC-B; and (3) FNMC-C 》351-22《.

Type B has a smooth inner surface having integral reinforcement within the conduit wall 》351-22(2)《. (Outside appearance is similar to liquidtight flexible metal conduit.)

Type B

The maximum approved electrical trade size for LFNC is 4 inches 》351-24《.

Type C has a corrugated internal and external surface having no integral reinforcement within the conduit wall 》351-22(3)《. (Similar to ENT, except flexible.)

Type C
Maximum Length, 6 Feet

The minimum electrical trade size for LFNC is ½ inch, unless a Section 351-24 provision applies.

Only with listed terminal fittings can be used with LFNC 》351-26《.

Angle connectors cannot be used for concealed raceway installations 》351-26《.

NOTE
Section 351-23 lists both permitted and non-permitted LFNC uses.

Straight Connectors

90° Angle Connectors

The equivalent of four quarter-bends (360° total) is the maximum allowed between pull points, e.g., conduit bodies and boxes 》351-30《.

■ Liquidtight Flexible Nonmetallic Conduit Installation

Liquidtight flexible nonmetallic conduit, as defined in Section 351-22(2), must be securely fastened at intervals not greater than 3 feet **》351-27《**.

LFNC must be secured within 12 inches of each connection to every outlet box, junction box, cabinet, or fitting **》351-27(a)《**.

Where flexibility is necessary, LFNC can be secured within 3 feet of the termination **》351-27(b)《**.

> **NOTE**
>
> *If an equipment grounding conductor is required for the circuits installed in LFNC, it can be installed either inside or outside of the conduit. Where installed on the outside, the length of the equipment grounding conductor must not exceed 6 feet and must be routed with the raceway or enclosure. Fittings and boxes must be bonded or grounded in accordance with Article 250 》351-28《.*

Strapping is not required for LFNC lengths of 6 feet, or less, from a fixture terminal connection (for tap conductors to light fixtures) as permitted in Section 410-67(c) **》351-27(b)《**.

Support is not required for fished LFNC **》351-27(b)《**.

Horizontal runs of liquidtight flexible nonmetallic conduit can be supported by openings through framing members (at intervals not greater than 3 feet), if securely fastened within 12 inches of each termination point **》351-27(c)《**.

OTHER RACEWAYS

■ Surface Nonmetallic Raceways

Surface nonmetallic raceway construction must be visibly distinguishable from other raceways. Surface nonmetallic raceways and their fittings must be designed so that sections can be mechanically coupled together, and installed without subjecting the wires to abrasion **》352-27《**.

The size and number of conductors installed in any raceway must not exceed the raceway's design limitations **》352-24 and 352-25《**.

Splices and taps are permitted within surface nonmetallic raceway having a removable cover which remains accessible once installed. The conductors, at the point of splices and taps, must not fill the raceway to more than 75 percent of its area. Splices and taps, in surface nonmetallic raceways without removable covers, must be made only in junction boxes. All splices and taps must be made using approved methods **》352-29《**.

Surface nonmetallic raceways can be installed in dry locations. Section 352-22 lists the locations where these raceways cannot be used.

> **NOTE**
>
> *Unbroken lengths of surface nonmetallic raceways can pass transversely through dry walls, partitions, and floors. Conductors must be accessible on both sides of the wall, partition, or floor 》352-28《.*

Where surface nonmetallic raceways are used in combination for both signaling and for lighting and power circuits, the different wiring systems must be run in separate compartments, easily identified by printed legend or by sharply contrasting interior finish colors while maintaining the same relative position of compartments throughout the premises **》352-26《**.

> **NOTE**
>
> *Where Article 250 requires equipment grounding, a separate equipment grounding conductor must be installed within the raceway 》352-23《.*

■ Surface Metal Raceways

Surface metal raceway construction must be visibly distinguishable from other raceways. Surface metal raceways and their fittings must be designed so that the sections can be electrically and mechanically coupled together, and installed without subjecting the wires to abrasion. Non-metallic covers and accessories can be used on surface metal raceways only if identified for such use ≫352-8 ≪.

Uses (permitted and not permitted) are located in Section 352-1.

The number and size of conductors installed in any raceway must not exceed the raceway's design limitations ≫352-3 and 352-4 ≪.

Section 310-15(b)(2)(a) derating factors do not apply to conductors installed in surface metal raceways where all of the following conditions are met: (1) the cross-sectional area of the raceway exceeds 4 square inches; (2) the current-carrying conductors do not exceed 30 in number; (3) the sum of the cross-sectional areas of all contained conductors does not exceed 20% of the interior cross-sectional area of the surface metal raceway ≫352-4 ≪.

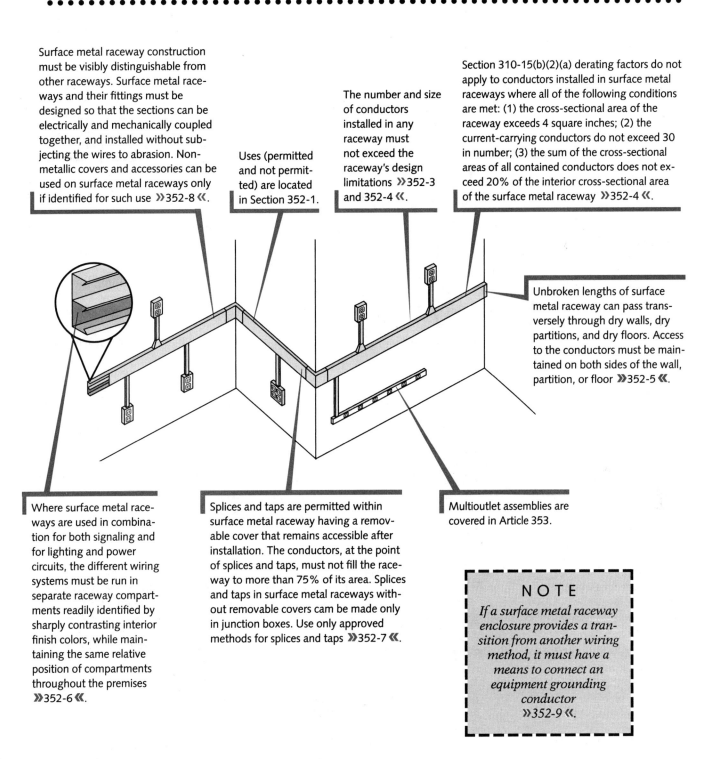

Unbroken lengths of surface metal raceway can pass transversely through dry walls, dry partitions, and dry floors. Access to the conductors must be maintained on both sides of the wall, partition, or floor ≫352-5 ≪.

Where surface metal raceways are used in combination for both signaling and for lighting and power circuits, the different wiring systems must be run in separate raceway compartments readily identified by sharply contrasting interior finish colors, while maintaining the same relative position of compartments throughout the premises ≫352-6 ≪.

Splices and taps are permitted within surface metal raceway having a removable cover that remains accessible after installation. The conductors, at the point of splices and taps, must not fill the raceway to more than 75% of its area. Splices and taps in surface metal raceways without removable covers cam be made only in junction boxes. Use only approved methods for splices and taps ≫352-7 ≪.

Multioutlet assemblies are covered in Article 353.

> **NOTE**
>
> *If a surface metal raceway enclosure provides a transition from another wiring method, it must have a means to connect an equipment grounding conductor ≫352-9 ≪.*

■ Strut-Type Channel Raceways

Part C of Article 352 applies to strut-type channel raceways, and accessories formed of metal, that are resistant to moisture or have corrosion protection and are judged suitable for the condition. These channel raceways can be steel (galvanized, stainless, enameled, or PVC-coated) or aluminum. Covers can be either metallic or nonmetallic
》352-40《.

Uses (permitted and not permitted), are covered in Sections 352-41 and 352-42.

Strut-type channel raceways can be suspended in air provided an approved method (designed for the purpose) is applied at intervals of not more than 10 feet
》352-47(b)《.

A strut-type channel raceway must be secured to the mounting surface with external retention straps at intervals not exceeding 10 feet, and within 3 feet of each raceway termination (outlet box, cabinet, junction box, etc.) 》352-47(a)《.

Splices and taps are permitted in raceways which provide access after installation via a removable cover. The conductors, at any point of splices or taps, must not fill the raceway to more than 75%. All splices and taps must be made by approved methods 》352-48《.

Strut-type channel raceway enclosures providing a transition to (or from) other wiring methods must accommodate the connection of an equipment grounding conductor. Strut-type channel raceway can be used as an equipment grounding conductor 》352-50《.

Use Table 352-45 to determine the maximum number of conductors permitted in strut-type channel raceway. Apply the appropriate Chapter 9 Tables for the outside diameter (O.D.) dimensions of the type and size of wire used 》352-45《.

Section 310-15(b)(2)(a) derating factors do not apply to conductors installed in strut-type channel raceways, where all of the following conditions are met: (1) the cross-sectional area of the raceway exceeds 4 square inches; (2) the current-carrying conductors do not exceed 30 in number; and (3) the sum of the cross-sectional areas of all contained conductors does not exceed 20% of the interior cross-sectional area of the strut-type channel raceway 》352-45《.

Conductors larger than that for which the strut-type channel raceway is listed must not be used 》352-44《.

> ### NOTE
> *Unbroken lengths of strut-type channel raceway can extend through walls, partitions, and floors provided closure strips are removable from either side and the portion within the wall, partition, or floor remains covered 》352-46《.*

■ Underfloor Raceways

Connections between raceways and distribution centers and wall outlets must be made using flexible metal conduit where not installed in concrete, rigid metal conduit, intermediate metal conduit, electrical metallic tubing, or approved fittings. Rigid nonmetallic conduit, electrical nonmetallic tubing, or liquidtight flexible nonmetallic conduit can be used where a metallic underfloor raceway system (not installed in concrete) provides for equipment grounding conductor termination »354-15«.

Underfloor raceways can be installed beneath a concrete (or other flooring material) surface. In office occupancies, installations flush with the concrete floor and covered with linoleum (or equivalent covering) are acceptable »354-2«.

The combined cross-sectional area of all conductors or cables must not exceed 40% of the interior cross-sectional area of the raceway »354-5«.

The size of conductors installed must not be larger than that for which the underfloor raceway is designed »354-4«.

Inserts must be leveled and sealed to prevent the entrance of concrete. Metal raceway inserts must also be metal and must be electrically continuous with the raceway »354-14«.

Install a suitable marker at, or near, each end of straight raceway runs to locate the last insert »354-9«.

Raceway dead ends must be closed »354-10«.

Seal and level junction boxes to the floor grade to prevent entrance of water or concrete. Underfloor metal raceway junction boxes must also be metal and must be electrically continuous with the raceway »354-13«.

Underfloor raceways must be laid so that a straight line from the center of one junction box, to the center to the next junction box, will coincide with the centerline of the raceway system. Raceways must be held firmly in place to prevent misalignment during construction »354-8«.

Splices and taps are acceptable only within junction boxes. Continuous, unbroken conductor connecting the individual outlets (so-called loop wiring) is not considered a splice or tap »354-6«.

Section 354-2(b) lists areas where underfloor raceways are not permitted.

CAUTION <<<<<<<<<<<<<<<<<
When an outlet is abandoned, discontinued, or removed, the supplying circuit conductors must be removed from the raceway. Splices or reinsulated conductors, such as would occur in the case of abandoned outlets on loop wiring, are not allowed in raceways »354-7«.

■ Underfloor Raceway Covering

Half-round and flat-top raceways not over 4 inches in width must be covered by ¾ inch or more of concrete (or wood) **》354-3(a)《**.

Linoleum
1/16"

3/4"   3/4"

4 Inches or Less

4 Inches or Less

4 Inches or Less

In office occupancies, approved flat-top metal raceways (of 4 inches or less in width), can be flush with the concrete floor surface, provided they are covered with a minimum ¹⁄₁₆-inch floor covering, such as linoleum. Where more than one (but fewer than four) single raceways are installed flush with the concrete, they must be contiguous with one another and be joined forming a rigid assembly **》354-3(d)《**.

Spacing at Least 1"

1"

Flat-top raceways, greater than 4 inches but not more than 8 inches wide, with a minimum of 1-inch spacing between the raceways, must be covered with at least 1 inch of concrete **》354-3(b)《**.

Over 4" Through 8"

Spacing Less Than 1":

1-1/2"

Flat-top raceways, greater than 4 inches but not more than 8 inches wide, which are spaced less than 1 inch apart must be covered with at least 1½ inches of **》354-3(b)《**.

Trench-type flush raceways having removable covers can be laid flush with the floor's surface. Such approved raceways shall be designed so that the cover plates provide adequate mechanical protection and rigidity equivalent to junction box covers **》354-3(c)《**.

Splices and taps are permitted in trench-type flush raceways whose removable covers are accessible after installation. The conductors, at the point of any splices or taps, must not fill more than 75% of the raceway's area **》354-6《**.

■ Cellular Metal Floor Raceways

● ●

The combined cross-sectional area of all conductors or cables must not exceed 40% of the cell or header's interior cross-sectional area »356-5 «.

Conductors larger than No. 1/0 can be installed only by special permission »356-4 «.

Inserts must be leveled to the floor grade and sealed against the entrance of concrete. Only metal inserts can be used with metal raceways and each must be electrically continuous with the raceway »356-10 «.

Connections between raceways and distribution centers and wall outlets must be made of flexible metal conduit (where not installed in concrete), rigid metal conduit, intermediate metal conduit, electrical metallic tubing, or approved fittings »356-11 «.

Junction boxes must be leveled to the floor grade and sealed against the entrance of water or concrete. Junction boxes used with these raceways must be metal and must be electrically continuous with the raceway »356-9 «.

For Article 356 purposes, a **cellular metal floor raceway** consists of the hollow spaces of cellular metal floors, together with suitable fittings, that may be approved as enclosures for electric conductors »356-1 «.

Splices and taps are made only in header access units or junction boxes. Continuous, unbroken conductor connecting the individual outlets (so-called loop wiring) is not considered a splice or tap »356-6 «.

A "cell" is a single, enclosed tubular space in a cellular metal floor member, whose axis is parallel to the axis of the metal floor member »356-1 «.

A "header" is a transverse raceway for electric conductors, providing access to predetermined cells of a cellular metal floor, accommodating the installation of electric conductors from a distribution center to the cells »356-1 «.

Section 356-2 lists areas where cellular metal floor raceways cannot be used.

CAUTION ‹‹‹‹‹‹‹‹‹‹‹‹‹‹‹‹‹‹‹‹‹‹‹‹‹‹‹‹‹‹‹‹‹‹‹‹‹‹‹

When an outlet is removed, abandoned, or discontinued, remove the section of circuit conductors supplying the outlet from the raceway. Raceways must not contain splices or reinsulated conductors as would occur with abandoned outlet loop wiring »356-7 «.

■ Cellular Concrete Floor Raceways

• •

Inserts must be leveled and sealed against the entrance of concrete. Inserts must be metal and must be fitted with grounded-type receptacles. A grounding conductor must connect the insert receptacles to a positive ground connection on the header »358-9 «.

The header must be installed in a straight line and at right angles to the cells »358-5 «.

Splices and taps are allowed only in header access units or junction boxes. Continuous, unbroken conductor connecting the individual outlets (so-called loop wiring) is not considered a splice or tap »358-12 «.

Mechanically secure the header to the top of the precast cellular concrete floor. The header must be electrically continuous throughout its entire length and must be electrically bonded to the distribution center enclosure »358-5 «.

A "header" is a transverse metal raceway for electric conductors that provides access to predetermined cells of a precast cellular concrete floor and permits the installation of conductors from a distribution center to the floor cells »358-2 «.

Conductor larger than No. 1/0 can be installed only by special permission »358-10 «.

Cellular concrete floor raceways are hollow spaces in floors constructed of precast cellular concrete slabs (together with suitable metal fittings) designed to provide access to the floor cells »358-1 «.

A "cell" is a single, enclosed tubular space in a floor made of precast cellular concrete slabs, whose direction is parallel to the direction of the floor member »358-2 «.

The combined cross-sectional area of all conductors or cables must not exceed 40% of the cell or header's cross-sectional area »358-11 «.

Section 358-4 lists areas where cellular concrete floor raceways cannot be used.

CAUTION ‹‹‹‹‹‹‹‹‹‹‹‹‹‹‹‹‹‹‹‹‹‹‹‹‹‹‹‹‹‹‹‹‹‹‹‹

When an outlet is removed, abandoned, or discontinued, remove the section of circuit conductors supplying the outlet from the raceway. Raceways must not contain splices or reinsulated conductors, such as loop wiring of abandoned outlets »358-13 «.

■ Wireways (Metal and Nonmetallic)

Wireways (sheet-metal and flame-retardant, nonmetallic) are troughs with hinged or removable covers for housing and protecting electric wires (and cables). The complete wireway system is installed before conductors are laid in place »362-1 and 14 «.

No conductor larger than that for which the wireway is designed shall be installed »362-4 and 18 «.

The sum of the cross-sectional area of all contained conductors must not exceed 20 percent of the wireway's interior cross-sectional area »362-5 and 362-19 «.

The maximum of current-carrying conductors in any metal wireway cross section is 30. Conductors for signaling circuits or controller conductors between a motor and its starter (used only for starting duty) are not considered as current-carrying conductors. Section 310-15(b)(2)(a) derating factors are not applicable to the 30 current-carrying conductors at 20 percent fill »362-5 «.

NOTE

Although metal and nonmetallic wireways can house more than 30 current-carrying conductors at any cross-section where Section 310-15(b)(2)(a) derating factors are applied, the sum of cross-sectional area of all contained conductors cannot exceed 20% of the wireway's interior cross-sectional area »362-5 Exception No. 1 and 362-19 «.

NOTE

The manufacturer's name (or trademark) on all wireways must remain visible after installation. In addition, the interior cross-sectional area (in square inches) must be marked on all nonmetallic wireways »362-12 and 27 «.

The 30-conductor limitation does not apply to metal wireways in theaters and similar locations »520-6 «.

The 20% metal wireway fill limitation does not apply to elevators and dumbwaiters »620-32 «.

CAUTION <<<<<<<<<<<<<<

Wireways containing No. 4 or larger conductors and used as junction or pull boxes must be sized according to Section 370-28 or 370-71.

■ Wireway Installation

Accessible splices and taps are permitted within a wireway. The conductors, at the point of any splice or tap, must not fill the wireway to more than 75% of its area ❯❯362-7 and 21 ❮❮.

Unbroken wireway lengths can pass transversely through walls. Access to the conductors must be maintained on both sides of the wall ❯❯362-9 and 24] ❮❮.

Metal wireways can be used: (1) for exposed work; (2) in concealed spaces only in accordance with Section 640-24; or (3) in hazardous (classified) locations as permitted by Sections 501-4(b) for Class 1, Division 2 locations; 502-4(b) for Class II, Division 2 locations; and 504-20 for intrinsically safe wiring ❯❯362-2(a) ❮❮.

Close all dead ends with listed fittings ❯❯362-10 and 25 ❮❮.

Where insulated conductors are deflected within a wireway, either at the ends or where conduits, fittings, or other raceways (or cables) enter (or leave) the wireway, or where the direction of the wireway is deflected greater than 30 degrees, Section 373-6 dimensions apply ❯❯362-6 and 20 ❮❮.

No. 4 or Larger Conductors

6 Times Larger Raceway

Metal wireways are not permitted where subject to severe physical damage or corrosive vapor ❯❯362-2(b) ❮❮.

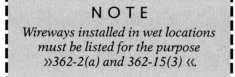

Extensions from wireways can be made with cord pendants (installed in accordance with Section 400-10) or with any Chapter 3 wiring method that includes an equipment grounding means. Where a separate equipment grounding conductor is employed, connection of the wiring method's equipment grounding conductors to the wireway must comply with Sections 250-8 and 250-12 ❯❯362-11 and 26 ❮❮.

Where No. 4 or larger insulated conductors enter a wireway through a raceway or cable, the distance between raceway and cable entries (enclosing the same conductor) must not be less than 6 times the trade diameter of the larger raceway or cable connector ❯❯362-6 ❮❮.

Listed **nonmetallic** wireways are permitted: (1) only for exposed work, except as permitted in accordance with Section 640-24; (2) where subject to corrosive vapors; and, (3) in wet locations where listed for the purpose ❯❯362-15 ❮❮.

Nonmetallic wireways shall not be used: (1) where subject to physical damage; (2) in any hazardous (classified) location, except as permitted in Section 504-20; (3) where exposed to sunlight unless listed and marked as suitable for the purpose; (4) where subject to ambient temperatures other than those for which nonmetallic wireway is listed; and, (5) for conductors whose insulation temperature limitations would exceed those for which the raceway is listed ❯❯362-16 ❮❮.

> **N O T E**
> *Wireways installed in wet locations must be listed for the purpose* ❯❯362-2(a) and 362-15(3) ❮❮.

CAUTION ❮❮❮❮❮❮❮❮❮❮❮❮❮❮❮❮❮
Extreme cold may cause nonmetallic wireways to become brittle and, therefore, more susceptible to damage from physical contact ❯❯362-15 FPN ❮❮.

■ Wireway Supports

Support intervals must not exceed 5 feet »362-8(a)«.

Metal

Horizontally run wire-ways must be supported at each end »362-8(a)«.

|← 5 Feet →|← 5 Feet →|

Individual Length **Metal**

Individual, metal wireway lengths (longer than 5 feet), run horizontally, must be supported at each end or joint, unless listed otherwise »362-8(a)«.

|← 10 Feet →|

The maximum distance between supports for individual lengths of metal wireway must not exceed 10 feet »362-8(a)«.

Nonmetallic

Horizontally run wireways must be supported at each end »362-22(a)«.

|← 3 Feet →|← 3 Feet →|← 3 Feet →|

Nonmetallic wireway support intervals must not exceed 3 feet »362-22(a)«.

Individual Length **Nonmetallic**

Horizontally run individual, nonmetallic wireway lengths of more than 3 feet must be supported at each end or joint »362-22(a)«.

|← 10 Feet →|

While individual, nonmetallic wireways can generally be supported as listed, in no case shall the distance between supports exceed 10 feet »362-22(a)«.

Nonmetallic wireway expansion fittings must be provided to compensate for thermal expansion and contraction where the change in length is expected to be 0.25 inch or more in a straight run »362-23«. Expansion characteristics of PVC nonmetallic wireway (which is the same as PVC rigid nonmetallic conduit) are listed in Table 347-9(A).

NOTE
Nonmetallic wireway vertical runs must be securely supported at intervals of 4 feet or less (except as otherwise listed), and can have only one joint between supports. Adjacent nonmetallic wireway sections must be securely fastened together to form a rigid joint »362-22(b)«.

NOTE
Metal wireway vertical runs must be securely supported at intervals not exceeding 15 feet, and are limited to one joint between supports. Adjacent wireway sections must be securely fastened together to form a rigid joint »362-8(b)«.

RACEWAY FILL

■ Same Size Conductors

Appendix C (located in the back of the *Code* book) can be used to find the maximum number of conductors permitted in a particular raceway (conduit or tubing). The conductor maximum is based on all conductors (in the raceway) being the same size (total cross-sectional area including insulation) **»** Chapter 9, Note1 **«** .

Tables C1 through 12 (in Appendix C) are based on the 40% fill for three or more conductors as permitted in Table 1 of Chapter 9.

Appendix C also lists the maximum number of fixture wires permitted in each type of conduit (or tubing).

■ Raceways Containing Multiconductor Cables

One Conductor   53%

A multiconductor cable of two or more conductors is treated as a single conductor for percentage conduit fill area calculations. To calculate the cross-sectional area of cables with elliptical cross sections, use the major diameter of the ellipse as a circle diameter **»** Chapter 9, Note 9 **«** .

> **N O T E**
> *While Table 1 applies only to complete conduit or tubing systems, it does not apply to sections of conduit used to protect exposed wiring from physical damage* **»** *Chapter 9, Note 2* **«** *.*

■ Different Size Conductors

Tables 4 and 5 (of Chapter 9) are used when a combination of different size conductors are installed in a single raceway **»** Chapter 9, Note 6 **«** .

Equipment Grounding Conductor

Use Table 4's 40% column when there are three or more conductors.

Use Tables 5 and 5A for dimensions (approximate area in square inches) of insulated conductors.

Include equipment grounding or bonding conductors, if any, when calculating raceway fill **»** Chapter 9, Note 3 **«** .

Table 4 contains trade size information for twelve different types of conduits and tubing. One column lists the internal diameters (in inches), the next lists the total area (in square inches), also referred to as the cross-sectional area. Table 4 also shows the maximum percent fill (in square inches) for tubing or conduits containing different numbers of conductors.

> **N O T E**
> *The approximate area (in square inches) for bare conductors are found in Table 8, column 6.*

■ Nipple Fill

A length of conduit (or tubing) measuring 24 inches or less is considered a nipple **»** Table 310-15(b)(2)(a) *Exception No. 3* and Chapter 9, Note 4 **«** .

|— 24 inches —|

For nipples, disregard Section 310-15(b)(2)(a) adjustment factors **»** Chapter 9, Note 4 **«** .

The square-inch area of conduit nipples must be calculated, since there is no table that provides that information. Somewhat confusing in its wording, Note 4 does not require a nipple fill of 60% of the cross-sectional area, but rather permits a 60% fill.

Installed between boxes, cabinets, and similar enclosures, conduit or tubing nipples can be filled to 60% of their total cross-sectional area **»** Chapter 9, Note 4 **«** .

> **N O T E**
> *When calculating the maximum number of same size conductors (total cross-sectional area including insulation) in a conduit or tubing, the next-higher whole number is used when the calculation results in a decimal of 0.8 or larger* **»** *Chapter 9, Note 7* **«** *.*

■ Raceway Fill Percentage

One Conductor **53%**	A conduit having a single conductor can be filled to 53% of the conduit's cross-sectional area.
Two Conductors **31%**	A conduit containing exactly 2 conductors can only be filled to 31% of the cross-sectional area of the conduit or tubing.
Three or More Conductors **40%**	A conduit containing 3 or more conductors can be filled to 40% of it's cross-sectional area.

NOTE

Tables 1 and 4 (in Chapter 9) lists maximum fill percentages for conduits and tubing.

When pulling three conductors or cables into a raceway, if the ratio of the raceway (inside diameter) to the conductor or cable (outside diameter) is between 2.8 and 3.2, jamming can occur. Jamming is less likely to occur when pulling four or more conductors or cables into a raceway » Chapter 9, Table 1, FPN No. 2«.

CAUTION <<<<<<<<<<<<<<<<<<<<<<<<<<<<

Table 1 is based on common conditions of proper cabling and alignment of conductors where the length of the pull and the number of bends are within reasonable limits. Be advised that, for certain conditions, a larger size conduit (or fewer conductors) should be considered » Chapter 9, Table 1, FPN No. 1«.

CONDUCTORS

■ Conductors in Parallel

Conductors of one phase, neutral, or grounded circuit conductors can have different physical characteristics from those of another phase, neutral, or grounded circuit conductors and still achieve balance. For example: neutral conductors do not have to be the same length as Phase A conductors, and Phase A conductors do not have to be the same length as Phase B conductors, etc. » 310-4«.

As a general rule, No. 1/0 and larger size conductors can be connected in parallel (electrically joined at both ends to form a single conductor) » 310-4«.

Neutrals   Phase   Phase   Phase
A   B   C

The paralleled conductors in each phase, neutral, or grounded circuit conductor must have the same characteristics » 310-4«. For example . . . all paralleled, Phase C conductors must:

(1) be of the same length,
(2) have the same conductor material,
(3) be the same size in circular mil area,
(4) have the same insulation type, and
(5) be terminated in the same manner.

NOTE

Equipment grounding conductors, used with conductors in parallel, must comply with Section 310-4 requirements, applying the sizing requirements of Section 250-122.

Conductors carrying alternating current, installed in metal enclosures or **metal raceways**, must be so arranged as to avoid heating the surrounding metal by induction » 300-20(a)«.

If run in separate raceways (or cables), the raceways (or cables), must have the same physical characteristics » 310-4«.

WARNING

Each metal raceway (housing paralleled conductors) must contain all phase conductors, the grounded conductor (where used) and all equipment grounding conductors » 300-20(a)«. For example: a Phase A conductor, a Phase B conductor, a Phase C conductor, and, if used, a neutral (or grounded) conductor and an equipment grounding conductor must be in each raceway.

Maximum Ampacities

Tables 310-16 through 310-20 list certain conductor maximum temperature ratings.

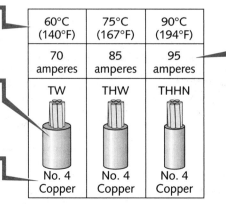

60°C (140°F)	75°C (167°F)	90°C (194°F)
70 amperes	85 amperes	95 amperes
TW	THW	THHN
No. 4 Copper	No. 4 Copper	No. 4 Copper

Table 310-13 contains specific conductor information, such as trade name, type letter, maximum operating temperature, application provisions, insulation, size, and outer covering.

Tables 310-16 through 310-21 list aluminum (or copper-clad aluminum) as well as copper conductors.

Maximum ampacities are listed in Tables 310-16 through 310-21. Other factors must be considered before using these ampacities: temperature limitations >> 110-14(c) <<, continuous loads >> 210-19(a) and 215-2(a) <<, ambient temperature >> Tables 310-16 through 310-21 <<, and the number of current-carrying conductors >> 310-15(b)(2)(a) <<, to name a few.

Conductor Temperature Limitations—100 Amperes or Less

Conductor temperature limitations can be compared to the strength of a chain. A chain is only as strong as its weakest link. One potential weak link for conductors is the termination (connection) point. The conductor ampacity cannot be higher than the lowest temperature rating of any connected termination, conductor, or device >> 110-14(c) <<.

Equipment termination provisions for circuits rated 100 amperes or less, or marked for Nos. 14 through 1 conductors, are used only for conductors rated 60°C >> 110-14(c)(1)(a) <<. Because the lowest temperature (weakest link) is 60°C, the ampacity of this conductor cannot exceed 70 amperes.

A No. 4 copper conductor with a temperature rating of 60°C (140EF) has an ampacity of 70 amperes >> Table 310-16 <<.

A temperature rating of 60/75°C means that the termination is suitable for 60° or 75°C conductors.

The ampacity of a No. 4 THHN copper conductor is 95 amperes >> Table 310-16 <<.

A No. 4 copper conductor with a temperature rating of 75°C (167°F) has an ampacity of 85 amperes >> Table 310-16 <<.

Equipment markings or listing information may further restrict the sizing and temperature ratings of connected conductors >> 110-14(c) FPN <<.

Equipment termination provisions for circuits rated 100 amperes or less, or marked for Nos. 14 through 1 conductors, can be used for conductors up to their maximum ampacities if the equipment is listed and identified for use with such conductors >> 110-14(c)(1) <<. The ampacity of this conductor now has a rating of 85 amperes because the lowest temperature (weakest link) is 75°C.

NOTE

When using conductors with temperature ratings higher than those specified for terminations, the higher ampacity can be used for ampacity adjustment, correction, or both >> 110-14(c) <<. For example, an adjustment factor of 80% must be applied to No. 4 THHN copper conductors because there will be more than three current-carrying conductors in the raceway. The connected load requires an overcurrent protection and conductor ampacity rating of at least 70 amperes, but the lowest temperature rated termination is 60°C. A 70-ampere overcurrent protective device can be installed with No. 4 THHN (90°C) copper conductors because the allowable ampacity (after the adjustment factor has been applied) is 76 amperes (95 × 80% = 76 amperes). Had derating not been possible (using the higher ampacity), the maximum overcurrent protective device would be 60 amperes (70 × 80% = 56 amperes).

■ Conductor Properties

A No. 6 (or smaller) insulated grounded conductor must be identified by a continuous white (or natural gray) outer finish or by three continuous white stripes on other than green insulation along its entire length »200-6(a) «.

An insulated grounded conductor larger than No. 6 shall be identified by either a continuous white (or natural gray) outer finish or by three continuous white stripes on other than green insulation along its entire length or, at the time of installation, by a distinctive white marking at its terminations. This marking must encircle the conductor (or insulation) »200-6(b) «.

> **N O T E**
>
> *No. 8 or larger conductors, installed in raceways, must be stranded* »310-3 «.

Table 310-13 lists conductor insulations and applications.

Ampacities of insulated conductors rated 0 through 2,000 volts, 60° through 90°C (140° through 194°F), with three or fewer current-carrying conductors installed in raceway, cable, or earth (directly buried) are found in Table 310-16. The ampacities listed there are based on an ambient temperature of 30°C (86°F). A variety of insulations are listed for both copper and aluminum (or copper-clad aluminum) conductors. Table 310-17 gives the same type of data as does Table 310-16, except for **single** conductors.

■ Conductor Temperature Limitations—Over 100 Amperes

When running *three* sets of paralleled copper conductors (for a 1,200 ampere overcurrent device), the minimum size is 600 kcmil, unless the equipment is listed and identified for use with 90°C (167°F) conductors.

Equipment termination provisions for circuits rated over 100 amperes, or marked for conductors larger than No. 1, are used only for conductors with higher temperature ratings provided the conductor ampacity is based on the 75°C (167°F) ampacity of the conductor size used »110-14(c)(2)(b) «.

Where the overcurrent device is rated over 800 amperes, the ampacity of the conductors it protects must be equal to, or greater than, the overcurrent device rating »240-3(c) «.

Switchboard is a large single panel, frame, or assembly of panels on which are mounted (on the face or back, or both) switches, overcurrent and other protective devices, buses, and, usually, instruments. Switchboards are generally accessible from both the front and the rear and are not intended for installation in cabinets »Article 100—Definitions «.

1,200 Ampere Switchboard

To 1,200 Ampere Overcurrent Device

Switchboard and panelboard provisions are found in Article 384.

When running *four* sets of paralleled copper conductors (for a 1,200 ampere overcurrent device), the minimum size is 350 kcmil, unless the equipment is listed and identified for use with 90°C (167°F) conductors.

> **N O T E**
>
> *The minimum size conductors (aluminum, copper-clad aluminum, or copper) that can be connected in parallel (electrically joined at both ends to form a single conductor) is size No. 1/0, unless meeting one of the four exceptions* »310-4 «.

> **CAUTION** ‹‹
>
> *Three paralleled sets of 500 kcmil, THHN, copper conductors do not meet NEC® provisions if the overcurrent device is 1,200 amperes, unless the equipment is listed and identified for use with 90°C (167°F) conductors. Because of Section 110-14(c)(2), the maximum ampacity (at 75°C) for these conductors is only 380 amperes. The total combined ampacity for each phase is, therefore, only 1,140 amperes. Section 240-3(c) requires a minimum conductor ampacity of 1,200 amperes.*

■ Ampacity Correction (Adjustment) Factors

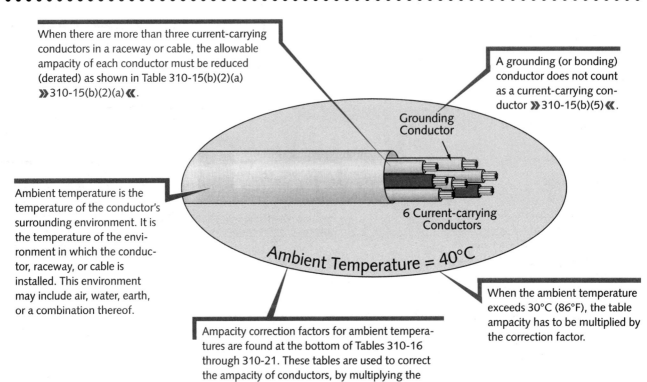

When there are more than three current-carrying conductors in a raceway or cable, the allowable ampacity of each conductor must be reduced (derated) as shown in Table 310-15(b)(2)(a) ≫ 310-15(b)(2)(a) ≪.

A grounding (or bonding) conductor does not count as a current-carrying conductor ≫ 310-15(b)(5) ≪.

Grounding Conductor

6 Current-carrying Conductors

Ambient temperature is the temperature of the conductor's surrounding environment. It is the temperature of the environment in which the conductor, raceway, or cable is installed. This environment may include air, water, earth, or a combination thereof.

Ambient Temperature = 40°C

Ampacity correction factors for ambient temperatures are found at the bottom of Tables 310-16 through 310-21. These tables are used to correct the ampacity of conductors, by multiplying the ampacity of the conductor by the appropriate factor.

When the ambient temperature exceeds 30°C (86°F), the table ampacity has to be multiplied by the correction factor.

The following are *not* counted as current-carrying conductors when using Table 310-15(b)(2)(a):

- Conductors of different systems, as provided in Section 300-3, unless the number of power and lighting conductors (current-carrying) exceeds three ≫ 310-15(b)(2)(a) *Exception No. 1* ≪.
- Conductors in cable trays, unless required by Section 318-11 ≫ 310-15(b)(2)(a) *Exception No. 2* ≪.
- Conductors in nipples of 24 inches or less ≫ 310-15(b)(2)(a) *Exception No. 3* ≪.
- Neutral conductors of normally balanced circuits containing three or more conductors ≫ 310-15(b)(4)(a) ≪.
- Grounding and bonding conductors ≫ 310-15(b)(5) ≪.

The following *shall* be counted as current-carrying conductors when using Table 310-15(b)(2)(a):

- The neutral conductor in a 3-wire circuit consisting of two phase (hot) wires and the neutral that is fed from a 4-wire, 3-phase wye-connected system ≫ 310-15(b)(4)(b) ≪.
- The neutral conductor of a 4-wire, 3-phase wye circuit where the major portion of the load consists of nonlinear loads, because harmonic currents are present in the neutral conductor ≫ 310-15(b)(4)(b) ≪. Examples: Electronic equipment (such as computers), electronic/electric-discharge lighting (such as fluorescent lights and lights with ballast), adjustable speed drive systems, and similar equipment. [Article 100—Definitions]
- A grounded conductor of any 2-wire circuit.

> **N O T E**
>
> *The maximum ampacities given in Tables 310-16 through 310-21 are based on a maximum of three current-carrying conductors and a maximum ambient temperature of 30°C (86°F). When these limitations are exceeded, the ampacities listed must be reduced, or **derated**.*

■ Continuous Loads

● ●

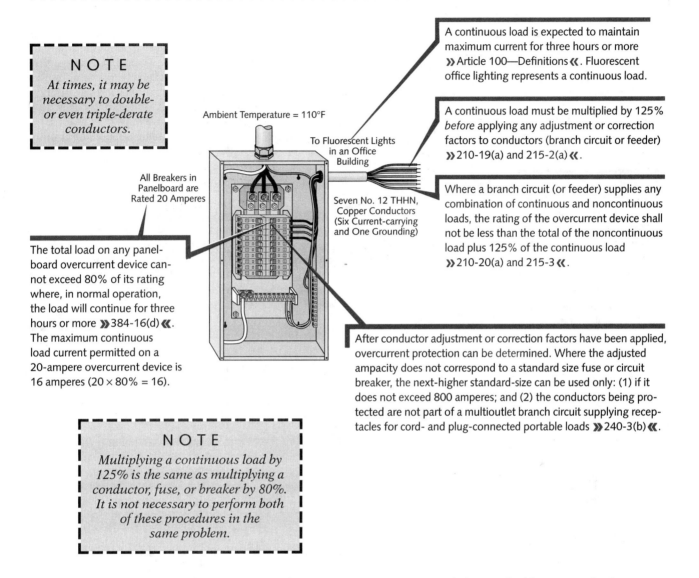

> **NOTE**
> *At times, it may be necessary to double- or even triple-derate conductors.*

Ambient Temperature = 110°F

To Fluorescent Lights in an Office Building

All Breakers in Panelboard are Rated 20 Amperes

Seven No. 12 THHN, Copper Conductors (Six Current-carrying and One Grounding)

A continuous load is expected to maintain maximum current for three hours or more ≫ Article 100—Definitions ≪. Fluorescent office lighting represents a continuous load.

A continuous load must be multiplied by 125% *before* applying any adjustment or correction factors to conductors (branch circuit or feeder) ≫ 210-19(a) and 215-2(a) ≪.

Where a branch circuit (or feeder) supplies any combination of continuous and noncontinuous loads, the rating of the overcurrent device shall not be less than the total of the noncontinuous load plus 125% of the continuous load ≫ 210-20(a) and 215-3 ≪.

The total load on any panel-board overcurrent device cannot exceed 80% of its rating where, in normal operation, the load will continue for three hours or more ≫ 384-16(d) ≪. The maximum continuous load current permitted on a 20-ampere overcurrent device is 16 amperes (20 × 80% = 16).

After conductor adjustment or correction factors have been applied, overcurrent protection can be determined. Where the adjusted ampacity does not correspond to a standard size fuse or circuit breaker, the next-higher standard-size can be used only: (1) if it does not exceed 800 amperes; and (2) the conductors being protected are not part of a multioutlet branch circuit supplying receptacles for cord- and plug-connected portable loads ≫ 240-3(b) ≪.

> **NOTE**
> *Multiplying a continuous load by 125% is the same as multiplying a conductor, fuse, or breaker by 80%. It is not necessary to perform both of these procedures in the same problem.*

To determine the maximum conductor ampacity and overcurrent protection, start with the actual Table ampacity for the type and size conductor, then derate as follows:

- No. 12 THHN, copper conductors have an ampacity of 30 amperes ≫ Table 310-16 ≪.
- Instead of multiplying the load by 125%, the conductor ampacity can be multiplied by 80% (the reciprocal of 125% is 80%). This calculation must be preformed before applying any adjustment or correction factors ≫ 210-19(a) ≪.

  30 amperes × 80% = 24 amperes

- Next, multiply the ampacity by 80%, because there are six current-carrying conductors in the conduit ≫ Table 310-15(b)(2)(a) ≪.

  24 amperes × 80% = 19.2 amperes

- Finally, multiply the ampacity by .87, because of the 110°F ambient temperature. This correction factor is found near the bottom of Table 310-16, in the 90°C column, in the row across from 105–113°F.

  19.2 amperes × .87 = 16.7 amperes

- The maximum ampacity for these No. 12 conductors is 16.7 amperes. The next-higher standard-size circuit breaker above 16.7 is 20 amperes.
- One last precaution: the load on the breaker cannot exceed 16 amperes because of the continuous load ≫ 384-16(d) ≪.

■ Support Requirements for Conductors in Vertical Raceways

Conductors in vertical raceways must be supported if the vertical rise exceeds Table 300-19(a) values ≫300-19(a)≪. While three specific methods for supporting cables are listed in Sections 300-19(b)(1), (2), and (3), other equally effective methods are also permitted.

Clamping devices constructed of, or employing, insulating wedges inserted in raceway ends is one type of conductor support. If clamping the insulation does not adequately support the cable, the conductor must also be clamped ≫300-19(b)(1)≪.

One cable support must be provided at the top of the vertical raceway, or as close to the top as practical. Intermediate supports must be provided so that supported conductor lengths do not exceed those specified In Table 300-19(a) ≫300-19(a)≪.

Wedges   Insulating Support   Insulator Used to Deflect Cable

Covered boxes can be inserted at the required intervals in which insulating supports are installed and secured in a satisfactory manner to withstand the weight of the conductors attached thereto ≫300-19(b)(2)≪.

Cables can be supported in junction boxes, by deflecting the cables at least 90 degrees and carrying them horizontally to a distance not less than twice the diameter of the cable, provided the cables are carried on at least two insulating supports, and are secured thereto by tie wires (if desired) ≫300-19(b)(3)≪.

In using this, cables must be supported at intervals not greater than 20% of the values in Table 300-19(a) ≫300-19(b)(3)≪.

Table 300-19(a)
Maximum Distance between Conductor Supports in Vertical Raceways (in feet)

AWG or Circular-Mil Size of Conductor	Aluminum or Copper-Clad Aluminum	Copper
No. 18 through No. 8	100	100
No. 6 through 1/0	200	100
2/0 through 4/0	180	80
250 through 350 kcmil	135	60
400 through 500 kcmil	120	50
600 through 750 kcmil	95	40
Over 750 kcmil	85	35

Summary

- Raceways cannot be supported by ceiling grid support wires, but can be supported by independent (additional) support wires.
- Raceways cannot support other raceways, cables, conductors, or non-electric equipment.
- The maximum total of conduit bends, between pull points, is four quarter-bends (360°).
- Rigid metal conduit is a heavy-walled metal raceway that can be threaded.
- Intermediate metal conduit has thinner walls than does rigid, and it can also be threaded.
- Unless made up with threaded couplings, IMC and RMC support requirements are the same as EMT.
- EMT, the thinnest walled classification of metal non-flexible raceways, provides protection from all but severe physical damage.
- Exposed rooftop installations of EMT must contain an equipment grounding conductor.

- Rigid nonmetallic conduit is commonly referred to as PVC.
- The distance between RNC supports increases as conduit size increases.
- Certain raceways, permitted as fixture whips, do not require support if installed in lengths of six feet or less.
- Wireways are troughs with hinged or removable covers for housing and protecting conductors and cables.
- Wireways can pass transversely through walls if the length passing through the wall is unbroken and if conductors can be accessed on both sides of the wall.
- Raceway fill can be calculated from Tables 4 and 5, located in Chapter 9.
- Conduit (and tubing) nipples can be filled to 60% of their cross-sectional area.

- Certain conductor properties are listed in Table 310-13.
- Conductor temperature limitations must be considered when determining overcurrent protection, especially an overcurrent protective device rated more than 800 amperes.

- Ambient temperature, the number of current-carrying conductors, and continuous loads can alter maximum conductor ampacity or overcurrent protection.

Unit 5 Competency Test

• •

NEC® Reference   **Answer**

_____ _____ 1. Where exposed to the weather, raceways shall be:

    a) rainproof and arranged to drain.

    b) watertight and arranged to drain.

    c) weatherproof and arranged to drain.

    d) raintight and arranged to drain.

_____ _____ 2. Most conduit and tubing type raceways have a maximum equivalent of four _____ bends between pull points, e.g., conduit bodies and boxes.

_____ _____ 3. The cross-sectional area of 2" electrical metallic tubing is _____ square inches.

_____ _____ 4. All cut ends of conduits shall be reamed or otherwise finished to remove rough edges. Where conduit is threaded in the field, a standard cutting die with a _____ inch taper per foot (1 in 16) shall be used.

_____ _____ 5. Multioutlet assemblies shall not be installed:

       I. in hoistways

      II. within dry partitions

     III. within dry locations

    a) I only   b) III only   c) I and II only   d) I, II, and III

_____ _____ 6. Splices and taps shall be permitted within a wireway provided they are accessible. The conductors, including splices and taps, shall not fill the wireway to more than _____ of its area at that point.

_____ _____ 7. A _____ shall be defined as a transverse raceway for electric conductors, providing access to predetermined cells of a cellular metal floor, thereby permitting the installation of electric conductors from a distribution center to the cells.

_____ _____ 8. In both exposed and concealed locations, where a cable or raceway-type wiring method is installed through bored holes in joist, rafters, or wood members, holes shall be bored so that the edge of the hole is not less than _____ from the nearest edge of the wood member.

_____ _____ 9. How many No. 14 THHN conductors, including an equipment grounding conductor, can be installed in a ⅜" flexible metal conduit using inside fittings?

_____ _____ 10. What is the maximum distance between supports for a straight run of 2" intermediate metal conduit made up with threaded couplings?

_____ _____ 11. The sum of cross-sectional areas of all contained conductors at any cross section of a wireway shall not exceed _____ fill.

_____ _____ 12. Where conductors carrying alternating current are installed in metal raceways, they shall be so arranged as to avoid heating the surrounding metal by _____.

NEC® Reference   Answer

_____ _____   13. A _____ is a type of surface or flush raceway designed to hold conductors and receptacles, assembled in the field or at the factory.

_____ _____   14. Where practicable, contact between dissimilar metals anywhere in the system shall be avoided to eliminate the possibility of _____.

_____ _____   15. What is the maximum number of No. 4 THHN conductors permitted in 1½" rigid metal conduit?

_____ _____   16. Where the same conductors enter and then exit a wireway through conduit or tubing, the distance between those raceway entries shall not be less than _____ times the trade diameter of the larger raceway.

_____ _____   17. All components of an exposed wiring system, installed where walls are frequently washed, must be mounted with a minimum of _____ airspace between components (boxes, fittings, conduits, etc.) and the wall (supporting surface).

_____ _____   18. Where an equipment bonding jumper is installed on the outside of a raceway, the length shall not exceed _____ and shall be routed with the raceway or enclosure.

_____ _____   19. Connections from headers to cabinets and other enclosures, in cellular concrete floor raceways, shall be made by means of:

        I. listed metal raceways

        II. listed nonmetallic raceways

        III. listed fittings

    a) I only   b) I and II only   c) I and III only   d) I, II, and III

_____ _____   20. Electrical nonmetallic tubing shall be secured at least every _____.

_____ _____   21. Intermediate metal conduit encased in a concrete slab, on the third floor of an office building, is considered a _____ location.

_____ _____   22. When an outlet from an underfloor raceway is discontinued, the sections of circuit conductors supplying the outlet:

    a) shall be permitted to be spliced

    b) shall be removed from the raceway.

    c) shall be permitted to be reinsulated.

    d) shall be handled like abandoned outlets on loop wiring.

_____ _____   23. Rigid metal conduit, elbows, couplings, and fittings shall be permitted to be installed in concrete, in direct contact with the earth, or in areas subject to severe corrosive influences where protected by _____ and judged suitable for the condition.

_____ _____   24. Where installed in raceways, conductors of size _____ and larger shall be stranded.

_____ _____   25. 3/0 AWG aluminum conductors in vertical raceways must be supported if the vertical rise is greater than _____ feet.

_____ _____   26. Where raceways containing ungrounded conductors _____ or larger enter a cabinet, box enclosure, or raceway, the conductors shall be protected by a substantial fitting providing a smoothly rounded insulating surface, unless the conductors are separated from the fitting or raceway by substantial insulating materials securely fastened in place.

_____ _____   27. Nonmetallic wireways shall be supported where run horizontally at intervals not to exceed _____, and at each end or joint, unless listed for other support intervals.

_____   _____   28. Fixtures shall not be used as a raceway for circuit conductors unless ____ for use as a raceway.

_____   _____   29. If a 1¼" EMT raceway containing three conductors is already filled to 20%, what is the raceway's cross-sectional area?

_____   _____   30. Metal wireways shall not contain more than ____ current-carrying conductors at any cross section.

_____   _____   31. Expansion fittings and telescoping sections of metal raceways shall be made electrically continuous by ____ or other means.

_____   _____   32. Rigid metal conduit shall be securely fastened within ____ of each outlet box, junction box, device box, cabinet, conduit body, or other conduit termination.

_____   _____   33. The maximum distance between supports for 4" rigid nonmetallic conduit is ____ feet.

_____   _____   34. When calculating the maximum number of conductors permitted in a conduit or tubing, all of the same size (total cross-sectional area including insulation), the next-higher whole number shall be used to determine the maximum number of conductors permitted when the calculation results in a decimal of ____ or larger.

_____   _____   35. Where insulated conductors are deflected within a wireway, either at the ends or where conduits, fittings, or other raceways or cables enter or leave the wireway, or where the direction of the wireway is deflected greater than ____, dimensions corresponding to Section 373-6 shall apply.

General Provisions

Objectives

After studying this unit, the student will:

- ✔ be able to calculate the minimum number of 15- and 20-ampere branch circuits in a one-family dwelling.
- ✔ know the requirements for single receptacles on individual branch circuits.
- ✔ understand the branch circuit ratings allowed for general-purpose receptacles.
- ✔ know how to layout general-purpose receptacles in a dwelling.
- ✔ know the receptacle ratings allowed on various size branch circuits.
- ✔ understand the requirements for receptacle boxes.
- ✔ have a good understanding of split-wire receptacles.
- ✔ know the requirements for wet bar receptacles.

- ✔ be familiar with the provisions concerning the minimum length of free conductors inside boxes.
- ✔ know the receptacle replacement requirements.
- ✔ understand the general requirements for lighting and switches.
- ✔ be familiar with the use of a white conductor as an ungrounded conductor.
- ✔ understand general requirements for devices and fixture boxes.
- ✔ know the provisions for outdoor receptacles.
- ✔ be familiar with the provisions concerning illuminating outdoor entrances and exits.
- ✔ understand the requirements for attaching receptacles and/or lighting to vegetation (such as trees).

Introduction

Unit 6 contains provisions concerning general areas, both inside and outside, of one-family dwellings. A one-family dwelling is a building consisting solely of one dwelling unit »Article 100«. Learning codes which pertain to one-family dwellings is of great importance, because these codes are the foundation for all residential wiring. One-family dwellings are built in all shapes and sizes. Some general codes apply to every dwelling, such as the placement of receptacles in habitable rooms. Other codes permit the installation of nonmetallic-sheathed cable in any level one-family (or two-family) dwelling, but prohibit its installation in multifamily dwellings exceeding three floors above grade. For this reason, wiring methods are explained in separate units. For information on codes pertaining to nonmetallic-sheathed cable, refer to Unit 4.

CAUTION <<<<<<<<<<<<<

Some state and local jurisdictions may not allow the use of nonmetallic-sheathed cable in any dwelling. Obtain a copy of any additional rules and regulations for your area.

Electrical blueprints (floor plans) may be optional on a one-family dwelling. Sometimes it is the responsibility of the electrician to design and install the electrical system. An electrical floor plan is shown below. The floor plan shows receptacles, lighting, switches, and other items which might be found on a residential plan. The *NEC*® references (shown here in red), are, of course, not present on general plans.

This chapter covers everything in a one-family dwelling from a single receptacle in an individual branch circuit to illumination of entrances and exits. Branch circuits will be explained briefly in this unit and in greater detail in later units. Receptacles have both general and specific codes. General codes are explained in this unit, while specific codes for rooms and areas with special receptacle requirements are explained in Unit 7. This unit also includes general requirements for lighting and switches. Specific requirements can be found in Unit 7.

ELECTRICAL FLOOR PLAN (BLUEPRINT)

Symbols Used on Floor Plan

⊖ Duplex Receptacle	⊖ 240 Volt Receptacle	S Single-Pole Switch	S_DS Dimmer Switch	◄ Telephone Outlet	Ceiling Paddle Fan	▭ Surface Fluorescent Fixture
⊖GFCI Ground-Fault Circuit-Interrupter Protected (GFCI) Receptacle	⊙ Floor Receptacle	S_3 Three-Way Switch	○ Ceiling Lighting Outlet	Ⓣ Thermostat		--- Switch Leg Indication
	⊙DW Dishwasher Outlet	S_4 Four-Way Switch	⊸○ Wall Lighting Outlet	(SD) Smoke Detector	⊢─┤ Fluorescent Strip	
⊖W.P.GFCI Weatherproof and GFCI Protected Receptacle	⊙GD Garbage Disposer Outlet	S_D Door Switch	Recessed Lighting Outlet	CH Chime		
	⊙TC Trash Compactor Outlet			⊡ Push Button	EF Exhaust Fan	

BRANCH CIRCUITS

■ General Purpose Branch Circuit

Branch circuit conductors shall have an ampacity not less than the maximum load served ≫210-19(a)≪.

General purpose branch circuit over-current protection shall not exceed 15 amperes for No. 14 copper; 20 amperes for No. 12 copper; and 30 amperes for No. 10 copper ≫240-3(d)≪.

A general purpose branch circuit may feed lights, receptacles, or both.

■ Required Branch Circuits

A minimum of two 20-ampere small appliance branch circuits are required ≫210-11(c)(1) and 210-52(b)≪.

At least one 20-ampere laundry branch circuit is required ≫210-11(c)(2) and 210-52(f)≪.

A minimum of one 20-ampere bathroom branch circuit is required ≫210-11(c)(3) and 210-52(d)≪.

■ Calculating 15-Ampere Branch Circuits

One-Family Dwelling
1,400 Square Feet
240/120 Volts

Calculation to find the minimum number of 15 ampere branch circuits:

► Calculate the general lighting and receptacle load by using Table 220-3(a).
► First, multiply 1,400 square feet by 3 volt-amperes. 1,400 × 3 = 4,200 volt-amperes
► Next, divide the total volt-amperes by 120 volts to find the total amperes. 4,200 ÷ 120 = 35 amperes
► Finally, divide 35 amperes by 15 (for a 15 ampere circuit) to find the minimum number of circuits. 35 ÷ 15 = 2.33 (Round up to the next whole number if the result of the calculation is not a whole number.)
► If 15 ampere circuits are installed for general lighting and receptacles, then, at least three circuits are required.

■ Calculating 20-Ampere Branch Circuits

One-Family Dwelling
1,400 Square Feet
240/120 Volts

Calculation to find the minimum number of 20 ampere branch circuits:

► Calculate the general lighting and receptacle load by using Table 220-3(a).
► First, multiply 1,400 square feet by 3 volt-amperes. 1,400 × 3 = 4,200 volt-amperes
► Next, divide the total volt-amperes by 120 volts to find the total amperes. 4,200 ÷ 120 = 35 amperes
► Finally, divide 35 amperes by 20 (for a 20 ampere circuit) to find the minimum number of circuits. 35 ÷ 20 = 1.75 (Round up to the next whole number if the result of the calculation is not a whole number.)
► If 20 ampere circuits are installed for general lighting and receptacles, then, at least two circuits are required.

> ### N O T E
> *At least four 20-ampere branch circuits are required (in addition to any installed for general lighting and receptacles): two small appliance branch circuits, one laundry branch circuit, and one bathroom branch circuit.*

Individual Branch Circuit for a Water Heater

Calculation to find the maximum overcurrent protective device:

▶ The overcurrent protection shall not exceed the rating marked on the appliance. If the rating is not marked, the overcurrent protection shall not exceed 150% of the rated current ≫ 422-11(e) ≪. $18.75 \times 150\% = 28.13$

▶ If the calculated rating does not correspond to a standard size fuse or breaker, as found in 240-6(a), the next higher standard rating shall be permitted ≫ 422-11(e) ≪. The next standard size fuse or breaker higher than the calculated rating of 28.13 amperes is 30 amperes.

▶ The maximum overcurrent protective device (fuse or breaker) is 30 amperes.

The minimum conductor size is No. 10 copper ≫ Table 310-16 ≪. No. 12 conductors are not permitted, because the overcurrent protection shall not exceed 20 amperes for No. 12 copper, which is insufficient for this application ≫ 240-3(d) ≪.

A branch circuit supplying a fixed storage-type water heater (120 gallon capacity or less) shall have a rating not less than 125% of the water heater nameplate rating ≫ 422-13 ≪.

Calculation to find the minimum circuit ampacity:

▶ First, divide the wattage by the voltage to find amperage. $4{,}500 \div 240 = 18.75$

▶ Next, multiply the amperage by 125% as required by Section 422-13. $18.75 \times 125\% = 23.44$

▶ The minimum ampacity required for this water heater is 23.44 amperes.

CAPACITY U.S. GAL. 50
VOLTAGE 240 AC, 1-PHASE
WATTAGE TOTAL 4,200

Ampere Rating of Receptacles

Receptacle branch circuits must have a rating of 20 amperes for laundry areas, bathrooms, kitchens (except for refrigeration equipment supplied from an individual branch circuit rated 15 amperes or greater), dining rooms, pantries, breakfast rooms, and similar areas. Throughout the remainder of a one-family dwelling, 15-ampere receptacle branch circuits are allowed.

Lighting and receptacles can share the same branch circuit, except for small-appliance receptacles, bathroom receptacles (unless the circuit supplies a single bathroom), and laundry receptacles ≫ 210-23(a) ≪. See Unit 7 for additional information concerning receptacles and lighting in these specific areas.

15-ampere overcurrent device

15-ampere Receptacles

20-ampere Receptacles

Receptacles installed on 15-ampere circuits must have an ampere rating not more than 15 amperes ≫ Table 210-21(b)(3) ≪.

A 20-ampere receptacle cannot be installed on a 15-ampere circuit.

20-ampere overcurrent device

15-ampere Receptacles

20-ampere Receptacles

Receptacles installed on 20-ampere circuits can have an ampere rating of either 15 or 20 ≫ Table 210-21(b)(3) ≪.

NOTE

Branch circuits for receptacles cannot be smaller than No. 14 ≫ 210-19(d) ≪. Generally, 15-ampere circuits require No. 14, and 20-ampere circuits require No. 12, copper conductors. Larger conductors may be needed to compensate for voltage drop, ambient temperature, or where more than three current-carrying conductors exist in a raceway or cable.

■ A Single Receptacle on an Individual Branch Circuit

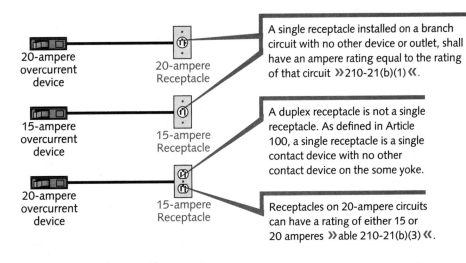

20-ampere overcurrent device

20-ampere Receptacle

A single receptacle installed on a branch circuit with no other device or outlet, shall have an ampere rating equal to the rating of that circuit »210-21(b)(1)«.

15-ampere overcurrent device

15-ampere Receptacle

A duplex receptacle is not a single receptacle. As defined in Article 100, a single receptacle is a single contact device with no other contact device on the some yoke.

20-ampere overcurrent device

15-ampere Receptacle

Receptacles on 20-ampere circuits can have a rating of either 15 or 20 amperes »able 210-21(b)(3)«.

RECEPTACLES

■ General Receptacle Placement

Wall space determines the number of receptacles, in a given dwelling. Receptacle outlets shall be installed in kitchens, family rooms, dining rooms, living rooms, parlors, libraries, dens, sun rooms, bedrooms, recreation rooms, or similar rooms or areas in accordance with the provisions specified in 210-52(a).

NOTE

Dwelling unit receptacle outlet general provisions (Section 210-52) apply to any part of a basement containing habitable rooms, such as a den, recreational room, etc.

6 feet   12 feet   6 feet   6 feet

6 feet

Receptacles shall be installed so that no point along the floor line in any wall space is more than 6 feet, measured horizontally, from an outlet in that space »210-52(a)(1)«.

The maximum distance between receptacles is 12 feet.

12 feet

6 feet

6 feet   12 feet

Whenever practicable, receptacle outlets shall be spaced equal distances apart »10-52(a)(1)«.

■ Wall Spaces 2 Feet in Width

Wall space 23" wide

No Receptacle Required

Wall space 24" wide

Receptacle Required

A receptacle is required for any wall space 2 feet or more in width (including space measured around corners), unbroken along the floor line by doorways, fireplaces, and similar openings »210-52(a)(2)(a)«.

Maximum Distance to a Receptacle

WARNING

All branch circuits that supply 125-volt, single-phase, 15- and 20-ampere receptacle outlets located in dwelling-unit bedrooms must be protected by an arc-fault circuit interrupter(s). This requirement becomes effective January 1, 2002 »210-12(b)«. An **arc-fault circuit interrupter** is a device intended to provide protection from the effects of arc faults by recognizing characteristics unique to arcing and by functioning to de-energize the circuit when an arc fault is detected »210-12(a)«.

An easy way to understand the placement of dwelling receptacles is to imagine having a floor lamp with a 6-foot cord. Anywhere this lamp is placed around the wall, there should be a receptacle within reach, without using as extension cord. Even when placed beside a door opening, an outlet should be within reach. If the lamp is placed next to a wall that is at least 24 inches wide, an outlet should be available within that wall space.

The maximum distance to any receptacle, along the floor line, measured horizontally, shall be 6 feet »210-52(a)(1)«.

Space Measured Around Corners

The maximum distance **between receptacles**, along the floor line, measured horizontally, shall be 12 feet »210-52(a)(1)«.

Wall space includes space measured around corners »210-52(a)(2)(a)«.

Fixed Panels

Glass door fixed panels in exterior walls are counted as wall space »210-52(a)(2)(b)«.

Sliding panels in exterior walls are not counted as wall space »210-52(a)(2)b«.

The maximum distance to any receptacle, along the floor line, measured horizontally, shall be 6 feet »210-52(a)(1)«.

Unit 6: General Provisions **101**

Fixed Room Dividers

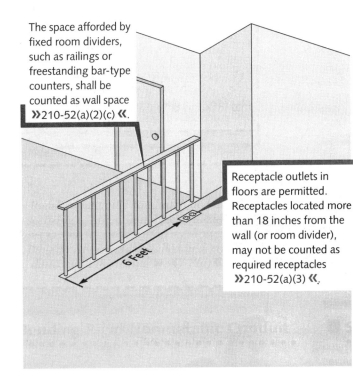

The space afforded by fixed room dividers, such as railings or freestanding bar-type counters, shall be counted as wall space ≫210-52(a)(2)(c)≪.

Receptacle outlets in floors are permitted. Receptacles located more than 18 inches from the wall (or room divider), may not be counted as required receptacles ≫210-52(a)(3)≪.

```
NOTE
Not all types of receptacle boxes are suit-
able for floor installation. Different types
of boxes may be required for different
types of floor construction, i.e., wood or
concrete. Receptacle floor boxes must be
listed specifically for the type of floor in
which it is installed ≫370-27(b)≪.
```

Balcony Handrail

Receptacle floor boxes must be listed specifically for the type of floor in which it is installed ≫370-27(b)≪.

A floor receptacle may be required if a balcony handrail is longer than 6 feet and the area is one listed in Section 210-52(a).

Miscellaneous Receptacle Requirements

While receptacles located more than 5½ feet above the floor are permitted, they are not counted as required receptacles ≫210-52≪.

6 feet

Although receptacle outlets within cabinets or cupboards are permitted, they are not counted as required receptacles ≫210-52≪.

```
NOTE
Height requirements for wall
receptacles are not defined.
```

Lighting fixtures and/or appliance with built-in receptacle(s) are permitted, but these receptacles do not count as required receptacles ≫210-52≪.

■ Electric Baseboard Heaters

A baseboard heater more than 12 feet in length does not eliminate the requirements of Section 210-52(a)(1). A receptacle is still required for a wall space containing a baseboard heater.

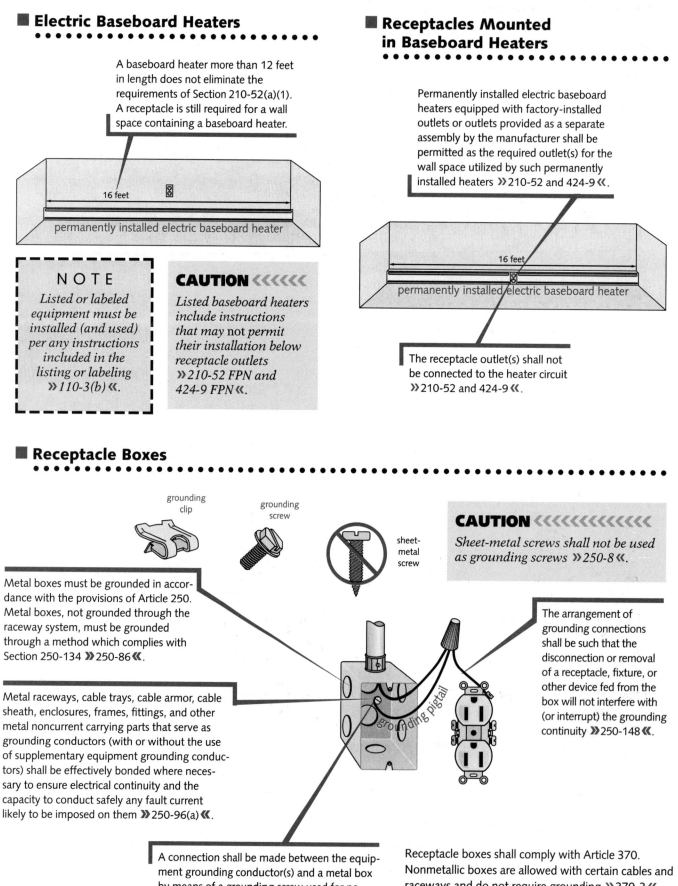

16 feet

permanently installed electric baseboard heater

NOTE
Listed or labeled equipment must be installed (and used) per any instructions included in the listing or labeling »110-3(b)«.

CAUTION «««««
Listed baseboard heaters include instructions that may not *permit their installation below receptacle outlets* »210-52 FPN and 424-9 FPN«.

■ Receptacles Mounted in Baseboard Heaters

Permanently installed electric baseboard heaters equipped with factory-installed outlets or outlets provided as a separate assembly by the manufacturer shall be permitted as the required outlet(s) for the wall space utilized by such permanently installed heaters »210-52 and 424-9«.

16 feet

permanently installed electric baseboard heater

The receptacle outlet(s) shall not be connected to the heater circuit »210-52 and 424-9«.

■ Receptacle Boxes

grounding clip

grounding screw

sheet-metal screw

CAUTION «««««««««««
Sheet-metal screws shall not be used as grounding screws »250-8«.

Metal boxes must be grounded in accordance with the provisions of Article 250. Metal boxes, not grounded through the raceway system, must be grounded through a method which complies with Section 250-134 »250-86«.

Metal raceways, cable trays, cable armor, cable sheath, enclosures, frames, fittings, and other metal noncurrent carrying parts that serve as grounding conductors (with or without the use of supplementary equipment grounding conductors) shall be effectively bonded where necessary to ensure electrical continuity and the capacity to conduct safely any fault current likely to be imposed on them »250-96(a)«.

grounding pigtail

The arrangement of grounding connections shall be such that the disconnection or removal of a receptacle, fixture, or other device fed from the box will not interfere with (or interrupt) the grounding continuity »250-148«.

A connection shall be made between the equipment grounding conductor(s) and a metal box by means of a grounding screw used for no other purpose. Other grounding devices listed for that purpose are allowed »250-148(a)«.

Receptacle boxes shall comply with Article 370. Nonmetallic boxes are allowed with certain cables and raceways and do not require grounding »370-3«. An expanded description of boxes, including box fill calculations, can be found in Unit 3 of this guidebook.

■ Grounding-Type Receptacles

Nongrounding-type receptacles may only be used to replace existing nongrounding receptacles (see "Receptacle Replacements" in this Unit).

Grounding-type receptacles shall be used on 15 and 20 ampere branch circuits ≫ 210-7(a)≪ .

For 15, 20, and 30 ampere circuits the equipment grounding conductor shall be the same size as the current-carrying conductors ≫ Table 250-122≪ . The size grounding conductor for a 15-ampere circuit is No. 14; a 20-ampere circuit is No. 12; and a 30-ampere circuit is No. 10, copper conductor.

N O T E

The grounding contact shall be effectively grounded by connection to the equipment grounding conductor of the circuit which supplies the receptacle ≫ 210-7(b) and (c)≪ .

No. 14 copper conductors are required for 15-ampere circuits, while 20-ampere circuits require No. 12 ≫ Table 310-16≪ .

■ Split-Wire Receptacles

A duplex receptacle, having two circuits on the same yoke, is considered a multiwire circuit. Under this condition, in dwelling units, the circuits are not thought of as separate. A means must be provided to simultaneously disconnect all ungrounded (hot) conductors ≫ 210-4(b)≪ . This is accomplished through the use of either one double-pole breaker or two single-pole breakers with tie handles.

A switch has been installed to control one of the multi-wire branch circuits.

A split-wire receptacle receiving power from a single source (one breaker or one fuse) is not a multi-wire receptacle.

WARNING

In multiwire branch circuits, the continuity of a grounded conductor shall not be dependent upon the device ≫ 300-13(b)≪ . If breaking the grounded conductor at the receptacle breaks the circuit down the line, then the grounded conductors must *not* be connected to the receptacle. Simply splice the grounded conductors and install a jumper wire to the receptacle.

Switch-controlled, split-wire duplex receptacle(s) are sometimes installed in lieu of a lighting outlet ≫ 210-70(a)(1) *Exception No. 1*≪ .

The tab has been removed so one receptacle can be controlled from the switch, and the other receptacle can remain live, whether the switch is on or off.

120 Volts

0 Volts

N O T E

For additional information concerning multiwire branch circuits, see the definition located in Unit 2.

A multiwire receptacle circuit consists of one (or more) duplex receptacles, one (or more) multiple receptacles, two (or more) single receptacles, or combinations thereof.

One half of the duplex receptacle is controlled by a wall switch, while the other half is a typical receptacle.

■ Wet Bar Receptacles

Ground-fault circuit interrupter (GFCI) protection is required where receptacles are installed to serve countertop surfaces *and* are located within 6 feet of the outside edge of the wet bar sink »210-8(a)(7)«.

Receptacles within 6 feet of the sink, which do not serve the countertop, do not require GFCI protection »210-8(a)(7)«.

For cord- and plug-connected appliances, an accessible, separable connector or an accessible plug and receptacle shall be permitted to serve as the disconnecting means. Where the separable connector (or plug and receptacle) are not accessible, cord- and plug-connected appliances shall have a disconnecting means in accordance with Section 422-31 »422-32(a)«.

If located outside the kitchen or dining room, wet bar countertop receptacle placement is not specified »210-52(c)«.

■ Room Air Conditioner Receptacle

If supplied by an individual branch circuit, the rating of a cord- and attachment plug-connected room air conditioner can be no more than 80% of the circuit's ampere rating »210-23(a) and 440-62(b)«.

Where other receptacles, lighting units, or appliances are also supplied, the rating of a cord- and attachment plug-connected room air conditioner shall not exceed 50% of the branch circuit ampere rating »210-23(a) and 440-62(c)«.

The attachment plug is permitted to serve as the disconnecting means »440-13«.

The length of the supply cord shall not exceed 10 feet for a 120-volt, or 6 feet for a 208- or 240-volt room air conditioner »440-64«.

> **NOTE**
> *For additional provisions pertaining to room air conditioners, see Part G of Article 440.*

■ Length of Free Conductor

At least 6 inches of free conductor, measured from the point in the box where it emerges from its raceway or cable sheath, shall be left at each outlet, junction, and switch point »300-14«.

Conductors must extend a minimum of 3 inches outside of the box opening where the opening to the box is less than 8 inches in any dimension »300-14«.

> **NOTE**
> *No minimum length is specified for conductors that are not spliced or terminated at the outlet, junction, or switch point »300-14 Exception«.*

RECEPTACLE REPLACEMENTS

■ Grounding Means Present or Installed in the Enclosure

A grounding-type receptacle shall be installed in any receptacle enclosure having a grounding means **》210-7(d)(1)《**.

A grounding-type receptacle shall be connected to the grounding conductor in accordance with Section 210-7(c) or Section 250-130(c).

Replace with...

If an equipment grounding conductor is installed and connected to the grounding electrode system in accordance with Section 250-130(c), a grounding-type receptacle shall be installed **》210-7(d)(1)《**.

NOTE

An existing two-wire receptacle may be attached to a metal receptacle enclosure. Just because the box is metal, does not guarantee that it is grounded to the grounding electrode system.

■ Receptacles Requiring GFCI Protection

Because of changes in GFCI protection requirements, some old receptacles must now be replaced with GFCI protected receptacles **》210-7(d)(2)《**.

Replace with...

All receptacles which serve kitchen countertop surfaces shall have ground-fault circuit-interrupter protection **》210-8(a)(6)《**.

■ No Grounding Means Present . . . Nongrounding-Type Replaces Nongrounding-Type

Existing Two-wire Circuit

Where no grounding means exists in the receptacle enclosure, a nongrounding-type receptacle can replace an existing nongrounding-type receptacle **》210-7(d)(3)(a)《**.

Replace with...

■ No Grounding Means Present . . . GFCI Replaces Nongrounding-Type

Existing Two-wire Circuit

Where no grounding means exists in the receptacle enclosure, a GFCI receptacle can replace the old receptacle **》210-7(d)(3)(b)《**.

No Equipment Ground

Replace with...

This GFCI receptacle must be labeled **"No Equipment Ground"** **》210-7(d)(3)(b)《**.

◼ No Grounding Means Present . . .
GFCI and Grounding-Type Replaces Nongrounding-Type

Existing
Two-wire
Circuit

Existing
Two-wire
Circuit

This GFCI receptacle must be labeled **"No Equipment Ground"** ⟫ 210-7(d)(3)(b) ⟪.

Replace with...

Existing two-wire receptacle is not being replaced

No
Equipment
Ground

Where no grounding means exists in the receptacle enclosure, a GFCI receptacle can replace an existing nongrounding-type receptacle ⟫ 210-7(d)(3)(b) ⟪.

No
Equipment
Ground

A grounding-type receptacle may replace a nongrounding-type receptacle where the circuit is ground-fault protected. These receptacles shall be labeled **GFCI Protected** and **No Equipment Ground** ⟫ 210-7(d)(3)(c) ⟪.

GFCI
Protected

Replace with...

An equipment grounding conductor shall not be connected to these receptacles ⟫ 210-7(d)(3)(c) ⟪.

> **CAUTION** ⟪⟪⟪⟪⟪⟪⟪⟪⟪
>
> *An equipment grounding conductor shall not be connected from the ground-fault circuit-interrupter-type receptacle to any outlet supplied from the GFCI receptacle ⟫ 210-7(d)(3)(b) ⟪.*

◼ No Grounding Means Present . . .
Grounding-Type Replaces Nongrounding-Type

Existing
Two-wire
Circuit

A GFCI breaker has been installed to feed the existing two-wire circuit

A grounding-type receptacle may replace a nongrounding-type receptacle where the circuit is ground-fault protected.

No
Equipment
Ground

GFCI
Protected

Replace with...

Existing two-wire receptacle is not being replaced

These receptacles shall be labeled **"GFCI Protected"** and **"No Equipment Ground"** ⟫ 210-7(d)(3)(c) ⟪.

No
Equipment
Ground

GFCI
Protected

Replace with...

An equipment grounding conductor shall not be connected to these receptacles ⟫ 210-7(d)(3)(c) ⟪.

> **N O T E**
>
> *Consumers should be alerted to the fact that a receptacle is not grounded. This is especially important as more goods (such as personal computers) require grounded receptacles.*

LIGHTING AND SWITCHING

■ Outdoor Entrances and Exits

At least one wall switch-controlled lighting outlet shall be installed in every habitable room of a dwelling »210-70(a)(1) «.

Entry

Front Porch

At least one wall switch-controlled lighting outlet is required to provide exterior side illumination of outdoor entrances or exits having grade level access »210-70(a)(2) «.

> **NOTE**
> *Remote, central, or automatic lighting control is permitted for outdoor entrances »210-70(a)(2) Exception «.*

■ General Lighting

Bedroom

Bathroom

Hallway

Closet

Bedroom

Except for kitchens and bathrooms, every habitable room is permitted one or more wall switch-controlled receptacle(s) in lieu of a lighting outlet »210-70(a)(1) *Exception No.1* «. Switch-controlled receptacles allow for cord-connected lighting.

At least one wall switch-controlled lighting outlet shall be installed in every bathroom of a dwelling »210-70(a)(1) «.

Dwelling hallways require a minimum of one wall switch-controlled lighting outlet »210-70(a)(2) «.

At least one wall switch-controlled lighting outlet shall be installed in every habitable room of a dwelling »210-70(a)(1) «.

All switches shall be located so that they may be operated from a readily accessible place, not more than 6 feet, 7 inches, from the floor to the center of the switch operating handle »380-8(a) «. No other height requirements are stipulated for wall switches.

> **NOTE**
> *Occupancy sensors are permitted to control lighting outlets in lieu of wall switches »210-70(a)(1) Exception No. 2 «. One of two requirements must be met in order to use occupancy sensors: either a regular wall switch must be present in addition to the sensor, or the sensor must be equipped with a manual override that allows the sensor to function as a switch. Either type of wall switch shall be located at a customary location.*

■ Ceiling-Suspended (Paddle) Fans

A listed box is required for a ceiling-suspended (paddle) fan where the box is the only support for the fan »370-27(c) «. Those boxes are permitted to support ceiling fans weighing no more than 35 pounds »422-18(a) «. Include any light kit or accessory when determining the total weight of the fan.

> **NOTE**
> *Ceiling-suspended (paddle) fans exceeding 35 pounds shall be supported independently of the outlet box »422-18(b) «.*

■ Cables Installed to Feed Switches

> **N O T E**
>
> *An outlet box can support a light fixture weighing no more than 50 pounds, unless the box is listed for a weight greater than the fixture. A separate means for supporting a light fixture (independent of the outlet box) is also permitted* **»410-16(a)«**.

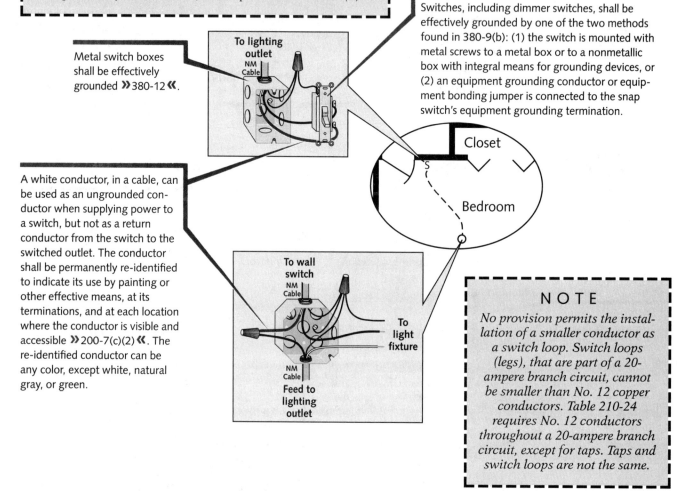

Metal switch boxes shall be effectively grounded **»380-12«**.

Switches, including dimmer switches, shall be effectively grounded by one of the two methods found in 380-9(b): (1) the switch is mounted with metal screws to a metal box or to a nonmetallic box with integral means for grounding devices, or (2) an equipment grounding conductor or equipment bonding jumper is connected to the snap switch's equipment grounding termination.

A white conductor, in a cable, can be used as an ungrounded conductor when supplying power to a switch, but not as a return conductor from the switch to the switched outlet. The conductor shall be permanently re-identified to indicate its use by painting or other effective means, at its terminations, and at each location where the conductor is visible and accessible **»200-7(c)(2)«**. The re-identified conductor can be any color, except white, natural gray, or green.

> **N O T E**
>
> *No provision permits the installation of a smaller conductor as a switch loop. Switch loops (legs), that are part of a 20-ampere branch circuit, cannot be smaller than No. 12 copper conductors. Table 210-24 requires No. 12 conductors throughout a 20-ampere branch circuit, except for taps. Taps and switch loops are not the same.*

OUTDOOR RECEPTACLES AND LIGHTING

■ Required Outdoor Receptacles

One-family dwellings require at least one receptacle outlet (accessible at grade level and not more than 6 feet, 6 inches above grade) at the front and back of the dwelling **»210-52(e)«**.

Each grade level unit of a two-family dwelling requires at least one receptacle outlet (accessible at grade level and not more than 6 feet, 6 inches above grade) at the front and back of the dwelling **»210-52(e)«**.

Receptacle Enclosures and Covers

A wet location receptacle where the product plugged into it will be attended while in use (e.g. portable tools, etc.) requires an enclosure that is weatherproof when the attachment plug is removed ≫410-57(b)(2)≪.

An outdoor receptacle outlet shall be located so that water accumulation is unlikely to contact the outlet cover or plate ≫410-57(f)≪.

A wet location receptacle where the product plugged into it is *not* attended while in use (e.g., sprinkler system controllers, landscape lighting, holiday lights, etc.) requires an enclosure that is weatherproof whether the attachment plug is inserted or removed ≫410-57(b)(1)≪.

The enclosure for a receptacle installed in a wall flush-mounted outlet box requires the use of a weatherproof faceplate assembly that provides a watertight seal between the plate and the wall ≫410-57(e)≪.

An outdoor receptacle, whether in a location protected from the weather or in other damp locations, requires a weatherproof receptacle enclosure when the receptacle is covered (attachment plug not inserted and receptacle cover closed). A receptacle location protected from the weather includes areas such as: under roofed, open porches, canopies, marquees, etc., which are not subjected to beating rain or water run-off ≫410-57(a)≪.

Receptacles More Than 6 Feet 6 Inches Above Grade

At least one receptacle outlet, accessible at grade level and not more than 6 feet, 6 inches above grade, shall be installed at the front *and* back of the dwelling ≫210-52(e)≪.

7 Feet

While receptacles are permitted more than 6 feet, 6 inches above grade, they do not count as required outdoor receptacles.

NOTE
Outdoor receptacles require GFCI protection ≫210-8(a)(3)≪.

NOTE
GFCI protection is not required for receptacles that are not readily accessible and are supplied by a dedicated branch circuit for electric snow-melting or de-icing equipment. The installation must comply with Article 426 ≫210-8(a)(3) Exception≪. (See Section 426-28 for ground-fault protection of equipment.) Outdoor receptacles that are not readily accessible (such as under an eve) require GFCI protection.

■ Support for Receptacle Enclosures

As a general rule, a single conduit cannot support an enclosure that contains a device ≫ 70-23(f)≪ .

Rigidly and securely fasten in place all boxes and raceways ≫ 300-11(a) and 370-23(a)≪ .

An enclosure containing devices can be supported by two conduits under the conditions found in Section ≫ 370-23(f)≪ : (1) The box shall not exceed 100 cubic inches; (2) The box has either threaded entries or hubs identified for the purpose; (3) The box shall be supported by two or more conduits threaded wrenchtight into the enclosure or hubs; (4) Each conduit shall be secured within 18 inches of the enclosure if all entries are on the same side.

Supporting an enclosure from grade requires rigid support in the form of conduit or a brace (metal, polymeric, or wood) ≫ 370-23(b)≪ . Wood brace cross sections must be at least 1 inch by 2 inches ≫ 370-23(b)(2)≪ . Wood braces must be treated to withstand the conditions when used in wet locations ≫ 370-23(b)(2)≪ .

> **NOTE**
> *Dwelling unit outdoor receptacles require GFCI protection for personnel ≫ 210-8(a)(3)≪ .*

■ Illumination of Entrances and Exits

Install at least one wall switch-controlled lighting outlet to provide illumination on the exterior side of outdoor entrances and exits (doorways) ≫ 210-70(a)(2)≪ . It is not required that the light fixture be located adjacent to the doorway, as long as illumination around the doorway is provided.

Install at least one receptacle outlet at the front and back of the dwelling ≫ 210-52(e)≪ . Outdoor receptacles require GFCI protection ≫ 210-8(a)(3)≪ .

■ Wiring in Trees

Outdoor lighting fixtures and associated equipment can be supported by trees ≫410-16(h)≪ .

Vegetation (such as trees) cannot be used to support overhead conductor spans ≫225-26≪ .

> **NOTE**
> *Direct buried conductors and cables, emerging from the ground, shall be protected by enclosures or raceways which extend from the minimum cover distance required by Section 300-5(a) below grade to a point 8 feet or more above finished grade ≫300-5(d)≪.*

Summary

- General-purpose branch circuits may feed lights, receptacles, or any combination thereof.
- An individual branch circuit feeds only one piece of equipment.
- The maximum distance to any receptacle, measured horizontally along the floor line, is 6 feet.
- A receptacle is required for any wall space 2 feet or wider.
- Twnety-ampere receptacles are not permitted on 15-ampere branch circuits.
- Both 15- and 20-ampere receptacles are permitted on 20-ampere branch circuits.
- GFCI-protected receptacles may replace nongrounded-type receptacles where the receptacle enclosure is without a grounding means.
- GFCI-protected receptacles, on nongrounded systems, must be marked in accordance with Section 210-7(d)(3) provisions.

- A means to simultaneously disconnect all ungrounded conductors must be provided where there are two circuits on the same yoke.
- A switch-controlled lighting outlet shall be installed in every habitable room of a dwelling, including bathrooms, hallways, and stairways.
- Boxes are permitted to support light fixtures weighing 50 pounds or less.
- Listed boxes are permitted to support ceiling fans weighing no more than 35 pounds.
- One-family dwellings require at least one exterior GFCI protected receptacle outlet at the front and back of the dwelling.
- Outdoor entrances and exits must have at least one wall switch-controlled lighting outlet.
- Although lighting and receptacles can be supported by trees, trees cannot support overhead conductor spans.

Unit 6 Competency Test

• •

NEC® Reference   Answer

_____   _____   1. A 20-ampere rated duplex receptacle may be installed on a _____ ampere branch circuit.

    a) 15 ampere   b) 20 ampere
    c) 15 or 20 ampere   d) 20 or 25 ampere

_____   _____   2. The continuity of a grounded conductor shall not be dependent upon the device in _____.

    a) branch circuits not having a grounded conductor

    b) multiwire branch circuits

    c) individual branch circuits

    d) branch circuits

_____   _____   3. Receptacles located _____ above the floor are not counted in the required number of receptacles along the wall.

_____   _____   4. At least _____ inches of free conductor, measured from the point in the box where it emerges from its raceway or cable sheath, shall be left at each outlet, junction, and switch point for splices or the connection of fixtures or devices.

_____   _____   5. In nonmetallic-sheathed cable, the equipment grounding conductor for 15-, 20-, and 30-ampere branch circuits _____.

    a) is required only with aluminum or copper-clad aluminum cable

    b) must be the same size as the insulated circuit conductors

    c) may be one size smaller than the insulated circuit conductors

    d) may be two sizes smaller than the insulated circuit conductors

_____   _____   6. Receptacles installed for the attachment of portable cords shall be rated at not less than 15 amperes, 125 volts, or _____ amperes, 250 volts, and shall be of a type not suitable for use as lampholders.

_____   _____   7. A fixture that weighs more than 6 pounds or exceeds _____ inches in any dimension shall not be supported by the screw shell of a lampholder.

_____   _____   8. Which of the following is not a standard classification for a branch circuit supplying several loads?

    a) 20 amperes   b) 25 amperes   c) 30 amperes   d) 50 amperes

_____   _____   9. A cord connector that is supported by a permanently-installed cord pendant shall be considered a(n) _____ outlet.

_____   _____   10. Grounding-type receptacles shall be installed only on circuits of the _____ for which they are rated, except as provided in Tables 210-21(b)(2) and (b)(3).

    I. voltage class

    II. wattage

    III. current

    a) I only   b) I and III only   c) I and II only   d) I, II, and III

_____   _____   11. Ceiling suspended fans that do not exceed _____ pounds in weight, with or without accessories, shall be permitted to be supported by outlet boxes identified for such use and supported in accordance with Sections 370-23 and 370-27.

_____   _____   12. Receptacle outlets in floors shall not be counted as part of the required number of receptacle outlets unless located within _____ of the wall.

NEC® Reference	Answer
_____	_____

13. Fixtures shall be wired with conductors having insulation suitable for the environmental conditions, _____, _____, and _____ to which the conductors will be subjected.

14. A light fixture that weighs more than 50 pounds shall be supported _____.

15. One 15-ampere rated duplex receptacle may be installed on a _____ ampere individual branch circuit.

16. A branch circuit that supplies only one utilization equipment is known as a(n) _____ branch circuit.

17. The maximum distance between receptacles in a one-family dwelling is _____ feet.

18. A nongrounding-type receptacle has been replaced with a new grounding-type duplex receptacle. The 15-ampere circuit breaker feeding the circuit has been replaced with a 15-ampere ground-fault circuit-breaker. The new receptacle must be marked _____.

    I. "Two-Wire Circuit"

    II. "No Equipment Ground"

    III. "GFCI Protected"

    a) III only   b) I and III only   c) II and III only   d) I, II, and III

19. An outlet where one or more receptacles are installed is called a _____.

20. What is the maximum distance that a receptacle wall box can be set back, from the finished surface of a 1/4 inch wood paneling wall?

    a) 0.0625 inch   b) 0.125 inch   c) 0.25 inch   d) 0.0 inch

21. Switch-controlled receptacles (in lieu of a lighting outlet) are permitted in all but which of the following?

    a) bedroom   b) library   c) dining room   d) hallway

22. A 50-ampere rated single receptacle may be installed on a _____ ampere individual branch circuit.

23. A _____ is a system or circuit conductor that is intentionally grounded.

24. The definition of a branch circuit is _____.

    a) the circuit conductors between the service and the sub-panel

    b) the circuit conductors prior to the final overcurrent device protecting the circuit

    c) the circuit conductors between the final overcurrent device protecting the circuit and the outlet(s)

    d) the circuit conductors between the final overload device protecting the circuit and the outlet(s

25. In dwelling units, a duplex receptacle with two circuits on the same yoke must have _____ as a means of disconnect.

    I. two single-pole circuit breakers without tie handles

    II. one double-pole circuit breaker

    III. two single-pole circuit breakers with tie handles

    a) I only   b) I or II only   c) II only   d) II or III only

UNIT 7 — Specific Provisions

Objectives

After studying this unit, the student will:

- ✔ know the required ampere rating for receptacles and branch circuits in kitchens, pantries, dining rooms, breakfast areas, and similar areas.
- ✔ be familiar with the requirements for kitchen countertop receptacle placement, including islands and peninsulas.
- ✔ know the minimum number of circuits required for kitchens, pantries, dining rooms, breakfast areas, and similar areas.
- ✔ understand requirements pertaining to permanently connected, as well as cord and plug connected appliances.
- ✔ be able to perform load calculations for appliance branch circuits.
- ✔ know the specific provisions for placement of GFCI-protected receptacles.
- ✔ understand the requirements for hallway and stairway receptacles and lighting.

- ✔ be able to identify which fixtures are, and which are not, permitted in closets, and the placement thereof.
- ✔ understand the comprehensive definition of a bathroom and know the requirements for receptacles, lighting, and fans.
- ✔ know the requirements for receptacles and lighting in attached garages, detached garages, and basements.
- ✔ understand the requirements for laundry (including clothes dryers) receptacles and branch circuits.
- ✔ be familiar with the requirements for attic and crawl space lighting and receptacles.
- ✔ understand lighting and receptacle requirements around HVAC equipment.

Introduction

Unit 6 contains provisions concerning general areas, both inside and outside, of one-family dwellings. The material in Unit 7 addresses more complex issues, requiring additional provisions for specific areas (such as kitchens, hallways, clothes closets, bathrooms, garages, basements, etc.). For instance, a kitchen may require the general provisions found in Unit 6, as well as specific provisions found in this unit. A dining room having no countertop surface requires receptacle placement by the general rule, but it must be fed from a 20-ampere branch circuit. Some areas require only one receptacle outlet, while bathroom receptacle requirements depend on the number and location of sinks. This unit also includes specific requirements for lighting outlets and switches. The pages that follow provide significant information on specific provisions for areas such as kitchens, dining rooms, breakfast areas, hallways, stairways, clothes closets, bathrooms, attached as well as detached garages, basements, attics, and crawl spaces.

KITCHENS, DINING ROOMS, AND BREAKFAST AREAS

■ Receptacle and Branch Circuit Rating

Receptacles installed on the 20-ampere small appliance circuits may be rated either 15 or 20 amperes ≫ Table 210-2(b)(3) ≪.

All receptacles installed to serve kitchen countertops shall have ground-fault circuit-interrupter protection ≫ 210-8(a)(6) ≪.

2' Max.   2' Max.   4' Max.

> **NOTE**
> *In kitchens, pantries, breakfast rooms, and dining areas, circuits which serve countertop surface receptacles and general purpose receptacles shall be 20-ampere small appliance branch circuits ≫ 210-52(b)(1) ≪. A minimum of two 20-ampere small appliance branch circuits are required ≫ 210-11(c)(1) ≪.*

Receptacle placement in a kitchen is determined by Section 210-52(a) and (c). All of the general receptacle placement provisions found in Section 210-52(a) apply, as well as additional requirements found in Section 210-52(c) for countertops.

Twenty-four (24) inches is the maximum distance to any receptacle, measured horizontally along the wall line ≫ 210-52(c)(1) ≪. An easy way to understand countertop receptacle placement is to imagine a toaster with a 24-inch cord. Anywhere this toaster is placed around the countertop wall, there should be a receptacle within reach.

■ Other Outlets Fed from Small Appliance Branch Circuits

A dedicated clock outlet, installed to provide power and support, is permitted on a small appliance branch circuit ≫ 210-52(b)(2) *Exception No. 1* ≪.

Outdoor receptacles are not permitted on small appliance branch circuits ≫ 210-52(b)(2) ≪.

> **NOTE**
> *A 120-volt receptacle installed for gas cooking equipment can be fed from a small appliance branch circuit ≫ 210-52(b)(2) Exception No. 2 ≪.*

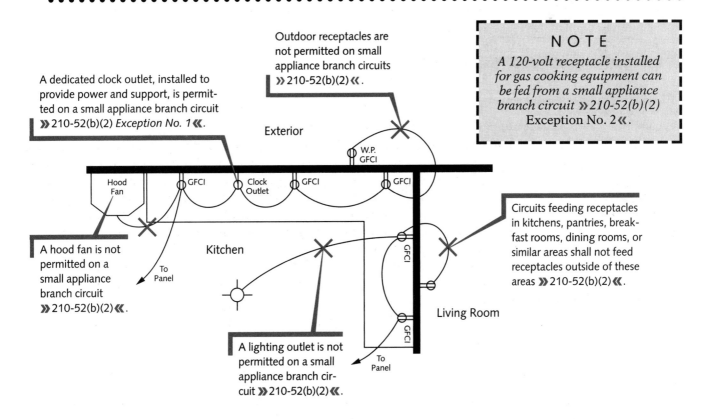

A hood fan is not permitted on a small appliance branch circuit ≫ 210-52(b)(2) ≪.

A lighting outlet is not permitted on a small appliance branch circuit ≫ 210-52(b)(2) ≪.

Circuits feeding receptacles in kitchens, pantries, breakfast rooms, dining rooms, or similar areas shall not feed receptacles outside of these areas ≫ 210-52(b)(2) ≪.

■ Kitchen Countertop Receptacle Placement

Twenty-four (24) inches is the maximum distance to any receptacle, along the wall line, measured horizontally ›› 210-52(c)(1)‹‹ .

A receptacle installed for refrigeration equipment may be fed from one of the 20-ampere small appliance branch circuits, or from an individual branch circuit rated 15 or 20 amperes ›› 210-52(b)(1) *Exception No 2*‹‹ .

Receptacles not serving kitchen countertops do not require GFCI protection.

A receptacle installed behind a refrigerator does not count as a required countertop receptacle ›› 210-52(c)(5)‹‹ .

Each wall counter space 12 inches or wider requires a receptacle ›› 210-52(c)(1)‹‹ .

All receptacles installed to serve kitchen countertop surfaces shall have ground-fault circuit-interrupter protection ›› 210-8(a)(6)‹‹ .

Branch circuits in kitchens shall be 20-ampere circuits ›› 210-52(b)(1)‹‹ .

A minimum of two small appliance branch circuits are required for receptacles serving kitchen countertops ›› 210-52(b)(3)‹‹ . Either, or both, of these small appliance branch circuits may feed other receptacles in the kitchen, pantry, breakfast room and/or dining area.

Receptacles located more than 18 inches above countertops are allowed but do not count as required receptacles ›› 210-52(c)(5)‹‹ .

■ Kitchen Countertop Receptacles

As a general rule, receptacle outlets shall be located above the countertop ›› 210-52(c)(5)‹‹ . Receptacles located more than 18 inches above countertops are allowed but do not count as required receptacles. Installation of receptacles below the countertop are allowed by some exceptions (see Section 210-52(c)(5) *Exception*).

Ground-fault circuit-interrupter protection is required for receptacles installed to serve kitchen countertops ›› 210-8(a)(6)‹‹ .

Receptacles located inside cabinets or cupboards, which do not serve kitchen countertop surfaces, do not require ground-fault circuit-interrupter protection ›› 210-8(a)(6)‹‹ .

GFCI protection is not needed for receptacles installed along the wall to meet Section 210-52(a) requirements.

Receptacles located inside cabinets or cupboards do not count as required kitchen countertop receptacles ›› 251-52‹‹ .

■ Island Receptacle Placement

At least one receptacle outlet shall be installed at each island counter space with a long dimension of 24 inches, or more, and a short dimension of 12 inches or more »210-52(c)(2)«.

A receptacle is required for each island counter space with at least a 12 by 24 inch area that is separated from other counter space because of range tops, refrigerators, or sinks »210-52(c)(4)«.

■ Peninsular Receptacle Placement

Counter spaces separated by range tops, refrigerators, or sinks require a receptacle for each 12 by 24 inch area »210-52(c)(4)«.

A receptacle is required for each peninsular counter space with at least a 12 by 24 inch area that is separated from other counter space because of range tops, refrigerators, or sinks »210-52(c)(3)«.

■ Receptacles Mounted Below Countertops

Where an island or peninsular countertop is flat across its entire surface, receptacle installation below the countertop is permitted under certain conditions. If a wall, backsplash, overhead cabinet, or similar area is available above the counter, the receptacle must be mounted above the countertop. Receptacles installed more than 18 inches above the countertop, such as in an overhead cabinet, are permitted but are not included as required countertop receptacles »210-52(c)(5) *Exception*«.

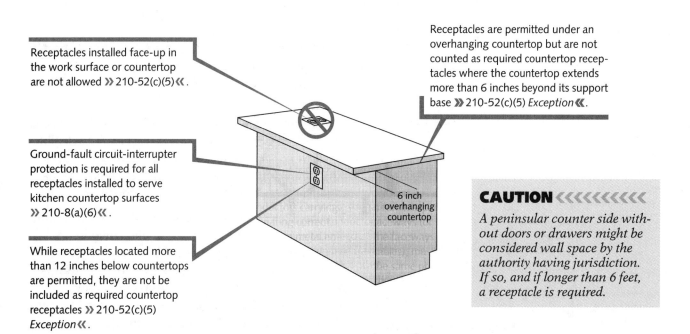

Receptacles installed face-up in the work surface or countertop are not allowed »210-52(c)(5)«.

Ground-fault circuit-interrupter protection is required for all receptacles installed to serve kitchen countertop surfaces »210-8(a)(6)«.

While receptacles located more than 12 inches below countertops are permitted, they are not be included as required countertop receptacles »210-52(c)(5) *Exception*«.

Receptacles are permitted under an overhanging countertop but are not counted as required countertop receptacles where the countertop extends more than 6 inches beyond its support base »210-52(c)(5) *Exception*«.

6 inch overhanging countertop

CAUTION ‹‹‹‹‹‹‹‹‹‹

A peninsular counter side without doors or drawers might be considered wall space by the authority having jurisdiction. If so, and if longer than 6 feet, a receptacle is required.

■ Permanently Connected Appliances

A disconnecting means is not required if the branch circuit switch or circuit breaker can be seen while working on the appliance ≫ 422-31(b) ≪.

A disconnecting means is not required if the branch circuit switch or circuit breaker can be locked in the open (off) position ≫ 422-31(b) ≪.

Water heaters shall not be connected with flexible cords ≫ 422-16(a) ≪.

Grounding must meet Section 250-134 requirements for equipment fastened-in-place, or connected by permanent wiring methods.

Adequately protect conductors where subject to physical damage ≫ 300-4 ≪.

NOTE
A permanently connected appliance, rated over 300 volt-amperes or ⅛ horsepower, must have a means to disconnect all ungrounded (hot) conductors from the appliance ≫ 422-30 ≪.

A permanently connected appliance, rated over 300 volt-amperes or ⅛ horsepower, must have a means to disconnect all ungrounded (hot) conductors from the appliance ≫ 422-30 ≪.

A disconnecting means is not required if the branch circuit switch or circuit breaker can be locked in the open (off) position ≫ 422-31(b) ≪.

A fastened-in-place appliance (dishwasher, garbage disposer, etc.) shall not be connected to a small appliance branch circuit ≫ 210-52(b)(2) ≪.

Grounding must meet Section 250-134 requirements for equipment fastened-in-place, or connected by permanent wiring methods.

■ Kitchen Waste (Garbage) Disposers

Some appliances (dishwashers, trash compactors, and kitchen garbage disposers) can be cord and plug connected. This arrangement provides a means of disconnect where accessible ≫ 422-32(a) ≪.

The switch controlling a kitchen waste (garbage) disposer serves as a disconnecting means. Also see Section 430-109(c).

Appliance receptacles installed inside cabinets, as permitted by Section 422-16, do not require GFCI protection.

Grounding must meet Section 250-138 requirements for cord and plug connected equipment.

Kitchen waste disposers can be cord and plug connected under certain conditions ≫ 422-16(b)(1) ≪. The flexible cord must be identified as suitable for the purpose by the appliance manufacture's installation instructions, and must meet *all* of the following conditions:

▶ The flexible cord shall be terminated with a grounding-type attachment plug. A listed kitchen disposer, distinctly marked as protected by a system of double insulation or its equivalent, does not require termination with a grounding-type attachment plug.
▶ The cord must be between 18 and 36 inches in length.
▶ The receptacle must be located so the flexible cord will not become damaged.
▶ The receptacle shall be accessible.

■ Built-In Dishwashers and Trash Compactors

Built-in dishwashers and trash compactors can be cord and plug connected under certain conditions ≫422-16(b)(2)≪. The flexible cord must be identified as suitable for the purpose by the appliance manufacturer's installation instructions, and must meet *all* of the following conditions:

▶ The flexible cord shall be terminated with a grounding-type attachment plug. A listed dishwasher or trash compactor, distinctly marked as protected by a system of double insulation, or its equivalent, does not require termination with a grounding-type attachment plug.

▶ The cord must be between 3 and 4 feet in length when measured from the face of the attachment plug to the plane of the rear of the appliance.

▶ The receptacle must be located so the flexible cord will not become damaged.

▶ The receptacle shall be accessible.

Where accessible, cord and plug arrangements may serve as a disconnecting means ≫422-32(a)≪. A receptacle installed under the sink (inside the cabinet) may serve as a means of disconnect.

■ Appliance Branch Circuit Rating

Branch circuits, rated 15 or 20 amperes, may feed lighting units, utilization equipment or a combination of both. The rating of any one cord and plug connected utilization equipment shall not exceed 80% of the branch circuit ampere rating ≫210-23(a)≪. Cord and plug connected equipment may be portable, or may be fastened-in-place equipment. Fastened-in-place utilization equipment shall not exceed 50% of the branch circuit ampere rating where the circuit also feeds lighting units, and/or non-fastened-in-place utilization equipment.

NOTE
Local codes may require an separate branch circuits for the dishwasher and the garbage disposer, i.e., one branch circuit for each.

This receptacle could be split-wired to accommodate two individual branch circuits.

The full load current for a single-phase motor is found in Table 430-148.

The total load shall not exceed the branch circuit rating ≫220-4≪. For motor-operated, fastened-in-place appliances with a motor larger than ⅛ horsepower, multiply the largest motor by 125% and add 100% of the other loads ≫220-4(a)≪.

■ Calculation for One Counter-Mounted Cooking Unit and One Wall-Mounted Oven on the Same Circuit

Calculation for one cooktop with a nameplate rating of 6.5 kW at 240 volts and one oven with a nameplate rating of 4 kW at 240 volts supplied from one branch circuit:

▶ Add the rating of the cooktop and oven and compute the branch circuit in accordance with Table 220-19.

▶ 6.5 kW + 4 kW = 10.5 kW

▶ Table 220-19, Column A for one unit less than 12 kW = 8 kW

▶ 8 kW = 8,000 watts
8,000 watts ÷ 240 volts = 33.3 amperes

▶ Overcurrent protective device = 35 amperes ≫240-3(b)≪.

▶ Conductor size = No. 8 ≫Table 310-16 and 110-14(c)≪.

■ Calculation for One Counter-Mounted Cooking Unit (Cooktop)

Branch circuit calculation for one cooktop with a nameplate rating of 6.5 kW at 240 volts:

▶ Use the nameplate rating to determine the branch circuit conductor size and the overcurrent protection for one counter-mounted cooking unit ≫ Table 220-19, Note 4 ≪.

▶ 6.5 kW = 6,500 watts
6,500 watts ÷ 240 = 27.1 amperes

▶ Overcurrent protective device = 30 amperes ≫ 240-3(b) ≪.

▶ Conductor size = No. 10 ≫ Table 310-16 and 110-14(c) ≪.

■ Calculation for One Wall-Mounted Oven

Branch circuit calculation for one oven with a nameplate rating of 4 kW at 240 volts:

▶ Use the nameplate rating to determine the branch circuit conductor size and the overcurrent protection for one wall-mounted oven ≫ Table 220-19, Note 4 ≪.

▶ 4 kW = 4,000 watts
4,000 watts ÷ 240 = 16.7 amperes

▶ Overcurrent protective device = 20 amperes ≫ 240-3(b) ≪.

▶ Conductor size = No. 12 ≫ Table 310-16 and 110-14(c) ≪.

■ Counter-Mounted Cooking Units (Cooktops) and Wall-Mounted Ovens

NOTE
A separable connector or a plug and receptacle combination shall not serve as the disconnecting means for ovens and cooktops ≫ 422-16(b)(3) ≪.

The nameplate rating of a counter-mounted cooking unit or a wall oven shall be the minimum branch circuit load ≫ Table 220-19, Note 4 ≪.

Wall-mounted ovens and counter-mounted cooking units may be connected permanently or by cord and plug ≫ 422-16(a)(3) ≪.

A disconnecting means is required ≫ 422-30 ≪.

■ Cooktop and Wall-Mounted Oven(s) on Same Branch Circuit

Where a single branch circuit feeds one cooktop *and* one or two ovens (in the same room), add the nameplate ratings of the appliances, treating the total kW as one range to be found in Table 220-19 ≫ Table 220-19, Note 4 ≪.

Tap conductors from a 50-ampere branch circuit must have an ampacity of at least 20 and not less than the ampacity of the individual unit served ≫ 210-19(c) *Exception No. 1* ≪.

No disconnecting means is required if the branch circuit switch or circuit breaker can be locked in the open (off) position ≫ 422-31(b) ≪.

It is permissible to feed one cooktop and one wall-mounted oven, or one cooktop and two wall-mounted ovens, with one branch circuit where all of the units are in the same room ≫ Table 220-19 Note 4 ≪.

■ Electric Ranges

> **N O T E**
> *Ranges connected to existing branch circuits can be grounded by the provisions found in Section 250-140.*

It is permissible to compute the branch circuit load for one range in accordance with Table 220-19 »Table 220-19 Note 4 «.

A disconnecting means is required [422-30]. An accessible separable connector or an accessible plug and receptacle can serve as the disconnecting means »422-32(a)«.

Grounding must meet the Section 250-138 requirements for cord and plug connected equipment.

Most household electric ranges are cord and plug connected. The attachment plug and receptacle can serve as the disconnecting means if they are accessible by removing a drawer on the front of the range »422-32(b)«.

> **N O T E**
> *The neutral conductor for a household electric range, a wall-mounted oven, or a counter-mounted cooking unit shall not be smaller than No. 10 copper »210-19(c) Exception No. 2«.*

New branch circuit conductors installed for a range must have an insulated grounded (neutral) conductor and a grounding means. Grounding must meet Section of 250-138 requirements for cord and plug connected equipment.

■ Calculation for Ranges Rated 8¾ kW Through 12 kW

Branch circuit calculation for one range with a nameplate rating of 12 kW at 240 volts:

- ▸ Compute the branch circuit load for one range in accordance with Table 220-19
- ▸ 12 kW = 8 kW (Column A)
  8 kW = 8,000 watts
  8,000 watts ÷ 240 = 33.3 amperes
- ▸ Overcurrent protective device = 40 amperes »240-3(b)«.
- ▸ Conductor size = No. 8 »Table 310-16 and 110-14(c)«.

> **N O T E**
> *The minimum branch circuit rating for a range rated 8¾ kW, or more, is 40 amperes »210-19(c)«.*

■ Calculation for Ranges Rated More Than 12 kW

Branch circuit calculation for one range with a nameplate rating of 15 kW at 240 volts:

- ▸ Compute the branch circuit load for one range in accordance with Table 220-19.
- ▸ Column A is for 12-kW and under ranges. Note 1 applies to ranges over 12 kW.
- ▸ First, subtract 12-kW from the 15-kW range.
  15 kW – 12 kW = 3 kW
- ▸ Next, multiply 3 kW by 5% = 15%
- ▸ Then, multiply 8 kW (Column A) by 15%.
  8 kW × 15% = 1.2 kW
- ▸ Finally, add 1.2 kW to 8 kW. 1.2 kW + 8 kW = 9.2 kW = 9,200 watts
- ▸ 9,200 watts ÷ 240 volts = 38.3 amperes
- ▸ Overcurrent protective device = 40 amperes »240-3(b)«.
- ▸ Conductor size = No. 8 »Table 310-16 and 110-14(c)«.

■ Kitchen, Dining Room, and Breakfast Area

A minimum of two small appliance branch circuits are required for receptacles serving kitchen countertops ≫210-52(b)(3)≪. Either, or both, of these small appliance branch circuits may feed other receptacles in the kitchen, pantry, breakfast room and/or dining area.

All receptacles in the kitchen, pantry, breakfast room, dining room, or similar area shall be supplied from 20-ampere branch circuits (except for refrigeration equipment) ≫210-52(b)(1)≪.

Non-countertop receptacle placement is determined by wall space ≫210-52(a)≪.

A minimum of two circuits are required for receptacles located in the kitchen, pantry, breakfast room, dining room, or similar areas ≫210-52(b)(1)≪.

The receptacle outlet installed for refrigeration equipment can be supplied from an individual branch circuit rated 15 amperes, or greater ≫210-52(b)(1) Exception No. 2≪.

Range hoods and lights are not permitted on small appliance branch circuits ≫210-52(b)(2)≪.

At least one wall switch-controlled lighting outlet shall be installed in dining rooms, kitchens, and breakfast areas ≫210-70(a)≪. One or more receptacles controlled by a wall switch are permitted in lieu of a lighting outlet in dining rooms and breakfast areas, but *not* in kitchens ≫210-70(a)(1)≪.

HALLWAYS AND STAIRWAYS

■ Hallway and Stairway Lighting

A minimum of one lighting outlet is required in each hallway and stairway ≫210-70(a)(2)≪. The light must be wall switch-controlled, except where controlled by remote, central, or automatic means ≫210-70(a)(2) Exception≪. A receptacle controlled by a wall switch is not permitted.

Interior stairways of 6 or more steps shall have a wall switch located at each floor level ≫210-70(a)(2)≪.

■ Hallway Receptacle Placement

Receptacles in hallways are not subject to the general provisions for placement determined by wall space.

Hallways measuring 10 feet or more in length must have at least one receptacle outlet ≫210-52(h)≪. Length is determined by measuring along the hallway center line, without passing through a doorway.

9 feet 11 inches

If length is less than 10 feet, no receptacle is required.

3 feet 6 inches

6 feet 6 inches

If length is 10 feet or more, a receptacle is required.

CLOTHES CLOSETS

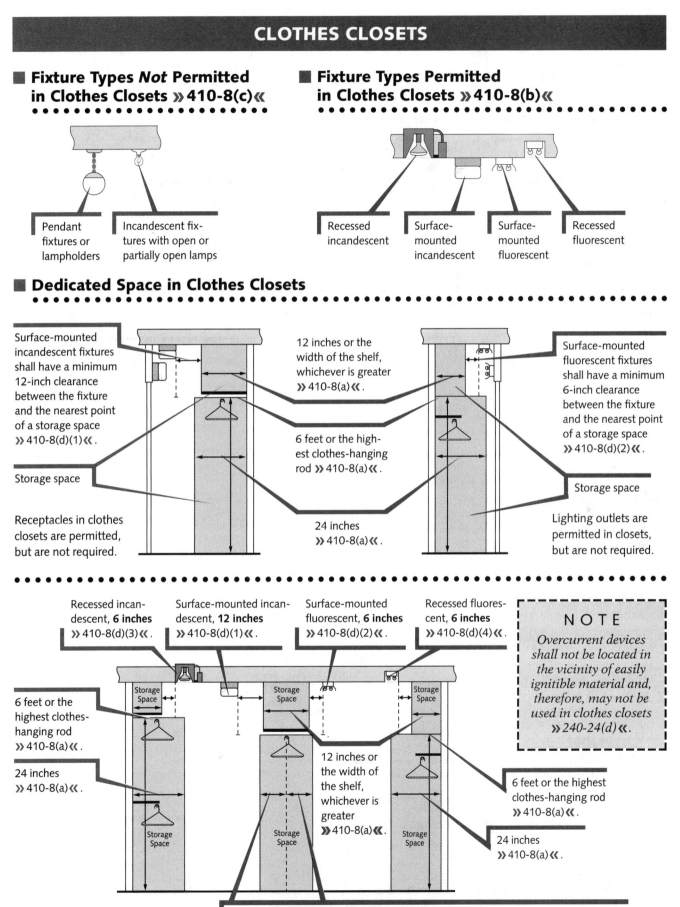

■ Fixture Types *Not* Permitted in Clothes Closets ≫ 410-8(c) ≪

- Pendant fixtures or lampholders
- Incandescent fixtures with open or partially open lamps

■ Fixture Types Permitted in Clothes Closets ≫ 410-8(b) ≪

- Recessed incandescent
- Surface-mounted incandescent
- Surface-mounted fluorescent
- Recessed fluorescent

■ Dedicated Space in Clothes Closets

Surface-mounted incandescent fixtures shall have a minimum 12-inch clearance between the fixture and the nearest point of a storage space ≫ 410-8(d)(1) ≪.

Storage space

Receptacles in clothes closets are permitted, but are not required.

12 inches or the width of the shelf, whichever is greater ≫ 410-8(a) ≪.

6 feet or the highest clothes-hanging rod ≫ 410-8(a) ≪.

24 inches ≫ 410-8(a) ≪.

Surface-mounted fluorescent fixtures shall have a minimum 6-inch clearance between the fixture and the nearest point of a storage space ≫ 410-8(d)(2) ≪.

Storage space

Lighting outlets are permitted in closets, but are not required.

Recessed incandescent, **6 inches** ≫ 410-8(d)(3) ≪.

Surface-mounted incandescent, **12 inches** ≫ 410-8(d)(1) ≪.

Surface-mounted fluorescent, **6 inches** ≫ 410-8(d)(2) ≪.

Recessed fluorescent, **6 inches** ≫ 410-8(d)(4) ≪.

6 feet or the highest clothes-hanging rod ≫ 410-8(a) ≪.

24 inches ≫ 410-8(a) ≪.

Storage Space

Storage Space

Storage Space

Storage Space

Storage Space

Storage Space

12 inches or the width of the shelf, whichever is greater ≫ 410-8(a) ≪.

6 feet or the highest clothes-hanging rod ≫ 410-8(a) ≪.

24 inches ≫ 410-8(a) ≪.

> **N O T E**
> *Overcurrent devices shall not be located in the vicinity of easily ignitible material and, therefore, may not be used in clothes closets* ≫ *240-24(d)* ≪.

For a closet with access to both sides of a hanging rod, the storage space shall include the volume below the highest rod extending 12 inches on either side of the rod on a plane horizontal to the floor and extending the entire length of the rod. 410-8(a)

BATHROOMS

■ Bathroom "Area"

Ground-fault circuit-interrupter (GFCI) protection for personnel is required for every 125-volt receptacle located in the bathroom area »210-8(a)(1)«.

Closet

Bathroom

GFCI

GFCI

SS

Bathroom Definition: An area including a basin (lavatory or sink) with one or more of the following: a toilet, a tub, or a shower [Article 100—Definitions]. A bathroom is an **area**—not necessarily a single room.

One or more lighting outlets, controlled by a wall switch, are required in bathrooms »210-70(a)(1)«.

The general provisions for receptacle placement by wall space do not apply to bathrooms.

N O T E

Service disconnecting means shall not be located in bathrooms »230-70(a)«. Overcurrent devices, other than supplementary overcurrent protection, shall not be installed in bathrooms »240-24(e)«. Supplementary overcurrent protection is not branch circuit overcurrent protection. It is an additional overcurrent protection usually installed within light fixtures, appliances and other equipment. It is not required that supplementary overcurrent protection be readily accessible »240-10]«.

■ Receptacle Within 36 Inches of Sink

At least one wall receptacle shall be located within 36 inches of the outside edge of each basin (lavatory or sink) »210-52(d)«.

3 Feet
GFCI Protected

■ Bathroom Branch Circuit Rating

Bathroom receptacle outlets shall be supplied by at least one 20-ampere branch circuit. »210-11(C)(3)«.

15-ampere overcurrent device

GFCI Protected

20-ampere overcurrent device

GFCI Protected

■ Bathroom Receptacles

CAUTION «««««

The receptacle outlet must be located on a wall that is adjacent to the basin location 210-52(d).

Receptacles are not permitted in a face-up position in the working surface or countertop »210-52(d)«.

At least one wall receptacle shall be located within 36 inches of the outside edge of each basin (lavatory or sink) »210-52(d)«.

GFCI Protected   GFCI Protected   GFCI Protected

Receptacles may have an individual rating of either 15 or 20 amperes, but must be supplied from a 20-ampere branch circuit »210-11(c)(3)«. Ground-fault circuit-interrupter (GFCI) protection for personnel is required for *all* bathroom receptacles »210-8(a)(1)«.

■ Bathroom

• •

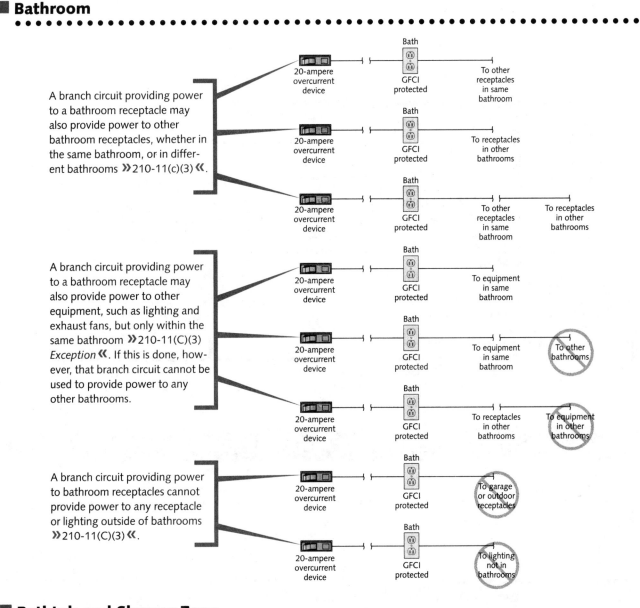

A branch circuit providing power to a bathroom receptacle may also provide power to other bathroom receptacles, whether in the same bathroom, or in different bathrooms ≫210-11-(c)(3)≪.

A branch circuit providing power to a bathroom receptacle may also provide power to other equipment, such as lighting and exhaust fans, but only within the same bathroom ≫210-11(C)(3) *Exception*≪. If this is done, however, that branch circuit cannot be used to provide power to any other bathrooms.

A branch circuit providing power to bathroom receptacles cannot provide power to any receptacle or lighting outside of bathrooms ≫210-11(C)(3)≪.

■ Bathtub and Shower Zone

• •

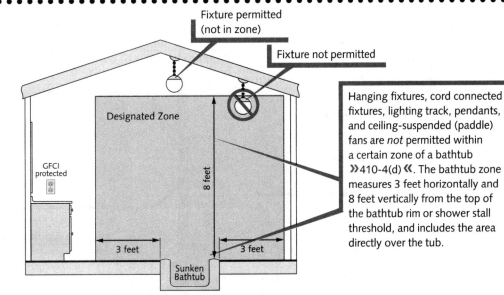

Hanging fixtures, cord connected fixtures, lighting track, pendants, and ceiling-suspended (paddle) fans are *not* permitted within a certain zone of a bathtub ≫410-4(d)≪. The bathtub zone measures 3 feet horizontally and 8 feet vertically from the top of the bathtub rim or shower stall threshold, and includes the area directly over the tub.

■ Bathtub and Shower Zone

Recessed lights, surface-mounted lights, and exhaust fans are permitted within the bathtub zone. Exposed metal parts must be grounded in accordance with Section 250-110.

Hanging fixtures, cord connected fixtures, lighting track, pendants, and ceiling-mounted (paddle) fans are permitted only *outside* the zone.

Exhaust Fan

Designated Zone

8 feet

GFCI protected

A receptacle shall not be installed within a bathtub or shower space ≫410-57(c) ≪.

Bathtub, Shower, or Hydromassage Bathtub

3 feet

Designated Zone

3 feet   3 feet

Bathtub, Shower, or Hydromassage Bathtub

3 feet   3 feet

Designated Zone

The bathtub zone measures 3 feet horizontally and 8 feet vertically from the top of the bathtub rim or shower stall threshold. The zone is all encompassing and includes the area directly over the tub or shower stall ≫410-4(d) ≪.

BASEMENTS AND GARAGES

■ Garage Receptacles

If a duplex receptacle is installed, GFCI protection is required. If a single receptacle is installed, GFCI protection is not required ≫210-8(a)(2) *Exception No. 2* ≪.

Laundry equipment receptacle(s) are not counted as required garage receptacle(s) ≫210-52(g) ≪.

A single receptacle installed on a 20-ampere branch circuit must have a rating of 20 amperes ≫210-21(b)(1) ≪.

Dryer   Washer

Sump Pump

Garage

Receptacle Located in Ceiling

At least one receptacle is required in an attached garage ≫210-52(g) ≪. (Receptacle placement is not determined by wall

Receptacles not readily accessible do not require GFCI protection ≫210-8(a)(2) *Exception No. 1* ≪.

Chest Freezer

Refrigerator

GFCI

All receptacles installed in garages must have ground-fault circuit-interrupter protection, unless installed under one of the exceptions found in 210- 8(a)(2).

If a duplex receptacle is installed for two appliances, GFCI protection is not required. The two appliances must be cord and plug connected and must be located within a dedicated space where in normal use they are not easily moved from one place to another ≫210-8(a)(2) *Exception No. 2* ≪.

Non-GFCI protected receptacles installed by an exception found in Section 210-8(a)(2) do not count toward the one garage receptacle requirement ≫210-8(a)(2) ≪.

■ GFCI Receptacles in Garage

If a duplex receptacle is installed, GFCI protection is required ≫210-8(a)(2)≪.

If a single receptacle is installed, GFCI protection is *not* required ≫210-8(a)(2) *Exception No. 2*≪.

■ Lighting

Basements and attached garages require at least one receptacle each ≫210-52(g)≪. (Receptacle placement is not determined by wall space.)

At least one wall switch-controlled lighting outlet is required for basements and attached garages ≫210-70(a)(2) and (3)≪.

At least one wall switch-controlled lighting outlet is required to provide illumination on the exterior side of an outside entrance or exit with grade level access ≫210-70(a)(2)≪.

Interior stairways with 6 steps, or more, shall have a wall switch located at each floor level ≫210-70(a)(2)≪.

A vehicle garage door is not considered an outside entrance or exit and therefore does not require a wall switch-controlled lighting outlet ≫210-70(a)(2)≪.

A lighting outlet shall be installed at, or near, equipment which may require servicing ≫210-70(a)(3)≪.

A lighting outlet installed to provide illumination for the general area (basement, garage, etc.) may also serve as the equipment lighting, provided it is located at, or near, the equipment.

N O T E

Unfinished basements are defined as portions or areas of the basement not intended as habitable rooms and limited to storage areas, work areas, and the like ≫210-8(a)(5)≪.

■ Disconnecting Means

• •

A disconnecting means is not required if the branch circuit switch or circuit breaker is within 50 feet and can be seen while working on the appliance 》422-31(b)《.

The branch circuit switch or circuit breaker cannot be seen while working on the appliance. A disconnecting means is required unless the branch circuit switch or circuit breaker can be locked in the open (off) position 》422-31(b)《.

A permanently connected appliance rated over 300 volt-amperes or 1/8 horsepower requires a means to disconnect all ungrounded (hot) conductors from the appliance 》422-30《.

■ Detached Garages

• •

> NOTE
>
> *Lighting and receptacles are not required in detached garages. There are no provisions to provide electric power to detached garages. However, when electric power is provided, the provisions for attached garages must be met 》210-52(g)《.*

At least one receptacle is required in each detached garage with electric power 》210-52(g)《.

All receptacles installed in detached garages must have ground-fault circuit-interrupter protection, unless installed under one of the exceptions found in 210-8(a)(2).

At least one wall switch-controlled lighting outlet is required for detached garages with electric power 》210-70(a)(2)《.

A detached garage, with electric power, requires at least one wall switch-controlled lighting outlet to provide illumination on the exterior side of an outside entrance or exit with grade-level access 》210-70(a)(2)《.

■ Accessory Buildings

All receptacles installed in accessory buildings having a floor located at, or below, grade level not intended as habitable rooms and limited to storage areas, work areas, and areas of similar use must have ground-fault circuit-interrupter protection, unless installed under one of the exceptions found in 210- 8(a)(2).

Lighting and receptacles are not required in accessory buildings. There are no provisions to provide electric power to detached garages.

LAUNDRY AREAS

■ Receptacles and Lighting

Laundry receptacle outlet(s) must be fed from a 20-ampere branch circuit ≫210-11(c)(2)≪.

The clothes dryer outlet must have an insulated grounded (neutral) conductor *and* an equipment grounding conductor ≫250-138≪.

To Panel   To Panel

At least one receptacle is required for the laundry area ≫210-52(f)≪.

Dryer   Washer

Laundry Area

A lighting outlet is required for spaces containing equipment which require servicing ≫210-70(a)(3)≪. A lighted area immediately adjacent to a laundry area (such as a hallway) may provide sufficient light for both areas.

Hall

S

NOTE

Electric clothes dryers connected to existing branch circuits are permitted to follow Section 250-140.

■ Receptacles and Lighting

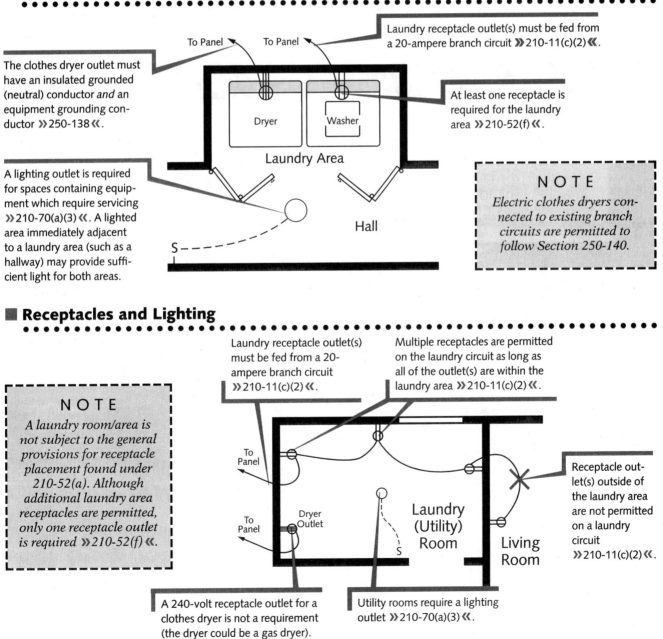

Laundry receptacle outlet(s) must be fed from a 20-ampere branch circuit ≫210-11(c)(2)≪.

Multiple receptacles are permitted on the laundry circuit as long as all of the outlet(s) are within the laundry area ≫210-11(c)(2)≪.

NOTE

A laundry room/area is not subject to the general provisions for receptacle placement found under 210-52(a). Although additional laundry area receptacles are permitted, only one receptacle outlet is required ≫210-52(f)≪.

To Panel

To Panel

Dryer Outlet

S

Laundry (Utility) Room

Living Room

Receptacle outlet(s) outside of the laundry area are not permitted on a laundry circuit ≫210-11(c)(2)≪.

A 240-volt receptacle outlet for a clothes dryer is not a requirement (the dryer could be a gas dryer).

Utility rooms require a lighting outlet ≫210-70(a)(3)≪.

■ Receptacles and Lighting

Laundry receptacle outlet(s) must be fed from a 20-ampere branch circuit ≫ 210-11(c)(2) ≪. A duplex receptacle installed on an individual 20 ampere branch circuit may have either a 15- or 20-ampere rating ≫ 210-21(b)(2) ≪.

Height requirements for laundry receptacle outlet(s) are not specified, except they cannot be located more than 5½ feet above the floor ≫ 210-52 ≪.

The clothes dryer outlet must have an insulated grounded (neutral) conductor *and* an equipment grounding conductor ≫ 250-138 ≪.

A receptacle outlet installed for a specific appliance, such as laundry equipment, shall be installed within 6 feet of the intended location of the appliance ≫ 210-50(c) ≪.

ATTIC AND CRAWL SPACES

■ Attic Spaces

Attic spaces used for storage or containing equipment which may require servicing must have at least one lighting outlet ≫ 210-70(a)(3) ≪.

A receptacle is not required for attic spaces that do not contain equipment which may require servicing.

Attic spaces not used for storage or attics that do not contain equipment which may require servicing, do not require lighting outlets.

At least one switch shall be located at the usual entry and exit to the attic space ≫ 210-70(a)(3) ≪. If the light fixture contains a switch and is located at the usual point of entry, a separate switch may not be required

Cables not installed within framing members may need protection up to a height of at least 7 feet ≫ 333-12(a), 334-10(h), and 336-6(d) ≪.

■ Crawl and Underfloor Spaces

Crawl spaces with lighting outlet(s) require at least one switch located at the usual entry and exit to the crawl space »210-70(a)(3) «. If the light fixture contains a switch and is located at the usual point of entry, a separate switch may not be required.

A 125-volt, single-phase, 15- or 20-ampere rated receptacle outlet shall be installed at an accessible location for the servicing of equipment. The receptacle shall be located on the same level and within 25 feet of the equipment. The receptacle shall not be connected to the load side of the equipment disconnecting means »210-63 «.

Receptacle outlet(s) require ground-fault circuit-interrupter protection where installed in a crawl space that is not above grade level »210-8(a)(4) «.

Underfloor spaces used for storage containing equipment which may require servicing must have at least one lighting outlet. A lighting outlet shall be located at or near equipment which may require servicing »210-70(a)(3) «.

A disconnecting means must be provided for HVAC equipment »440-3(b), 422-30, and 424-19 «.

A ground-fault protected receptacle located outside of the crawl space may meet the requirements of Section 210-63 if the receptacle is on the same level and within 25 feet of the equipment.

No lighting outlet(s) are required for underfloor spaces which are not used for storage or contain no equipment which may require servicing

Summary

- Receptacle branch circuits rated 20 amperes are required for kitchens, pantries, dining rooms, breakfast areas, and similar areas.
- Twenty-four (24) inches is the maximum distance to any countertop receptacle, along the wall line, measured horizontally.
- General provisions for countertop receptacle placement do not apply to islands and peninsulas.
- A minimum of two circuits is required for receptacles serving kitchen countertops.
- Receptacles located in kitchens, pantries, dining rooms, breakfast areas, and similar areas, also require a minimum of two circuits.

- All receptacles installed to serve kitchen countertops shall have GFCI protection.
- Some appliances can be cord and plug connected.
- Permanently connected as well as cord and plug connected appliances require a disconnecting means.
- Branch circuit conductors installed for ranges and clothes dryers must have an insulated grounded (neutral) conductor and a grounding means.
- Hallways measuring 10 feet or more in length require only one receptacle.

- Clothes closets have dedicated space where light fixtures are not permitted.
- Only certain types of light fixtures are permitted in the dedicated space within bathrooms.
- Receptacles in bathrooms must be GFCI protected and be supplied by 20 ampere branch circuit(s).
- As a general rule, garage and basement receptacles require GFCI protection.
- Lighting is required at or near equipment which may require servicing.

- At least one laundry receptacle outlet is required and must be fed from a 20 ampere branch circuit.
- Under certain conditions, attics and crawl spaces may require a lighting outlet and receptacle.
- A receptacle is required on the same level and within 25 feet of HVAC equipment which may require servicing.

Unit 7 Competency Test

NEC® Reference   **Answer**

_____ _____ 1. The maximum distance between kitchen countertop receptacles is _____ feet.

_____ _____ 2. Hanging fixtures, located directly above any part of the bathtub, shall be installed so that the fixture is not less than _____ feet above the top of the bathtub rim.

_____ _____ 3. A 15-ampere rated duplex receptacle may be installed on a _____ branch circuit.

        a) 15-ampere   b) 20-ampere
        c) 15- or 20-ampere   d) 15-, 20- or 25-ampere

_____ _____ 4. The minimum-size THHN copper neutral conductor permitted for an 8¾ kW, 240-volt, 3-wire household electric range is _____.

        a) No. 12   b) No. 10   c) No. 8   d) No. 6

_____ _____ 5. Fixtures shall be constructed, or installed, so that adjacent combustible material will not be subjected to temperatures in excess of _____ degrees Fahrenheit.

_____ _____ 6. A one-family dwelling has a front entrance on the east side of the house, an attached garage which has two 10' wide doors on the south side, and a back entrance door on the west side of the house. This one-family dwelling requires _____ lighting outlet(s) for these outdoor entrances.

_____ _____ 7. Overcurrent devices shall not be located in the vicinity of easily ignitible material such as in _____.

_____ _____ 8. A single receptacle installed on an individual branch circuit shall have an ampere rating not less than _____% of the rating of the branch circuit.

        a) 125   b) 100   c) 80   d) 50

_____ _____ 9. Appliance receptacle outlets installed for specific appliances, such as laundry equipment, shall be placed within _____ feet of the intended location of the appliance.

_____ _____ 10. Tap conductors for household cooking equipment supplied from a 50-ampere branch circuit shall have an ampacity of not less than _____ amperes.

_____ _____ 11. One cooktop and two ovens are connected to one branch circuit in a kitchen of a one-family dwelling. The cooktop has a nameplate rating of 4 kW at 240 volts, while each oven has a nameplate rating of 5.5 kW at 240 volts. The minimum load calculation for this branch circuit is _____ kW.

_____ _____ 12. The rating of a single cord and plug connected utilization equipment shall not exceed _____ % of the branch circuit ampere rating.

_____ _____ 13. A recessed incandescent fixture with a completely enclosed lamp installed in the ceiling of a clothes closet must have a minimum clearance of _____ inches between the fixture and the nearest point of the storage space.

_____ _____ 14. Where the *NEC*® specifies that one equipment shall be "within sight" of another equipment, the specified equipment shall be visible and not more than _____ feet distant from the other.

_____ _____ 15. In a dwelling unit, a hallway measuring 21½ feet in length requires a minimum of _____ receptacle outlet(s).

_____ _____ 16. A permanently connected dishwasher has a nameplate rating of 780 volt-amperes. The circuit breaker is permitted to serve as the disconnecting means where it is:

  I. within sight from the dishwasher.

  II. capable of being locked in the open position.

  III. capable of being locked in the closed position.

  a) I only   b) I or III only   c) I or II only   d) I, II, or III

_____ _____ 17. Interior stairway lighting outlets shall be controlled by a wall switch at each floor level, where the difference between levels is _____ steps or more.

_____ _____ 18. Supplementary overcurrent devices shall not be required to be _____.

_____ _____ 19. The maximum-length cord permitted on an electrically operated trash compactor is _____ inches.

_____ _____ 20. For a one-family dwelling, at least one receptacle outlet shall be installed in each:

  I. basement

  II. attached garage

  III. detached garage

  a) I only   b) I and II only   c) II and III only   d) I, II and III

_____ _____ 21. A refrigerator receptacle can be fed from a:

  I. 15-ampere individual branch circuit.

  II. 20-ampere individual branch circuit.

  III. 20-ampere small appliance branch circuit.

  a) II only   b) III only   c) II or III   d) I, II, or III

_____ _____ 22. A receptacle installed on the laundry branch circuit, having no other receptacles, can be a:

  I. 15-ampere single receptacle.

  II. 15-ampere duplex receptacle.

  III. 20-ampere duplex receptacle

  a) III only   b) I or III   c) II or III   d) I, II, or III

_____ _____ 23. At least one GFCI-protected receptacle shall be located within _____ feet from the outside edge of each bathroom sink.

_____ _____ 24. Receptacles are permitted under an overhanging countertop, but are not counted as required countertop receptacles, where the countertop extends more than _____ beyond its support base.

_____ _____ 25. A branch circuit providing power to a bathroom receptacle may also provide power to:

      I. two other receptacles, two light fixtures, and an exhaust fan in the same bathroom.

      II. two other receptacles in the same bathroom and receptacles in two additional bathrooms.

      III. one light fixture in the same bathroom and one light fixture in another bathroom.

    a) I only   b) I or II only   c) II or III only   d) I, II, or III

_____ _____ 26. Inside a bathroom, _____ receptacle can be installed between two and three feet horizontally from the bathtub rim.

    a) a weatherproof   b) a GFCI-protected

    c) a GFCI-protected, weatherproof   d) no

Load Calculations

Objectives

After studying this unit, the student will:

- ✔ be able to calculate the general lighting load in a one-family dwelling.
- ✔ know the minimum volt-ampere requirements for small appliance and laundry branch circuits.
- ✔ know how to apply demand factors to the general lighting load.
- ✔ be able to apply demand factors to fastened-in-place appliances.
- ✔ know how to calculate feeder-demand loads for household cooking equipment
- ✔ be able to calculate feeder-demand loads for household cooking equipment.
- ✔ be able to calculate heating and air-conditioning feeder-demand loads.

- ✔ be able to calculate a one-family dwelling service and/or feeder (using the standard method.)
- ✔ be able to calculate a one-family dwelling service and/or feeder (using the optional method.)
- ✔ know how to size service and feeder conductors.
- ✔ be able to calculate and choose the appropriate size neutral conductor.
- ✔ understand how the grounding electrode conductor is selected.

Introduction

Load calculations must be performed in order to determine the size of services and feeders. Service conductors are the conductors between the power provider and the disconnecting means, whether a main breaker or main lug panel. A feeder is defined as all circuit conductors between the service disconnecting means, and the panelboard containing fuses or breakers which feed branch circuits. Additional load calculations will be needed for dwellings with panelboard(s) which are not part of the service. For example, a separate panelboard might be installed near the middle of the house to help eliminate voltage drop in branch circuit conductors. Since the remote panelboard (sub-panel) does not carry the total load, an additional calculation is needed to size the feeder and panelboard.

This Unit simplifies the standard, as well as optional, load calculation methods for one-family dwellings. A blank form is provided at the beginning of each method. The floor plan of the one-family dwelling being calculated is found on page 96 while the specifications are on page 137. Both calculation methods will use the same specifications. Each line on each form is explained in detail using that information. The completed form follows the explanations, showing all of the information, together in one place.

COMPILING INFORMATION ESSENTIAL TO LOAD CALCULATIONS

■ Total Floor Area Square Footage

Open porches are not included in the calculated floor area of dwelling units ≫ 220-3(a) ≪. Although this porch has a roof, it is considered open, and therefore, is not included in the calculation.

Scale rulers are also useful in determining the location of outlets (receptacles, lights, switches, etc.).

Before a scale ruler can be used to determine dimensions, the scale used to produce the blueprint (floor plan) must be known. (Then the area of an exterior porch, for example, is simple to calculate.)

Front Door

Front Porch

10 feet

Scale: ½ inch = 1 foot

Scale rulers are useful in determining the total square footage of a structure (house, building, etc.) that has been drawn on a set of blueprints (floor plans).

A regular tape measure (ruler, yard stick, etc.) can be used, but the use of a scale ruler makes the job much easier.

■ Gathering Information

What is the dwelling's total square-foot area, using the outside dimensions?

55'

35'

Front Porch
7' x 10'

How many small appliance branch circuits will be installed?

How many laundry circuit(s) will be installed?

If an electric clothes dryer will be installed, what is the volt-ampere load?

What household cooking equipment will be installed and what is the volt-ampere load of each?

What fastened-in-place appliances will be installed and what is the volt-ampere load of each?

Gathering certain information is necessary to accurately perform load calculations. All the essential data can be identified by simply answering the questions found on the load calculation form, either standard or optional. Enter the information on the appropriate line of the load calculation forms provided with this text book.

Will there be an air-conditioning system? If so, what is the total volt-ampere load, including the compressor and fan motors? If an electric heating system will be installed, what is the total volt-ampere load, including the strip heat and blower motor?

STANDARD METHOD: ONE-FAMILY DWELLINGS

Standard Method Load Calculation for One-Family Dwellings

1 General Lighting and Receptacle Loads 220-3(a)
Do not include open porches, garages, unused or
unfinished spaces not adaptable for future use. $3 \times$ _____ = **1**
(sq. ft. using outside dimensions)

2 Small Appliance Branch Circuits 220-16(a)
*At least **two** small appliance branch circuits*
must be included. 210-11(c)(1) $1,500 \times$ _____ = **2**
(minimum of two)

3 Laundry Branch Circuit(s) 220-16(b)
*At least **one** laundry branch circuit*
must be included. 210-11(c)(2) $1,500 \times$ _____ = **3**
(minimum of one)

4 Add lines 1, 2, and 3 | **4** | | Lines 5 through 8 utilize the demand factors found in Table 220-11.

5 _____ $- 3,000 =$ **5**
(line 4) (if 117,000 or less, skip to line 8)

6 _____ $- 117,000 =$ **6**
(line 5, if more than 117,000)

7 _____ $\times 25\% =$ **7**
(line 6)

8 _____ $\times 35\% =$ **8**
(smaller of line 5 or 117,000)

9 Total General Lighting and Receptacle Load $3,000 +$ _____ $+$ _____ $=$ **9**
(line 7) (line 8)

10 Fastened-In-Place Appliances 220-17
Use the nameplate rating.
Do not include electric ranges,   water heater /   dishwasher /   disposer /
clothes dryers, space-heating   compactor /   /   /
equipment, or air-conditioning
equipment.   /   /   /

If fewer than four units, put total volt-amperes on line 10.
If four or more units, multiply total volt-amperes by 75%.   _____ $\times 75\% =$ **10**
(volt-amps of four or more)

11 Clothes Dryers 220-18
(If present, otherwise skip to line 12.) Use 5,000 watts or the nameplate rating, whichever is larger.
(The neutral demand load is 70% for feeders. 220-22) | **11** |

12 Ranges, Ovens, Cooktops, and Other Household Cooking Appliances Over 1,750 Watts 220-19
(If present, otherwise skip to line 13.) Use Table 220-19 and all of the applicable Notes.
(The neutral demand load is 70% for feeders. 220-22) | **12** |

13 Heating or Air-Conditioning System (Compare the heat and A/C, and omit the smaller.) 220-21
Include the air handler when using either one. For heat pumps, include the compressor and the
maximum amount of electric heat which can be energized while the compressor is running. | **13** |

14 Largest Motor (one motor only) 220-14 and 430-24
Multiply the volt-amperes of the largest motor by 25%.   _____ $\times 25\% =$ **14**
(volt-amps of largest motor)

15 Total Volt-Ampere Demand Load: *Add lines 9 through 14 to find the minimum required volt-amperes.* | **15** |

16 Minimum Amperes
Divide the total
volt-amperes by   _____ $\div$ _____ $=$ **16**
the voltage   (line 15) (voltage) (minimum amperes) | **17** Minimum Size Service and/or Feeder 240-6(a) | **17** |

18 Size the Service and/or Feeder Conductors.
Use Section 310-15(b)(6) to find the service conductors up to 400 amperes.
Ratings in excess of 400 amperes shall comply with Table 310-16.   **Minimum Size Conductors** | **18** |
310-15(b)(6) also applies to feeder conductors serving as the main power feeder.

19 Size the Neutral Conductor 220-22
Section 310-15(b)(6) states that the neutral service and/or feeder conductor
can be smaller than the ungrounded (hot) conductors, provided the   **Minimum Size Neutral Conductor** | **19** |
requirements of Sections 215-2, 220-22, and 230-42 are met.
Section 250-24(b)(1) states that the neutral cannot be smaller than the
required grounding electrode conductor specified in Table 250-66.

20 Size the Grounding Electrode Conductor Table 250-66
Using line 18 (the service and/or feeder conductor size), find the same size
in Table 250-66. (Pay close attention to copper and aluminum conductors.)   **Minimum Size Grounding Electrode Conductor** | **20** |
Follow that row across to either copper or aluminum grounding electrode
conductor and select the correct size.

Sample One-Family Dwelling Load Calculation

> **NOTE**
>
> *The information to the right will be used throughout the remainder of this unit to demonstrate the use of the load calculation forms, both standard and optional.*

Dwelling outside dimensions . . . . 35' × 55'
Front porch (included within outside dimensions) . . . 7' × 10'
Small appliance branch circuits . . . . four
Laundry branch circuits . . . . two
Water heater . . . . 4.5 kW, 240 volt
Dishwasher . . . . 10-amperes, 120 volt
Waste (garbage) disposer . . . . ½ HP, 115 volt
Trash compactor . . . . 7.5-amperes, 120 volt
Three attic fans (4.2-amperes each) . . . . 12.6-amperes, 120 volt
Clothes dryer . . . . 5.5 kW, 240 volt
Counter-mounted cooking unit (cook top) . . . . 7 kW, 240 volt
Wall-mounted oven . . . . 6 kW, 240 volt
Electric heat (3 banks @ 5 kW each) . . . . 15 kW, 240 volt
Air handler (blower motor) . . . . 3.2-amperes, 115 volt
Air-conditioner compressor . . . . 16.6-amperes, 230 volt
Condenser fan motor . . . . 2-amperes, 115 volt

Assume water heater, clothes dryer, counter-mounted cooking unit, wall-mounted oven, and electric heat kW ratings equivalent to kVA.

Line 1—General Lighting and Receptacle Loads

In Table 220-3(a) the general lighting loads are listed according to the type of occupancy and are determined by requiring a minimum lighting load for each square foot of floor area.

The lighting load for residential dwelling units is 3 volt-amperes (VA) per square foot.

The dwelling floor plan on page 135 has an outside dimension of 35' × 55' (1,925 sq. ft.), but there is an open porch with a dimension of 7' × 10' (70 sq. ft.). Since open porches are not included in the total area, 70 is deducted from 1,925, leaving 1,855 square feet.

1 General Lighting and Receptacle Loads 220-3(a)
Do not include open porches, garages, unused or unfinished spaces not adaptable for future use.

$3 \times \underline{\quad 1,855 \quad}$ (sq. ft. using outside dimensions) $= \boxed{1} \quad 5,565$

> **NOTE**
>
> *All 15 or 20 ampere general-use receptacle outlets (except for small appliance and laundry), are considered part of the general lighting load. The single asterisk (*) footnote at the bottom of Table 220-3(a) refers to Section 220-3(b)(10). which states that no additional load calculations are required for these outlets.*

The floor area for each floor shall be computed using the outside dimensions. The total floor area does not include open porches, garages, or unused (including unfinished) spaces which are not adaptable for future use »220-3(a)«.

Line 2—Small Appliance Branch Circuits

Small appliance branch circuits must be included when calculating a dwelling unit service »220-16(a)«.

Each small appliance branch circuit is calculated at no less than 1,500 volt-amperes »220-16(a)«.

These loads can be included with the general lighting load and subjected to the demand factors provided in Table 220-11 »220-16(a)«.

2 Small Appliance Branch Circuits 220-16(a)
*At least **two** small appliance branch circuits must be included. 210-11(c)(1)*

$1,500 \times \underline{\quad 4 \quad}$ (minimum of two) $= \boxed{2} \quad 6,000$

A dwelling unit must have at least two small appliance branch circuits »210-11(c)(1)«.

> **NOTE**
>
> *If a dwelling has more than one feeder, a separate load calculation is needed for each. It is not necessary to include this step if a feeder does not supply small appliance branch circuit(s). A load calculation for a feeder supplying small appliance branch circuit(s) must include at least 1,500 volt-amperes for each 2-wire circuit »220-16(a)«.*

> **NOTE**
>
> *An individual branch circuit for refrigeration equipment (permitted by Section 210-52(b)(1) Exception No. 2) can be excluded from this calculation »220-16(a) Exception«.*

Line 3—Laundry Branch Circuit(s)

Laundry branch circuit(s) must be included when calculating a dwelling unit service »220- 16(b) «.

Each laundry branch circuit is calculated at no less than 1,500 volt-amperes »220-16(b) «.

| 3 | Laundry Branch Circuit(s) 220-16(b) *At least **one** laundry branch circuit must be included. 210-11(c)(2)* | | 1,500 × ___2___ (minimum of one) = | 3 | 3,000 |

At least one laundry branch circuit is required per dwelling unit »210-11(c)(2) «.

These loads can be included with the general lighting load and subjected to the demand factors provided in Table 220-11 »220-16(b) «.

Lines 4 Through 8—Applying Demand Factors Found in Table 220-11

Line 4 is simply the total of lines 1 through 3.

Lines 5 through 8 outline the procedure that derates by using demand factors provided in Table 220-11.

| 4 | Add lines 1, 2, and 3 | 4 | 14,565 | Lines 5 through 8 utilize the demand factors found in Table 220-11. |

Line 5 is the result of subtracting 3,000 from the number in Line 4.

The first 3,000 (or less) volt-amperes must remain at 100% »Table 220-11 «.

If Line 5 is 117,000 or less, then Lines 6 and 7 can be skipped.

If Line 4 is 120,000 or less, then Lines 6 and 7 will remain empty.

| 5 | 14,565 (line 4) − 3,000 = | 5 | 11,565 (if 117,000 or less, skip to line 8) | 6 | ---- (line 5, if more than 117,000) − 117,000 = | 6 | ---- |
| 7 | ---- (line 6) × 25% = | 7 | ---- | 8 | 11,565 (smaller of line 5 or 117,000) × 35% = | 8 | 4,048 |

> **NOTE**
>
> *Generally, one-family dwelling load calculations will not use Lines 6 and 7. (For example: a one-family dwelling with a total floor area of 25,000 sq. ft., 20 small appliance branch circuits, and 10 laundry circuits will not need Lines 6 and 7.)*

Insert the smaller of Line 5 or 117,000.

The next 117,000 (from 3,001 to 120,000) is multiplied by 35% »Table 220-11 «.

The result found on line 8 has been rounded up from 4,047.75 to the next whole number, thus eliminating the decimal.

Line 9—Total General Lighting (and Receptacle) Load

Line 9 is the total of general lighting, and general-use receptacles, small appliance branch circuits, and laundry branch circuit(s) after derating.

If Line 7 was skipped (left blank), leave this blank also.

This shaded block draws attention to the first of six lines (boxes) that will be added together to give the total volt-ampere demand load for the one-family dwelling.

| 9 | Total General Lighting and Receptacle Load | 3,000 + ___----___ (line 7) + ___4,048___ (line 8) = | 9 | 7,048 |

If performing a load calculation for a feeder (and panel) that supplies less than 1,000 sq. ft. of floor area and does not supply small appliance or laundry branch circuits, the result could be less than 3,000. For example, if a feeder is installed to supply only 800 sq. ft. of floor area, and will not supply any small appliance or laundry branch circuits, then Line 9 will be 2,400 (3 VA per sq. ft. × 800 sq. ft. = 2,400).

■ Line 10—Fastened-In-Place Appliances

The term **appliance** designates utilization equipment commonly built in standardized types and sizes, installed as a unit to perform specific function(s) ≫ Article 100 ≪.

For the purpose of this load calculation, a kW rating will be the same as a kVA rating.

Multiplying voltage by amperage produces volt-amperes (VA).

Horsepower ratings must be converted to volt-amperes (VA) before they can be added to this form. The full load current ratings for single-phase motors are found in Table 430-148. Multiply the amperes (9.8) by the rated voltage (115) to find the volt-amperes (1,127).

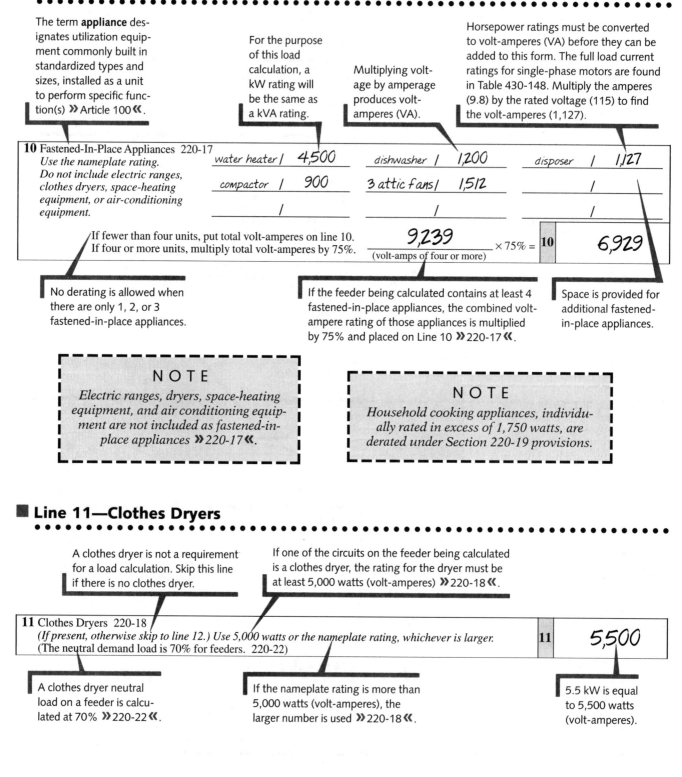

10 Fastened-In-Place Appliances 220-17
Use the nameplate rating.
Do not include electric ranges, clothes dryers, space-heating equipment, or air-conditioning equipment.

water heater / 4,500   dishwasher / 1,200   disposer / 1,127
compactor / 900   3 attic fans / 1,512   /
            /   /   /

If fewer than four units, put total volt-amperes on line 10.
If four or more units, multiply total volt-amperes by 75%.   9,239   × 75% =   **10**   6,929
(volt-amps of four or more)

No derating is allowed when there are only 1, 2, or 3 fastened-in-place appliances.

If the feeder being calculated contains at least 4 fastened-in-place appliances, the combined volt-ampere rating of those appliances is multiplied by 75% and placed on Line 10 ≫ 220-17 ≪.

Space is provided for additional fastened-in-place appliances.

> **N O T E**
> *Electric ranges, dryers, space-heating equipment, and air conditioning equipment are not included as fastened-in-place appliances ≫ 220-17 ≪.*

> **N O T E**
> *Household cooking appliances, individually rated in excess of 1,750 watts, are derated under Section 220-19 provisions.*

■ Line 11—Clothes Dryers

A clothes dryer is not a requirement for a load calculation. Skip this line if there is no clothes dryer.

If one of the circuits on the feeder being calculated is a clothes dryer, the rating for the dryer must be at least 5,000 watts (volt-amperes) ≫ 220-18 ≪.

11 Clothes Dryers 220-18
(If present, otherwise skip to line 12.) Use 5,000 watts or the nameplate rating, whichever is larger.
(The neutral demand load is 70% for feeders. 220-22)   **11**   5,500

A clothes dryer neutral load on a feeder is calculated at 70% ≫ 220-22 ≪.

If the nameplate rating is more than 5,000 watts (volt-amperes), the larger number is used ≫ 220-18 ≪.

5.5 kW is equal to 5,500 watts (volt-amperes).

■ Line 12—Household Cooking Appliances

Household cooking appliances (ranges, wall-mounted ovens, counter-mounted cooking units, etc.) are not required in a load calculation. Skip this line if there are no cooking appliances rated over 1¾ kW.

Individual household cooking appliances rated more than 1,750 watts (volt-amperes) can be derated by using Table 220-19 demand factors (and all of the applicable notes).

12 Ranges, Ovens, Cooktops, and Other Household Cooking Appliances Over 1,750 Watts 220-19 *(If present, otherwise skip to line 13.) Use Table 220-19 and all of the applicable Notes.* (The neutral demand load is 70% for feeders. 220-22)	**12**	*8,450*

The neutral load on a feeder is calculated at 70% for household electric ranges, wall-mounted ovens, and counter-mounted cooking units »220-22«.

A 7 kW counter-mounted cooking unit (cook top) and a 6 kW wall-mounted oven are included in this calculation. Both units fall within the parameters of Column C. The demand factor percentage found in that column for 2 units is 65%. Multiply the total kW (13) by the demand factor percentage (65%) to find the derated load (8.45 kW). 8.45 kW is equal to 8,450 watts (volt-amperes).

■ Line 13—Heating or Air-Conditioning System

The smaller of two (or more) **noncoincident loads** can be omitted, as long as they will never be energized simultaneously »220-21«. By definition, noncoincident means not occurring at the same time. A dwelling's electric heating system and air conditioning system may be noncoincident loads.

The air handler (blower motor) has a rating of 368 volt-amperes ($3.2 \times 115 = 368$). Since both loads will be energized simultaneously, add the air handler to the 15 kW of electric heat.

Fixed electric space heating loads are computed at 100% of the total connected load »220-15«.

13 Heating or Air-Conditioning System (Compare the heat and A/C, and omit the smaller.) 220-21 *Include the air handler when using either one. For heat pumps, include the compressor and the* *maximum amount of electric heat which can be energized while the compressor is running.*	**13**	*15,368*

The air handler (blower) works with either the heating or the air conditioning. Add it to both calculations.

The air-conditioner compressor ($16.6 \times 230 = 3,818$ VA), condenser fan motor ($2 \times 115 = 230$ VA), and blower motor ($3.2 \times 115 = 368$ VA) have a combined load of 4,416 volt-amperes. Because this is less than the heating load omit the A/C compressor and the condenser fan motor.

> **N O T E**
>
> *A heat pump with supplementary heat is not considered a noncoincident load. Add the compressor full-load current to the maximum amount of heat that can be energized while the compressor is running. Remember to include all associated motor(s) (blower, condenser, etc.).*

■ Line 14—Largest Motor

Multiply the largest motor in the dwelling calculation by 25% »220-14 and 430-24«.

The largest motor is usually the air conditioner compressor. If the air conditioner load has been omitted, use the second largest motor. In this case, use the ½-HP-garbage disposer motor.

The result of 281.75 has been rounded up, thus eliminating the decimal.

14 Largest Motor (one motor only) 220-14 and 430-24 *Multiply the volt-amperes of the largest motor by 25%.*	$\dfrac{\textit{1,127}}{\text{(volt-amps of largest motor)}} \times 25\% =$ **14**	*282*

■ Line 15—Total Demand Load

Add lines 9 through 14 to find the service's minimum volt-ampere load.

15 Total Volt-Ampere Demand Load: *Add lines 9 through 14 to find the minimum required volt-amperes.*	**15**	*43,577*

■ Lines 16 and 17—Minimum Service and/or Feeder

Place the total volt-ampere amount found in Line 15 here.

Write down the source voltage that will supply the feeder.

Fractions of an ampere 0.5 and higher are rounded up, while fractions less than 0.5 are dropped ≫ 220-2(b) ≪.

The service (or feeder) overcurrent protection must be higher than the number found in Line 16. Section 240-6(a) provides a list of standard ampere ratings for fuses and circuit breakers.

16 Minimum Amperes *Divide the total volt-amperes by the voltage*	$\dfrac{43,577}{\text{(line 15)}} \div \dfrac{240}{\text{(voltage)}} =$	**16**	*182* (minimum amperes)	**Minimum Size 17 Service and/or Feeder 240-6(a)**	**17**	*200*

Find the amperage by dividing the volt-amperes by the voltage.

This is the minimum amperage rating required for the service and/or feeder, being calculated.

The next standard size fuse or breaker above 182 amperes is 200 ≫ 240-6(a) ≪.

■ Line 18—Minimum Size Conductors

Use Section 310-15(b)(6) to find service entrance and service lateral conductors.

Section 310-15(b)(6) also applies to feeder conductors that serve as the main power feeder to a dwelling unit and are installed in a raceway or cable (with or without an equipment grounding conductor). It is not required that the feeder conductors be larger than the service-entrance conductors ≫ 310-15(b)(6) ≪.

The overcurrent device rating must be 400 amperes, or less.

Table 310-16 requires 3/0 (75°C) copper conductors for 200 amperes. However, since this is a dwelling service, use Table 310-15(b)(6), which allows 2/0 copper conductors.

18 Size the Service and/or Feeder Conductors. *Use Section 310-15(b)(6) to find the service conductors up to 400 amperes. Ratings in excess of 400 amperes shall comply with Table 310-16. 310-15(b)(6) also applies to feeder conductors serving as the main power feeder.*	**Minimum Size Conductors**	**18**	*2/0 copper 4/0 aluminum*

The main power feeder is the feeder(s) that supplies the lighting and appliance branch circuit panelboard(s) ≫ 310-15(b)(6) ≪.

Table 310-15(b)(6) provides sizes for both copper and aluminum conductors. Copper-clad aluminum conductors are listed in the column with aluminum.

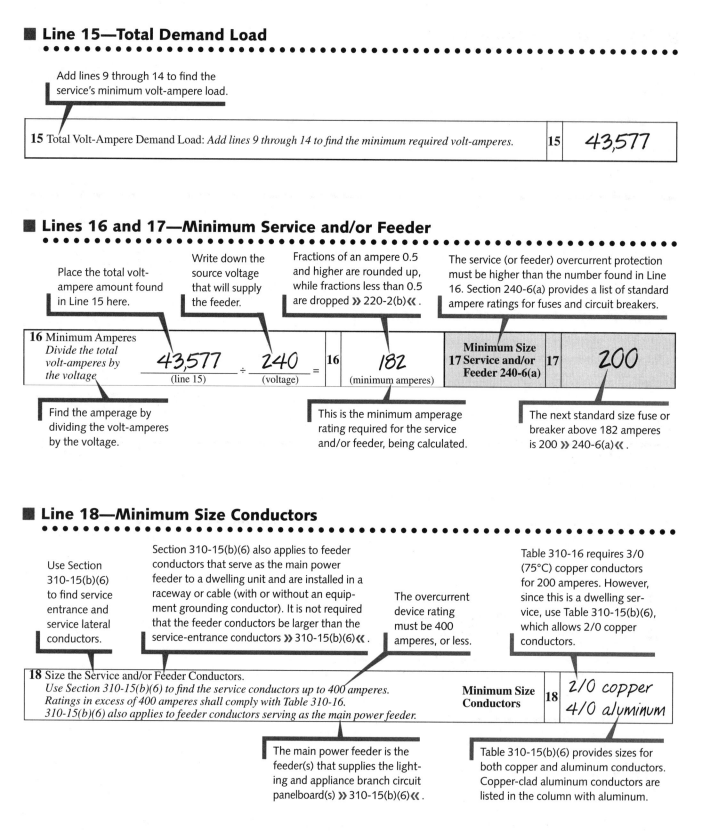

■ Line 19—Neutral Conductor

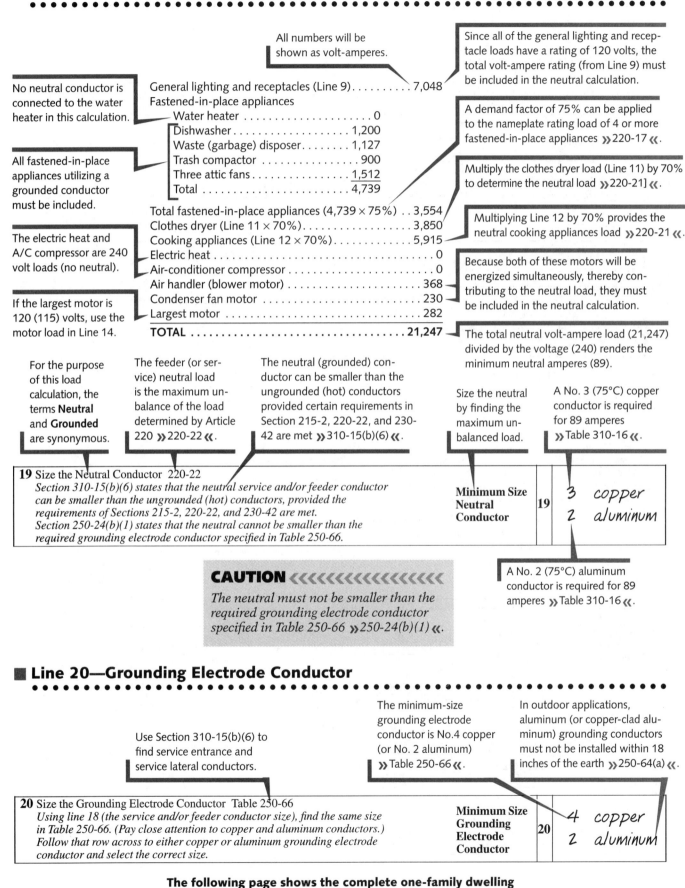

All numbers will be shown as volt-amperes.

No neutral conductor is connected to the water heater in this calculation.

Since all of the general lighting and receptacle loads have a rating of 120 volts, the total volt-ampere rating (from Line 9) must be included in the neutral calculation.

General lighting and receptacles (Line 9) 7,048
Fastened-in-place appliances

| Water heater . . . . 0 |
| Dishwasher . . . . 1,200 |
| Waste (garbage) disposer . . . . 1,127 |
| Trash compactor . . . . 900 |
| Three attic fans . . . . 1,512 |
| Total . . . . 4,739 |

All fastened-in-place appliances utilizing a grounded conductor must be included.

A demand factor of 75% can be applied to the nameplate rating load of 4 or more fastened-in-place appliances »220-17 «.

Multiply the clothes dryer load (Line 11) by 70% to determine the neutral load »220-21] «.

Total fastened-in-place appliances (4,739 × 75%) . . 3,554
Clothes dryer (Line 11 × 70%) 3,850
Cooking appliances (Line 12 × 70%) 5,915
Electric heat 0
Air-conditioner compressor 0
Air handler (blower motor) 368
Condenser fan motor 230
Largest motor 282
TOTAL **21,247**

The electric heat and A/C compressor are 240 volt loads (no neutral).

If the largest motor is 120 (115) volts, use the motor load in Line 14.

Multiplying Line 12 by 70% provides the neutral cooking appliances load »220-21 «.

Because both of these motors will be energized simultaneously, thereby contributing to the neutral load, they must be included in the neutral calculation.

The total neutral volt-ampere load (21,247) divided by the voltage (240) renders the minimum neutral amperes (89).

For the purpose of this load calculation, the terms **Neutral** and **Grounded** are synonymous.

The feeder (or service) neutral load is the maximum unbalance of the load determined by Article 220 »220-22 «.

The neutral (grounded) conductor can be smaller than the ungrounded (hot) conductors provided certain requirements in Section 215-2, 220-22, and 230-42 are met »310-15(b)(6) «.

Size the neutral by finding the maximum unbalanced load.

A No. 3 (75°C) copper conductor is required for 89 amperes »Table 310-16 «.

19 Size the Neutral Conductor 220-22			
Section 310-15(b)(6) states that the neutral service and/or feeder conductor can be smaller than the ungrounded (hot) conductors, provided the requirements of Sections 215-2, 220-22, and 230-42 are met. Section 250-24(b)(1) states that the neutral cannot be smaller than the required grounding electrode conductor specified in Table 250-66.	**Minimum Size Neutral Conductor**	**19**	3 *copper* 2 *aluminum*

A No. 2 (75°C) aluminum conductor is required for 89 amperes »Table 310-16 «.

CAUTION «««««««««««««««
The neutral must not be smaller than the required grounding electrode conductor specified in Table 250-66 »250-24(b)(1) «.

■ Line 20—Grounding Electrode Conductor

Use Section 310-15(b)(6) to find service entrance and service lateral conductors.

The minimum-size grounding electrode conductor is No.4 copper (or No. 2 aluminum) »Table 250-66 «.

In outdoor applications, aluminum (or copper-clad aluminum) grounding conductors must not be installed within 18 inches of the earth »250-64(a) «.

20 Size the Grounding Electrode Conductor Table 250-66			
Using line 18 (the service and/or feeder conductor size), find the same size in Table 250-66. (Pay close attention to copper and aluminum conductors.) Follow that row across to either copper or aluminum grounding electrode conductor and select the correct size.	**Minimum Size Grounding Electrode Conductor**	**20**	4 *copper* 2 *aluminum*

The following page shows the complete one-family dwelling load calculation according to the standard method.

Standard Method Load Calculation for One-Family Dwellings

1 General Lighting and Receptacle Loads 220-3(a) *Do not include open porches, garages, unused or* *unfinished spaces not adaptable for future use.*		$3 \times$ ___1,855___ (sq. ft. using outside dimensions)	= **1**	**5,565**
2 Small Appliance Branch Circuits 220-16(a) *At least **two** small appliance branch circuits* *must be included. 210-11(c)(1)*		$1,500 \times$ ___4___ (minimum of two)	= **2**	**6,000**
3 Laundry Branch Circuit(s) 220-16(b) *At least **one** laundry branch circuit* *must be included. 210-11(c)(2)*		$1,500 \times$ ___2___ (minimum of one)	= **3**	**3,000**

4 Add lines 1, 2, and 3	**4** *14,565*	Lines 5 through 8 utilize the demand factors found in Table 220-11.

5 ___*14,565*___ $- 3,000 =$ **5** (line 4)	*11,565* (if 117,000 or less, skip to line 8)	**6** ___----___ $- 117,000 =$ **6** (line 5, if more than 117,000)	----
7 ___----___ $\times 25\% =$ **7** (line 6)	----	**8** ___*11,565*___ $\times 35\% =$ **8** (smaller of line 5 or 117,000)	*4,048*

9 Total General Lighting and Receptacle Load $3,000 +$ ___----___ $+$ ___*4,048*___ $=$ **9** (line 7) (line 8)		**7,048**

10 Fastened-In-Place Appliances 220-17
Use the nameplate rating.
Do not include electric ranges,
clothes dryers, space-heating
equipment, or air-conditioning
equipment.

water heater / *4,500*	dishwasher / *1,200*	disposer / *1,127*
compactor / *900*	3 attic fans / *1,512*	/
/	/	/

If fewer than four units, put total volt-amperes on line 10.
If four or more units, multiply total volt-amperes by 75%.   ___*9,239*___ $\times 75\% =$ **10**   **6,929**
(volt-amps of four or more)

11 Clothes Dryers 220-18 *(If present, otherwise skip to line 12.) Use 5,000 watts or the nameplate rating, whichever is larger.* (The neutral demand load is 70% for feeders. 220-22)	**11**	**5,500**
12 Ranges, Ovens, Cooktops, and Other Household Cooking Appliances Over 1,750 Watts 220-19 *(If present, otherwise skip to line 13.) Use Table 220-19 and all of the applicable Notes.* (The neutral demand load is 70% for feeders. 220-22)	**12**	**8,450**
13 Heating or Air-Conditioning System (Compare the heat and A/C, and omit the smaller.) 220-21 *Include the air handler when using either one. For heat pumps, include the compressor and the* *maximum amount of electric heat which can be energized while the compressor is running.*	**13**	**15,368**
14 Largest Motor (one motor only) 220-14 and 430-24 *Multiply the volt-amperes of the largest motor by 25%.* ___*1,127*___ $\times 25\% =$ **14** (volt-amps of largest motor)	**14**	**282**
15 Total Volt-Ampere Demand Load: *Add lines 9 through 14 to find the minimum required volt-amperes.*	**15**	**43,577**

16 Minimum Amperes *Divide the total* *volt-amperes by* *the voltage*	___*43,577*___ $\div$ ___*240*___ $=$ **16** (line 15) (voltage)	*182* (minimum amperes)	**Minimum Size** **17 Service and/or** **Feeder 240-6(a)** **17**	**200**

18 Size the Service and/or Feeder Conductors. *Use Section 310-15(b)(6) to find the service conductors up to 400 amperes.* *Ratings in excess of 400 amperes shall comply with Table 310-16.* *310-15(b)(6) also applies to feeder conductors serving as the main power feeder.*	**Minimum Size** **Conductors** **18**	*2/0 copper* *4/0 aluminum*
19 Size the Neutral Conductor 220-22 *Section 310-15(b)(6) states that the neutral service and/or feeder conductor* *can be smaller than the ungrounded (hot) conductors, provided the* *requirements of Sections 215-2, 220-22, and 230-42 are met.* *Section 250-24(b)(1) states that the neutral cannot be smaller than the* *required grounding electrode conductor specified in Table 250-66.*	**Minimum Size** **Neutral** **Conductor** **19**	*3 copper* *2 aluminum*
20 Size the Grounding Electrode Conductor Table 250-66 *Using line 18 (the service and/or feeder conductor size), find the same size* *in Table 250-66. (Pay close attention to copper and aluminum conductors.)* *Follow that row across to either copper or aluminum grounding electrode* *conductor and select the correct size.*	**Minimum Size** **Grounding** **Electrode** **Conductor** **20**	*4 copper* *2 aluminum*

OPTIONAL METHOD: ONE-FAMILY DWELLINGS

Optional Method Load Calculation for One-Family Dwellings

1 General Lighting and Receptacle Loads 220-30(b)(2) *Do not include open porches, garages, unused or* *unfinished spaces not adaptable for future use.* $3 \times$ _____ = **1** (sq. ft. using outside dimensions)	
2 Small Appliance Branch Circuits 220-30(b)(1) *At least **two** small appliance branch circuits* $1,500 \times$ _____ = **2** *must be included. 210-11(c)(1)* (minimum of two)	
3 Laundry Branch Circuit(s) 220-30(b)(1) *At least **one** laundry branch circuit* $1,500 \times$ _____ = **3** *must be included. 210-11(c)(2)* (minimum of one)	

4 Appliances 220-30(b)(3) and (4)
*Use the nameplate rating of **ALL** appliances (fastened-in-place, permanently connected, or connected to a specific circuit), ranges, ovens, cook tops, motors, and clothes dryers.*
Convert any nameplate rating given in amperes to volt-amperes by multiplying the amperes by the rated voltage.

Do not include any heating or air-conditioning equipment in this section.

Total volt-amperes of all appliances **LISTED BELOW**   **4**

water heater /	clothes dryer /	range /
dishwasher /	disposer /	/
/	/	/
/	/	/
/	/	/

5 Apply Section 220-30(b) demand factor to the total of lines 1 through 4.   **5**

_____ $- 10,000 =$ _____ $\times 40\% =$ _____ $+ 10,000 =$
(total of lines 1 through 4)

6 Heating and/or Air-Conditioning System 220-30(c)
*Use the nameplate rating(s) in volt-amperes for all applicable systems in lines **a** through **e**.*

a) Air-conditioning and cooling system(s), including heat pump compressors and supplemental heating, unless the controller prevents the compressor and supplemental heating from operating at the same time.
_____ $\times 100\% =$ **a)**

b) Electric thermal storage and other heating systems where the usual load is expected to be continuous at full nameplate value: *Systems qualifying under this selection shall not be figured under any other selection in 220-30(c).*
_____ $\times 100\% =$ **b)**

c) Central electric space heating equipment, including integral supplemental heating in heat pumps where the controller prevents the compressor and supplemental heating from operating at the same time.
_____ $\times 65\% =$ **c)**

d) Electric space heating equipment, if less than four separately controlled units:
_____ $\times 65\% =$ **d)**

e) Electric space heating equipment, if four or more separately controlled units:
_____ $\times 40\% =$ **e)**

7 Total Volt-Ampere Demand Load: _____ $+$ _____ $=$ **7**
(largest VA rating from line 6a through 6e)   (line 5)

8 Minimum Amperes
Divide the total volt-amperes by the voltage   _____ $\div$ _____ $=$ **8**
(line 15)   (voltage)   (minimum amperes)

9 Minimum Size Service and/or Feeder 240-6(a)   **9**   (minimum is 100 amperes)

10 Size the Service and/or Feeder Conductors.
Use Section 310-15(b)(6) to find the service conductors up to 400 amperes.
Ratings in excess of 400 amperes shall comply with Table 310-16.
310-15(b)(6) also applies to feeder conductors serving as the main power feeder.   **Minimum Size Conductors**   **10**

11 Size the Neutral Conductor 220-22
Note: There is no optional method for calculating the neutral conductor.
Section 310-15(b)(6) states that the neutral service and/or feeder conductor can be smaller than the ungrounded (hot) conductors, provided the requirements of Sections 215-2, 220-22, and 230-42 are met.
Section 250-24(b)(1) states that the neutral cannot be smaller than the required grounding electrode conductor specified in Table 250-66.   **Minimum Size Neutral Conductor**   **11**

12 Size the Grounding Electrode Conductor Table 250-66
Using line 18 (the service and/or feeder conductor size), find the same size in Table 250-66. (Pay close attention to copper and aluminum conductors.)
Follow that row across to either copper or aluminum grounding electrode conductor and select the correct size.   **Minimum Size Grounding Electrode Conductor**   **12**

Lines 1 Through 3—General Lighting and Receptacle Loads

In the **Optional Method**, Lines 1, 2 and 3 are calculated exactly as described in the **Standard Method**. (See explanations on pages 137 and 138).

1	General Lighting and Receptacle Loads 220-30(b)(2) *Do not include open porches, garages, unused or* *unfinished spaces not adaptable for future use.*	3 × __1,855__ (sq. ft. using outside dimensions)	= **1** **5,565**
2	Small Appliance Branch Circuits 220-30(b)(1) *At least two small appliance branch circuits* *must be included. 210-11(c)(1)*	1,500 × __4__ (minimum of two)	= **2** **6,000**
3	Laundry Branch Circuit(s) 220-30(b)(1) *At least one laundry branch circuit* *must be included. 210-11(c)(2)*	1,500 × __2__ (minimum of one)	= **3** **3,000**

Line 4—Appliances

This Section includes appliances (fastened-in-place, permanently connected, or connected to a specific circuit), ranges, wall-mounted ovens, counter-mounted cooking units, clothes dryers, and water heaters ≫220-30(b)(3)≪.

Heating and air-conditioning systems are *not* included in this computation.

Use the exact nameplate rating of the clothes dryer, even if less than 5,000 volt-amperes.

4	Appliances 220-30(b)(3) and (4) *Use the nameplate rating of ALL* *appliances (fastened-in-place,* *permanently connected, or* *connected to a specific circuit),* *ranges, ovens, cook tops, motors,* *and clothes dryers.* *Convert any nameplate rating* *given in amperes to volt-amperes* *by multiplying the amperes* *by the rated voltage.*	Total volt-amperes of all appliances LISTED BELOW **4** **27,739**

Do not include any heating or air-conditioning equipment in this section.

water heater /	4,500	clothes dryer /	5,500	range /	
dishwasher /	1,200	disposer /	1,127	cook top /	7,000
compactor /	900	3 attic fans /	1,512	oven /	6,000
/		/		/	
/		/		/	

Do not apply the demand factors in Table 220-19 to household cooking equipment (ranges, cook tops, and ovens). List the nameplate ratings as they appear.

Line 5—Applying Demand Factors to Lines 1 Through 4

This demand factor is found in Section 220-30(b).

This is the newly calculated load which includes everything except the heating and air-conditioning systems.

5 Apply Section 220-30(b) demand factor to the total of lines 1 through 4.

__42,304__ − 10,000 = __32,304__ × 40% = __12,922__ + 10,000 = **5** **22,922**
(total of lines 1 through 4)

(5,565 + 6,000 + 3,000 + 27,739)

Because the first 10,000 volt-amperes are calculated at 100%, subtract 10,000 here.

The remainder of the load is calculated at 40%.

The result on Line 5 has been rounded up from 12,921.6.

■ Line 6—Heating and/or Air-Conditioning Systems

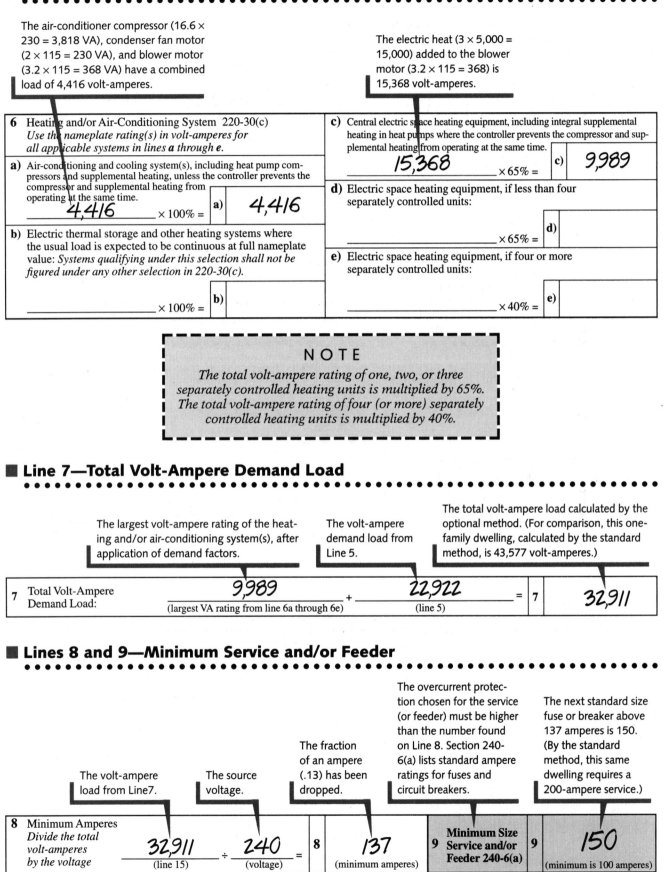

The air-conditioner compressor (16.6 × 230 = 3,818 VA), condenser fan motor (2 × 115 = 230 VA), and blower motor (3.2 × 115 = 368 VA) have a combined load of 4,416 volt-amperes.

The electric heat (3 × 5,000 = 15,000) added to the blower motor (3.2 × 115 = 368) is 15,368 volt-amperes.

6 Heating and/or Air-Conditioning System 220-30(c)
*Use the nameplate rating(s) in volt-amperes for all applicable systems in lines **a** through **e**.*

a) Air-conditioning and cooling system(s), including heat pump compressors and supplemental heating, unless the controller prevents the compressor and supplemental heating from operating at the same time.

_____4,416_____ × 100% =   **a)**   4,416

b) Electric thermal storage and other heating systems where the usual load is expected to be continuous at full nameplate value: *Systems qualifying under this selection shall not be figured under any other selection in 220-30(c).*

_____ × 100% =   **b)**

c) Central electric space heating equipment, including integral supplemental heating in heat pumps where the controller prevents the compressor and supplemental heating from operating at the same time.

_____15,368_____ × 65% =   **c)**   9,989

d) Electric space heating equipment, if less than four separately controlled units:

_____ × 65% =   **d)**

e) Electric space heating equipment, if four or more separately controlled units:

_____ × 40% =   **e)**

> ### NOTE
> *The total volt-ampere rating of one, two, or three separately controlled heating units is multiplied by 65%. The total volt-ampere rating of four (or more) separately controlled heating units is multiplied by 40%.*

■ Line 7—Total Volt-Ampere Demand Load

The largest volt-ampere rating of the heating and/or air-conditioning system(s), after application of demand factors.

The volt-ampere demand load from Line 5.

The total volt-ampere load calculated by the optional method. (For comparison, this one-family dwelling, calculated by the standard method, is 43,577 volt-amperes.)

7 Total Volt-Ampere Demand Load: _____9,989_____ + _____22,922_____ = **7**   32,911
(largest VA rating from line 6a through 6e)   (line 5)

■ Lines 8 and 9—Minimum Service and/or Feeder

The volt-ampere load from Line 7.

The source voltage.

The fraction of an ampere (.13) has been dropped.

The overcurrent protection chosen for the service (or feeder) must be higher than the number found on Line 8. Section 240-6(a) lists standard ampere ratings for fuses and circuit breakers.

The next standard size fuse or breaker above 137 amperes is 150. (By the standard method, this same dwelling requires a 200-ampere service.)

8 Minimum Amperes *Divide the total volt-amperes by the voltage*   _____32,911_____ ÷ _____240_____ = **8**   137   **9** Minimum Size Service and/or Feeder 240-6(a) **9**   150
(line 15)   (voltage)   (minimum amperes)   (minimum is 100 amperes)

■ Line 10—Minimum Size Conductors

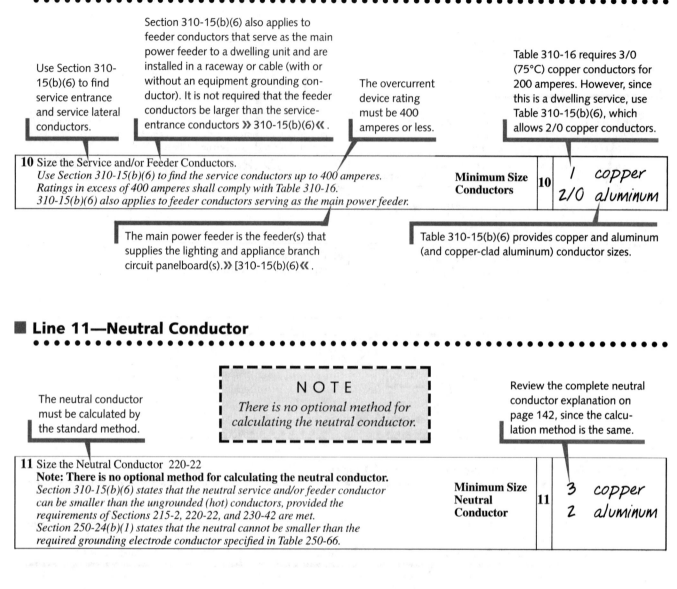

Use Section 310-15(b)(6) to find service entrance and service lateral conductors.

Section 310-15(b)(6) also applies to feeder conductors that serve as the main power feeder to a dwelling unit and are installed in a raceway or cable (with or without an equipment grounding conductor). It is not required that the feeder conductors be larger than the service-entrance conductors ≫ 310-15(b)(6) ≪.

The overcurrent device rating must be 400 amperes or less.

Table 310-16 requires 3/0 (75°C) copper conductors for 200 amperes. However, since this is a dwelling service, use Table 310-15(b)(6), which allows 2/0 copper conductors.

| 10 Size the Service and/or Feeder Conductors. *Use Section 310-15(b)(6) to find the service conductors up to 400 amperes. Ratings in excess of 400 amperes shall comply with Table 310-16. 310-15(b)(6) also applies to feeder conductors serving as the main power feeder.* | Minimum Size Conductors | 10 | 1 copper 2/0 aluminum |

The main power feeder is the feeder(s) that supplies the lighting and appliance branch circuit panelboard(s). ≫ [310-15(b)(6) ≪.

Table 310-15(b)(6) provides copper and aluminum (and copper-clad aluminum) conductor sizes.

■ Line 11—Neutral Conductor

The neutral conductor must be calculated by the standard method.

> **NOTE**
> *There is no optional method for calculating the neutral conductor.*

Review the complete neutral conductor explanation on page 142, since the calculation method is the same.

| 11 Size the Neutral Conductor 220-22 **Note: There is no optional method for calculating the neutral conductor.** *Section 310-15(b)(6) states that the neutral service and/or feeder conductor can be smaller than the ungrounded (hot) conductors, provided the requirements of Sections 215-2, 220-22, and 230-42 are met. Section 250-24(b)(1) states that the neutral cannot be smaller than the required grounding electrode conductor specified in Table 250-66.* | Minimum Size Neutral Conductor | 11 | 3 copper 2 aluminum |

■ Line 12—Grounding Electrode Conductor

The minimum size grounding electrode conductor is No. 6 copper (or No. 4 aluminum) ≫ Table 250-66 ≪.

| 12 Size the Grounding Electrode Conductor Table 250-66 *Using line 18 (the service and/or feeder conductor size), find the same size in Table 250-66. (Pay close attention to copper and aluminum conductors.) Follow that row across to either copper or aluminum grounding electrode conductor and select the correct size.* | Minimum Size Grounding Electrode Conductor | 12 | 6 copper 4 aluminum |

On the following page is the complete one-family dwelling load calculation according to the optional method.

Optional Method Load Calculation for One-Family Dwellings

1	General Lighting and Receptacle Loads 220-30(b)(2) *Do not include open porches, garages, unused or unfinished spaces not adaptable for future use.*	$3 \times$ ___**1,855**___ (sq. ft. using outside dimensions)	= **1**	**5,565**
2	Small Appliance Branch Circuits 220-30(b)(1) *At least **two** small appliance branch circuits must be included. 210-11(c)(1)*	$1,500 \times$ ___**4**___ (minimum of two)	= **2**	**6,000**
3	Laundry Branch Circuit(s) 220-30(b)(1) *At least **one** laundry branch circuit must be included. 210-11(c)(2)*	$1,500 \times$ ___**2**___ (minimum of one)	= **3**	**3,000**

4 Appliances 220-30(b)(3) and (4)
*Use the nameplate rating of **ALL** appliances (fastened-in-place, permanently connected, or connected to a specific circuit), ranges, ovens, cook tops, motors, and clothes dryers. Convert any nameplate rating given in amperes to volt-amperes by multiplying the amperes by the rated voltage.*

Do not include any heating or air-conditioning equipment in this section.

Total volt-amperes of all appliances LISTED BELOW **4** **27,739**

water heater / **4,500**	clothes dryer / **5,500**	range /
dishwasher / **1,200**	disposer / **1,127**	cook top / **7,000**
compactor / **900**	3 attic fans/ **1,512**	oven / **6,000**
/	/	/
/	/	/

5 Apply Section 220-30(b) demand factor to the total of lines 1 through 4.

___**42,304**___ (total of lines 1 through 4) $- 10,000 =$ ___**32,304**___ $\times 40\% =$ ___**12,922**___ $+ 10,000 =$ **5** **22,922**

6 Heating and/or Air-Conditioning System 220-30(c)
*Use the nameplate rating(s) in volt-amperes for all applicable systems in lines **a** through **e**.*

a) Air-conditioning and cooling system(s), including heat pump compressors and supplemental heating, unless the controller prevents the compressor and supplemental heating from operating at the same time.
___**4,416**___ $\times 100\% =$ **a)** **4,416**

b) Electric thermal storage and other heating systems where the usual load is expected to be continuous at full nameplate value: *Systems qualifying under this selection shall not be figured under any other selection in 220-30(c).*
_____ $\times 100\% =$ **b)**

c) Central electric space heating equipment, including integral supplemental heating in heat pumps where the controller prevents the compressor and supplemental heating from operating at the same time.
___**15,368**___ $\times 65\% =$ **c)** **9,989**

d) Electric space heating equipment, if less than four separately controlled units:
_____ $\times 65\% =$ **d)**

e) Electric space heating equipment, if four or more separately controlled units:
_____ $\times 40\% =$ **e)**

7 Total Volt-Ampere Demand Load: ___**9,989**___ (largest VA rating from line 6a through 6e) $+$ ___**22,922**___ (line 5) $=$ **7** **32,911**

8	Minimum Amperes *Divide the total volt-amperes by the voltage*	___**32,911**___ (line 15) $\div$ ___**240**___ (voltage) $=$ **8** **137** (minimum amperes)	**9** Minimum Size Service and/or Feeder 240-6(a)	**9** **150** (minimum is 100 amperes)

10	Size the Service and/or Feeder Conductors. *Use Section 310-15(b)(6) to find the service conductors up to 400 amperes. Ratings in excess of 400 amperes shall comply with Table 310-16. 310-15(b)(6) also applies to feeder conductors serving as the main power feeder.*	Minimum Size Conductors	**10**	**1 copper** **2/0 aluminum**
11	Size the Neutral Conductor 220-22 **Note: There is no optional method for calculating the neutral conductor.** *Section 310-15(b)(6) states that the neutral service and/or feeder conductor can be smaller than the ungrounded (hot) conductors, provided the requirements of Sections 215-2, 220-22, and 230-42 are met. Section 250-24(b)(1) states that the neutral cannot be smaller than the required grounding electrode conductor specified in Table 250-66.*	Minimum Size Neutral Conductor	**11**	**3 copper** **2 aluminum**
12	Size the Grounding Electrode Conductor Table 250-66 *Using line 18 (the service and/or feeder conductor size), find the same size in Table 250-66. (Pay close attention to copper and aluminum conductors.) Follow that row across to either copper or aluminum grounding electrode conductor and select the correct size.*	Minimum Size Grounding Electrode Conductor	**12**	**6 copper** **4 aluminum**

Summary

● ●

- A dwelling unit load calculation, for general lighting and receptacles, is determined by total square feet.
- For this purpose, Table 220-3(a) requires a unit load of 3 volt-amperes per square foot.
- At least 1 laundry and 2 small appliance branch circuits are required when performing a service (or main power feeder) calculation.
- Laundry and small appliance branch circuits are calculated using a rating of at least 1,500 volt-amperes each.
- Table 220-11 demand factors are applied to the general lighting, small appliance, and laundry loads.
- The total volt-ampere rating of four or more fastened-in-place appliances is multiplied by 75%.

- Demand loads for household cooking appliances (over 1¾ kW rating) are found in Table 220-19.
- The heating and air-conditioning loads are compared, and the larger of the two is used.
- For heat pumps, include the compressor and the maximum amount of electric heat which can be energized while the compressor is running.
- Two load calculation methods are provided in Article 220, **Standard** and **Optional**.
- The neutral conductor can be smaller than the ungrounded (hot) conductors, provided the requirements of Sections 215-2, 220-22, and 230-42 are met.
- Table 250-66 is used to size grounding electrode conductors.

Unit 8 Competency Test

● ●

NEC® Reference   Answer

_____   _____   1. It shall be permissible to apply a demand factor of _____ percent to the nameplate rating load of _____ or more appliances fastened in place, other than electric ranges, clothes dryers, space-heating equipment, or air-conditioning equipment, that are served by the same feeder in a one-family dwelling.

_____   _____   2. What is the lighting demand load (before derating) for a house with outside dimensions of 30' × 48' on the first floor and 22' × 42' on the second floor?

_____   _____   3. A homeowner wants to add twelve general-purpose receptacle outlets to the electrical drawings, before the house is built. What additional load, in volt-amperes, will this add to the service-load calculation?

_____   _____   4. In each dwelling unit, the feeder load shall be computed at _____ volt-amperes for each 2-wire small appliance branch circuit required by Section 210-11(c)(1).

_____   _____   5. What is the demand load for the service for two 3½ kW wall-mounted ovens and one 5 kW counter-mounted cooking unit? All of the appliances are supplied from a single branch circuit and are located in the same room of a one-family dwelling.
      a) 12 kW   b) 9.6 kW   c) 8 kW   d) 6.6 kW

_____   _____   6. A 4.5 kVA, 240 volt clothes dryer will contribute _____ amperes to the neutral load, when calculating the service by the standard method.

_____   _____   7. A one-family dwelling contains the following appliances: a 1 kVA, 115-volt dishwasher; a ¼ HP, 115-volt waste (garbage) disposer, and a ⅓ HP, 115-volt trash compactor. Using the standard method, what is the load contribution, in volt-amperes, for these appliances?

_____   _____   8. For one-family dwellings, the service disconnecting means must have a rating of at least _____ amperes, 3-wire.

_____   _____   9. A one-family dwelling has a 175-ampere, 240-volt service that is fed with 75°C conductors. What is the minimum size copper grounding electrode conductor required?

_____   _____   10. A one-family dwelling has six 3 kW electric wall heaters (with individual thermostats) and five room air-conditioners. Two air-conditioners are rated 10.5 amperes at 230 volts, and the others are rated 7.2 amperes at 115 volts. A service calculation is being performed, using the optional method. How many volt-amperes should be included (after demand factors) for the heating and air-conditioning? (Assume heater kW ratings equivalent to kVA.)

_____   _____   11. In each dwelling unit, a feeder load of at least _____ volt-amperes must be included for each 2-wire laundry branch circuit installed as required by Section 210-11(c)(2).

_____   _____   12. A 12 kW, 240-volt range will contribute _____ watts to the load, when calculating the service by the standard method.

Questions 13 through 20 are based on a 120/240 volt, 3-wire, single-phase one-family dwelling containing the following:

Floor area . . . . 2,500 sq. ft.	Clothes dryer. . . . . 4 kW, 240 volt
Range . . . . 12 kW, 240 volt	Electric heat (2 banks @ 3 kW each) . . 6 kW, 240 volt
Water heater. . . . . 4 kW, 240 volt	Air handler (blower motor). . . . . ¼ HP, 115 volt
Dishwasher. . . . . 1.5 kW, 120 volt	Air-conditioner compressor . . . . 5 HP, 230 volt
Waste (garbage) disposer . . ⅓ HP, 115 volt	Condenser fan motor. . . . . ⅙ HP, 115 volt

Assume water heater, clothes dryer, range, and electric heat kW ratings equivalent to kVA.

Using Part B (the standard method), of Article 220:

_____   _____   13. What is the minimum ampere rating for the service?

_____   _____   14. What is minimum size service?

_____   _____   15. What is the minimum size THWN copper conductor that can be installed as the service (ungrounded) conductor?

_____   _____   16. What is the minimum size grounding electrode conductor?

Using the Optional Method:

_____   _____   17. What is the minimum ampere rating for the service?

_____   _____   18. What is minimum size service?

_____   _____   19. What is the minimum size THWN copper conductor that can be installed as the service (ungrounded) conductor?

_____   _____   20. What is the minimum size grounding electrode conductor?

_____   _____   21. What is the minimum size THWN copper conductor that can be installed as the neutral conductor?

Services and Electrical Equipment

Objectives

After studying this unit, the student will:

- ✔ be able to determine adequate strength for a mast supporting service-drop conductors.
- ✔ understand service-entrance cable provisions.
- ✔ know the definition of a service lateral and understand the applicable provisions.
- ✔ be familiar with minimum wiring clearance requirements (vertical, horizontal, etc.) for service and outside wiring.
- ✔ have a good understanding of the amount of three-dimensional working space required around service equipment, panelboards, electric equipment, etc.
- ✔ thoroughly understand panelboards, cabinets, and cutout boxes.
- ✔ be familiar with the multitude of provisions relating to circuit breakers and fuses.
- ✔ know and understand different aspects of service equipment and panelboard installations.

- ✔ be able to identify the appropriate Table (310-15[b][6] or 310-16) for conductor sizing.
- ✔ understand that electric equipment must be installed in a neat and workmanlike manner.
- ✔ have a detailed understanding of grounding and bonding procedures.
- ✔ be able to properly position grounded and grounding conductors in remote panelboards (subpanels).
- ✔ have an extensive understanding of the grounding system, as a whole.
- ✔ understand how to ground and bond panelboards on the supply side, as well as the load side, of the service disconnecting means.
- ✔ be able to prevent objectionable current flow in grounding conductors.
- ✔ be familiar with the different grounding possibilities for panelboards installed in separate buildings or structures.

Introduction

Because the service is the heart of the electrical system, the importance of understanding this unit is enormous. Many of the regulations learned in this unit will be applicable in all types of occupancies, not merely single-family dwellings. Unit 9 is divided into five sections. The first, **Service-entrance Wiring Methods**, addresses both overhead and underground service entrance provisions. The section entitled **Service and Outside Wiring Clearances** covers the clearances required above roofs and final grade. Other clearances (horizontal, vertical, and diagonal) are shown for windows, doors, balconies, platforms, etc. Next, the **Working Space Around Equipment** section illustrates common working zone requirements. **Service Equipment and Panelboards** explains panelboards, cutout boxes, and cabinets. Specific provisions pertaining to circuit breakers, fuses, and circuit breaker panels are also covered. **Grounding**, the last section, explains in detail Article 250 provisions relating to service equipment and panelboards. Refer to Unit 9 often when studying (or installing) services, whether one-family, multi-family dwellings, commercial, or even industrial buildings.

Since many local jurisdictions and electric utility companies have supplemental requirements, exercise caution in compiling a complete set of service installation requirements. Examples of these requirements include, but are not limited to: restrictions on the type of conduit used for service mast; minimum conduit trade size; specific strapping methods; restrictions on the use of service-entrance cables; special underground service provisions; limitations on the maximum distance service-entrance conductors can run within a building or structure; and various grounding provisions. Obtain a copy of these additional rules (if any) for your area.

SERVICE-ENTRANCE WIRING METHODS

■ Service Mast as Support

Service-drop and service-entrance conductors must be arranged so that water cannot enter the service raceway or equipment »230-54(g) «.

Only power service-drop conductors can be attached to a service-mast »230-28 «.

It is not required (where approved) that rigid (and intermediate) metal conduit be securely fastened within 3 feet of the service head for above-the-roof termination of a mast »345-12(a) and 346-12(a) «.

2½-inch Rigid Metal Conduit

Open conductors must be attached to fittings identified as suitable for use with service conductors or to noncombustible, nonabsorbent insulators securely attached to the building or other structure »230-27 «.

A service mast used to support service-drop conductors must be of adequate strength or be supported by braces (or guys) to safely withstand the strain imposed by the service drop »230-28 «.

■ Service Heads (Weather Heads)

Service raceways must be equipped with a raintight service head at the point of connection to service-drop conductors »230-54(a) «.

Conductors of different potential must be brought out of service heads through separately bushed openings »230-54(e) «.

An insulated grounded conductor, larger than No. 6, must be identified either by a continuous white or natural gray outer finish or by three continuous white stripes on other than green insulation along its entire length or by a distinctive white marking at its terminations during installation. This marking must completely encircle the conductor or insulation »200-6(b) «.

Raceway-type service masts require raceway fittings which are identified for use with service masts »230-28 «.

Conductors must have adequate mechanical strength and sufficient ampacity to carry the current for the load (as computed in accordance with Article 220) »230-23(a) «.

■ Service Mast Attachments

The service drop consists of the overhead conductors, inclusive of any splices, extending from the last pole (or other aerial support) to the service-entrance conductors at the building (or similar structure) »Article 100 «.

Service Drop

Telephone Wire

T.V. Cable

Other Cable or Wiring

Nothing, except electrical service-drop conductors, can be attached to the service mast »230-28 «.

■ Service Drops Attached to Buildings

Open conductors must be attached to fittings identified specifically for use with service conductors, or to noncombustible, nonabsorbent insulators which are securely attached to the building (or other structure) ≫230-27≪.

Service heads must be located above the point where service-drop conductors are attached to the building or other structure ≫230-54(c)≪.

Where it is impracticable to locate the service head above the point of attachment, the service head must be located within 24 inches of that point ≫230-54(c) *Exception*≪.

2-inch EMT

Since this service mast does not support service-drop conductors, another wiring method (EMT) is used. Other allowable service-entrance methods are listed in Section 230-43.

EMT (electrical metallic tubing) couplings and connectors shall be made up tight. If installed in wet locations, they must be of the raintight type ≫348-10≪.

Electric equipment must be firmly secured to the surface on which it is mounted. Wooden plugs driven into holes in masonry, concrete, plaster, etc., are not acceptable ≫110-13(a)≪.

■ Support of Overhead Conductors

Vegetation (such as trees) cannot be used to support overhead conductor spans, except for temporary wiring in accordance with Article 305 ≫225-26≪.

Drip loops must be formed on each individual conductor. To prevent moisture penetration, service-entrance conductors are connected to the service-drop conductors either (1) below the level of the service head, or (2) below the level of the termination of the service-entrance cable sheath ≫230-54(f)≪.

■ Goosenecks in Service-Entrance Cables

Bends, and all other handlings, in service-entrance cable must be made so that the cable's protective covering remains undamaged. In addition, the radius of the curve of the inner edge of any bend must be at least five times the diameter of the cable ≫338-6≪.

Goosenecks in service-entrance cables must be located above the point of attachment of the service-drop conductors to the building (or other structure) ≫230-54(c)≪.

Type SE cable can be formed in a gooseneck and taped with a self-sealing weather resistant thermoplastic ≫230-54(b) *Exception*≪.

Cable Strap

Drip loops must be formed on all individual conductors ≫230-54(f)≪.

Service-Entrance Cable

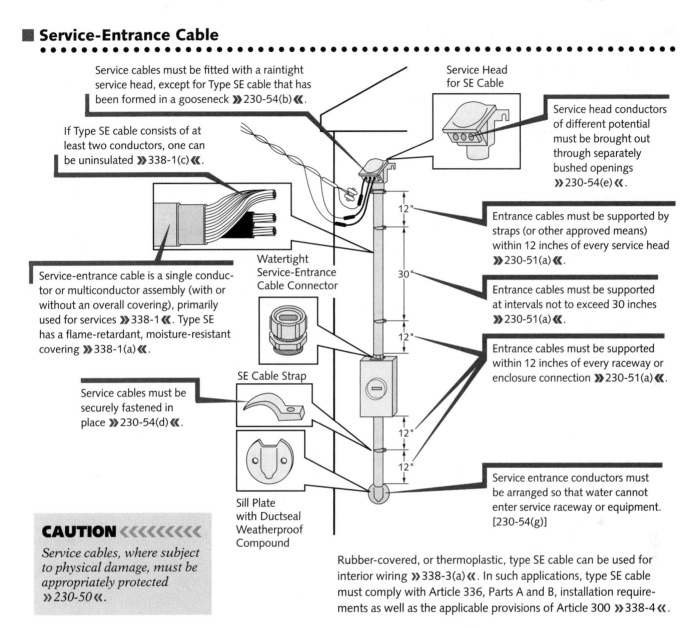

Service cables must be fitted with a raintight service head, except for Type SE cable that has been formed in a gooseneck »230-54(b)«.

If Type SE cable consists of at least two conductors, one can be uninsulated »338-1(c)«.

Service-entrance cable is a single conductor or multiconductor assembly (with or without an overall covering), primarily used for services »338-1«. Type SE has a flame-retardant, moisture-resistant covering »338-1(a)«.

Watertight Service-Entrance Cable Connector

Service cables must be securely fastened in place »230-54(d)«.

SE Cable Strap

Sill Plate with Ductseal Weatherproof Compound

Service Head for SE Cable

Service head conductors of different potential must be brought out through separately bushed openings »230-54(e)«.

Entrance cables must be supported by straps (or other approved means) within 12 inches of every service head »230-51(a)«.

Entrance cables must be supported at intervals not to exceed 30 inches »230-51(a)«.

Entrance cables must be supported within 12 inches of every raceway or enclosure connection »230-51(a)«.

Service entrance conductors must be arranged so that water cannot enter service raceway or equipment. [230-54(g)]

CAUTION «««««««««

Service cables, where subject to physical damage, must be appropriately protected »230-50«.

Rubber-covered, or thermoplastic, type SE cable can be used for interior wiring »338-3(a)«. In such applications, type SE cable must comply with Article 336, Parts A and B, installation requirements as well as the applicable provisions of Article 300 »338-4«.

Conductors Considered Outside of a Building

CAUTION «««««««««

Although not stipulated in the NEC®, some local jurisdictions have adopted specific distance requirements for service conductors. (For example, the length for each service conductor between the meter socket enclosure and the service disconnecting means could be limited to only six feet.)

Service Disconnecting Means

Building or structure interior wall

An interior service disconnecting means must be installed as close as possible to the point of entrance of the service conductors and must be readily accessible. A raceway, beneath a 2-inch concrete slab, lies outside the building. Under this condition, the service disconnecting means could be installed on (or in) an interior wall of the building. (See Caution.)

Concrete Slab

To Meter Socket Enclosure

Outside of Building

Conductors installed under 2 or more inches of concrete beneath a building (or other structure) are considered outside »230-6«.

■ Underground Services

Service-lateral conductors must be insulated for the applied voltage **»** 230-30 **«**. Grounded conductors, qualifying under Section 230-30 *Exception*, can be uninsulated (bare).

Metal raceways containing service conductors must be grounded **»** 250-80 **«**.

Wiring methods, permitted for services, are listed in Section 230-43.

Service lateral consists of the underground service conductors between the street main and the first point of connection to the service-entrance conductors in a terminal box, meter base, or other enclosure whether inside or outside the building wall. Any risers at a pole (or other structure) or from transformers are included **»** Article 100 **«**.

Rigid Metal Conduit

Table 300-5

> **N O T E**
> *Service-lateral conductors must have adequate mechanical strength and sufficient ampacity to carry the current for the load (as computed in accordance with Article 220)* **»** *230-31(a)* **«**.

Direct buried conduit (or other raceways) must be installed in adherence with Table 300-5 minimum cover requirements **»** 300-5(a) **«**.

■ Underground Service-Entrance Cable

Although Type USE cable identified for underground use has a moisture-resistant covering, a flame-retardant covering is not required **»** 338-1(b) **«**.

Underground service-lateral conductors must be protected against damage in accordance with Section 300-5. Service-lateral conductors entering a building must be installed in accordance with Section 230-6 or must be protected by an approved raceway wiring method identified in Section 230-43 **»** 230-32 **«**.

Type USE cable, used for service laterals, can emerge aboveground at outside terminations (in meter bases or other enclosures) provided Section 300-5(d) protection requirements are met **»** 338-2 **«**.

If Type USE cable contains at least two conductors, one can be uninsulated **»** 338-1(c) **«**.

Table 300-5

When installed, direct buried cable must meet Table 300-5 minimum cover requirements **»** 300-5(a) **«**.

Where the conductors or cables emerge as a direct burial wiring method, a bushing (or fitting) must be installed to protect conductors from abrasion on the end of the conduit (or tubing) that terminates underground. A seal providing the same level of protection can be used instead of a bushing **»** 300-5(h) and 300-15(c) **«**.

Cabled, single-conductor, Type USE constructions, recognized for underground use, may have a bare copper conductor cabled with the assembly. Underground approved, Type USE single, parallel, or cabled conductor assemblies, may have a bare copper concentric conductor applied. An outer overall covering is not required for these constructions **»** 338-1(b) **«**.

Wiring Methods

● ●

> ### N O T E
> *The wiring method of choice must be installed according to provisions for that particular method. For example, rigid metal conduit must be installed according to Article 346 provisions.*

Service entrance conductors must be installed according to applicable *Code* requirements for the type of wiring method used and limited to the following methods: (1) open wiring on insulators; (2) Type IGS cable; (3) rigid metal conduit; (4) intermediate metal conduit; (5) electrical metallic tubing; (6) electrical nonmetallic tubing (ENT); (7) service-entrance cables; (8) wireways; (9) busways; (10) auxiliary gutters; (11) rigid nonmetallic conduit; (12) cablebus; (13) Type MC cable; (14) mineral-insulated, metal-sheathed cable; (15) flexible metal conduit (or liquidtight flexible metal conduit) not longer than 6 feet between raceways, or between raceway and service equipment, having equipment bonding jumper routed with the flexible metal conduit (or liquidtight flexible metal conduit) according to Section 250-102(a), (b), (c), and (e) provisions; or (16) liquidtight flexible nonmetallic conduit ≫230-43 ≪.

SERVICE AND OUTSIDE WIRING CLEARANCES

Accessibility to Pedestrians

● ●

The point of attachment of the service-drop conductors to a building (or other structure) must meet Section 230-24 minimum clearances. In all cases this point of attachment must be at least 10 feet above finished grade ≫230-26 ≪.

Overhead conductors, accessible only to pedestrian traffic, must have a minimum 10-foot vertical clearance above finished grade, sidewalks, platforms, or projections from which they might be reached. The voltage must not exceed 150 to ground ≫230-24(b) and 225-18 ≪.

The 3-foot minimum requirement does not apply to conductors run above the top level of a window ≫230-9 *Exception* and 225-19(d) *Exception* ≪.

The 3-foot clearance from windows, found in Section 230-9, does not apply to raceways and service equipment.

A drip loop's lowest point has a minimum vertical clearance of 10 feet ≫230-24(b) ≪.

10 Feet

10 Feet

■ Vertical Clearances

A 10-foot minimum vertical clearance is required from any platform or projection, from which overhead conductors (of 150 volts, or less, to ground) might be reached **》230-24(b) and 225-18 《**.

A minimum 12-foot vertical clearance is required over residential property and driveways, and over commercial areas (not subject to truck traffic) provided conductor voltage does not exceed 300 volts to ground **》230-24(b) and 225-18 《**.

Conductor voltage exceeding 300 volts to ground increases the minimum vertical clearance from 12 feet to 15 feet **》230-24(b) and 225-18 《**.

Public streets, alleys, roads, parking areas subject to truck traffic, driveways on non-residential property, and other land traversed by vehicles (such as cultivated, grazing, forest, and orchard) require a minimum 18 foot vertical clearance **》230-24(b) and 225-18 《**.

10 Feet

12 Feet

15 Feet

18 Feet

> **N O T E**
> *Swimming pool overhead conductor clearances are listed in Section 680-8.*

■ Clearance from Windows and Accessible Areas

Permitted

Non-opening Window

Permitted

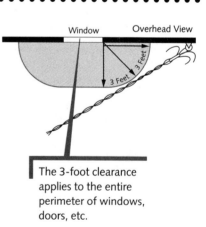

Window   Overhead View

3 Feet

Service conductors, installed as open conductors or multiconductor cable (having no overall outer jacket), must have a clearance of at least 3 feet from windows (designed to open), doors, porches, balconies, ladders, stairs, fire escapes, and similar locations **》230-9 and 225-19(d) 《**.

The 3-foot clearance requirement does not apply to conductors run above the top level of a window **》230-9 *Exception* and 225-19(d) *Exception* 《**.

The 3-foot clearance is not required for windows which are not designed to open.

The 3-foot clearance applies to the entire perimeter of windows, doors, etc.

■ Conductors Obstructing Openings

> **NOTE**
>
> *A minimum 18-foot vertical clearance is required over land routinely traversed by vehicles (such as cultivated, grazing, forest, and orchard)* »230-24(b) and 225-18 «.

Overhead conductors must not obstruct building entrances and must not be installed beneath openings designed for passage of materials, such as openings in farm and commercial buildings »230-9 and 225-19(d) «.

■ Conductors Above Overhanging Portions of Roofs

A maximum of 6 feet of service-drop conductors pass over the roof.

Conductors terminate in a through-the-roof raceway (or other approved support).

The vertical clearance for service-drop conductors can be reduced to 18 inches provided all Section 230-24(a) *Exception No. 3* requirements are met.

The voltage between conductors does not exceed 300 volts.

The horizontal service-drop conductor measurement cannot exceed 4 feet.

This reduction in clearance applies to conductors passing above roof overhang only.

18 Inch Minimum

4 Feet

Overhanging Portion of the Roof

> **NOTE**
>
> *If all the requirements of Section 225-19(a) Exception No. 3 are met, the vertical clearance for the overhead spans of open conductors, including multiconductor cables, can be reduced to 18 inches.*

■ Conductor Clearance Above Flat Roofs

Service-drop conductor vertical clearances are based on the conductor temperature of 60°F (15°C), no wind, with final unloaded sag in the wire, conductor, or cable »230-24 «.

While service-drop conductors must not be readily accessible, they must comply with Section 230-24(a) through (d) for services not over 600 volts, nominal »230-24 «.

Conductors must have a vertical clearance of at least 8 feet above the roof surface »230-24(a) and 225-19(a) «.

8 Foot Minimum

Path of Service Drop Conductors Above Roof (More than 4 Feet)

Guy Wire

8 Feet

3 Foot Minimum

8 Feet

3 Foot Minimum

A service mast supporting service-drop conductors must be of adequate strength or be supported by braces (or guys) to safely withstand the strain imposed by the service drop »230-28 «.

The vertical clearance above roof level must be maintained for a distance of at least 3 feet from all edges of the roof »230-24(a) and 225-19(a) «.

■ Conductors Above Sloped Roofs

The voltage between the ungrounded (hot) conductors is 240 volts.

Where the voltage between conductors does not exceed 300, and the roof has a slope of not less than 4 inches in 12, a 3 foot clearance is permitted ≫ 230-24(a) *Exception No. 2* and 225-19(a) *Exception No. 2* ≪ .

Roof Slope   12
              6

This roof, as denoted by this symbol, has a slope (or slant) of 6 inches in 12.

Unlike Section 230-24(b) which stipulates that "vertical clearance" also includes the drip loop, Section 230-24 provisions mention only the service-drop conductor. Because the end of the service-drop conductors could be part (or all) of the drip loop, caution as advised.

The horizontal service-drop measurement is less than 4 feet. However, the conductors pass over more than just the overhanging portion of this roof. Therefore, the clearance cannot be reduced to 18 inches ≫ 230-24(a) *Exception No. 3* and 225-19(a) *Exception No. 3* ≪ .

Service equipment rated at 600 volts, or less, must be marked as suitable for use as service equipment. Individual meter socket enclosures are not considered service equipment ≫ 230-66 ≪ .

3 Foot Minimum

Path of Service Drop Conductors above Roof (Less than 4 Feet)

■ Various Roof Designs

Service-drop conductors above a roof with a slope of at least 4 inches in 12 has a minimum clearance of 3 feet ≫ 230-24(a) *Exception No. 2* ≪ .

This roof has a slope of 4 inches in 12.

The point of attachment is usually the lowest point of the service-drop. If, however the conductors run above the roof's highest point (ridge or peak), a taller service mast may be required.

This roof has a slope of 6 inches in 12 (more than 4 in 12).

Service Drop Above Roof (More than 4 Feet)

The supply voltage for all three service-drop conductor illustrations is 240 volts.

Service Drop Above Roof (More than 4 Feet)

A service mast supporting service-drop conductors must be of adequate strength or be supported by braces (or guys) to safely withstand the strain imposed by the service drop ≫ 230-28 ≪ .

Guy Wire

3 Feet

12
4

8 Feet

Service Drop Above Roof (More than 4 Feet)

3 Feet

12
6

12
3

Service-drop conductors above a roof with a slope less than 4 inches in 12 have a minimum clearance of 8 feet ≫ 230-24(a) ≪ .

This roof has a slope of 3 inches in 12 (less than 4 in 12).

> **NOTE**
> Section 225-19 lists the vertical clearance for the overhead span of open conductors and open multiconductor cables.

> **NOTE**
> Because each of the illustrated service-drops has a horizontal measurement of more than 4 feet, the amount of slope determines the minimum clearance requirement (either 3 feet or 8).

■ Clearance Reduction

No more than 6 linear feet of service-drop conductors pass over the roof.

Conductors terminate in a through-the-roof raceway (or other approved support.)

Voltage between conductors does not exceed 300 volts.

The service-drop's minimum vertical clearance is 18 inches where all Section 230-24(a) *Exception No. 3* requirements have been met.

The horizontal service-drop conductor measurement does not exceed 4 feet.

18 Inch Minimum

Path of Service Drop Conductors Above Roof (Less than 4 Feet)

Overhanging Portion of the Roof

The reduction in clearance (to 18 inches) applies to sloped roofs as well as flat roofs where the service-drop passes above the overhang portion of the roof only.

NOTE
If all the requirements of Section 225-19(a) Exception No. 3 are met, the vertical clearance for the overhead spans of open conductors, including multiconductor cables, can be reduced to 18 inches.

WORKING SPACE AROUND EQUIPMENT

■ Headroom and Height

Overcurrent devices cannot be located in the vicinity of easily ignitable material (such as clothes closets) ≫ 240-24(d) ≪ . Although mentioned specifically, clothes closets are only one type of area that may contain easily ignitable materials.

A minimum of 6½ feet of working space headroom is required around service equipment, switchboards, panelboards, or motor control centers ≫ 110-26(e) ≪ .

A clear work space must extent from the grade, floor, or platform to the height required by Section 110-26(e) ≫ 110-26(a)(3) ≪ .

Work space illumination must be provided for all indoor installations of service equipment, panelboards, etc. Supplemental light fixtures are not required where the work space is illuminated by an adjacent light source ≫ 110-26(d) ≪ .

Sufficient working space access must be provided and maintained about all electric equipment to permit ready and safe equipment operation and maintenance ≫ 110-26 ≪ .

6½ Feet

Heating Equipment

NOTE
Compliance with the minimum headroom of working space is optional for 200 amperes, or less, service equipment or panelboards, in existing dwellings ≫ 110-26(e) Exception ≪ .

Working space required by Section 110-26 must not be used for storage ≫ 110-26(b) ≪ .

NOTE
Electric equipment work space must have at least one entrance of sufficient area to provide access ≫ 110-26(c) ≪ .

■ Maximum Depth of Associated Equipment

Surrounding space (equal to the width and depth of the equipment) must be clear of foreign systems, unless protection is provided against damage from condensation, leaks, or breaks in such foreign systems. This zone extends from the top of the equipment to the structural ceiling. A dropped, suspended, or similar ceiling that does not strengthen the building is not considered a structural ceiling. Sprinkler protection is permitted within the dedicated space, where the piping complies with Section 110-26 »*110-26(f)(1)(b)-(d)* «.

Wireway

Wireways are troughs (sheet-metal or non-metallic) with hinged or removable covers used to house and protect electric wires and cables. Conductors are laid in place after the wireway has been installed as a complete system »362-1 and 362-14 «.

Article 362 outlines wireway provisions.

6 Inches

12 Inches

In an electrical installation, other associated equipment located above or below the electrical equipment cannot extend more than 6 inches beyond the front of that electrical equipment »110-26(a)(3) «.

■ Working Space Width

The frontal work space width of electric equipment must be the greater of the width of the equipment or 30 inches »110-26(a)(2) «.

Section 110-26 does not stipulate that the panel (or equipment) be located in the center of the work space. It can be located anywhere within the required space.

30 Inches

OR

30 Inches

OR

30 Inches

A panel can be located in the middle, with equal amounts of space on each side, or . . .

A panel can be located on the far left side, with the additional space on the right, or . . .

A panel can be located on the far right, with the additional space on the left side.

The work space extends to a height of 6½ feet or the height of the equipment, whichever if greater. Above 6½ feet or the height of the equipment, the dedicated space must equal the width and depth of the equipment »*110-26(a), (e) and (f)* «.

N O T E

The working space must always permit at least a 90-degree opening of equipment doors or hinged panels »*110-26(a)(2)* «.

■ Overlapping Work Space

The work space width required in front of electric equipment is the greater of the width of the equipment or 30 inches ≫ 110-26(a)(2) ≪ .

■ Equipment Behind Doors

Equipment can be located behind a door provided a 3-foot working space is achieved by closing the door.

Working Space for Equipment

30 Inch Working Space for Left Panel

In this configuration, it is not necessary that each panel have it's own dedicated working space.

A panel (or equipment) can share the dedicated work space of other equipment associated with the electrical installation.

30 Inch Working Space for Right Panel

Overcurrent devices must be readily accessible unless one of the conditions in Section 240-24(a)(1) through (4) apply ≫ 240-24(a) (The term "readily accessible" is defined in Unit 2 of this book.) ≪ .

■ Working Space Depth

Concrete, brick, or tile walls are recognized as grounded ≫ Table 110-26(a) Note 2 ≪ .

Except as elsewhere required or permitted, the work space area for equipment operating at 600 volts, nominal (or less) to ground, which is likely to require examination, adjustment, servicing, or maintenance while energized, must comply with (1), (2), and (3) of Section 110-26(a) ≫ 110-26(a) ≪ .

Table 110-26(a) specifies working space depth requirements in the direction of access to live parts. Distances are measured from the live parts (if exposed) or from the front or opening (if enclosed) ≫ 110-26(a)(1) ≪ .

$6\frac{1}{2}$ Feet

$6\frac{1}{2}$ Feet

3 Feet

30 inches

The minimum headroom is 6½ feet, except for existing dwellings whose service does not exceed 200 amperes ≫ 110-26(e) ≪ .

$6\frac{1}{2}$ Feet

$6\frac{1}{2}$ Feet

3 Feet

30 inches

The minimum acceptable width of working space is 30 inches ≫ 110-26(a)(2) ≪ .

The minimum work space depth for equipment operating at 150 volts, nominal (or less) to ground is 3 feet, regardless of the conditions found in Table 110-26(a). For equipment operating between 151 and 600 volts, to ground, refer to Table 110-26(a).

Electric equipment must be firmly secured to the surface on which it is mounted. Wooden plugs driven into holes in masonry, concrete, plaster, or similar materials are not acceptable ≫ 110-13(a) ≪ .

> **NOTE**
> *Installation and working space requirements for equipment operating over 600 volts, nominal are found in Sections 110-30 through 110-40.*

Dedicated Space

Surrounding space (equal to the width and depth of the equipment) must be clear of foreign systems, unless protection is provided against damage from condensation, leaks, or breaks in such foreign systems. This zone extends from the top of the equipment to the structural ceiling. A dropped, suspended, or similar ceiling, that does not strengthen the building is not considered a structural ceiling »110-26(f)(1) «.

Depth of Panel   Width of Panel

Motor control centers and equipment within the scope of Article 384, must be located in dedicated spaces and must be protected from damage as covered in (1) and (2) of Section 110-26(f).

> **NOTE**
> *Outdoor clear work space includes the zone described in Section 110-26(a). No architectural appurtenances or other equipment can occupy in this zone* »110-26(f)(2) «.

Outdoor Service Equipment

Outdoor electrical equipment must be installed in suitable enclosures and must be protected from accidental piping system spills and leaks, and from accidental contact by unauthorized personnel, or vehicular traffic »110-26(f)(2) «.

6½ Feet

3 Feet

30 Inches or Width of the Equipment

Panelboards must be mounted in dead front enclosures (cabinets, cutout boxes, etc.) specifically designed for that purpose »384-18 «. Dead front means having no exposed live parts on the operating side of the equipment »Article 100 «.

The zone described in Section 110-26(a) must be included in the minimum clear working space. No architectural appurtenance or other equipment can occupy this zone »110-26(f)(2) «.

> **NOTE**
> *Panelboards in damp or wet locations must comply with Section 373-2(a)* »384-17 «.

Electrical Equipment

> **NOTE**
> *Equipment grounding conductors are found in Article 250, Part F.*

The disconnecting means for air-conditioning or refrigerating equipment must be located within sight from the equipment and must also be readily accessible »440-14 «.

All electric equipment must be surrounded by sufficient, unobstructed space to allow ready and safe equipment operation and maintenance »110-26 «. Working space is mandatory for equipment likely to require examination, adjustment, servicing, or maintenance while energized »110-26(a) «.

Installation of the disconnecting means can be on, or within, the air-conditioning or refrigerating equipment »440-14 «. If installed within the equipment, it must be readily accessible. ("Accessible, Readily" is defined in Article 100 of the *NEC*® and Unit 2 of this book.)

Article 440 details air-conditioning and refrigerating equipment provisions. General requirements, disconnecting means, overcurrent protection, conductor sizing, and room air conditioners are a few of the covered topics.

SERVICE EQUIPMENT AND PANELBOARDS

■ Panelboards

A panelboard (1) consists of one, or more, panel units designed for assembly, thereby forming a single panel; (2) includes buses and automatic overcurrent devices; (3) may contain switches for controlling circuits (light, heat, or power); (4) are designed for mounting in a cabinet or cutout box situated in, or against, a wall (or partition); and (5) are accessible only from the front »Article 100 «. (Panelboards include, but are not limited to, circuit breaker panels and fused or non-fused switches (disconnects).

Main Breaker (MB) Panelboard

Main Lug (MLO) Panelboard

Panelboards include overcurrent devices.

Panelboards include buses.

Panelboards are mounted in cabinets, or cutout boxes, which are accessible only from the front.

Panelboards can be single or multi-panel units.

■ Enclosures

A cabinet is an enclosure which by design may be surface or flush mounted, having a frame, mat, or trim attached to which are (or can be) swinging door(s) »Article 100 «.

Another type of enclosure, a cutout box, is designed for surface mounting and has swinging doors (or covers) secured directly to and extending from one of the enclosure walls »Article 100 «.

Type 1 Enclosure

Cabinets can be either flush or surface mounted.

Cabinets

Cabinets include trims with swinging doors.

Cutout boxes have swinging doors or covers attached directly to the enclosure.

Type 1 Enclosure

Type 1 enclosures, intended for indoor use, primarily provide protection against contact with the enclosed equipment in locations not subject to unusual service conditions.

Type 3R Enclosure

Cutout boxes are surface mounted.

Dead front is defined as having no exposed live parts on the operating side of the equipment »Article 100 «.

Cutout Boxes

Type 1 Enclosure

Type 3R Enclosure

Type 3R enclosures, intended for outdoor use, primarily provide protection against falling precipitation and must remain undamaged by external formation of ice. They are not intended to provide protection against conditions such as dust, internal condensation, or internal icing.

■ Maximum Number of Overcurrent Devices

When counting the number of over-current devices within a panelboard, the main breaker is not added.

A 2-pole circuit breaker counts as two overcurrent devices for the purpose of Section 384-15 » 384-15 《.

The purpose of a panelboard's circuits (and circuit modifications) must be legibly identified on a circuit directory affixed to the face (or inside) of the panel door(s) » 384-13 《. The marking must be durable enough to withstand the surrounding environment » 110-22 《.

A hand-operable circuit breaker (equipped with a lever or handle), or a power-operated circuit breaker (capable of being opened by hand in the event of a power failure), can serve as a switch depending on the number of poles. Note: See Sections 240-81 and 240-83 provisions » 380-11 《.

Twin (dual) circuit breakers count as two overcurrent devices. Similarly, quad circuit breakers count as four.

A maximum of 42 overcurrent devices (exclusive of those provided for in the mains) can be installed in any one cabinet or cutout box of a lighting and appliance branch circuit panelboard » 384-15 《.

41 42

WARNING

Lighting and appliance branch circuit panelboards physically inhibit the installation of excess overcurrent devices. These panelboards accommodate only that number for which the panelboard was designed, rated, and approved » 384-15 《.

■ Unused Openings

The total load on any panelboard overcurrent device must not exceed 80 percent of its rating where, in normal operation, the load will continue for three hours or more » 384-16(d) 《.

The purpose of all panelboard circuits (and circuit modifications) must be legibly identified on a circuit directory affixed to the face or inside of the panel door(s) » 384-13 《. The marking must survive environmental conditions intact » 110-22 《.

Filler Plate

Busbars, both insulated and bare, must be rigidly mounted » 384-31 《.

Within the scope of Article 373 (Cabinets, Cutout Boxes, and Meter Socket Enclosures), unused openings must be closed in a manner which affords protection equivalent to that of the enclosure » 373-4 and 110-12(a) 《.

■ Circuit Breakers

A **circuit breaker** is a device designed to both open and close a circuit by nonautomatic means, and to open the circuit automatically when subjected to a predetermined overcurrent without being damaged (when properly applied within its rating) »Article 100«.

A circuit breaker's **Inverse time** is a qualifying term indicating an intentional delay in the tripping action of the breaker. The length of the delay decreases as the magnitude of the current increases »Article 100«.

All circuit breakers having an interrupting rating other than 5,000 amperes must be appropriately marked. The interrupting rating is not required on circuit breakers used for supplementary protection »240-83(c)«. (Note: The ampere interrupting rating (AIR) is not the current rating of the circuit breaker.)

Ampere Interrupting Rating 10,000 (AIR) RMS Symmetrical 120/240 Vac

A **nonadjustable** circuit breaker does not allow the value of current which triggers the tripping action, or the time required for its operation, to be altered »Article 100«.

Single-pole

Circuit breakers must have a voltage rating (clearly marked) no less than the nominal system voltage which indicates their ability to interrupt fault current (between phases or phase to ground) »240-83(e)«.

A circuit breaker's open (off) or closed (on) positions must be clearly identified »240-81 and 380-7«.

For vertically-operated circuit breakers, the "up" position of the handle must be the "on" position »384-81«.

Two-pole

Three-pole

Circuit breakers and switches (general-use and motor-circuit), enclosure mounted according to Section 380-3, must clearly identify the open (off) or closed (on) positions »380-7«.

The ampere rating of any circuit breaker rated at 100 amperes or less, and 600 volts or less, must be molded, stamped, etched, or similarly marked into the handle or escutcheon area »240-83(b)«.

> **NOTE**
>
> *Circuit breakers used as switches in 120-volt and 277-volt fluorescent lighting circuits must be listed and be marked "SWD" »240-83(d)«.*

A circuit breaker's ampere rating must be durably marked and remain visible after installation. When necessary to ensure visibility, removal of a trim or cover is permitted »240-83(a)«.

Single-Pole
Grounded Fault
Circuit Interrupter

Twin or Dual

> **NOTE**
>
> *Circuit breakers must be trip free as well as manually operable in both the closed and opened positions. While a circuit breaker may normally operate by electrical or pneumatic means, a means of manual operation must also be provided »240-80«.*

The standard ampere ratings for fuses and inverse time circuit breakers are: 15, 20, 25, 30, 35, 40, 45, 50, 60, 70, 80, 90, 100, 110, 125, 150, 175, 200, 225, 250, 300, 350, 400, 450, 500, 600, 700, 800, 1000, 1200, 1600, 2000, 2500, 3000, 4000, 5000, and 6000 amperes. Fuses and inverse time circuit breakers with nonstandard ampere ratings can be used. The standard ampere ratings for fuses are: 1, 3, 6, 10, and 601 »240-6(a)«.

CAUTION «««««

Circuit breakers and fuses can be connected in parallel provided they are factory assembled and listed as a unit. Individual fuses, circuit breakers, or combinations cannot otherwise be connected in parallel »240-8«.

■ Back-Fed Devices

> **N O T E**
>
> *A back-fed device functioning as a service disconnect must be permanently marked to identify it as a service disconnect* » *230-70(b)* «.

Feeder Conductors

The purpose of all panelboard circuits (and circuit modifications) must be legibly identified on a circuit directory affixed to the face or inside of the panel door(s) » 384-13 «.

These remote panelboard (subpanel) feeder conductors originated from a main panelboard. The conductors are connected to the load side terminals of the circuit breaker, instead of connecting to the main lugs. In this configuration (the reverse of normal operation), the conductor feeds the breaker, thus the term back-fed device.

Retaining Clip

Back-fed plug-in-type over-current protection devices or plug-in-type main lug assemblies, used to terminate field installed ungrounded supply conductors, must be secured to the panelboard by an additional fastener mounting means » 384-16(g) «.

Since a bolt-on breaker is not a plug-in type overcurrent protection device, no additional fastener (screw, clip, etc.) is needed.

Bolt-on Breaker

■ Installation

Enclosures installed in wet locations must be weather-proof » 373-2(a) «.

The service conductors must be attached to the service disconnecting means by pressure connectors, clamps, or other approved means » 230-81 «.

Cabinets and cutout boxes must have sufficient space to prevent crowding of installed conductors » 373-7 «.

In damp or wet locations, surface-type enclosures (within the scope of Article 373) must be located or equipped so that moisture or water cannot enter or accumulate within the cabinet or cutout box. A minimum of ¼-inch airspace between the enclosure and the wall or other supporting surface is required » 373-2(a) «.

Panelboards in damp or wet locations must be installed in compliance with Section 373-2(a) » 384-17 «.

Concrete, brick, and tile walls are considered to be grounded » Table 110-26(a) Note 2 «.

Electric equipment must be firmly secured to the surface on which it is mounted. Wooden plugs driven into holes in masonry, concrete, plaster, or similar materials are not acceptable » 110-13(a) «.

■ Grouping of Service Disconnects

• •

Energized parts of service equipment must be enclosed as specified in 230-62(a), or guarded as specified in 230-62(b) » 230-62« .

Two to six circuit breakers, or sets of fuses, are acceptable as the overcurrent device to provide overload protection » 230-90(a) *Exception No. 3*« .

A service disconnecting means consisting of more than one switch or circuit breaker, as permitted by Section 230-71, must have a combined rating of all switches or circuit breakers used, not less than the rating required by Section 230-79 » 230-80« .

Multiple service disconnects (from two to six) as permitted in Section 230-71, must be grouped » 230-72(a)« .

Both Service Disconnects on the Outside

Both Service Disconnects on the Inside

One Service Disconnect on the Outside, One Service Disconnect on the Inside

> **N O T E**
>
> *Equipment whose purpose is to interrupt current at fault levels must have an interrupting rating sufficient for the nominal circuit voltage, as well as the current available at the equipment's line terminals. Equipment intended to interrupt current at other than fault levels must have an interrupting rating at nominal circuit voltage sufficient for current interruption » 110-9« .*

■ Location of Service Disconnecting Means

• •

The service disconnecting means shall be installed at a readily accessible location either outside of a building or structure, or inside as near as possible to the point of entrance of the service conductors » 230-70(a)« .

Outside   Inside

> **CAUTION** ‹‹‹‹‹‹‹‹‹‹‹‹‹‹‹‹‹‹
>
> *Neither service disconnecting means nor overcurrent devices (other than supplementary overcurrent protection) can be located in bathrooms as defined in Article 100 » 230-70(a) and 240-24(e)« .*

> **N O T E**
>
> *Maximum service conductor lengths are not specified within a building or structure. The provision merely states that the disconnecting means must be located nearest the point of service conductor entrance. Since the length of service conductors can alter available short-circuit current, the length should be kept to a minimum. Some local jurisdictions define a maximum acceptable length for service conductors.*

■ Service Equipment

Because individual meter socket enclosures are not considered service equipment, a "suitable for use as service equipment" marking is not required ≫230-66 ≪.

WARNING

Local utility companies often set parameters (height, grounding, etc.) for metering equipment installation. Local jurisdictions may also have special regulations concerning service equipment. Obtain copies of all applicable provisions for the area in which the equipment will be installed.

Suitable for Use as Service Equipment

MAIN

100 Amperes

Service equipment rated 600 volts, or less, must be clearly marked as suitable for use as service equipment ≫230-66 ≪.

Each service disconnect must be permanently marked identifying it as a service disconnect ≫230-70(b) ≪.

A one-family dwelling service disconnecting means must have a minimum rating of 100 amperes, 3-wire ≫230-79(c) ≪.

A means to disconnect all conductors in a building (or other structure) from the service-entrance conductors must be provided ≫230-70 ≪.

■ Multiple Service Disconnects

Section 230-90 requires that each ungrounded (hot) service conductor have overload protection. Such protection shall be provided by an overcurrent device in series with each ungrounded service conductor having a rating (or setting) not higher than the allowable conductor ampacity ≫230-90(a) ≪. Under 230-90(a), one of the five exceptions, allow more than one overcurrent device for overload protection.

The service disconnecting means for each service (permitted by Section 230-2), or for each set of service-entrance conductors (permitted by Section 230-40, Exception No. 1 or Exception No. 3) shall consist of not more than six switches or six circuit breakers whether mounted in a single enclosure, in a group of separate enclosures; in or on a switchboard. No more than six disconnects per service can be grouped in any one location ≫230-71(a) ≪.

One 2-Pole Breaker   One 1-Pole Breaker   Two 1-Pole Breakers

Handle Tie

Each Is Considered One Operation of the Hand

Two or three single-pole switches or breakers, individually operable, are permitted on multiwire circuits (one pole for each ungrounded conductor). The multipole disconnect must be equipped with "handle ties" or a "master handle" to disconnect all service conductors. Six operations of the hand is the maximum allowed ≫230-71(b) ≪.

Two to six circuit breakers, or sets of fuses, can serve as the overcurrent device, providing overload protection ≫230-90(a) *Exception No. 3* ≪.

Single-pole circuit breakers, grouped in accordance with Section 230-71(b), constitute one protective device ≫230-90(a) ≪.

The total circuit breaker (or fuse) ratings can exceed the service conductors ampacity, provided the calculated load (in accordance with Article 220) does not exceed the ampacity of the service conductors ≫230-90(a) *Exception No. 3* ≪.

The service disconnecting means must have a rating not less than the load to be carried, as determined in accordance with Article 220. In no case shall the rating be lower than specified in Section 230-79(a), (b), (c), or (d) ≫230-79 ≪.

Each service disconnect must be permanently marked as a service disconnect ≫230-70(b) ≪. Each individual service disconnect must be marked. A single marking for multiple-service disconnects is not permitted.

NOTE
For information on service equipment in panelboards, see Section 384-16(a).

NOTE
A set of fuses shall be considered all the fuses required to protect all of a circuit's ungrounded conductors ≫230-90(a) ≪.

Power Panelboards

Power Panelboard

An individual overcurrent protective device is not required for a power panelboard, used as service equipment, with multiple disconnecting means in accordance with Section 230-71 **»**384-16(b) *Exception* **«**.

According to Article 384, panelboards are classified as (1) lighting and appliance branch-circuit or (2) power panelboards **»**384-14 **«**.

A power panelboard has no more than 10% of its overcurrent devices protecting lighting and appliance branch circuits **»**384-14(b) **«**.

> **NOTE**
> *A power panelboard (with multiple disconnecting means) not used as service equipment, having neutral connections available, and more than 10% of its overcurrent devices protecting branch circuits rated 30 amperes or less, must be protected on the supply side by an overcurrent protective device having a rating not exceeding that of the panelboard »384-16(b) «.*

Lighting and Appliance Branch-Circuit Panelboards

Raceway containing feeder conductors

Ungrounded (hot) Conductors

Panelboards are classified as (1) lighting and appliance branch-circuit or (2) power panelboards **»**384-14 **«**.

Grounding Conductor

The front edge of cabinets (panelboards) situated in walls constructed of noncombustible material (concrete, tile, etc.) must be within ¼ inch of the finished surface **»**373-3 **«**.

Neutral Conductor

Lighting and Appliance Panelboard

Individual protection for a lighting and appliance panelboard is not required if the feeder has overcurrent protection not greater than the panelboard's rating **»**384-16(a) *Exception No. 1* **«**.

A lighting and appliance branch-circuit panelboard has more than 10% of its overcurrent devices protecting lighting and appliance branch circuits **»**384-14(a) **«**.

A lighting and appliance branch circuit has a connection to the panelboard neutral, and has overcurrent protection of 30 amperes or less in one or more conductors **»**384-14(a) **«**.

A grounding connection must not be made to any grounded circuit conductor on the load side of the service disconnecting means, except as otherwise permitted in Article 250 **»**250-24(a)(5) **«**.

> **NOTE**
> *Each lighting and appliance branch circuit panelboard must be individually protected on the supply side by no more than two main circuit breakers (or two sets of fuses) having a combined rating not greater than the panelboard's rating »384-16(a) «. (See the two exceptions to this main rule.)*

> **NOTE**
> *Panelboards used as service equipment in pre-existing individual residential occupancy installations, do not require individual protection for lighting and appliance branch-circuit panelboards »384-16(a) Exception No. 2 «.*

> **CAUTION** ‹‹‹‹‹‹‹‹‹‹‹‹‹
> *Cabinets in walls constructed of combustible material (wood, etc.) must be flush with or extend beyond the finished surface »373-3 «.*

■ Common-Trip and Single-Pole Circuit Breakers

In addition to common-trip 2-pole circuit breakers, individual single-pole breakers (with or without handle ties) are permitted as the protection for each ungrounded conductor for single-phase circuits (or 3-wire, direct current circuits) used to feed line-to-line loads, in grounded systems ≫240-20(b)(2) ≪.

Individual single-pole circuit breakers, with or without approved handle ties, are permitted as protection for each ungrounded conductor in multiwire branch circuits serving only single-phase, line-to-neutral loads, except where limited by Section 210-4(b) ≫240-20(b)(1) ≪.

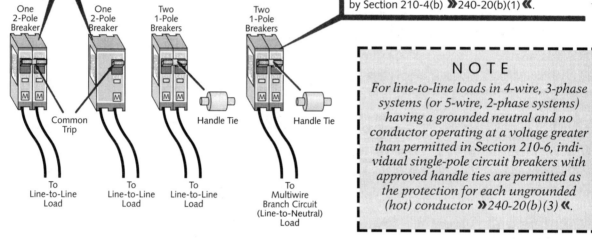

One
2-Pole
Breaker

One
2-Pole
Breaker

Two
1-Pole
Breakers

Two
1-Pole
Breakers

Common Trip

Handle Tie

Handle Tie

To
Line-to-Line
Load

To
Line-to-Line
Load

To
Line-to-Line
Load

To
Multiwire
Branch Circuit
(Line-to-Neutral)
Load

> **NOTE**
>
> *For line-to-line loads in 4-wire, 3-phase systems (or 5-wire, 2-phase systems) having a grounded neutral and no conductor operating at a voltage greater than permitted in Section 210-6, individual single-pole circuit breakers with approved handle ties are permitted as the protection for each ungrounded (hot) conductor ≫240-20(b)(3) ≪.*

■ Cartridge Fuses and Fuseholders

Fuse-holders are designed so that it will be difficult to put a fuse of any given class into a fuseholder intended for a current lower, or voltage higher, than that of the class to which the fuse belongs ≫240-60(b) ≪.

Fuses must be plainly marked (by printing on or a label attached to the fuse barrel), showing the following: (1) ampere rating, (2) voltage rating, (3) interrupting rating if other than 10,000 amperes, (4) "current-limiting" where applicable, and (5) the name or trademark of the manufacturer. The interrupting rating is not required on fuses used for supplementary protection ≫240-60(c) ≪.

Knife-Blade Type
Cartridge Fuses

Ferrule Type
Cartridge Fuses

100 250 VAC 10,000 AIR RMS Sym.

60 250 VAC 10,000 AIR RMS Sym.

60 250 VAC 200,000 AIR RMS Sym.

100 250 VAC 200,000 AIR RMS Sym.

Non-Current
Limiting
Fuses

Current
Limiting
Fuses

Fuseholders for current-limiting fuses must not permit insertion of noncurrent-limiting fuses ≫240-60(b) ≪.

Cartridge fuses and fuseholders are classified according to voltage and amperage ranges. Fuses rated 600 volts, nominal or less, can be used for equal or lesser voltages ≫240-61 ≪.

> **NOTE**
>
> *Cartridge fuses, and fuseholders with a 300-volt rating, can be used in the following: (1) circuits not exceeding 300 volts between conductors, and (2) single-phase, line-to-neutral circuits supplied from a 3-phase, 4-wire solidly grounded neutral source with line-to-neutral voltage not exceeding 300 volts ≫240-60(a) ≪.*

■ Conductor Sizing

Table 310-15(b)(6) lists conductors which can be used as service-entrance, service-lateral, and feeder conductors serving as the main power feeder to a dwelling unit and being installed in raceway or cable (with or without an equipment grounding conductor) **》310-15(b)(6)《**.

The minimum size service-entrance conductors for this 200-ampere disconnect are 2/0 copper or 4/0 aluminum (or copper-clad aluminum) **》Table 310-15(b)(6)《**.

Feeder conductors (not used as the main power feeder) are found in Table 310-16.

200-Ampere Main Breaker

100-Ampere Breakers

100-Ampere Disconnect

To 100-Ampere Panelboard

> **N O T E**
> *Dwelling unit feeder conductors are not required to be larger than the service-entrance conductors* **》310-15(b)(6)《**.

The minimum size feeder conductors for the 100-ampere remote panelboard are No. 4 copper or No. 2 aluminum (or copper-clad aluminum). As Section 310-15(b)(6) specifies, the main power feeder is that feeder(s) supplying the lighting and appliance branch-circuit panelboard(s) **》310-15(b)(6)《**.

Article 440 outlines provisions relating to disconnecting means, branch circuit overload protection, and branch circuit conductors for air-conditioning and refrigeration equipment.

> **N O T E**
> *The grounded (neutral) conductor can be smaller than the ungrounded (hot) conductors, provided Section 215-2, 220-22, and 230-42 requirements are met* **》310-15(b)(6)《**. *The grounded conductor must not be smaller than the required grounding electrode conductor specified in Table 250-66* **》250-24(b)(1)《**.

■ Installation

Cabinets and cutout boxes must have space to sufficiently accommodate (without crowding) all installed conductors **》373-7《**.

Electric equipment must be installed in a neat and workmanlike manner **》110-12《**. "Workmanlike," by definition, implies skillful and characteristic of a good workman.

Conductor sizes are expressed in American Wire Gage (AWG) or in circular mils **》110-6《**.

All electric equipment internal parts (busbars, wiring terminals, insulators, etc.) must remain undamaged and uncontaminated by foreign materials (paint, plaster, cleaners, abrasives, corrosive residues, etc.). Equipment parts which are broken, bent, cut, corroded, impaired due to chemical action, heat, or otherwise damaged in such a way as to adversely affect safe operation or mechanical strength must not be used **》110-12(c)《**.

Sections 373-10 and 373-11 itemize construction specifications for cabinets, cutout boxes, and meter socket enclosures.

> **N O T E**
> *Unless adequate space is available, enclosures for switches or overcurrent devices must not be used as junction boxes, auxiliary gutters, or raceways for conductors feeding through or (tapping off) to other switches or overcurrent devices. Conductors must not fill more than 40% of a cross-sectional area of the space; and conductors, splices, and taps must not fill the wiring space at any cross section to more than 75% of the cross-sectional area* **》373-8《**.

■ Plug Fuses, Fuseholders, and Adapters

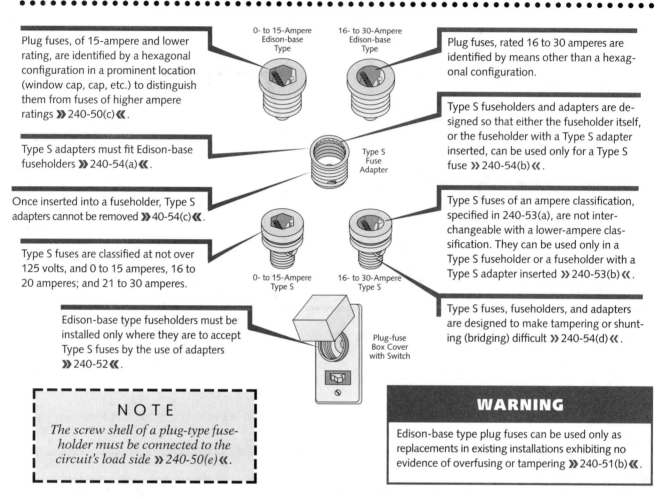

Plug fuses, of 15-ampere and lower rating, are identified by a hexagonal configuration in a prominent location (window cap, cap, etc.) to distinguish them from fuses of higher ampere ratings ≫ 240-50(c) ≪.

Type S adapters must fit Edison-base fuseholders ≫ 240-54(a) ≪.

Once inserted into a fuseholder, Type S adapters cannot be removed ≫ 40-54(c) ≪.

Type S fuses are classified at not over 125 volts, and 0 to 15 amperes, 16 to 20 amperes; and 21 to 30 amperes.

Edison-base type fuseholders must be installed only where they are to accept Type S fuses by the use of adapters ≫ 240-52 ≪.

0- to 15-Ampere Edison-base Type

16- to 30-Ampere Edison-base Type

Type S Fuse Adapter

0- to 15-Ampere Type S

16- to 30-Ampere Type S

Plug-fuse Box Cover with Switch

Plug fuses, rated 16 to 30 amperes are identified by means other than a hexagonal configuration.

Type S fuseholders and adapters are designed so that either the fuseholder itself, or the fuseholder with a Type S adapter inserted, can be used only for a Type S fuse ≫ 240-54(b) ≪.

Type S fuses of an ampere classification, specified in 240-53(a), are not interchangeable with a lower-ampere classification. They can be used only in a Type S fuseholder or a fuseholder with a Type S adapter inserted ≫ 240-53(b) ≪.

Type S fuses, fuseholders, and adapters are designed to make tampering or shunting (bridging) difficult ≫ 240-54(d) ≪.

N O T E

The screw shell of a plug-type fuseholder must be connected to the circuit's load side ≫ 240-50(e) ≪.

WARNING

Edison-base type plug fuses can be used only as replacements in existing installations exhibiting no evidence of overfusing or tampering ≫ 240-51(b) ≪.

GROUNDING

■ Grounded (Neutral) Conductor in the Service Disconnecting Means

Grounding and bonding jumpers must be connected by exothermic welding, or by listed means, such as pressure connectors and clamps ≫ 250-8 ≪.

N O T E

The bonding jumper (screw, strap, etc.) is installed only in service equipment, and not in subpanels.

Meter Socket Enclosure

Bonding Jumper

Service Disconnectings Means

Neutral Bar

Main Bonding Jumper (Screw)

Bonding Jumper

A grounded system must have an unspliced main bonding jumper connecting the equipment grounding conductor(s) and the service-disconnect enclosure to the grounded conductor of the system (within each service disconnect enclosure) ≫ 250-28 ≪.

A single bar in the service-disconnect enclosure may serve as the neutral bar (or bus) and the equipment grounding terminal bar (or bus).

An ac system grounded at any point and operating at less than 1,000 volts must have a grounded conductor that is: (1) run to each service disconnecting means; (2) that is routed with the phase conductors; and (3) bonded to each service disconnecting means enclosure ≫ 250-24(b) ≪.

The grounded conductor must be routed with the phase conductors; cannot be smaller than the required grounding electrode conductor specified in Table 250-66; and is not required to be larger than the largest ungrounded service-entrance conductor ≫ 250-24(b)(2) ≪.

■ Main Bonding Jumper

A main bonding jumper is the connection between the grounded circuit conductor (neutral) and the equipment grounding conductor at the service **»** Article100 **«** .

Screw

In cases where the main bonding jumper is a screw only, it must be identified with a green finish which remains visible after installation **»** 250-28(b) **«** .

Main bonding jumpers must be attached by exothermic welding, or other listed means such as pressure connectors or clamps. Sheet-metal screws are not acceptable **»** 250-28(c) **«** .

Neutral Bar

Strap

Main bonding jumpers must be made of copper or other corrosion-resistant material. A wire, bus, screw, or similar suitable conductor is acceptable as a main bonding jumper **»** 250-28(a) **«** .

A grounded system must have an unspliced main bonding jumper connecting the equipment grounding conductor(s) and the service-disconnect enclosure to the grounded conductor of the system (within each service disconnect enclosure) **»** 250-28 **«** .

CAUTION «««««««««««««««««««««
Panelboards, listed as suitable for use as service equipment, are furnished with a main bonding jumper that has not yet been installed. When using the panelboard as service equipment, install the main bonding jumper (screw, strap, etc.). When the panelboard is used for other purposes, such as a subpanel, do not *install the main bonding jumper.*

N O T E
The main bonding jumper cannot be smaller than the sizes shown in Table 250-66 for grounding electrode conductors **»** *250-28(d)* **«** *. (Listed main bonding jumpers furnished with panelboards meet the provisions of Section 250-28.)*

■ Remote Panelboards (Subpanels)

Unless grounded by connection to the grounded circuit conductor (permitted by Sections 250-32, 250-140, and 250-142), noncurrent-carrying metal parts of equipment, raceways, and other enclosures must be grounded (where required) by one of the following methods: (a) by any of the equipment grounding conductors permitted by Section 250-118, or (b) by an equipment grounding conductor contained within the same raceway, cable, or otherwise run with the circuit conductors **»** 250-134 **«** .

Equipment grounding conductors can be bare, covered, or insulated. Individually covered or insulated equipment grounding conductors must have a continuous outer finish, either green or green with yellow stripe(s), except as otherwise permitted by Article 250, Part F **»** 250-119 **«** .

A grounded circuit conductor cannot be used for grounding noncurrent-carrying metal parts of equipment on the load side of the service disconnecting means, unless an exception has been met **»** 250-142(b) **«** .

Lighting and Appliance Panelboard

Neutral Terminal Bar

Equipment Grounding Terminal Bar

Panelboard cabinets and panelboard frames, if metal, must be in physical contact with one another and must be grounded. Grounding conductors must not be connected to a terminal bar provided for grounded conductors (may be a neutral) unless the bar is identified for that purpose and is located where interconnection between equipment grounding conductors and grounded circuit conductors comply with Article 250 **»** 384-20 **«** .

N O T E
The bonding jumper (screw, strap, etc.) is installed only in the service disconnecting means and not in subpanels.

Only in service disconnecting means can the neutral terminal bar be connected to the enclosure or equipment grounding terminal bar. At this point, the neutral bar is isolated from the equipment grounding system.

A grounding connection must *not* be made to any grounded circuit conductor on the load side of the service disconnecting means except as otherwise allowed in Article 250 **»** 250-24(a)(5) **«** .

■ Grounding Electrode Conductor Connections

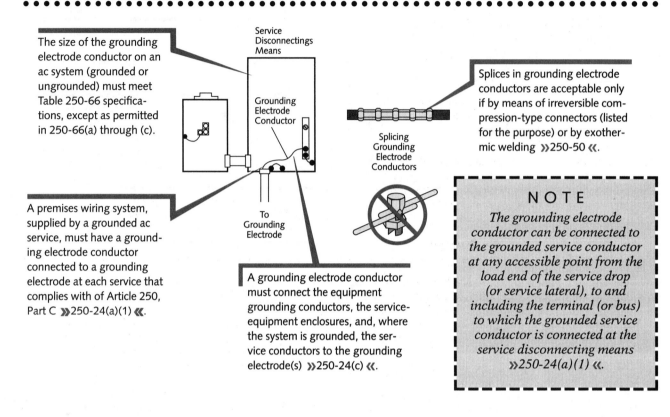

The size of the grounding electrode conductor on an ac system (grounded or ungrounded) must meet Table 250-66 specifications, except as permitted in 250-66(a) through (c).

A premises wiring system, supplied by a grounded ac service, must have a grounding electrode conductor connected to a grounding electrode at each service that complies with of Article 250, Part C »250-24(a)(1) «.

Service Disconnectings Means

Grounding Electrode Conductor

To Grounding Electrode

A grounding electrode conductor must connect the equipment grounding conductors, the service-equipment enclosures, and, where the system is grounded, the service conductors to the grounding electrode(s) »250-24(c) «.

Splicing Grounding Electrode Conductors

Splices in grounding electrode conductors are acceptable only if by means of irreversible compression-type connectors (listed for the purpose) or by exothermic welding »250-50 «.

NOTE
The grounding electrode conductor can be connected to the grounded service conductor at any accessible point from the load end of the service drop (or service lateral), to and including the terminal (or bus) to which the grounded service conductor is connected at the service disconnecting means »250-24(a)(1) «.

■ Methods of Bonding at the Service

One method is to bond equipment to the grounded service conductor in a Section 250-8 approved manner »250-94(1) «.

Connections utilizing threaded couplings or threaded bosses on enclosures must be wrenchtight »250-94(2) «.

Threaded Couplings

Hubs or Threaded Bosses

Bonding Bushing Bonding Jumper

Threadless Connector

Threadless couplings and connectors are approved where made up tight for metal raceways and metal-clad cables. Standard locknuts and bushings cannot be used to bond these items »250-94(3) «.

Concentric Knockouts Eccentric Knockouts

Other approved devices: such as bonding type locknuts and bushings »250-94(4) «.

Bonding Locknut

Bonding Bushing

Bonding Wedge

Regular Locknut

Standard locknuts or bushings cannot be the sole means for bonding as required by Section 250-94 »250-94 «.

Bonding jumpers, which meet other Article 250 requirements, must be used around concentric (or eccentric) knockouts that are punched (or otherwise formed) in a way that impairs the electrical connection to ground »250-94 «.

■ Bonding at the Service

Bonding is the permanent joining of metallic parts to form an electrically conductive path thereby ensuring continuity and capacity to safely conduct any current likely to be imposed »Article 100 «.

Hubs (or threaded bosses) are acceptable when bonding service equipment enclosures to threaded conduits or threadless connectors »250-94(2) and (3) «.

All service equipment noncurrent-carrying metal parts (raceways, enclosures containing service conductors, meter fittings, metal raceway, or armor enclosing a grounding electrode conductor, etc.) must be effectively bonded together. »250-92(a) «.

Metal Raceway

Service Disconnect

Metal Raceway

Bonding, where necessary, must be provided not only to ensure electrical continuity but also ample capacity to safely conduct any likely fault current »250-90 «.

Electrical continuity at service equipment, service raceways, and service conductor enclosures must be accomplished by one of the approved methods in Section 250-94.

■ Grounding Electrode System

If available on the premises of each building or structure being served, each item in Section 250-50(a) through (d), and any made electrodes in accordance with Sections 250-52(c) and (d), shall be bonded together to form the grounding electrode system »250-50 «.

The metal frame of the building or structure, where effectively grounded, must be bonded »250-50(b) «.

Metal Frame

No. 2 Bare Copper

A ground ring, which encircles the building or structure in direct contact with the earth at a depth of at least 2½ feet below the surface, consisting of at least 20 feet of bare copper conductor not smaller than No. 2 is allowed »250-50(d) «.

A metal underground water pipe must be in direct contact with the earth for a minimum of 10 feet (including any metal well casing effectively bonded to the pipe) and electrically continuous (or made so by bonding around insulating joints or sections or insulating pipe) to the points where the grounding electrode conductor and the bonding conductors connect »250-50(a) «.

An electrode encased by at least 2 inches of concrete, located within and near the bottom of a concrete foundation or footing in direct contact with the earth, consisting of one or more bare or zinc galvanized or other electrically conductive coated steel reinforcing bars each at least ½ inch in diameter and 20 feet continuous length, or consisting of at least 20 feet of bare copper conductor not smaller than No. 4, is acceptable »250-50(c) «.

Water Meter

Bonding Jumper

Grounding path continuity, or the bonding connection to interior piping, must not rely on water meters, filtering devices, or similar equipment »250-50(a)(1) «.

Bonding jumper(s) shall be installed in accordance with Sections 250-64(a),(b), and (e), shall be sized in accordance with 250-66, and shall be connected as specified in Section 250-70 »250-50 «.

Grounding Electrode Conductor

Metal Water Pipe

Underground

5 Feet

10 Feet

Interior metal water piping (located more than 5 feet from the point of entrance to the building) cannot be used as part of the grounding electrode system, nor as a conductor to interconnect electrodes that are part of the grounding electrode system, unless the exception for industrial and commercial buildings is met »250-50 «.

1/2" Steel Reinforcing Bar

OR

No. 4 Bare Copper

20 Feet

20 Feet

NOTE

An unspliced grounding electrode conductor can be run to any convenient grounding electrode available in the system, or to grounding electrode(s) individually. It must be sized for the largest grounding electrode conductor required among all the electrodes being connected »250-50 «.

CAUTION 〈〈〈〈〈〈〈〈〈〈〈〈〈〈〈〈

An underground metal water pipe requires an additional electrode of a type specified in Section 250-50 or in Section 250-52. For dwellings, the supplemental electrode can be bonded to the grounding electrode conductor, the grounded service-entrance conductor, the nonflexible grounded service raceway, or to any grounded service enclosure »250-50(a)(2) «.

Rod and Pipe Electrodes

The electrode's upper end must be flush with, or below, ground level unless the aboveground end and the grounding electrode conductor attachment are adequately protected against physical damage as specified in Section 250-10 »250-52(c)(3)«.

At least 8 feet of electrode length must be in contact with the soil.

Where rock bottom is encountered, the electrode can be driven at an oblique angle not to exceed 45 degrees from vertical »250-52(c)(3)«.

Where rock bottom is encountered, the electrode can be buried in a trench that is at least 2½ feet deep »250-52(c)(3)«.

Electrodes of iron or steel rods must be at least ⅝ inch in diameter. Stainless steel rods (less than ⅝ inch in diameter), nonferrous rods, or their equivalent must be listed, and ½ inch or more in diameter »250-52(c)(2)«.

In multiple electrode configurations, each electrode of one grounding system (including that used for lightning rods) must be at least 6 feet from any other electrode of another grounding system. Two or more grounding electrodes, effectively bonded together, are considered a single grounding electrode system »250-52«.

Each plate electrode must expose at least 2 square feet of surface to exterior soil. (One square foot of surface area on each side of a 1' × 1' plate equals 2 square feet.)

A single electrode consisting of a rod, pipe, or plate that does not have a resistance to ground of 25 ohms (or less) shall be augmented by an additional electrode of any type specified in Section 250-50 or 250-52. Multiple electrodes installed to meet the requirements of Section 250-56 must be at least 6 feet apart »250-56«.

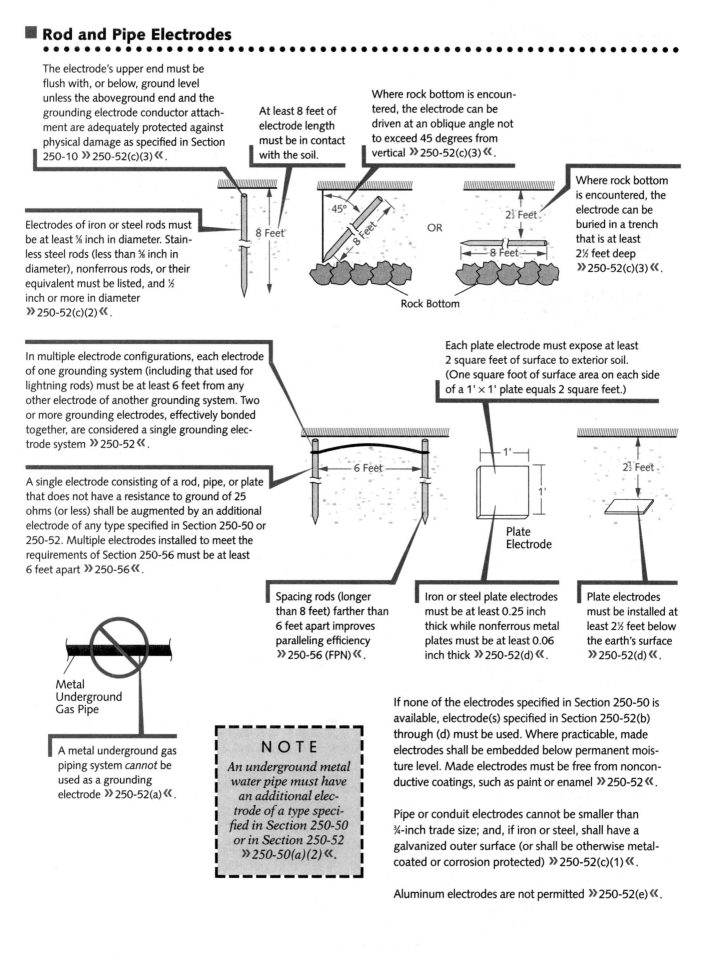

Metal Underground Gas Pipe

A metal underground gas piping system *cannot* be used as a grounding electrode »250-52(a)«.

Spacing rods (longer than 8 feet) farther than 6 feet apart improves paralleling efficiency »250-56 (FPN)«.

Iron or steel plate electrodes must be at least 0.25 inch thick while nonferrous metal plates must be at least 0.06 inch thick »250-52(d)«.

Plate electrodes must be installed at least 2½ feet below the earth's surface »250-52(d)«.

NOTE

An underground metal water pipe must have an additional electrode of a type specified in Section 250-50 or in Section 250-52 »250-50(a)(2)«.

If none of the electrodes specified in Section 250-50 is available, electrode(s) specified in Section 250-52(b) through (d) must be used. Where practicable, made electrodes shall be embedded below permanent moisture level. Made electrodes must be free from nonconductive coatings, such as paint or enamel »250-52«.

Pipe or conduit electrodes cannot be smaller than ¾-inch trade size; and, if iron or steel, shall have a galvanized outer surface (or shall be otherwise metal-coated or corrosion protected) »250-52(c)(1)«.

Aluminum electrodes are not permitted »250-52(e)«.

■ Grounding Electrode Conductor

A grounding electrode conductor must be used to connect the equipment grounding conductors, the service-equipment enclosures, and, in a grounded system, service conductors to the grounding electrode ≫250-24(c) ≪.

The grounding electrode conductor must be installed in one continuous length (without a splice or joint), unless spliced by an irreversible means ≫250-50 and 250-64(c) ≪.

A grounding electrode conductor (or its enclosure) must be securely fastened to the surface on which it is carried ≫250-64(b) ≪.

Where a raceway is used as protection for a grounding conductor, the installation must comply with appropriate raceway article requirements ≫250-64(e) ≪.

Bonding Locknut or Bushing

Metal Raceway

Metal raceways which enclose grounding electrode conductors must be bonded on each end ≫250-64(e) ≪.

Only one conductor can be connected to the grounding electrode by a single clamp or fitting, unless it is listed for multiple conductors ≫250-70 ≪.

Metal enclosures, enclosing grounding electrode conductors, must be electrically continuous from the point of cabinet (or equipment) attachment to the grounding electrode, and must be fastened securely to the ground clamp (or fitting). Metal enclosures, not physically continuous from cabinet (or equipment) to the grounding electrode, must be made electrically continuous by bonding each end to the grounding conductor ≫250-64(e) ≪.

The grounding conductor must be connected to the grounding electrode by exothermic welding, or by listed means, such as lugs, pressure connectors, clamps, etc. Connections depending on solder cannot be used ≫250-70 ≪.

The grounding electrode conductor must be made of copper, aluminum, or copper-clad aluminum. The material selected shall be inherently resistant to corrosive conditions existing at the site of installation, or shall be suitably protected against corrosion. The conductor can be solid (or stranded), insulated, covered, or bare ≫250-62 ≪.

Aluminum or copper-clad aluminum grounding conductors (insulated or bare) must not be used in direct contact with masonry, or earth, or where subject to corrosive conditions. If used outside, aluminum or copper-clad aluminum grounding conductors must not be installed within 18 inches of the earth ≫250-64(a) ≪.

One variety of ground-rod clamp listed for direct burial

Ground clamps must be listed for the grounding electrode and grounding electrode conductor materials; and, if used on pipe (rod or other buried electrodes) must also be listed for direct soil burial ≫250-70 ≪.

■ Grounding Electrode Conductor Sizing

While the grounding electrode conductor must be sized in accordance with Section 250-66, the tap conductors can be sized in accordance with the grounding electrode conductors specified in Section 250-66 for the largest conductor serving respective enclosures. The tap conductors must be connected to the grounding electrode conductor so that the grounding electrode conductor remains without a splice or joint ≫ 250-64(d)≪ .

A No. 4 copper (or No. 2 aluminum or copper-clad aluminum) grounding electrode conductor is required if the service-entrance conductors are 2/0 or 3/0 copper ≫ Table 250-66≪ .

To Service Disconnecting Means

Metal Water Pipe

Ground Rod

If the grounding electrode conductor is connected to made electrodes as permitted in Section 250-52(c) or (d), that portion of the conductor serving as the sole connection to the grounding electrode is required to be no larger than No. 6 copper wire, or No. 4 aluminum wire ≫ 250-66(a)≪ .

Interior metal water piping, located more than 5 feet from the point of entrance to the building, cannot be used as part of the grounding electrode system nor as a conductor to interconnect electrodes that are part of the system ≫ 250-50≪ .

Concrete Encased Electrode

To Ground Ring

Where the grounding electrode conductor is connected to a concrete encased electrode as permitted in Section 250-50(c), that portion of the conductor serving as the sole connection to the grounding electrode is not required to be larger than No. 4 copper wire ≫ 250-66(b)≪ .

To ensure a permanent and effective ground for a metal piping system used as a grounding electrode, effective bonding must be provided as necessary around insulated joints (and sections) as well as around any equipment likely to be disconnected for repairs or replacement ≫ 250-68(b)≪ .

If, as Section 250-50(d) permits, the grounding electrode conductor is connected to a ground ring, that portion of the conductor serving as the sole connection to the grounding electrode does not have to be larger than the ground ring conductor ≫ 250-66(c)≪ .

A No. 4 or larger conductor (copper or aluminum) must be protected if exposed to severe physical damage. A No. 6 grounding conductor, not exposed to physical damage, can run along the surface of the building construction without metal covering or protection provided it is securely fastened to the construction. Otherwise, it must be housed in rigid metal conduit, intermediate metal conduit, rigid nonmetallic conduit, electrical metallic tubing, or cable armor ≫ 250-64(b)≪ .

No. 8 and smaller grounding electrode conductors must be installed in rigid metal conduit, intermediate metal conduit, rigid nonmetallic conduit, electrical metallic tubing, or cable armor ≫ 250-64(b)≪ .

■ Equipment Bonding Jumper on the Supply Side of the Service

● ●

> **NOTE**
>
> *Read Section 250-102(c) for service-entrance conductors larger than 1,100 kcmil copper, or 1,750 kcmil aluminum.*

> **NOTE**
>
> *The interior metal water piping system must be bonded either to the service equipment enclosure, the grounded conductor at the service, the grounding electrode conductor (if of sufficient size), or to the grounding electrode(s) used. The bonding jumper must be sized in accordance with Table 250-66, and installed in accordance with Section 250-64(a),(b), and (e). The bonding jumper's point of attachment must be accessible ≫ 250-104(a)≪.*

Service Disconnecting Means

Main Bonding Jumper

Bonding Bushing

Bonding Locknut

Grounding electrode conductor and equipment bonding jumper sizes, on the supply side of the service, are determined by service-entrance conductor sizes.

Unit 8 explains grounded (neutral) conductor sizing.

Bonding jumpers, on the supply side of the service, cannot be smaller than the sizes shown in Table 250-66 for grounding electrode conductors ≫ 250-102(c)≪.

The grounding electrode conductor must be sized in accordance with Section 250-66 ≫ 250-64(d)≪.

> **CAUTION** ⟨⟨⟨⟨⟨⟨⟨⟨⟨⟨⟨⟨
>
> *Interior metal piping with the potential to become energized must be bonded to the service equipment enclosure, the grounded conductor at the service, the grounding electrode conductor (if of sufficient size), or to the grounding electrode(s) used. The bonding jumper must be sized in accordance with Table 250-122 by using the rating of the circuit which may energize the piping. The equipment grounding conductor of this circuit can serve as the bonding means ≫ 250-104(c)≪.*

■ Equipment Bonding Jumper on the Load Side of the Service

● ●

A panelboard (on the load side of the service overcurrent devices) protected by a 100 ampere overcurrent protection device (circuit breaker or fuse), requires a No. 8 copper equipment grounding conductor ≫ Table 250-122≪.

Wire type copper, aluminum, or copper-clad aluminum equipment grounding conductors must not be less than shown in Table 250-122, but are not required to be larger than the circuit conductors supplying the equipment. When a raceway, cable armor, or sheath is used as the equipment grounding conductor (as provided in Section 134(a) and 250-118), it must comply with Section 250-2(d) ≫ 250-122≪.

Metal panelboard cabinets and frames must be grounded and in physical contact with each other. Where the panelboard is used with nonmetallic raceway (or cable), or where separate grounding conductors are provided, a terminal bar for the grounding conductors must be secured within the cabinet. The terminal bar must be bonded to the metal cabinet and panelboard frame. Otherwise it shall be connected to the grounding conductor that is run with the conductors feeding the panelboard ≫ 384-20≪.

To 100 Ampere Feeder

Main Bonding Jumper **Not** Installed

Lighting and Appliance Panelboard

Equipment grounding conductor and equipment bonding jumper sizes (on the load side of the service) are determined by the size of the overcurrent device which protects the equipment.

Unit 8 explains grounded (neutral) conductor sizing.

Equipment bonding jumpers (on the load side of the service overcurrent devices) must be: (1) sized, as a minimum, in accordance with Table 250-122; but (2) are not required to be larger than the circuit conductors supplying the equipment; and (3) cannot be smaller than No. 14 copper ≫ 250-102(d)≪.

> **NOTE**
>
> *An equipment grounding conductor run with (or enclosing) the circuit conductors must be one of those listed in Section 250-118.*

> **NOTE**
>
> *Where conductor sizes are adjusted to compensate for voltage drop, equipment grounding conductors, if required, must be adjusted proportionately according to circular mil area ≫ 250-122≪.*

■ Separate Buildings or Structures

Where multiple buildings (or structures) are supplied from a common ac service by a feeder (or more than one branch circuit), each building must have a grounding electrode (as described in Part C) connected to the metal enclosure of the building disconnecting means. If there are no electrodes, a grounding electrode must be installed which meets Part C requirements »250-32(a)«.

A remote panelboard (subpanel) installed in a detached garage (or structure) is not considered a service disconnecting means.

Panelboard

An equipment grounding conductor (described in Section 250-118) must be run with the supply conductors and connected to the building (or structure) disconnecting means and to the grounding electrode(s). The equipment grounding conductor must be used for grounding or bonding or equipment, structures, or frames required to be grounded or bonded. Any installed grounded conductor must not be connected to the equipment grounding conductor or the grounding electrode(s) »250-32(b)(1)«.

Where (1) an equipment grounding conductor is *not* run with the supply to the building or structure; and (2) there are no continuous metallic paths bonded to the grounding system in both buildings or structures involved; and (3) ground-fault protection of equipment has not been installed on the common ac service, the grounded circuit conductor run with the supply to the building or structure must be connected to the building or structure disconnecting means and to the grounding electrode(s) and must be used for grounding or bonding equipment, structures, or frames requiring bonding or grounding »250-32(b)(2)«.

Three Conductor (**With** Ground) Feeder

Either Or

Three Conductor (**Without** Ground) Feeder

Equipment Grounding Conductor

Main Bonding Jumper **Not** Installed

Main Bonding Jumper Installed

Grounding Electrode Conductor   Bonding Locknut

The grounding conductor installation must comply with Article 250, Part C »250-32(f)«.

The size of the grounding conductor to the grounding electrode(s) must not be less than given in Table 250-122, and does not have to be larger than the largest ungrounded supply conductor »250-32(f)«.

> ### NOTE
> *Where one or more disconnecting means supply one or more additional buildings (or structures) under single management, and where these disconnecting means are located remote from those buildings (or structures) according to Section 225-32, Exception No. 1 and 2 provisions, all of Section 230-32(d)(1) through (3) conditions must be met »230-32(d)«.*

A grounding electrode is not required at separate buildings (or structures) where only one branch circuit is supplied »250-32(a) *Exception* «.

■ Preventing Objectionable Current

In a system that must be grounded, electrical equipment must be connected to the supply system grounded conductor so that an effective return path for fault current is established »250-2(b)«.

If multiple grounding connections result in an objectionable flow of current, one or more of the following alterations in Section 250-6(b)(1) through (4) are permitted, provided that the requirements of Section 250-2(d) are met: (1) discontinue one or more, but not all, of such grounding connections; (2) change the location of the grounding connections; (3) interrupt the continuity of the conductor (or conductive path) linking the grounding connections; and (4) take other suitable remedial action approved by the authority having jurisdiction »250-6(b)«.

The grounding of electric systems, circuit conductors, surge arresters, and conductive noncurrent-carrying materials (and equipment) must be installed and arranged to prevent an objectionable flow of current over the grounding conductors or grounding paths »250-6(a)«.

Service Disconnecting Means

Meter Socket Enclosure

Neutral Conductor

Nonmetallic Raceways

To Grounding Electrode

CAUTION <<<<<<<<<<<<<<<<<<<<<<<<

Connecting a grounded (neutral) and grounding conductor in parallel creates objectionable current flow on the grounding conductor, because the unbalanced current flows through both the grounded and grounding conductor. Should the grounded conductor lose its continuity (become disconnected), the grounding conductor would then carry all of the unbalanced (neutral) current. A hazard would exist if the grounding conductor's ampacity rating is less than the current flowing through it.

Electrical conductive materials, such as metal water piping, metal gas piping, and structural steel members, which are likely to become energized, must be bonded to the supply system grounded conductor »250-2(c)«.

Temporary currents resulting from accidental conditions, such as ground-fault currents, that occur only while the grounding conductors are performing their intended protective functions, are not classified as objectionable current for the purpose specified in 250-6(a) and (b) »250-6(c)«.

Summary

- A service mast must safely withstand the strain imposed by service-drop conductors.
- Nothing, except electrical service-drop conductors, can be attached to the service mast.
- Wiring methods, permitted for services, are listed in Section 230-43.
- Service-drop conductor vertical ground clearances are found in Section 230-24(b)
- Service-drop conductor roof clearances are in Section 230-24(a).
- A minimum of 6½ feet of work-space headroom is required around service equipment and panelboards.

- A minimum frontal work space width is 30 inches for service equipment and panelboards.
- The depth of this work space must be at least 3 feet.
- Panelboards are mounted in cabinets or cutout boxes.
- A maximum of 42 overcurrent devices can be installed in any one panelboard.
- Each service disconnect must be permanently marked identifying it as a service disconnect.
- The service disconnecting means must be installed at a readily accessible location whether inside or outside.

- Two to six circuit breakers, or sets of fuses, can serve as the overcurrent device, providing overload protection.
- Table 310-15(b)(6) can be used to size conductors that serve as the main power feeder to a dwelling.
- Conductors not serving as the main power feeder are sized from Table 310-16.
- Electric equipment must be installed in a neat and workmanlike manner.
- Grounding and bonding conductors must be connected by listed means.
- The grounded (neutral) conductor is isolated from the equipment grounding conductor in subpanels.

- Grounding electrode conductors and bonding jumpers on the service supply side are sized from Table 250-66.
- Equipment grounding conductors and bonding jumpers, on the load side of the service, are sized according to Table 250-122.
- All noncurrent-carrying metal parts of service equipment must be effectively bonded together.
- Grounding electrode systems (and conductors) are located in Part C of Article 250.
- Panelboards, in separate buildings or structures, must be installed according to Section 250-32 provisions.

Unit 9 Competency Test

NEC® Reference	Answer
_____	_____

1. The point of attachment of the service-drop conductors to a building must not be less than _____ feet above finish grade.

2. An equipment bonding jumper on the supply side of the service must not be smaller than the sizes shown in Table _____.

3. Service conductors run above the top level of a window shall be:

    a) 8 feet above window.

    b) accessible.

    c) 3 feet above window.

    d) considered out of reach.

4. Every circuit breaker having an interrupting rating other than _____ amperes shall have its interrupting rating shown on the circuit breaker.

5. Energized parts of service equipment that are not enclosed shall be installed on a _____ and guarded in accordance with Sections 110-18 and 110-27.

    I. switchboard

    II. panelboard

    III. control board

    a) II only   b) I or II only   c) II or III only   d) I, II, or III

6. A lighting and appliance branch-circuit panelboard is one having more than _____ % of its overcurrent devices protecting lighting and appliance branch circuits.

7. Service entrance cables shall be supported by straps or other approved means within _____ of every service head, gooseneck, or connection to a raceway or enclosure.

    a) 1'   b) 6'   c) 12'   d) 30'

8. Bonding jumpers meeting the other requirements of this article shall be used around _____ knockouts that are punched or otherwise formed so as to impair the electrical connection to ground.

NEC® Reference   Answer

_____ _____   9. Residential 240 volt, servicedrop conductors must have a vertical clearance above a driveway of at least _____.

_____ _____   10. A grounding ring encircling the building or structure, in direct contact with the earth at a depth below earth's surface of not less than 2½ feet, consisting of at least _____ of bare copper conductor not smaller than No. 2.

_____ _____   11. Plug fuses of 15-ampere and lower rating shall be identified by a _____ configuration of the window, cap, or other prominent part to distinguish them from fuses of higher ampere ratings.

_____ _____   12. Where more than one grounding electrode is used, each electrode of one grounding system (including that used for lightning rods) shall not be less than _____ from any other electrode of another grounding system.

_____ _____   13. Not more than _____ overcurrent devices (other than those provided for in the mains) of a lighting and appliance branch-circuit panelboard shall be installed in any one cabinet or cutout box.

_____ _____   14. Service heads and _____ in service entrance cables shall be located above the point of attachment of the service-drop conductors to the building or other structure.

_____ _____   15. The width of the working space in front of the electric equipment shall be the width of the equipment or _____ inches, whichever is greater.

_____ _____   16. Which of the following is not a true statement concerning an equipment grounding conductor?

   a) Under certain conditions, equipment grounding conductors shall be required to be larger than circuit conductors supplying the equipment.

   b) One equipment grounding conductor may serve multiple circuits.

   c) Under certain conditions, equipment grounding conductors may be run in parallel.

   d) The size of equipment grounding conductors shall be increased to compensate for voltage drop.

_____ _____   17. The minimum headroom of working spaces about service equipment, switchboards, panelboards, or motor control centers shall be _____.

_____ _____   18. Service-drop conductors shall have a vertical clearance of not less than _____ above the roof surface.

_____ _____   19. In damp or wet locations, surface-type cabinets, cutout boxes, and meter socket enclosures shall be so placed or equipped as to prevent moisture or water from entering and accumulating within them, and mounted so there is at least _____ airspace between the enclosure and the wall or other supporting surface.

_____ _____   20. Other equipment associated with the electrical installation located above or below the electrical installation shall be permitted to extend not more than _____ beyond the front of the electrical equipment.

   a) 0 inches   b) 6 inches   c) 12 inches   d) 24 inches

_____ _____   21. What size copper equipment bonding jumper is required for a 200 ampere panelboard, located on the load side of the service overcurrent device? (The conductors feeding the panelboard are 2/0 copper conductors.)

_____ _____   22. Plug fuses of the Edison-base type shall be classified at not over 125 volts and _____ amperes.

NEC® Reference	Answer
_____	_____

23. A _____ is used to connect the grounding electrode to the equipment grounding conductor, to the grounded conductor, or to both, of the circuit at the service equipment.

24. A single electrode consisting of a rod, pipe, or plate that does not have a resistance to ground of _____ or less shall be augmented by one additional electrode.

25. A disconnecting means serving a hermetic refrigerant motor-compressor shall be selected on the basis of the nameplate rated-load current or branch-circuit selection current, whichever is greater, and locked-rotor current, respectively, of the motor-compressor. The ampere rating shall be at least _____ of the nameplate rated-load current or branch-circuit selection current, whichever is greater.

26. Circuit breakers rated _____ or less and 600 volts or less shall have the ampere rating molded, stamped, etched, or similarly marked into their handles or escutcheon areas.

27. A dwelling has a 175-ampere service that is fed with THW copper conductors. What is the minimum size copper grounding electrode conductor?

28. The _____ is the connection between the grounded circuit conductor and the equipment grounding conductor at the service.

29. Which of the following is not a standard ampere rating for fuses?

      a) 25 amperes   b) 50 amperes   c) 75 amperes   d) 601 amperes

30. The paralleling efficiency of ground rods longer than _____ feet is improved by spacing greater than 6 feet.

31. Where the service-entrance phase conductors are larger than 1,100 kcmil copper, the bonding jumper shall have an area not less than _____ % of the area of the largest phase conductor.

32. Service conductors are the conductors from the service point to the _____.

33. The minimum depth of working space for a 120/240 volt panelboard, with exposed live parts on one side and grounded parts on the other, shall be at least _____ feet.

34. Each plate electrode shall expose not less than _____ sq. ft. of surface to exterior soil and shall be installed not less than _____ below the surface of the earth.

35. Where installed on the outside of a raceway, the length of the equipment bonding jumper shall not exceed _____ and shall be routed with the raceway or enclosure.

Comprehensive Provisions

Objectives

• •

After studying this unit, the student will:

✔ be aware that general receptacle and lighting installation provisions for single-family dwellings are almost identical to those for multi-family structures.

✔ know that certain exceptions permit the installation of more than one service on a multi-family building.

✔ thoroughly understand the provisions limiting the number of service disconnecting means to only six.

✔ be familiar with the provision requiring that each multi-family dwelling occupant have access to their service disconnecting means.

✔ understand that each service supplying a building must be connected to the same grounding electrode.

✔ be familiar with service conductor vertical clearance requirements.

✔ be able to determine when Table 310-15(b)(6) can be used instead of Table 310-16 to size service and feeder conductors.

✔ know that outdoor receptacles are permitted, but not required in multi-family dwellings.

✔ be able to choose the appropriate voltage drop formula, insert the proper data, and compute the correct answer.

Introduction

• •

A multi-family dwelling is a building containing three or more dwelling units. Apartments or condominiums are examples of multi-family dwellings. They can be any size, from one story structures containing three units to high-rise buildings containing hundreds of units. One-family dwelling units (covered in Section 2 of this book) contain fundamental provisions essential to all dwelling units, whether one-, two-, or multi-family. Most of the general installation requirements (such as receptacle and switch placement), as well as most of the specific provisions (such as kitchen and bathroom regulations) pertain to all dwelling occupancies. A few requirements, such as outdoor receptacle outlets, are specific to one- and two-family dwellings. Service provisions pertaining to multi-family dwellings, such as the maximum number of service disconnecting means are explained. This unit also includes a brief description and explanation of voltage drop formulas. A thorough understanding of one-family dwelling requirements is essential to understanding those for multi-family dwellings.

PLANS (BLUEPRINTS)

■ Partial Site Plan

• •

Dwelling unit conductors, as listed in Table 310-15(b)(6), are permitted as 120/240, 3-wire, single-phase service-entrance conductors, service lateral conductors, and feeder conductors serving as the main power feeder to a dwelling unit and installed in raceway or cable with or without an equipment grounding conductor ≫ [310-15(b)(6)≪.

In a multiple-occupancy building, each occupant must have access to the unit's service disconnecting means, unless electric service and electrical maintenance are provided by building management ≫ 230-72(c)≪.

General receptacle installation provisions for multi-family dwellings are almost identical to those for one-family dwellings ≫ 210-52≪.

Outdoor outlets are not required for multi-family dwellings ≫ 210-52(e)≪.

Use Part B of Article 220 to perform standard method load calculations for both feeders and services.

Use Section 220-32 to perform multi-family dwelling optional load calculations for feeders or services consisting of 3 or more units.

Bldg. D

A laundry receptacle is not required, if one of the applicable exceptions is met ≫ 210-52(f)≪.

Bldg. C

With few exceptions, general lighting provisions are the same for both single- and multi-family dwellings ≫ 210-70≪.

Use Section 220-30 to perform optional method load calculations for individual feeders in each unit of multi-family dwellings.

Swimming Pool

Clubhouse

Bldg. E

Swimming pool provisions are contained in Article 680.

Swimming pool motor loads are computed in accordance with Sections 430-22 and 430-24 requirements.

Clubhouse and swimming pool load calculations are determined in accordance with Article 220, Part B provisions.

The clubhouse is exempt from dwelling unit(s) provisions.

If any portion of the clubhouse is designed, or intended, for the assembly of more than 99 persons, that portion must comply with Article 518 provisions.

Air-conditioning and refrigeration equipment provisions are found in Article 440.

> ### NOTE
> *House loads must be computed in accordance with Article 220, Part B and must be in addition to the dwelling unit loads computed by Table 220-32 demand factors ≫ 220-32(b)≪.*

■ Electrical Plan

Equipment servicing receptacle fed from house panel.

Photo-cell controlled exterior hallway and stairway lighting connected to house panel.

Floor Plan
Building D
Building Type II
First Floor - Electrical

SERVICES

■ Maximum Number of Services

Vertical service-drop conductor clearances are covered in Unit 9 of this book.

A building (or other structure) shall be supplied by only one service unless permitted in Section 230-2(a) through (d) »230-2 «.

Each service disconnecting means, permitted by Section 230-2, whether for a service or for a set of service-entrance conductors permitted by Section 230-40 (*Exception No. 1* or *Exception No. 3*) shall consist of not more than six switches (or circuit breakers) mounted in a single enclosure, a group of separate enclosures, or a switchboard configuration »230-71(a) «.

Auxiliary gutter provisions are located in Article 374.

NOTE
Unit 9 explains equipment working space, Section 110-26.

The conductors and equipment which deliver energy from the electricity supply system to the wiring system of the premises is called the service »Article 100 «.

Wiring spaces can be supplemented by auxiliary gutters at meter centers, distribution centers, switchboards, and similar points and may enclose conductors or busbars. Auxiliary gutters, however, must not be used to enclose switches, overcurrent devices, appliances, or similar equipment »374-1 «.

■ Maximum Number of Disconnects Per Service

Section 230-40 basic requirements state that each service drop (or lateral) must supply only one set of service-entrance conductors. However, several exceptions exist.

Two to six service disconnecting means in separate enclosures, grouped at one location and supplying separate loads from one service drop or lateral, can be supplied by one set of service-entrance conductors »230-40 *Exception No. 2* «.

There shall be no more than six disconnects per service grouped in a single location »230-71(a)«.

N O T E
For the purpose of Section 230-71, disconnecting means used solely for the control circuit of the ground-fault protection system, installed as part of the listed equipment, is not considered a service disconnecting means »*230-71(a)*«.

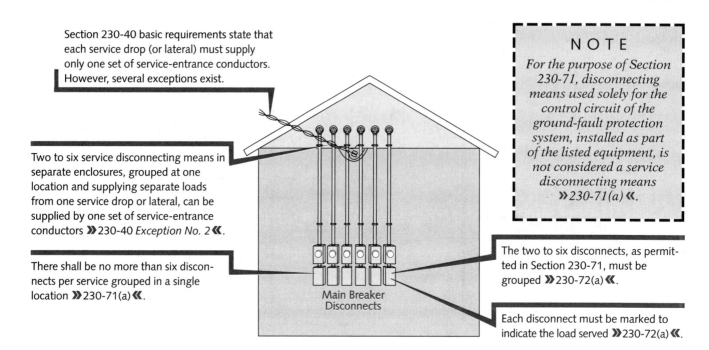

Main Breaker Disconnects

The two to six disconnects, as permitted in Section 230-71, must be grouped »230-72(a)«.

Each disconnect must be marked to indicate the load served »230-72(a)«.

■ Two Disconnects Supplying Twelve Units

Conductors must have adequate mechanical strength and sufficient ampacity to carry the current for the load as computed in accordance with Article 220 »230-23(a)«.

Many possible combinations exist for these twelve units: 1 main disconnect feeding all 12 units; 3 main disconnects feeding 12 units (4 units per disconnect); and 6 main disconnects feeding 12 units (2 units per disconnect), to name a few. The arrangement of service disconnects is not critical, as long as no more than six operations of the hand disconnect all of the service conductors.

There must be no more than six disconnects per service grouped in a single location »230-71(a)«.

Service Disconnect

If the service disconnecting means consists of more than one switch or circuit breaker, as permitted by Section 230-71, the combined ratings of all the switches or circuit breakers used must meet the rating requirements of Section 230-79 »230-80«.

In a multiple-occupancy building, each occupant must have access to their unit's service disconnecting means »230-72(c)«.

Each service disconnect must be permanently marked to identify it as a service disconnect »230-70(b)«.

N O T E
In a multiple-occupancy building where electric service and electrical maintenance are provided by building management, under continuous supervision, the service disconnecting means supplying more than one occupancy can be accessible only to authorized management personnel »*230-72(c)* Exception«.

■ More Than One Service

If the load requirements of a single-phase installation are greater than the serving agency normally supplies through one service, multiple services are permitted ≫230-2(c)(2)≪.

Service 1 of 2 services.
Service 2 is located on East end of this building, supplying the East half of this building.

Service 2 of 2 services.
Service 1 is located on West end of this building, supplying the West half of this building.

Six is the maximum number of service disconnects for any one location ≫230-71(a)≪.

A building or other structure can have another service for fire pumps where a separate service is required ≫230-2(a)(1)≪.

A building or other structure can have another service for emergency, legally required standby, optional standby, or parallel power production systems where a separate service is required ≫230-2(a)(2) through (5)≪.

By special permission, in multiple-occupancy buildings where there is no space available, service equipment does not have to be accessible to all occupants 230-2(b)(1). Special permission is the written consent of the authority having jurisdiction ≫Article 100≪.

Multiple services are permitted where the capacity requirements are in excess of 2,000 amperes at a supply voltage of 600 volts or less ≫230-2(c)(1)≪.

Where a building or structure is supplied by multiple services, or by any combination of branch circuits, feeders, and services, a permanent plaque (or directory) must be installed at each service disconnect location denoting all other services, feeders, and branch circuits supplying that building as well as the area served by each ≫230-2(e)≪.

> **NOTE**
> *One set of service-entrance conductors can supply the circuits of a two-family or multi-family dwelling as covered in Section 210-25 ≫230-40 Exception No. 4≪.*

> **WARNING**
> If a building supplied by separate services requires connection to a grounding electrode, the same grounding electrode must be used for all services. Two or more grounding electrodes that are effectively bonded together constitute a single grounding electrode system in this sense ≫250-58≪.

■ One Service Lateral—One Location

One set of service-entrance conductors can supply two to six service disconnecting means where located in separate enclosures, grouped at one location, and supply separate loads from one service lateral. The one set of service-entrance conductors can supply each or several such service equipment enclosures ≫230-40 *Exception No. 2*≪.

Maximum of six disconnects

One Service Lateral

All Conductors 1/0 or Larger

To Utility Transformer

Only as applied to Section 230-40, *Exception No. 2*, underground sets of conductors, size 1/0 and larger, running to the same location and connected together at their supply end (but not at their load end) constitutes one service lateral ≫230-2≪.

The **service lateral** is the underground service conductors between the main and the first point of connection to the service entrance conductors in a terminal box, meter base, etc. (inside or outside the building wall). This includes any risers at a pole, from transformers, etc. ≫Article 100≪.

■ One Service Lateral—Six Disconnects Maximum

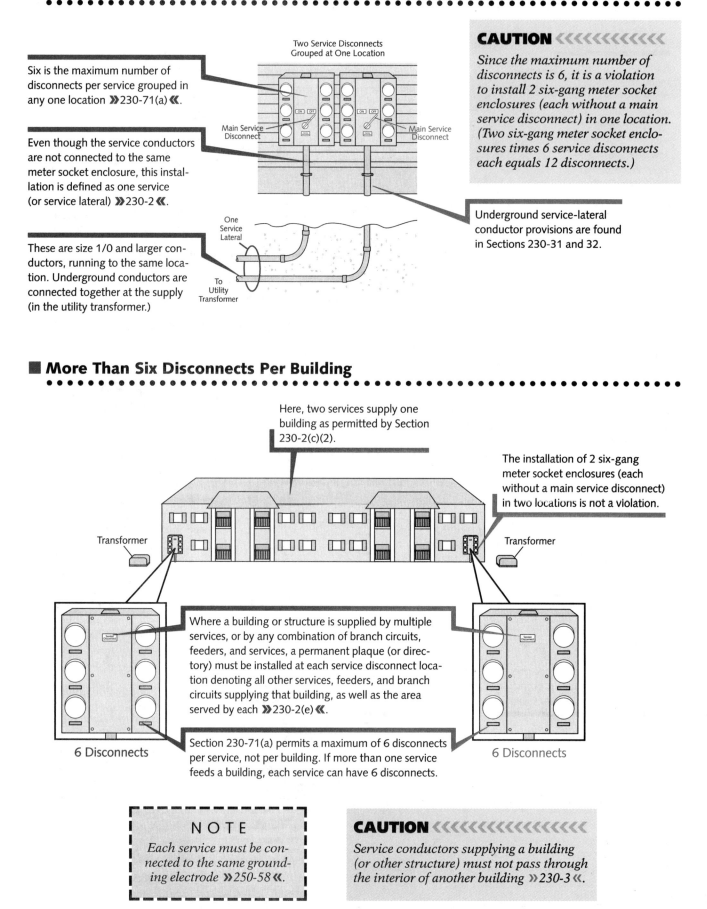

Six is the maximum number of disconnects per service grouped in any one location ≫230-71(a)≪.

Even though the service conductors are not connected to the same meter socket enclosure, this installation is defined as one service (or service lateral) ≫230-2≪.

These are size 1/0 and larger conductors, running to the same location. Underground conductors are connected together at the supply (in the utility transformer.)

Two Service Disconnects Grouped at One Location

Main Service Disconnect

Main Service Disconnect

ON OFF

ON OFF

One Service Lateral

To Utility Transformer

CAUTION ≪≪≪≪≪≪≪≪≪
Since the maximum number of disconnects is 6, it is a violation to install 2 six-gang meter socket enclosures (each without a main service disconnect) in one location. (Two six-gang meter socket enclosures times 6 service disconnects each equals 12 disconnects.)

Underground service-lateral conductor provisions are found in Sections 230-31 and 32.

■ More Than Six Disconnects Per Building

Here, two services supply one building as permitted by Section 230-2(c)(2).

The installation of 2 six-gang meter socket enclosures (each without a main service disconnect) in two locations is not a violation.

Transformer

Transformer

Service Disconnect

Service Disconnect

6 Disconnects

6 Disconnects

Where a building or structure is supplied by multiple services, or by any combination of branch circuits, feeders, and services, a permanent plaque (or directory) must be installed at each service disconnect location denoting all other services, feeders, and branch circuits supplying that building, as well as the area served by each ≫230-2(e)≪.

Section 230-71(a) permits a maximum of 6 disconnects per service, not per building. If more than one service feeds a building, each service can have 6 disconnects.

```
NOTE
Each service must be con-
nected to the same ground-
ing electrode ≫250-58≪.
```

CAUTION ≪≪≪≪≪≪≪≪≪≪≪≪
Service conductors supplying a building (or other structure) must not pass through the interior of another building ≫230-3≪.

■ Service Conductors

Table 310-15(b)(6) can be used to size 120/240, 3-wire, single-phase service-entrance conductors and service lateral conductors, up to a rating of 400 amperes ≫310-15(b)(6) ≪.

This book explains standard and optional multi-family load calculations in Unit 11.

According to Table 310-15(b)(6), the minimum size THHN copper service conductors for a 400-ampere rating is 400 kcmil. Table 310-16 requires a 500-kcmil conductor to attain the same rating.

Two main service disconnecting means supplying power to twelve units must be computed on the basis of six units per service. Performing a service calculation on twelve units and then dividing by two is incorrect.

Service Disconnect

SERVICE WIRING CLEARANCES

■ Conductor Clearance Above Flat Roofs

A service mast supporting service-drop conductors must be strong enough to safely withstand the strain imposed by the service drop. Supplemental braces or guys may be necessary ≫230-28 ≪.

Service-drop vertical clearances are based on the conductor temperature of 60°F (15°C), no wind, with final unloaded sag in the conductor. Service-drop conductors must not be readily accessible ≫230-24 ≪.

Guy Wire

8 Feet

Conductors must have a minimum vertical clearance of 8 feet above the roof surface ≫230-24(a) ≪.

The vertical clearance above roof level must be maintained for a distance of at least 3 feet from all edges of the roof ≫230-24(a) ≪.

> ### N O T E
> *Vertical clearances of service drop conductors, covered in Section 230-24, pertains to every type of occupancy. Refer to Unit 9 (of this book) for detailed illustrations explaining vertical clearances.*

■ Clearance from Openings

Open service conductors or multiconductor cable (having no overall outer jacket), must have a clearance of at least 3 feet from windows (designed to open), doors, porches, balconies, ladders, stairs, fire escapes, and similar locations ≫230-9 ≪.

3 Feet

Public streets, alleys, roads, parking areas subject to truck traffic, and non-residential driveways require a minimum vertical clearance of 18 feet ≫230-24(b) ≪.

18 Feet

> ### N O T E
> *Where buildings exceed three stories or 50 feet in height, branch circuit and feeder overhead lines must be arranged, where practicable, so that a clear area at least 6 feet wide will be left adjacent to the buildings or beginning within 8 feet to facilitate the raising of fire-fighting ladders ≫225-19(e) ≪.*

■ Vertical Clearance Above Platforms

A drip loop's lowest point has a minimum vertical clearance of 10 feet **》230-24(b)《**.

Overhead conductors, accessible only to pedestrian traffic, must have a minimum 10-foot vertical clearance above finished grade, sidewalks, platforms, or projections from which they might be reached. The voltage must not exceed 150 to ground **》230-24(b)《**.

10 Feet

Vertical clearance of final spans above platforms, projections, or surfaces from which conductors might be reached, must be maintained in accordance with Section 230-24(b) **》230-9《**.

Here, the 3-foot clearance requirement does not apply to conductors run above the top level of a window **》230-9 *Exception*《**. Because of the fire escape platform, the 10-foot vertical clearance takes precedence over the window clearance.

PANELBOARDS AND EQUIPMENT

■ Access to Overcurrent Devices

Each occupant must have ready access to all overcurrent devices protecting the conductors supplying their particular occupancy **》240-24(b)《**.

Service Disconnect

Unit 101

Unit 102

In damp or wet locations, surface-type meter socket enclosures must be so situated and/or equipped to prevent moisture from accumulating within the enclosure. There must be at least ¼-inch airspace between the enclosure and the wall **》373-2(a)《**.

Where electric service and electrical maintenance are provided by building management, service and feeder overcurrent devices supplying more than one occupancy can be accessible only to authorized management personnel in the following: (1) in multiple occupancy buildings, and (2) for guest rooms of hotels and motels that are intended for transient occupancy **》240-24(b)《**.

Table 310-15(b)(6) lists conductors which can be used as service-entrance, service-lateral, and feeder conductors serving as the dwelling unit(s) main power feeder. The conductors can be in raceways or cables with (or without) equipment grounding conductors. These conductor ampacities can be used only in 120/240 volt, 3-wire, single-phase dwelling services and feeders **》Table 310-15(b)(6)《**.

CAUTION 《《《《《《《《《

Overcurrent devices must not be located in close proximity to easily ignitible material (such as in clothes closets) **》240-24(d)《**.

N O T E

Where the service overcurrent devices are locked, sealed, or otherwise not readily accessible to the occupant, branch-circuit overcurrent devices must be: (1) installed on the load side of the service overcurrent device; (2) mounted in a readily accessible location; and (3) of lower ampere rating than the service overcurrent device **》230-92《**.

■ Panelboards Supplying Individual Units

The front edge of cabinets (panelboards) situated in walls constructed of noncombustible material (concrete, tile, etc.) must be within ¼ inch of, or extend beyond, the finished surface ≫ 373-3 ≪.

Remote Panelboard (Subpanel)

Noncurrent-carrying metal parts of equipment, raceways, and other enclosures, must be grounded (where required) by one of the following methods: (1) by any of the equipment grounding conductors permitted by Section 250-118; or (2) by an equipment grounding conductor contained within the same raceway, cable, or otherwise run with the circuit conductors ≫ 250-134 ≪.

If all of the qualifications are met, Table 310-15(b)(6) can be used to size feeder conductors.

Main Bonding Jumper **Not** Installed

Install the bonding jumper (screw, strap, etc.) only in the service disconnecting means, not in subpanels.

At this point, the neutral bar is isolated from the equipment grounding system.

A grounded circuit conductor cannot be used for grounding noncurrent-carrying metal parts of equipment on the load side of the service disconnecting means, unless an exception has been met ≫ 250-142(b) ≪.

A grounding connection must *not* be made to any grounded circuit conductor on the load side of the service disconnecting means except as otherwise allowed in Article 250 ≫ 250-24(a)(5) ≪.

■ Rooftop Heating and Air-Conditioning Equipment

GFCI-Protected Receptacle

25 Feet

25 Feet

All 125-volt, single-phase, 15- and 20-ampere, dwelling unit, outdoor receptacles must have ground-fault circuit-interrupter protection ≫ 210-8(a)(3) ≪.

A 125-volt, single-phase, 15- or 20-ampere-rated receptacle outlet must be installed at an accessible location for the servicing of heating, air-conditioning, and refrigeration equipment on rooftops, in attics, and in crawl spaces ≫ 210-63 ≪.

The receptacle must be located on the same level, and within 25 feet, of the heating, air-conditioning, and refrigeration equipment ≫ 210-63 ≪. The provision does not stipulate that a receptacle must be located at every unit. One centrally located receptacle could meet the Section 210-63 requirement for all pieces of equipment located within the 25-foot radius.

Article 440 contains provisions pertaining to air-conditioning and refrigeration equipment.

■ Feeder Conductors

> **NOTE**
>
> *Where separate services, requiring grounding electrodes, supply a building, the same grounding electrode must be used* »250-58 «.

Table 310-15(b)(6) can be used to size 120/240, 3-wire, single-phase feeder conductors serving as the main power feeder to a dwelling unit »310-15(b)(6) «.

If each unit contains a 100-ampere panelboard, the feeders can be No. 4 AWG copper (or No. 2 AWG aluminum) conductors »Table 310-15(b)(6) «.

> **CAUTION** «««««««««
>
> *On long runs it may be necessary to increase the feeder conductors size, to compensate for voltage drop* »215-2(d) FPN No. 2 «.

Load calculations are computed for each individual unit having its own panelboard. Although located in a multi-family dwelling, each unit is calculated by one of the single-family methods. It is incorrect to perform a load calculation on multiple units and then divide by the number of units.

Standard and optional single-family load calculations are explained in Unit 8 of this text.

The authority having jurisdiction may require a diagram showing feeder details prior to installation. Such a diagram must show the area in square feet of the building (or other structure) supplied by each feeder, the total connected load before applying demand factors, the demand factors used, the computed load after applying demand factors, and the size and type of conductors that will be used »215-5 «.

BRANCH CIRCUITS

■ Optional Receptacles

Multi-family dwelling outdoor receptacle outlets, while not required, must be GFCI-protected, if installed »210-52(e) and 210-8(a)(3) «.

Dwelling-unit branch circuits shall supply only loads within, or associated only with, that dwelling unit »210-25 «.

In a multi-family dwelling where laundry facilities are provided on the premises, available to all building occupants, a laundry receptacle is not required in each individual unit »210-52(f) *Exception No. 1* «.

In other than one-family dwellings without laundry facilities, a laundry receptacle is not required »210-52(f) *Exception No. 2* «.

Receptacles

Twenty-four (24) inches is the maximum distance to any receptacle, measured horizontally, along the wall line »210-52(c)(1) «.

Dwelling-unit branch circuits shall supply only loads within, or associated only with, that dwelling unit »210-25 «.

Receptacles must be installed so that every 6 feet of wall space, measured horizontally along the floor line, contains a receptacle outlet »210-52(a)(1) «.

Whenever practicable, receptacle outlets should be spaced equal distances apart »210-52(a)(1) «.

All receptacles serving kitchen countertops must have ground-fault circuit-interrupter protection »210-8(a)(6) «.

Kitchen receptacles not serving countertops (such as receptacles behind refrigerators) do not require GFCI protection.

All receptacles in the kitchen, pantry, breakfast room and dining room, or similar areas must be supplied from 20-ampere small appliance branch circuits (except for refrigeration equipment) »210-52(b)(1) «.

Circuits feeding receptacles in kitchens, pantries, breakfast rooms, dining rooms, or similar areas must not feed receptacles outside these areas »210-52(b)(2) «.

A receptacle is required for each peninsular counter space (at least 12 by 24 inch in size) which is separated from other counter space by a range top, refrigerator, or sink »210-52(c)(3) «.

A circuit providing power to bathroom receptacles cannot provide power to any receptacles (or lighting) outside the bathroom, unless meeting an exception »210-11(c)(3) «.

GFCI protection is required for every 125-volt receptacle located in a bathroom area »210-8(a)(1) «.

A receptacle is required for any wall space, 2 feet or more in width (including space measured around corners), unbroken along the floor line by doorways, fireplaces, and similar openings »210-52(a)(2)(a) «.

Locate at least one wall receptacle within 36 inches of the outside edge of each basin (lavatory or sink). The receptacle outlet must be located on a wall adjacent to the basin location »210-52(d) «.

NOTE
Only a few differences exist between one-family and multi-family dwelling receptacle requirements. Refer to Units 6 and 7 for more information pertaining to general as well as specific receptacle provisions.

WARNING

All branch circuits that supply 125-volt, single-phase, 15- and 20-ampere receptacle outlets located in dwelling unit bedrooms must be protected by an arc-fault circuit interrupter(s). This requirement becomes effective January 1, 2002 »210-12(b) «. An *arc-fault circuit interrupter* is a device intended to provide protection from the effects of arc faults by recognizing characteristics unique to arcing and by functioning to de-energize the circuit when an arc fault is detected »210-12(a) «.

■ General Lighting

At least one wall switch-controlled lighting outlet must be installed in every habitable room of a dwelling »210-70(a)(1)«.

Clothes closet lighting outlet provisions are found in Section 410-8.

At least one wall switch-controlled lighting outlet must be installed in every dwelling bathroom »210-70(a)(1)«.

A minimum of one wall switch-controlled lighting outlet is required for dwelling hallways »210-70(a)(1)«.

Except for kitchens and bathrooms, every habitable room is allowed one or more wall switch-controlled receptacle(s) in lieu of a lighting outlet »210-70(a)(1) *Exception No. 1*«.

In hallways, stairways, and outdoor entrances, control of lighting can be remote, central, or automatic »210-70(a)(2) *Exception*«.

Branch circuits required for the purpose of lighting, central alarm, signal, communication, or other needs for public or common areas of a two-family or multi-family dwelling cannot be provided from equipment that supplies an individual dwelling unit »210-25«.

■ Appliance Disconnecting Means

A permanently connected appliance, rated over 300 volt-amperes or ⅛ horsepower, must have a means to disconnect all ungrounded (hot) conductors from the appliance »422-30«.

An appliance switch marked with an "off" position which effectively disconnects all ungrounded conductors can serve as the disconnecting means required by Article 422 where other means for disconnection are provided in 422-33(a) through (d) »422-33«.

A disconnecting means is not required if the branch circuit switch or circuit breaker can be locked in the open (off) position »422-31(b)«.

In multi-family dwellings, the other appliance disconnecting means (as described in Section 422-33) must be within the dwelling unit, or on the same floor as the dwelling unit in which the appliance is installed, and can also control lamps and other appliances »422-33(a)«.

A branch circuit supplying a fixed storage type water heater with a capacity of 120 gallons or less must have a rating of at least 125 percent of the water heater's nameplate rating »422-13«.

■ Heating and Air-Conditioning Equipment

For a hermetic refrigerant motor-compressor, the rated-load current marked on the equipment nameplate in which the motor-compressor is employed must be used in determining the rating or ampacity of the disconnecting means, the branch circuit conductors, the controller, the branch-circuit short-circuit and ground-fault protection, and the separate motor overload protection ≫ 440-6(a)≪.

A hermetic refrigerant motor-compressor consists of a compressor and motor, both enclosed in the same housing, with no external shaft or shaft seals, and the motor operating in the refrigerant ≫ 440-2≪.

The disconnecting means for air-conditioning or refrigerating equipment must be readily accessible and within sight of the equipment ≫ 440-14≪.

All electric equipment must be surrounded by sufficient, unobstructed space to allow ready and safe equipment operation and maintenance ≫ 110-26≪. Working space is mandatory for equipment likely to require examination, adjustment, servicing, or maintenance while energized ≫ 110-26(a)≪.

On long runs it may be necessary to increase the branch circuit conductors size, to compensate for voltage drop ≫ 210-19(a) FPN No. 4≪.

The ampacity of branch circuit conductors and the rating or setting of overcurrent protective devices that supply fixed electric space-heating equipment (consisting of resistance elements with or without a motor) must be at least 125% of the total load of the motors and heaters ≫ 424-3(b)≪.

Means must be provided to disconnect all ungrounded conductors from the heater, motor controller(s), and supplementary overcurrent protective device(s) of all fixed electric space-heating equipment ≫ 424-19≪.

In multi-family dwellings, the other disconnecting means, as described in Section 424-19(c), must be within the dwelling unit, or on the same floor as the dwelling unit in which the fixed heater is installed, and can also control lamps and other appliances ≫ 424-19(c)(1)≪.

CU-6
W.P.
W.P.
CU-4   CU-2
W.P.
AHU-2
Dryer
Washer
Electric Panel   SS

N O T E
Fixed heater switch(es) marked with an "off" position which effectively disconnects all ungrounded conductors can serve as the disconnecting means required by Article 424 where other means for disconnection are provided in 424-19(c)(1) through (4) ≫ 424-19(c)≪.

N O T E
The size of the branch circuit conductors and the overcurrent protective devices supplying fixed electric space-heating equipment, including a hermetic refrigerant motor-compressor with or without resistance units, must be computed in accordance with Sections 440-34 and 440-35 ≫ 424-3(b)≪.

VOLTAGE DROP

■ Conductor Properties (Chapter 9, Tables 8 and 9)

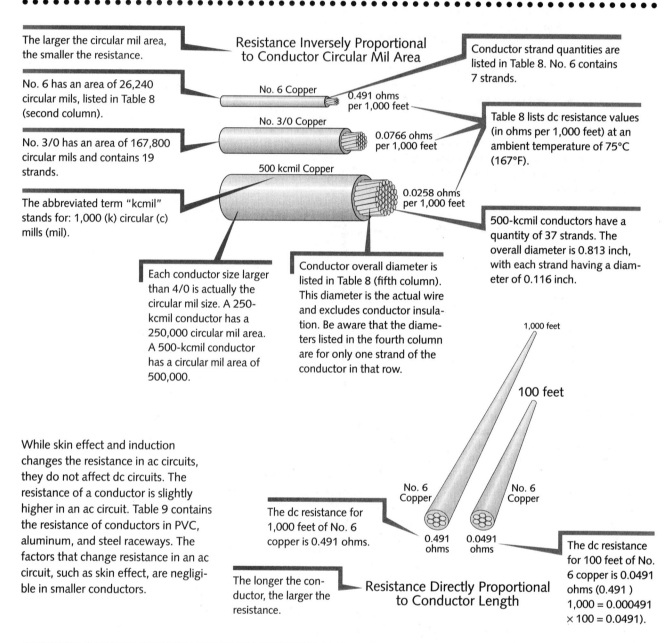

The larger the circular mil area, the smaller the resistance.

No. 6 has an area of 26,240 circular mils, listed in Table 8 (second column).

No. 3/0 has an area of 167,800 circular mils and contains 19 strands.

The abbreviated term "kcmil" stands for: 1,000 (k) circular (c) mills (mil).

Resistance Inversely Proportional to Conductor Circular Mil Area

No. 6 Copper — 0.491 ohms per 1,000 feet

No. 3/0 Copper — 0.0766 ohms per 1,000 feet

500 kcmil Copper — 0.0258 ohms per 1,000 feet

Conductor strand quantities are listed in Table 8. No. 6 contains 7 strands.

Table 8 lists dc resistance values (in ohms per 1,000 feet) at an ambient temperature of 75°C (167°F).

500-kcmil conductors have a quantity of 37 strands. The overall diameter is 0.813 inch, with each strand having a diameter of 0.116 inch.

Each conductor size larger than 4/0 is actually the circular mil size. A 250-kcmil conductor has a 250,000 circular mil area. A 500-kcmil conductor has a circular mil area of 500,000.

Conductor overall diameter is listed in Table 8 (fifth column). This diameter is the actual wire and excludes conductor insulation. Be aware that the diameters listed in the fourth column are for only one strand of the conductor in that row.

While skin effect and induction changes the resistance in ac circuits, they do not affect dc circuits. The resistance of a conductor is slightly higher in an ac circuit. Table 9 contains the resistance of conductors in PVC, aluminum, and steel raceways. The factors that change resistance in an ac circuit, such as skin effect, are negligible in smaller conductors.

1,000 feet

100 feet

No. 6 Copper

No. 6 Copper

0.491 ohms

0.0491 ohms

The dc resistance for 1,000 feet of No. 6 copper is 0.491 ohms.

The longer the conductor, the larger the resistance.

Resistance Directly Proportional to Conductor Length

The dc resistance for 100 feet of No. 6 copper is 0.0491 ohms (0.491) 1,000 = 0.000491 × 100 = 0.0491).

N O T E

The resistance values in Chapter 9, Tables 8 and 9 are valid only for the parameters given. Using conductors having coated strands, different stranding type, and especially, other temperatures, changes the resistance. The formula for temperature change, at the bottom of Table 8, is used to find the resistance of conductors at other than 75°C (167°F). For each 15°C change in temperature, the resistance will change approximately 5% (0.05). A lower temperature will result in a lower resistance; a higher temperature will result in a higher temperature. For example: the resistance for 1,000 feet of No. 6 copper at 30°C (86°F) is approximately 0.417 ohms. The difference between 75°C and 30°C is 45°C, which is approximately 15% (5% × 3). To find the change in resistance, multiply the resistance (0.491) by 15% (0.491 × 15% = .07365 = .074). Since the temperature is lower, the resistance will be lower (0.491 – 0.074 = 0.417).

■ Voltage Drop Recommendation

Voltage drop considerations are only recommendations; compliance is optional.

The conductor voltage that is dropped is a percentage of the source voltage.

Third Floor

The recommended voltage drop for a branch circuit is 3% or less. For example: the maximum allowable voltage drop for a 120-volt branch circuit is 3.6 volts (120 × 3%), and the maximum allowable for a 240-volt branch circuit is 7.2 volts (240 × 3%).

The recommended maximum voltage drop for the combined feeder and branch circuit is 5%. However, the voltage drop for feeders is not necessarily 2%. If the branch circuit is 3%, the feeder is 2%, but if the branch circuit is 1%, the feeder can be 4%. Any combination is possible, as long as the branch circuit does not exceed 3%, and the combined branch circuit and feeder does not exceed 5%.

Branch Circuit

Second Floor

First Floor

Not all electrical installations have feeder conductors.

Feeder

There is no recommended maximum voltage drop for service conductors.

The higher the resistance through a conductor, the higher the voltage drop. Such a conductor voltage drop could result in a reduction of voltage supplying the load.

A single-phase feeder (or branch circuit) conductor will have a total resistance twice that of one conductor. For example: the length of one conductor must be doubled in order to find the total resistance of the circuit. The total length for a three-phase circuit is found by multiplying the length of one conductor by the square root of 3 (1.732).

Branch circuit conductors, as defined in Article 100, sized to prevent a voltage drop exceeding 3 percent at the farthest outlet of power, heating, and lighting loads (or combinations of such loads) and where the maximum total voltage drop on both feeders and branch circuits to the farthest outlet does not exceed 5 percent, will provide reasonable efficient operation ≫210-19(a) FPN No. 4 and 215-2(d) FPN No. 2 ≪.

■ Single-Phase Voltage Drop Formulas

Formula Definitions:

V_D = Volts dropped from a circuit.

2 = Multiplying factor for single-phase (the 2 represents the conductor length in a single-phase circuit.)

K = Resistivity of the conductor material. The resistance value (measured in ohms) is based on a mil foot at a given temperature. A mil foot is a piece of wire 1 foot long and one mil in diameter. Each conductor resistivity (K) factor can be found by multiplying the circular mil area (cmil) by the resistance (R) for one foot. For example: the K factor for a No. 6 copper conductor at 75°C (167°F) is 12.88384, because 26,240 (area circular mills) × 0.000491 (ohms per foot) = 12.88384. At 75°C, the approximate K for copper is 12.9; and for aluminum is 21.2.

L = Length, **ONE WAY** only. The distance from the voltage source to the load.

I = Actual current used by the load. Do not use 125% for motors and continuous loads.

cmil = Circular mil area of the conductor (Chapter 9, Table 8).

R = Conductor resistance for one foot. **Note**: The resistance values, listed in Chapter 9, Tables 8 and 9, are for 1,000 feet. To find the resistance for one foot, simply divide the resistance by 1,000. Example: the resistance for one foot of No. 6 copper (at 75°C) is 0.000491 (0.491 ÷ 1,000 = 0.000491).

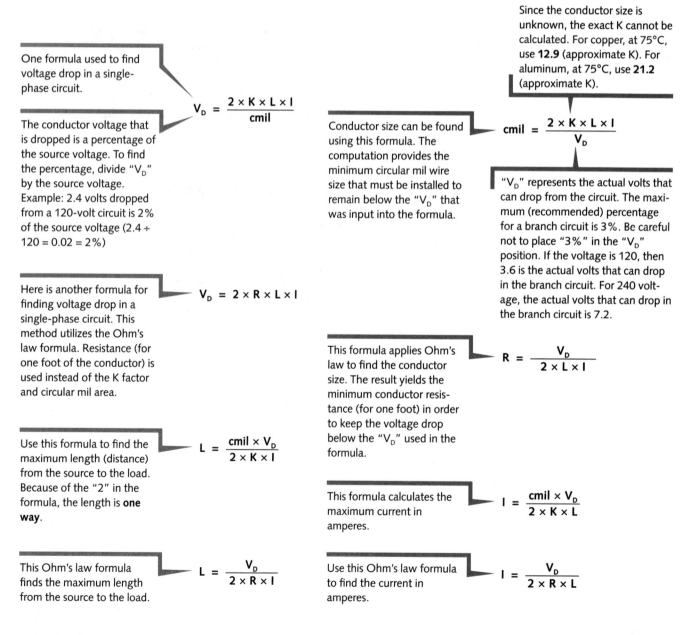

One formula used to find voltage drop in a single-phase circuit.

The conductor voltage that is dropped is a percentage of the source voltage. To find the percentage, divide "V_D" by the source voltage. Example: 2.4 volts dropped from a 120-volt circuit is 2% of the source voltage (2.4 ÷ 120 = 0.02 = 2%)

$$V_D = \frac{2 \times K \times L \times I}{cmil}$$

Here is another formula for finding voltage drop in a single-phase circuit. This method utilizes the Ohm's law formula. Resistance (for one foot of the conductor) is used instead of the K factor and circular mil area.

$$V_D = 2 \times R \times L \times I$$

Use this formula to find the maximum length (distance) from the source to the load. Because of the "2" in the formula, the length is **one way**.

$$L = \frac{cmil \times V_D}{2 \times K \times I}$$

This Ohm's law formula finds the maximum length from the source to the load.

$$L = \frac{V_D}{2 \times R \times I}$$

Since the conductor size is unknown, the exact K cannot be calculated. For copper, at 75°C, use **12.9** (approximate K). For aluminum, at 75°C, use **21.2** (approximate K).

Conductor size can be found using this formula. The computation provides the minimum circular mil wire size that must be installed to remain below the "V_D" that was input into the formula.

$$cmil = \frac{2 \times K \times L \times I}{V_D}$$

"V_D" represents the actual volts that can drop from the circuit. The maximum (recommended) percentage for a branch circuit is 3%. Be careful not to place "3%" in the "V_D" position. If the voltage is 120, then 3.6 is the actual volts that can drop in the branch circuit. For 240 voltage, the actual volts that can drop in the branch circuit is 7.2.

This formula applies Ohm's law to find the conductor size. The result yields the minimum conductor resistance (for one foot) in order to keep the voltage drop below the "V_D" used in the formula.

$$R = \frac{V_D}{2 \times L \times I}$$

This formula calculates the maximum current in amperes.

$$I = \frac{cmil \times V_D}{2 \times K \times L}$$

Use this Ohm's law formula to find the current in amperes.

$$I = \frac{V_D}{2 \times R \times L}$$

Three-Phase Voltage Drop Formulas

"1.732" is the multiplying factor for three-phase (the square root of 3 represents the conductor length in a three-phase circuit). The only difference between the single-phase and three-phase formulas, is that "1.732" has replaced "2".

N O T E
The formula definitions for three-phase correspond with the single-phase definitions.

To find voltage drop in a three-phase circuit:

$$V_D = \frac{1.732 \times K \times L \times I}{cmil}$$

$$V_D = 1.732 \times R \times L \times I$$

To compute the conductor size in a three-phase circuit:

$$V_D = \frac{1.732 \times K \times L \times I}{V_D}$$

To determine the resistance of the conductor in a three-phase circuit:

$$R = \frac{V_D}{1.7322 \times L \times I}$$

To find the maximum length (distance) from the source to the load in a three-phase circuit: (Because of the "1.732" in the formula, the length is **one way**.)

$$L = \frac{cmil \times V_D}{1.732 \times K \times I}$$

$$L = \frac{V_D}{1.732 \times R \times I}$$

To calculate the maximum current in amperes in a three-phase circuit:

$$I = \frac{cmil \times V_D}{1.732 \times K \times L}$$

$$I = \frac{V_D}{1.732 \times R \times L}$$

Variations of three-phase formulas

$$V_D = \frac{2 \times K \times L \times I \times .866}{cmil}$$

$$L = \frac{cmil \times V_D}{2 \times K \times I \times .866}$$

The single-phase formula can be change to a three-phase formula by simply placing ".866" on the line containing the "2". The purpose of putting ".866" on the same line as "2" is simple: $2 \times .866 = 1.732$.

■ Single-Phase Voltage Drop Example

• •

At 75°C, what size conductors are required to feed a 2-HP, 230-volt, single-phase motor that is 150 feet from the source? (Do not exceed *NEC* ® recommendations.)

Both the supply and load side of this equipment disconnecting means are branch circuit conductors; therefore, the maximum recommended voltage drop is 3% of the source voltage.

To Source Voltage

Branch Circuit 150 Feet

2 HP, 230 volt Single-phase Motor

Table 430-148 lists full-load current (in amperes) for single-phase ac motors.

Voltage drop formula used to find the minimum size conductor

Approximate K for copper at 75°C

Current in amperes for a 2-HP, 230-volt, single-phase motor » Table 430-148 «.

This is the minimum circular mil area required. Notice that a No. 12 conductor (6,530 circular mils) is too small for this installation.

$$\text{cmil} = \frac{2 \times K \times L \times I}{V_D} = \frac{2 \times 12.9 \times 150 \times 12}{6.9} = 6,730 = \text{No. 10 copper}$$

The minimum size conductor recommended for this installation.

Maximum amount of voltage drop on a 230-volt branch circuit that remains within the 3% recommendation (230 × 3% = 6.9)

An alternative formula equally effective in determining the minimum size conductor

In multiplying the 1 foot value by 1,000, the resistance values will correspond with Tables 8 and 9, which contains ohms per 1,000 feet values.

$$R = \frac{V_D}{2 \times L \times I} = \frac{6.9}{2 \times 150 \times 12} = 0.001917 \times 1,000 = 1.917 = \text{No. 10 copper}$$

The maximum conductor resistance for 1 foot

The conductor chosen cannot exceed this resistance number. Both solid and stranded No. 12 conductors have a higher resistance; therefore, No. 10 conductors (solid or stranded) must be used.

> **N O T E**
> *If the total distance of the branch circuit was less than 145 feet, a No. 12 copper conductor could be installed.*

Summary

● ●

- Overall receptacle installation requirements, with few exceptions, are the same for both one-family and multi-family dwellings.
- While outdoor receptacles are not required for multi-family dwellings, GFCI protection must be provided, if installed.
- Under certain conditions, a laundry receptacle is not required.
- A multi-family clubhouse (or portion thereof), designed for the assembly of more than 99 persons, must comply with Article 518 provisions.
- Generally, a building is supplied by only one service.
- There can be no more than six disconnects per service grouped in any single location.
- Where separate services supply a building, and a grounding electrode is required, the same electrode must be used.

- Separate service laterals, grouped in one location, constitute one service.
- Use Table 310-15(b)(6) to size 120/240, 3-wire, single-phase service-entrance conductors, service lateral conductors, and feeder conductors serving as the dwelling's main power feeder.
- Service and open conductor clearances pertain to all dwellings.
- Conductor properties are found in Chapter 9, Tables 8 and 9.
- Voltage drop compliance is discretionary.
- The recommended voltage drop for a branch circuit is 3% or less.
- The recommended voltage drop for combined feeder and branch circuits is 5% or less.
- Voltage drop formulas, complete with explanations, are found near the end of this unit.

Unit 10 Competency Test

NEC® Reference   **Answer**

_____ _____ 1. _____ is a general term including material, fittings, devices, appliances, fixtures, apparatus, and the like used as a part of, or in connection with, an electrical installation.

_____ _____ 2. Where a building is supplied by more than one service, or combination of branch circuits, feeders, and services, a permanent _____ or _____ shall be installed at each service disconnect location denoting all other services, feeders, and branch circuits supplying that building and the area served by each.

_____ _____ 3. Where the overcurrent device is rated over _____ amperes, the ampacity of the conductors it protects shall be equal to or greater than the rating of the overcurrent device.

_____ _____ 4. A(n) _____ is a device, or group of devices, or other means by which the conductors of a circuit can be disconnected from their source of supply.

_____ _____ 5. A(n) _____ is a device intended to provide protection from the effects of arc faults by recognizing characteristics unique to arcing and by functioning to de-energize the circuit when an arc fault is detected.

_____ _____ 6. Service drop conductors shall have a vertical clearance of not less than _____ above the roof surface.

_____ _____ 7. What size copper equipment grounding conductor is required for a circuit having a 40-ampere overcurrent device?

_____ _____ 8. Where a change occurs in the size of the _____, a similar change shall be permitted to be made in the size of the grounded conductor.

_____ _____ 9. _____ in dwelling units shall supply only loads within that dwelling unit or loads associated only with that dwelling unit.

_____ _____ 10. A metal elbow that is installed in an underground installation of rigid non-metallic conduit and is isolated from possible contact by a minimum cover of _____ to any part of the elbow shall not be required to be grounded.

NEC® Reference   Answer

_____ _____   11. Conductors shall be considered outside of a building or other structure where installed under not less than _____ of concrete beneath a building or other structure.

_____ _____   12. What is the service demand load (in kW) for an apartment complex containing ten each of the following: 15-kW ranges, 13-kW ranges, 11-kW ranges, and 9-kW ranges? (The service is 120/240 single-phase.)

_____ _____   13. In _____, remote, central, or automatic control of lighting shall be permitted.

_____ _____   14. Service conductors installed as open conductors shall have a clearance of not less than _____ from fire escapes.

_____ _____   15. A grounded circuit conductor shall be permitted to ground noncurrent-carrying metal parts of equipment on the _____ side of the ac service disconnecting means.

_____ _____   16. Each unit of a thirty-unit apartment building has a 4.5-kW clothes dryer. What is the service demand load contribution (in kW) for these clothes dryers? (The service is 120/240 single-phase.)

_____ _____   17. The maximum distance between receptacle outlets in a multi-family dwelling is _____.

_____ _____   18. Overcurrent devices shall not be located in the vicinity of _____, such as in clothes closets.

_____ _____   19. Cabinets and cutout boxes shall have _____ to accommodate all conductors installed in them without crowding.

_____ _____   20. Where conductors carrying alternating current are installed in metal enclosures or metal raceways, they shall be arranged so as to avoid heating the surrounding metal by _____.

_____ _____   21. _____ is the highest current at rated voltage that a device is intended to interrupt under standard test conditions.

_____ _____   22. Where conductors are adjusted in size to compensate for voltage drop, equipment grounding conductors, where installed, shall be adjusted proportionately according to:

_____ _____   23. Enclosures for overcurrent devices shall be mounted in a _____ position.

_____ _____   24. A metal pole supporting a lighting fixture, 18 feet above grade, must have a handhole of at least _____.

_____ _____   25. Fittings and connectors shall be used only with the specific wiring methods for which they are _____.

_____ _____   26. _____ is a combination consisting of a compressor and motor, both of which are enclosed in the same housing, with no external shaft or shaft seals, the motor operating in the refrigerant.

_____ _____   27. For permanently connected appliances rated over _____ volt-amperes or _____ horsepower the branch circuit switch or circuit breaker shall be permitted to serve as the disconnecting means where the switch or circuit breaker is within sight from the appliance or is capable of being locked in the open position.

_____ _____   28. The ampacity of the branch circuit conductors and the rating or setting of overcurrent protective devices supplying fixed electric space-heating equipment consisting of resistance elements with or without a motor shall not be less than _____ percent of the total load of the motors and the heaters.

NEC® Reference   Answer

_____ _____ 29. _____ is the necessary equipment, usually consisting of a circuit breaker(s) or switch(es) and fuse(s) and their accessories, connected to the load end of service conductors to a building or other structure, or an otherwise designated area, and intended to constitute the main control and cutoff of the supply.

_____ _____ 30. No hanging fixture parts can be located within a zone measured _____ feet vertically from the top of the bathtub rim and _____ feet vertically from the top of the shower stall threshold.

_____ _____ 31. What is the ac resistance of 500 feet of 1,000 kcmil aluminum wire in a steel conduit?

_____ _____ 32. At 75°C, what size copper conductors are required (without exceeding *NEC*® recommendations) to feed a ¼ horsepower, 115 volt, single-phase motor located 230 feet from the source?

_____ _____ 33. Interior metal water piping located more than _____ from the point of entrance to the building shall not be used as a part of the grounding electrode system.

_____ _____ 34. What is the minimum neutral demand load (in kW) for twelve apartments, each containing an 8-kW range?

_____ _____ 35. What is the voltage drop percentage on two No. 10 THW copper, stranded, branch circuit conductors, 120 feet long, supplying a 21-ampere, 240-volt load?

Load Calculations

Objectives

After studying this unit, the student will:

- ✔ know that the number of dwelling units is essential in performing a multi-family calculation.
- ✔ be able to calculate the general lighting load in a multi-family dwelling.
- ✔ correctly identify minimum volt-ampere requirements for small appliance branch circuits.
- ✔ thoroughly understand the exceptions pertaining to optional laundry branch circuits.
- ✔ know how to apply demand factors to the general lighting load, as well as to fastened-in-place appliances.
- ✔ be able to calculate demand loads for household clothes dryers, cooking equipment, and heating and air-conditioning systems.

- ✔ understand the standard method for calculating a multi-family dwelling service.
- ✔ successfully calculate a multi-family dwelling service (using the optional method.)
- ✔ be able to calculate a feeder for one unit in a multi-family dwelling, using either method.
- ✔ know how to size service and feeder conductors.
- ✔ be familiar with paralleled conductor provisions.
- ✔ correctly calculate and select the appropriate size neutral conductor.
- ✔ understand how grounding electrode conductors, as well as an equipment grounding conductors, are selected.

Introduction

Strong similarities exist between multi-family dwelling load calculations and one-family load calculations. Both can be performed by a standard method as well as an optional method. In fact, the similarities are so strong that this is reflected in the very design of the load calculation forms. It is, therefore, obviously beneficial to have a clear understanding of one-family dwelling load calculations before studying this unit (see Unit 8). It is absolutely essential to these calculations, to know the total number of dwelling units being supplied by the service. To illustrate this point: two main service disconnecting means supplying power to twelve units must be computed on the basis of six units per service. Performing a service calculation on twelve units and then dividing the final figure by two, would render inaccurate results.

The multi-family dwelling load calculations performed in this unit are based on the floor plan found in Unit 10. The statistical information for each individual dwelling unit remains unchanged for the purpose of load calculation. This data can be found in the floor-plan diagram immediately following this introduction. Examples include a multi-family dwelling load calculation for both a 12-unit and a 6-unit service. Each will be calculated by the standard method as well as the optional. Feeder and panelboard sizing for individual units is demonstrated by a standard one-family dwelling calculation. For comparison, the optional one-family dwelling is shown after the optional multi-family dwelling computation.

COMPILING LOAD CALCULATION INFORMATION

■ Load Calculation Information for Each Unit

The smaller of two (or more) non-coincident loads can be omitted, as long as they are never energized simultaneously ≫ 220-21 ≪ .

Calculate the floor area using the outside dimensions ≫ 220-3(a) ≪ .

Include household cooking appliances rated over 1¾ kW.

Include all fastened-in-place appliances.

30 Feet

Air-conditioner Compressor and Condenser Fan Motor located outside

Air-handler Unit with Electric Heat

Dryer

Laundry Branch Circuit

Washer

Electric Panel

Dish-washer

Range

Waste Disposer

Two Small Appliance Branch Circuits

35 Feet

Although a clothes dryer is not requirement for a load calculation, be sure to include a dryer load in units containing clothes dryer outlets.

A laundry branch circuit is optional in multi-family dwellings ≫ 210-52(f) *Exception No. 1 and 2* ≪ .

Include at least 2 small appliance branch circuits ≫ 210-11(c)(1) ≪ .

Gather all load calculation information needed to compute each service. For a service consisting of 12 units (apartment or condominium), compile the essential information for all 12 units. If the service consists of only 6 units, gather the data for those 6 units. Since most multi-family dwellings have a limited number of floor plans, compile the information pertaining to each "Unit Type" (style, floor plan, etc.). Be sure that each unit grouped in a unit type contains the same information, i.e. total square feet, small appliance branch circuits, laundry circuits, appliances, etc. For ease of demonstration in this section, each unit is identical in size and content.

Dwelling outside dimensions	30' × 35'
Small appliance branch circuits	two
Laundry branch circuit	one
Water heater	4.5 kW, 240 volt
Dishwasher	10 amperes, 120 volt
Waste (garbage) disposer	7.5 amperes, 120 volt
Clothes dryer	5.5 kW, 240 volt
Range	12 kW, 240 volt
Electric heat	5 kW, 240 volt
Air handler (blower motor)	3 amperes, 115 volt
Air-conditioner compressor	14 amperes, 230 volt
Condenser fan motor	2 amperes, 115 volt

Assume water heater, clothes dryer, range, and electric heat kW ratings equivalent to kVA.

STANDARD METHOD: MULTI-FAMILY DWELLINGS

■ Load Calculation for Twelve Units

The first multi-family load calculation form is based on 12 units. Each unit is identical, as shown above.

In a multiple-occupancy building, each occupant must have access to their unit's service disconnecting means ≫ 230-72(c) ≪ .

There must not be more than six disconnects (switches or circuit breakers) per service grouped in a single location ≫ 230-71(a) ≪ .

One service disconnecting switch controls all 12 units.

Each unit's feeder (and overcurrent protection) must be calculated individually, using the one-family dwelling form (standard or optional method).

Standard Method Load Calculation for Multi-Family Dwellings

1 General Lighting and Receptacle Loads 220-3(a)
Do not include open porches, garages, unused or unfinished spaces not adaptable for future use.

$3 \times$ _____ $\times$ _____ $=$ **1**
(sq. ft. outside dimensions) (number of units)

2 Small Appliance Branch Circuits 220-16(a)
*At least **two** small appliance branch circuits must be included. 210-11(c)(1)*

$1{,}500 \times$ _____ $\times$ _____ $=$ **2**
(minimum of two) (number of units)

3 Laundry Branch Circuit(s) 220-16(b)
*Include at least **one** laundry branch circuit ***unless** meeting 210-52(f) Exception No. 1 or 2. 210-11(c)(2)*

$1{,}500 \times$ _____ $\times$ _____ $=$ **3**
(minimum of two) (number of units)

4 Add lines 1, 2, and 3   **4** | Lines 5 through 8 utilize the demand factors found in Table 220-11.

5 _____ $- 3{,}000 =$ **5**
(line 4)
(if 117,000 or less, skip to line 8)

6 _____ $- 117{,}000 =$ **6**
(line 5, if more than 117,000)

7 _____ $\times 25\% =$ **7**
(line 6)

8 _____ $\times 35\% =$ **8**
(smaller of line 5 or 117,000)

9 Total General Lighting and Receptacle Load $3{,}000 +$ _____ $+$ _____ $=$ **9**
(line 7) (line 8)

10 Fastened-In-Place Appliances 220-17
*Use the nameplate rating.
Do not include electric ranges, clothes dryers, space-heating equipment, or air-conditioning equipment.*

water heaters / _____   dishwashers / _____   disposers / _____

_____ / _____   _____ / _____   _____ / _____

_____ / _____   _____ / _____   _____ / _____

If fewer than four units, put total volt-amperes on line 10.
If four or more units, multiply total volt-amperes by 75%.   _____ $\times 75\% =$ **10**
(volt-amps of four or more)

11 Clothes Dryers 220-18
(If present, otherwise skip to line 12.) Use 5,000 watts or the nameplate rating, whichever is larger. (The neutral demand load is 70% for feeders. 220-22)   **11**

12 Ranges, Ovens, Cooktops, and Other Household Cooking Appliances Over 1,750 Watts 220-19
(If present, otherwise skip to line 13.) Use Table 220-19 and all of the applicable Notes. (The neutral demand load is 70% for feeders. 220-22)   **12**

13 Heating or Air-Conditioning System (Compare the heat and A/C, and omit the smaller.) 220-21
Include the air handler when using either one. For heat pumps, include the compressor and the maximum amount of electric heat which can be energized while the compressor is running.   **13**

14 Largest Motor (one motor only) 220-14 and 430-24
Multiply the volt-amperes of the largest motor by 25%.   _____ $\times 25\% =$ **14**
(volt-amps of largest motor)

15 Total Volt-Ampere Demand Load: *Add lines 9 through 14 to find the minimum required volt-amperes.*   **15**

16 Minimum Amperes
Divide the total volt-amperes by the voltage   _____ $\div$ _____ $=$ **16**
(line 15) (voltage) (minimum amperes)

17 Minimum Size Service and/or Feeder 240-6(a)   **17**

18 Size the Service and/or Feeder Conductors.
Use Section 310-15(b)(6) to find the service conductors up to 400 amperes. Ratings in excess of 400 amperes shall comply with Table 310-16. 310-15(b)(6) also applies to feeder conductors serving as the main power feeder.   **Minimum Size Conductors**   **18**

19 Size the Neutral Conductor 220-22
Section 310-15(b)(6) states that the neutral service and/or feeder conductor can be smaller than the ungrounded (hot) conductors, provided the requirements of Sections 215-2, 220-22, and 230-42 are met. Section 250-24(b)(1) states that the neutral cannot be smaller than the required grounding electrode conductor specified in Table 250-66.   **Minimum Size Neutral Conductor**   **19**

20 Size the Grounding Electrode Conductor (for Service) 250-66
Using line 18 to find the grounding electrode conductor in Table 250-66.
Size the Equipment Grounding Conductor (for Feeder) 250-122
Use line 17 to find the equipment grounding conductor in Table 250-122. Equipment grounding conductor types are listed in Section 250-118.   **Minimum Size Grounding Electrode Conductor . . . or . . . Equipment Grounding Conductor**   **20**

■ Line 1—General Lighting and Receptacle Loads

The residential dwelling unit's lighting load is 3 volt-amperes (VA) per square foot »Table 220-3(a) «.

Each multi-family dwelling unit has an outside dimension of 30' × 35' (1,050 sq. ft.).

Insert the number of units (apartments or condominiums) which constitute one service. Additional space will be needed for computations if the units have different square foot dimensions.

1 General Lighting and Receptacle Loads 220-3(a) *Do not include open porches, garages, unused or unfinished spaces not adaptable for future use.*	$3 \times \underline{\quad 1,050 \quad} \times \underline{\quad 12 \quad} = $ **1** $\boxed{37,800}$
	(sq. ft. outside dimensions) (number of units)

NOTE

All 15- or 20-ampere general-use receptacle outlets (except for small appliance and laundry), are considered part of the general lighting load. The "a" footnote at the bottom of Table 220-3(a) refers to Section 220-3(b)(10), which states that no additional load calculations are required for these outlets.

Each unit's floor area shall be computed using the outside dimensions. The total floor area does not include open porches, garages, or unused (including unfinished) spaces which are not adaptable for future use »220-3(a) «.

■ Line 2—Small Appliance Branch Circuits

Include small appliance branch circuits when calculating a dwelling unit service »220-16(a) «.

Each small appliance branch circuit is calculated at no less than 1,500 volt-amperes »220-16(a) «.

If any unit(s) has a different number of small appliance branch circuits, calculate each separately.

2 Small Appliance Branch Circuits 220-16(a) *At least **two** small appliance branch circuits must be included. 210-11(c)(1)*	$1,500 \times \underline{\quad 2 \quad} \times \underline{\quad 12 \quad} = $ **2** $\boxed{36,000}$
	(minimum of two) (number of units)

NOTE

A refrigeration equipment individual branch circuit (permitted by Section 210-52[b][1] Exception No. 2) can be excluded from this calculation »220-16(a) Exception «.

Each dwelling unit must have at least two small appliance branch circuits »210-11(c)(1) «.

These loads can be included with the general lighting load and are subject to Table 220-11 demand factors »220-16(a) «.

■ Line 3—Laundry Branch Circuit(s)

Include laundry branch circuit(s) when calculating a dwelling unit service »220-16(b) «.

Each laundry branch circuit is calculated at no less than 1,500 volt-amperes »220-16(b) «.

Include only the number of units having a laundry branch circuit. If only 6 of 12 units contains a laundry branch circuit, put 6 as the number of units. In this calculation, each unit has a laundry branch circuit.

3 Laundry Branch Circuit(s) 220-16(b) *Include at least **one** laundry branch circuit *unless* meeting 210-52(f) Exception No. 1 or 2. 210-11(c)(2)*	$1,500 \times \underline{\quad 1 \quad} \times \underline{\quad 12 \quad} = $ **3** $\boxed{18,000}$
	(minimum of two) (number of units)

In a multi-family building where laundry facilities are provided on the premises, available to all occupants, a laundry receptacle is not required in each individual unit »210-52(f) Exception No. 1 «.

A laundry receptacle is still not required, even in multi-family dwellings without common laundry facilities »210-52(f) Exception No. 2 «.

Unless one of the exceptions is met, include at least one laundry branch circuit »210-11(c)(2) «.

These loads can be included with the general lighting load and are subject to the Table 220-11 demand factors »220-16(b) «.

■ Lines 4 through 8—Applying Demand Factors Found in Table 220-11

Line 4 represents the total of lines 1 through 3.

Lines 5 through 8 outline the derating procedure, using Table 220-11 demand factors.

| 4 | Add lines 1, 2, and 3 | 4 | *91,800* | Lines 5 through 8 utilize the demand factors found in Table 220-11. |

Line 5 is the result of subtracting 3,000 from the number in Line 4.

The first 3,000 (or less) volt-amperes must remain at 100% » Table 220-11«.

If Line 5 is 117,000 or less, skip Lines 6 and 7.

If Line 4 is 120,000 or less, then Lines 6 and 7 remain empty.

| 5 | *91,800* (line 4) − 3,000 = | 5 | *88,800* (if 117,000 or less, skip to line 8) | 6 | ---- (line 5, if more than 117,000) − 117,000 = | 6 | ---- |
| 7 | ---- (line 6) × 25% = | 7 | ---- | 8 | *88,800* (smaller of line 5 or 117,000) × 35% = | 8 | *31,080* |

Insert the smaller of Line 5 or 117,000.

Multiply the next 117,000 (from 3,001 to 120,000) by 35% » Table 220-11«.

■ Line 9—Total General Lighting (and Receptacle) Load

Line 9 is the total of general lighting, general-use receptacles, small appliance branch circuits, and laundry branch circuit(s) after derating.

If Line 7 is blank, leave this blank also.

Insert the result of line 8, not the number from Line 5.

| 9 | Total General Lighting and Receptacle Load | 3,000 + ---- (line 7) + *31,080* (line 8) = | 9 | *34,080* |

This shaded block draws attention to the first of six lines (boxes) which will be added together resulting in the total volt-ampere demand load for the one-family dwelling.

■ Line 10—Fastened-in-Place Appliances

The term **appliance** designates utilization equipment (commonly built in standardized types and sizes) installed as a unit to perform specific function(s) » Article 100«.

For the purpose of this load calculation, a kW rating is the same as a kVA rating.

Additional space may be needed to calculate total appliance volt-amperes. (1,200 × 12 = 14,400) and (900 × 12 = 10,800)

| 10 | Fastened-In-Place Appliances 220-17
Use the nameplate rating.
Do not include electric ranges, clothes dryers, space-heating equipment, or air-conditioning equipment. | water heaters / *54,000*
microwave ovens / *14,400*
/ | dishwashers / *14,400*
/
/ | disposers / *10,800*
/
/ |

If fewer than four units, put total volt-amperes on line 10. If four or more units, multiply total volt-amperes by 75%.

93,600 (volt-amps of four or more) × 75% = | 10 | *70,200*

Derating is allowed for a total of 4 (or more) fastened-in-place appliances. It is not necessary that the 4 fastened-in-place appliances be of different types. This derating factor can be used even if all units do not contain 4 appliances.

> **NOTE**
> *Household cooking appliances, individually rated in excess of 1,750 watts, are derated under Section 220-19 provisions.*

If the calculation contains at least 4 fastened-in-place appliances, the combined volt-ampere rating of those appliances is multiplied by 75% and placed on Line 10 » 220-17«.

> **NOTE**
> *Do not include electric ranges, dryers, space-heating equipment, and air conditioning equipment as fastened-in-place appliances » 220-17«.*

▪ Line 11—Clothes Dryers

A clothes dryer is not a requirement for a load calculation. Skip this line if there is no clothes dryer.

The dryer rating must be at least 5,000 watts (volt-amperes) **》220-18《**.

Table 220-18 can be used to derate five (or more) clothes dryer loads. First, find the demand factor percent across from the number of dryers, in this case the percent is 45. Next, multiply the total dryer load by that percentage and put the result in Line 11. (5,000 × 12 = 60,000 × 45% = 27,000)

11 Clothes Dryers 220-18
(If present, otherwise skip to line 12.) Use 5,000 watts or the nameplate rating, whichever is larger.
(The neutral demand load is 70% for feeders. 220-22)

11 *27,000*

A clothes dryer neutral load on a feeder is calculated at 70% **》220-22《**.

If the nameplate rating exceeds 5,000 watts (volt-amperes), use the larger number **》220-18《**.

> **N O T E**
> *For 4 (or fewer) clothes dryers the demand is 100%.*

▪ Line 12—Household Cooking Appliances

Household cooking appliances (ranges, wall-mounted ovens, counter-mounted cooking units, etc.) are not required in a load calculation. If no cooking appliances are rated over 1¾ kW, skip this line.

Individual household cooking appliances rated more than 1,750 watts (volt-amperes) can be derated using Table 220-19 demand factors (and all applicable notes).

The demand for twelve 12 kW ranges comes from Table 220-19, Column A. First, look at number of appliances column and find the number twelve. Next, follow that row across to the second column (Column A) and find the kW demand load, which is 27. Finally, multiply this number by 1,000 and place in Line 12. Since each range is 12 kW, Columns B and C cannot be used.

12 Ranges, Ovens, Cooktops, and Other Household Cooking Appliances Over 1,750 Watts 220-19
(If present, otherwise skip to line 13.) Use Table 220-19 and all of the applicable Notes.
(The neutral demand load is 70% for feeders. 220-22)

12 *27,000*

For household electric ranges, wall-mounted ovens, and counter-mounted cooking units the neutral load on a feeder is calculated at 70% **》220-22《**.

▪ Line 13—Heating or Air-Conditioning System

The smaller of two (or more) **noncoincident loads** can be omitted, as long as they are never energized simultaneously **》220-21《**. By definition, noncoincident means not occurring at the same time. A dwelling's electric heating system and air conditioning system, may be noncoincident loads.

The air handler (blower motor) has a rating of 345 volt-amperes (3 × 115 = 345). Since both loads are energized simultaneously, add the air handler to the 5 kW of electric heat.

Fixed electric space heating loads are computed at 100% of the total connected load **》220-15《**.

13 Heating or Air-Conditioning System (Compare the heat and A/C, and omit the smaller.) 220-21
Include the air handler when using either one. For heat pumps, include the compressor and the maximum amount of electric heat which can be energized while the compressor is running.

13 *64,140*

Because the air handler (blower) works with either the heating or the air conditioning, add it to both calculations.

> **N O T E**
> *A heat pump with supplementary heat is not considered a noncoincident load. Add the compressor full load current to the maximum amount of heat that can be energized while the compressor is running. Remember to include all associated motor(s), i.e., blower, condenser, etc.*

Each unit contains an air-conditioner compressor (14 × 230 = 3,220 VA), condenser fan motor (2 × 115 = 230 VA), and blower motor (3 × 115 = 345 VA). The combined load is 3,795 volt-amperes. Because this is less than the heating load omit the A/C compressor and the condenser fan motor. Multiply each unit's heat load by the number of units. (5,345 × 12 = 64,140)

■ Line 14—Largest Motor

Multiply the largest motor in the dwelling calculation by 25% ≫ 220-14 and 430-24 ≪ .

This means a total of one motor; not one motor per unit.

The largest motor is usually the air conditioner compressor. If the air conditioner load has been omitted, use the second largest motor. In this case, use the 7.5 ampere garbage disposer motor. (7.5 × 120 = 900)

14 Largest Motor (one motor only) 220-14 and 430-24 *Multiply the volt-amperes of the largest motor by 25%.*	$\dfrac{900}{\text{(volt-amps of largest motor)}} \times 25\% =$	**14**	*225*

■ Line 15—Total Demand Load

Find the minimum volt-ampere load by adding Lines 9 through 14.

15 Total Volt-Ampere Demand Load: *Add lines 9 through 14 to find the minimum required volt-amperes.*	**15**	*222,645*

■ Lines 16 and 17—Minimum Service and/or Feeder

Place the total volt-ampere figure found in Line 15 here.

Write down the source voltage supplying the service equipment.

Fractions of an ampere 0.5 and higher are rounded up, while fractions less than 0.5 are dropped ≫ 220-2(b) ≪ .

The service (or feeder) overcurrent protection must be higher than the number in Line 16. Section 240-6(a) provides a list of standard ampere ratings for fuses and circuit breakers.

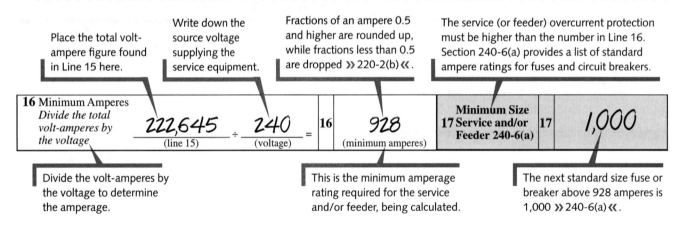

16 Minimum Amperes *Divide the total volt-amperes by the voltage*	$\dfrac{222,645}{\text{(line 15)}} \div \dfrac{240}{\text{(voltage)}} =$	**16**	*928* (minimum amperes)	**Minimum Size** **17** Service and/or Feeder 240-6(a)	**17**	*1,000*

Divide the volt-amperes by the voltage to determine the amperage.

This is the minimum amperage rating required for the service and/or feeder, being calculated.

The next standard size fuse or breaker above 928 amperes is 1,000 ≫ 240-6(a) ≪ .

■ Line 18—Minimum Size Conductors

*Conductors carrying alternating current, installed in metal enclosures or **metal raceways**, must be arranged to avoid heating of the surrounding metal by induction* »*300-20(a)* ‹‹.

When running **three** sets of paralleled copper conductors (for a 1,000 ampere overcurrent device), the minimum size is 400 kcmil, unless the equipment is listed and identified for use with 90°C (194°F) conductors. Even if a 90°C conductor (such as THHN) is installed, the ampacity rating cannot exceed the 75°C (167°F) ampacity which is 335 »110-14(c)(2) ‹‹. Since the overcurrent protection is more than 800 amperes, the conductor ampacity must match (or exceed) the 1,000 ampere rating (1,000 ÷ 3 = 333.3 minimum amperes, each conductor) »240-3(c) ‹‹.

Other parallel conductor combinations are possible, such as four sets of 250 kcmil Cu, four sets of 350 kcmil Al, five sets of No. 3/0 Cu, five sets of 250 kcmil Al, etc. The smallest conductor that can be run in parallel is No. 1/0 »310-4 ‹‹.

Use Section 310-15(b)(6) to find service entrance and service lateral conductors.

Section 310-15(b)(6) also applies to feeder conductors serving as the dwelling unit's main power feeder »310-15(b)(6) ‹‹.

Use Table 310-15(b)(6) to size conductors up to a 400 ampere rating.

---

18 Size the Service and/or Feeder Conductors.
Use Section 310-15(b)(6) to find the service conductors up to 400 amperes.
Ratings in excess of 400 amperes shall comply with Table 310-16.
310-15(b)(6) also applies to feeder conductors serving as the main power feeder.

| **Minimum Size Conductors** | **18** | *400 kcmil Cu* *600 kcmil Al* *three paralleled sets* |

NOTE

The paralleled conductors in each phase, neutral, or grounded circuit conductor must share the same characteristics »*310-4* ‹‹*. For example, all paralleled, Phase A conductors must:*

(1) be of the same length,
(2) have the same conductor material,
(3) be the same size in circular mil area,
(4) have the same insulation type,
(5) be terminated in the same manner.

Since the rating is more than 400 amperes, use Table 310-16 to find the minimum size conductors.

The number and type of paralleled conductor sets is a design consideration; not necessarily a *Code* issue. Exercise extreme care when designing a paralleled conductor installation without violating provisions, such as, 110-14(a), 110-14(c), 240-3(c), 250-66, 250-122(f) 300-20(a), 310-4, etc.

No. 1/0 and larger conductors can be connected in parallel (electrically joined at both ends to form a single conductor) »310-4 ‹‹.

■ Line 19—Neutral Conductor

For the purpose of this load calculation, the terms **Neutral** and **Grounded** are synonymous.

The feeder (or service) neutral load is the maximum unbalance of the load determined by Article 220 »220-22 ‹‹.

The neutral (grounded) conductor can be smaller than the ungrounded (hot) conductors provided certain requirements in Section 215-2, 220-22, and 230-42 are met »310-15(b)(6) ‹‹.

Find the maximum unbalanced load in order to size the neutral.

Although the neutral conductor requires only 151 amperes each (for three paralleled sets), it cannot be smaller than the appropriate grounding electrode conductor (see Line 20).

---

19 Size the Neutral Conductor 220-22
Section 310-15(b)(6) states that the neutral service and/or feeder conductor can be smaller than the ungrounded (hot) conductors, provided the requirements of Sections 215-2, 220-22, and 230-42 are met.
Section 250-24(b)(1) states that the neutral cannot be smaller than the required grounding electrode conductor specified in Table 250-66.

| **Minimum Size Neutral Conductor** | **19** | *3/0 Cu* *250 kcmil Al* *three paralleled sets* |

NOTE
The smallest paralleled grounded (neutral) conductor is 1/0 »*250-24(b)(2)* ‹‹*.*

The neutral must not be smaller than the required grounding electrode conductor specified in Table 250-66 »*250-24(b)(1)* ‹‹*.*

■ Line 19—Neutral Conductor (*continued*)

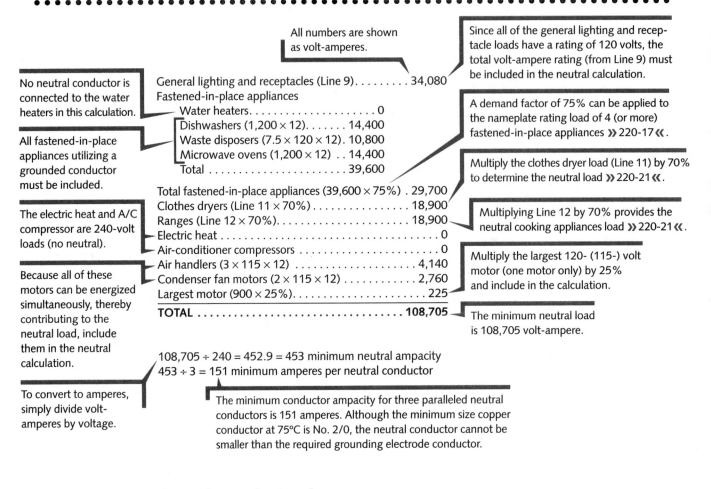

All numbers are shown as volt-amperes.

Since all of the general lighting and receptacle loads have a rating of 120 volts, the total volt-ampere rating (from Line 9) must be included in the neutral calculation.

No neutral conductor is connected to the water heaters in this calculation.

All fastened-in-place appliances utilizing a grounded conductor must be included.

A demand factor of 75% can be applied to the nameplate rating load of 4 (or more) fastened-in-place appliances »220-17«.

Multiply the clothes dryer load (Line 11) by 70% to determine the neutral load »220-21«.

Multiplying Line 12 by 70% provides the neutral cooking appliances load »220-21«.

The electric heat and A/C compressor are 240-volt loads (no neutral).

Multiply the largest 120- (115-) volt motor (one motor only) by 25% and include in the calculation.

Because all of these motors can be energized simultaneously, thereby contributing to the neutral load, include them in the neutral calculation.

General lighting and receptacles (Line 9)........ 34,080
Fastened-in-place appliances
    Water heaters.................... 0
    Dishwashers (1,200 × 12)...... 14,400
    Waste disposers (7.5 × 120 × 12). 10,800
    Microwave ovens (1,200 × 12) .. 14,400
    Total 39,600
Total fastened-in-place appliances (39,600 × 75%) . 29,700
Clothes dryers (Line 11 × 70%)................ 18,900
Ranges (Line 12 × 70%)..................... 18,900
Electric heat 0
Air-conditioner compressors 0
Air handlers (3 × 115 × 12) 4,140
Condenser fan motors (2 × 115 × 12) 2,760
Largest motor (900 × 25%)..................... 225
TOTAL **108,705**

The minimum neutral load is 108,705 volt-ampere.

108,705 ÷ 240 = 452.9 = 453 minimum neutral ampacity
453 ÷ 3 = 151 minimum amperes per neutral conductor

To convert to amperes, simply divide volt-amperes by voltage.

The minimum conductor ampacity for three paralleled neutral conductors is 151 amperes. Although the minimum size copper conductor at 75°C is No. 2/0, the neutral conductor cannot be smaller than the required grounding electrode conductor.

■ Line 20—Grounding Electrode Conductor

Table 250-66 has two headings; service conductor size and grounding electrode conductor size. Each heading contain two columns: one for copper and one for aluminum (or copper-clad aluminum). Because of the four columns, take extreme care in selecting the correct grounding electrode conductor.

The grounding electrode conductor is determined by the largest service-entrance conductor (or the equivalent area) if the service-entrance conductors are paralleled. The equivalent area for three parallel sets in this calculation is 1,200 kcmil copper (3 × 400) *or* 1,800 kcmil aluminum (3 × 600).

20 Size the Grounding Electrode Conductor (for Service) 250-66 *Using line 18 to find the grounding electrode conductor in Table 250-66.* Size the Equipment Grounding Conductor (for Feeder) 250-122 *Use line 17 to find the equipment grounding conductor in Table 250-122.* *Equipment grounding conductor types are listed in Section 250-118.*	**Minimum Size Grounding Electrode Conductor ... or ... Equipment Grounding Conductor**	**20**	*3/0 Cu* *250 kcmil Al*

The minimum size grounding electrode conductor is No. 3/0 copper (or 250 kcmil aluminum) »Table 250-66«.

The following page shows the completed standard method load calculation for a 12-unit, multi-family dwelling, having all units combined in a single service.

Standard Method Load Calculation for Multi-Family Dwellings

1 General Lighting and Receptacle Loads 220-3(a) *Do not include open porches, garages, unused or unfinished spaces not adaptable for future use.*	$3 \times \underline{1,050}$ (sq. ft. outside dimensions) $\times \underline{12}$ (number of units) =	**1**	**37,800**
2 Small Appliance Branch Circuits 220-16(a) *At least **two** small appliance branch circuits must be included. 210-11(c)(1)*	$1,500 \times \underline{2}$ (minimum of two) $\times \underline{12}$ (number of units) =	**2**	**36,000**
3 Laundry Branch Circuit(s) 220-16(b) *Include at least **one** laundry branch circuit ***unless*** meeting 210-52(f) Exception No. 1 or 2. 210-11(c)(2)*	$1,500 \times \underline{1}$ (minimum of two) $\times \underline{12}$ (number of units) =	**3**	**18,000**

4 Add lines 1, 2, and 3	**4**	**91,800**	Lines 5 through 8 utilize the demand factors found in Table 220-11.

5 $\underline{91,800}$ (line 4) $- 3,000 =$	**5**	**88,800** (if 117,000 or less, skip to line 8)	**6** $\underline{----}$ (line 5, if more than 117,000) $- 117,000 =$	**6**	**----**
7 $\underline{----}$ (line 6) $\times 25\% =$	**7**	**----**	**8** $\underline{88,800}$ (smaller of line 5 or 117,000) $\times 35\% =$	**8**	**31,080**

9 Total General Lighting and Receptacle Load	$3,000 + \underline{----}$ (line 7) $+ \underline{31,080}$ (line 8) $=$	**9**	**34,080**

10 Fastened-In-Place Appliances 220-17 *Use the nameplate rating. Do not include electric ranges, clothes dryers, space-heating equipment, or air-conditioning equipment.*	water heaters / **54,000** dishwashers / **14,400** disposers / **10,800** microwave ovens / **14,400** / / / / /		
If fewer than four units, put total volt-amperes on line 10. If four or more units, multiply total volt-amperes by 75%.	$\underline{93,600}$ (volt-amps of four or more) $\times 75\% =$	**10**	**70,200**

11 Clothes Dryers 220-18 *(If present, otherwise skip to line 12.) Use 5,000 watts or the nameplate rating, whichever is larger. (The neutral demand load is 70% for feeders. 220-22)*	**11**	**27,000**
12 Ranges, Ovens, Cooktops, and Other Household Cooking Appliances Over 1,750 Watts 220-19 *(If present, otherwise skip to line 13.) Use Table 220-19 and all of the applicable Notes. (The neutral demand load is 70% for feeders. 220-22)*	**12**	**27,000**
13 Heating or Air-Conditioning System (Compare the heat and A/C, and omit the smaller.) 220-21 *Include the air handler when using either one. For heat pumps, include the compressor and the maximum amount of electric heat which can be energized while the compressor is running.*	**13**	**64,140**

14 Largest Motor (one motor only) 220-14 and 430-24 *Multiply the volt-amperes of the largest motor by 25%.*	$\underline{900}$ (volt-amps of largest motor) $\times 25\% =$	**14**	**225**

15 Total Volt-Ampere Demand Load: *Add lines 9 through 14 to find the minimum required volt-amperes.*	**15**	**222,645**

16 Minimum Amperes *Divide the total volt-amperes by the voltage*	$\underline{222,645}$ (line 15) $\div \underline{240}$ (voltage) $=$	**16** **928** (minimum amperes)	**Minimum Size** **17** **Service and/or Feeder 240-6(a)**	**17** **1,000**

18 Size the Service and/or Feeder Conductors. *Use Section 310-15(b)(6) to find the service conductors up to 400 amperes. Ratings in excess of 400 amperes shall comply with Table 310-16. 310-15(b)(6) also applies to feeder conductors serving as the main power feeder.*	**Minimum Size** **Conductors**	**18** **400 kcmil Cu** **600 kcmil Al** *three paralleled sets*
19 Size the Neutral Conductor 220-22 *Section 310-15(b)(6) states that the neutral service and/or feeder conductor can be smaller than the ungrounded (hot) conductors, provided the requirements of Sections 215-2, 220-22, and 230-42 are met. Section 250-24(b)(1) states that the neutral cannot be smaller than the required grounding electrode conductor specified in Table 250-66.*	**Minimum Size** **Neutral** **Conductor**	**19** **3/0 Cu** **250 kcmil Al** *three paralleled sets*
20 Size the Grounding Electrode Conductor (for Service) 250-66 *Using line 18 to find the grounding electrode conductor in Table 250-66.* Size the Equipment Grounding Conductor (for Feeder) 250-122 *Use line 17 to find the equipment grounding conductor in Table 250-122. Equipment grounding conductor types are listed in Section 250-118.*	**Minimum Size** **Grounding Electrode** **Conductor . . . or . . .** **Equipment Grounding** **Conductor**	**20** **3/0 Cu** **250 kcmil Al**

SIX UNIT MULTI-FAMILY DWELLING CALCULATION

■ Two Meter Centers, Each Supplying Six Units

Each service drop (or lateral) supplies only one set of service-entrance conductors, unless two to six service-disconnecting means (in separate enclosures) are grouped at one location and supply separate loads from one service drop **》230-40** *Exception No. 2* **《**.

Because service equipment installations can vary greatly, the arrangement must be known before a load calculation is performed. In this configuration, each service (conductors and equipment) must be able to carry the calculated load for six dwelling units.

The following multi-family dwelling, standard method, load calculation is based on 6 units with identical configurations. Turn to page 209 for dwelling unit details.

Service-entrance conductors must be able to carry the ampacity determined in accordance with Article 220 **》230-42(a)《**.

No more than six disconnects per service can be grouped in one location **》230-71(a)《**.

■ Lines 1 through 9—General Lighting and Receptacle Load

Fill in the first three lines using six dwelling units instead of 12.

1 General Lighting and Receptacle Loads 220-3(a) *Do not include open porches, garages, unused or unfinished spaces not adaptable for future use.*	$3 \times$ ___*1,050*___ (sq. ft. outside dimensions) $\times$ ___*6*___ (number of units) =	**1**	*18,900*
2 Small Appliance Branch Circuits 220-16(a) *At least **two** small appliance branch circuits must be included. 210-11(c)(1)*	$1,500 \times$ ___*2*___ (minimum of two) $\times$ ___*6*___ (number of units) =	**2**	*18,000*
3 Laundry Branch Circuit(s) 220-16(b) *At least **one** laundry branch circuit must be included. 210-11(c)(2)*	$1,500 \times$ ___*1*___ (minimum of two) $\times$ ___*6*___ (number of units) =	**3**	*9,000*

4 Add lines 1, 2, and 3	**4** *45,900*	Lines 5 through 8 utilize the demand factors found in Table 220-11.	
5 ___*45,900*___ (line 4) $-3,000 =$	**5** *42,900* (if 117,000 or less, skip to line 8)	**6** ___----___ (line 5, if more than 117,000) $-117,000 =$ **6**	----
7 ___----___ (line 6) $\times 25\% =$ **7**	----	**8** ___*42,900*___ (smaller of line 5 or 117,000) $\times 35\% =$ **8**	*15,015*
9 Total General Lighting and Receptacle Load	$3,000 +$ ___----___ (line 7) $+$ ___*15,015*___ (line 8) $=$ **9**		*18,015*

Insert the result of Line 8; not the number from Line 5.

The general lighting and receptacle load for these six dwelling units is 18,015 volt-amperes. Compare this Line 9 with the Line 9 for twelve units (34,080). Obviously, this is not half of Line 9 for twelve units; therefore, a calculation must be performed for each different application.

Multiply the result of Line 5 by 35% and place on Line 8.

■ Lines 11 and 12—Clothes Dryers and Ranges

Use Table 220-18 to derate five (or more) clothes dryer loads. First, find the demand factor percent across from the number of dryers, in this case 70%. Next, multiply the total dryer load by that percent and record the result on Line 11. (5,000 × 6 = 30,000 × 70% = 21,000)

11 Clothes Dryers 220-18 *(If present, otherwise skip to line 12.) Use 5,000 watts or the nameplate rating, whichever is larger.* (The neutral demand load is 70% for feeders. 220-22)	**11**	*21,000*
12 Ranges, Ovens, Cooktops, and Other Household Cooking Appliances Over 1,750 Watts 220-19 *(If present, otherwise skip to line 13.) Use Table 220-19 and all of the applicable Notes.* (The neutral demand load is 70% for feeders. 220-22)	**12**	*21,000*

The demand for six 12 kW ranges comes from Table 220-19, Column A. First, look at the number of appliances column and find the number six. Next, follow that row across to Column A and find the kW demand load, which is 21. Finally, multiply this number by 1,000 and place in Line 12. Since each range is 12 kW, Columns B and C cannot be used.

■ Lines 15 Through 17—Minimum Amperes

The minimum volt-ampere load for these six dwelling units is 127,410.

15 Total Volt-Ampere Demand Load: *Add lines 9 through 14 to find the minimum required volt-amperes.*	**15**	*127,410*

| **16** Minimum Amperes
Divide the total volt-amperes by the voltage | $\dfrac{127,410}{\text{(line 15)}} \div \dfrac{240}{\text{(voltage)}} =$ | **16** | *531*
(minimum amperes) | **Minimum Size**
17 Service and/or
Feeder 240-6(a) | **17** | *600* |

The service-entrance conductors and equipment must have a rating of at least 531 amperes.

The next standard size fuse or circuit breaker above 531 is 600 amperes » 240-6(a) «.

■ Line 18—Minimum Size Conductors

At 75°C, the minimum size copper conductors, for two parallel sets, matching or exceeding 531 amperes, are 300 kcmil. Since the overcurrent protection is less than 800 amperes, the conductor ampacity can be less than the 600-ampere rating. Where the conductor ampacity does not correspond to a standard size fuse or breaker, and 240-3(b)1 through 3 conditions are met, use the next higher standard overcurrent device » 240-3(b) «. The combined rating of two 300 kcmil copper conductors is 570 amperes. 250 kcmil copper conductors cannot be used, because their combined rating of only 510 amperes is below the minimum required conductor ampacity of 531.

18 Size the Service and/or Feeder Conductors. *Use Section 310-15(b)(6) to find the service conductors up to 400 amperes.* *Ratings in excess of 400 amperes shall comply with Table 310-16.* *310-15(b)(6) also applies to feeder conductors serving as the main power feeder.*	**Minimum Size** **Conductors**	**18**	*300 kcmil Cu* *400 kcmil Al* *two paralleled sets*

Other parallel conductor combinations are possible, such as three sets of No. 3/0 Cu and three sets of No. 4/0 Al.

■ Line 19—Neutral Conductor

General lighting and receptacles (Line 9). 18,015
Fastened-in-place appliances
    Water heaters. 0
    Dishwashers (1,200 × 6). 7,200
    Waste disposers (7.5 × 120 × 6). . . 5,400
    Microwave ovens (1,200 × 6) 7,200
    Total 19,800

Total fastened-in-place appliances (19,800 × 75%) . 14,850
Clothes dryers (Line 11 × 70%) 14,700
Ranges (Line 12 × 70%). 14,700
Electric heat 0
Air-conditioner compressors 0
Air handlers (3 × 115 × 6). 2,070
Condenser fan motors (2 × 115 × 6) 1,380
Largest motor (900 × 25%). 225
TOTAL **65,940**

> **NOTE**
> *Even with a lower neutral ampacity, the conductors must remain this size, because the neutral conductor cannot be smaller than the required grounding electrode conductor* **》250-24(b)(1)《**.

The minimum neutral load is 65,940 volt-ampere.

65,940 ÷ 240 = 274.75 = 275 minimum neutral ampacity
275 ÷ 2 = 137.5 = 138 minimum amperes per neutral conductor

The minimum conductor ampacity for two paralleled neutral conductors is 138 amperes. At 75°C, the minimum size conductor is No. 1/0 copper *or* No. 3/0 aluminum.

19 Size the Neutral Conductor 220-22 *Section 310-15(b)(6) states that the neutral service and/or feeder conductor can be smaller than the ungrounded (hot) conductors, provided the requirements of Sections 215-2, 220-22, and 230-42 are met.* *Section 250-24(b)(1) states that the neutral cannot be smaller than the required grounding electrode conductor specified in Table 250-66.*	**Minimum Size Neutral Conductor**	**19**	*1/0 Cu* *3/0 Al* *two paralleled sets*

■ Line 20—Grounding Electrode Conductor

The equivalent area for two parallel sets of service entrance conductors, in this calculation, is 600 kcmil copper (2 × 300) . . . or . . . 800 kcmil aluminum (2 × 400).

20 Size the Grounding Electrode Conductor (for Service) 250-66 *Using line 18 to find the grounding electrode conductor in Table 250-66.* Size the Equipment Grounding Conductor (for Feeder) 250-122 *Use line 17 to find the equipment grounding conductor in Table 250-122.* *Equipment grounding conductor types are listed in Section 250-118.*	**Minimum Size Grounding Electrode Conductor . . . or . . . Equipment Grounding Conductor**	**20**	*1/0 Cu* *3/0 Al*

The minimum size grounding electrode conductor is No.1/0 copper *or* No. 3/0 aluminum **》Table 250-66《**.

The following page shows the complete standard method load calculation for a six-unit, multi-family dwelling.

Standard Method Load Calculation for Multi-Family Dwellings

1 General Lighting and Receptacle Loads 220-3(a) *Do not include open porches, garages, unused or unfinished spaces not adaptable for future use.*	3 × __1,050__ × __6__ = **1** (sq. ft. outside dimensions) (number of units)	**18,900**
2 Small Appliance Branch Circuits 220-16(a) *At least two small appliance branch circuits must be included. 210-11(c)(1)*	1,500 × __2__ × __6__ = **2** (minimum of two) (number of units)	**18,000**
3 Laundry Branch Circuit(s) 220-16(b) *At least one laundry branch circuit must be included. 210-11(c)(2)*	1,500 × __1__ × __6__ = **3** (minimum of two) (number of units)	**9,000**

4 Add lines 1, 2, and 3	**4** __45,900__	Lines 5 through 8 utilize the demand factors found in Table 220-11.

5 __45,900__ − 3,000 = **5** (line 4)	__42,900__ (if 117,000 or less, skip to line 8)	**6** __----__ − 117,000 = **6** (line 5, if more than 117,000)	__----__
7 __----__ × 25% = **7** (line 6)	__----__	**8** __42,900__ × 35% = **8** (smaller of line 5 or 117,000)	**15,015**

9 Total General Lighting and Receptacle Load	3,000 + __----__ + __15,015__ = **9** (line 7) (line 8)	**18,015**

10 Fastened-In-Place Appliances 220-17 *Use the nameplate rating.* *Do not include electric ranges, clothes dryers, space-heating equipment, or air-conditioning equipment.* water heaters / 27,000 dishwashers / 7,200 disposers / 5,400 microwave ovens / 7,200 / / / / /	
If fewer than four units, put total volt-amperes on line 10. If four or more units, multiply total volt-amperes by 75%. __46,800__ × 75% = **10** (volt-amps of four or more)	**35,100**

11 Clothes Dryers 220-18 *(If present, otherwise skip to line 12.) Use 5,000 watts or the nameplate rating, whichever is larger.* *(The neutral demand load is 70% for feeders. 220-22)* **11**	**21,000**
12 Ranges, Ovens, Cooktops, and Other Household Cooking Appliances Over 1,750 Watts 220-19 *(If present, otherwise skip to line 13.) Use Table 220-19 and all of the applicable Notes.* *(The neutral demand load is 70% for feeders. 220-22)* **12**	**21,000**
13 Heating or Air-Conditioning System (Compare the heat and A/C, and omit the smaller.) 220-21 *Include the air handler when using either one. For heat pumps, include the compressor and the maximum amount of electric heat which can be energized while the compressor is running.* **13**	**32,070**
14 Largest Motor (one motor only) 220-14 and 430-24 *Multiply the volt-amperes of the largest motor by 25%.* __900__ × 25% = **14** (volt-amps of largest motor)	**225**
15 Total Volt-Ampere Demand Load: *Add lines 9 through 14 to find the minimum required volt-amperes.* **15**	**127,410**

16 Minimum Amperes *Divide the total volt-amperes by the voltage*	__127,410__ ÷ __240__ = **16** (line 15) (voltage)	**531** (minimum amperes)	**Minimum Size 17 Service and/or Feeder 240-6(a)** **17** **600**
18 Size the Service and/or Feeder Conductors. *Use Section 310-15(b)(6) to find the service conductors up to 400 amperes. Ratings in excess of 400 amperes shall comply with Table 310-16. 310-15(b)(6) also applies to feeder conductors serving as the main power feeder.*			**Minimum Size Conductors** **18** **300 kcmil Cu 400 kcmil Al** *two paralleled sets*
19 Size the Neutral Conductor 220-22 *Section 310-15(b)(6) states that the neutral service and/or feeder conductor can be smaller than the ungrounded (hot) conductors, provided the requirements of Sections 215-2, 220-22, and 230-42 are met. Section 250-24(b)(1) states that the neutral cannot be smaller than the required grounding electrode conductor specified in Table 250-66.*			**Minimum Size Neutral Conductor** **19** **1/0 Cu 3/0 Al** *two paralleled sets*
20 Size the Grounding Electrode Conductor (for Service) 250-66 *Using line 18 to find the grounding electrode conductor in Table 250-66.* Size the Equipment Grounding Conductor (for Feeder) 250-122 *Use line 17 to find the equipment grounding conductor in Table 250-122. Equipment grounding conductor types are listed in Section 250-118.*			**Minimum Size Grounding Electrode Conductor ... or ... Equipment Grounding Conductor** **20** **1/0 Cu 3/0 Al**

STANDARD LOAD CALCULATION FOR EACH UNIT OF A MULTI-FAMILY DWELLING

■ Calculating Feeders for Individual Dwelling Units

A load calculation, on an individual unit in a multi-family dwelling, must be computed in accordance with one of the one-family-dwelling load-calculation methods (standard or optional).

A load calculation must be performed on each unique unit, i.e., different from any unit already calculated.

A grounded circuit conductor (neutral) cannot be used for grounding noncurrent-carrying metal equipment on the load side of the service disconnecting means »250-142(b)«. In other words, the neutral must be isolated from the equipment ground in a remote panelboard (sub-panel).

The service disconnecting means must have a rating of at least the calculated load, determined in accordance with Article 220 »230-79«.

Use Table 310-15(b)(6) to size the dwelling unit's main power feeder(s) »310-15(b)(6)«.

If the service disconnecting means consists of more than one switch or circuit breaker (permitted by Section 230-71), the combined ratings of all switches or circuit breakers used must not be less than that required by section 230-79 »230-80«.

Standard Method—One Unit

Each identically configured unit, i.e., sq. ft. dimensions, small appliance and laundry branch circuits, fastened-in-place appliances, etc., results in the same size feeder.

Although located in a multi-family dwelling, this dwelling unit contains one feeder and one panelboard. It is therefore computed in accordance with one of the one-family dwelling methods.

A load calculation must be performed on each dissimilar unit.

Standard Method Load Calculation for One-Family Dwellings

1	General Lighting and Receptacle Loads 220-3(a) *Do not include open porches, garages, unused or unfinished spaces not adaptable for future use.*	3 × **1,050** (sq. ft. using outside dimensions) = 1	**3,150**
2	Small Appliance Branch Circuits 220-16(a) *At least **two** small appliance branch circuits must be included. 210-11(c)(1)*	1,500 × **2** (minimum of two) = 2	**3,000**
3	Laundry Branch Circuit(s) 220-16(b) *At least **one** laundry branch circuit must be included. 210-11(c)(2)*	1,500 × **1** (minimum of one) = 3	**1,500**
4	Add lines 1, 2, and 3 4 **7,650**	Lines 5 through 8 utilize the demand factors found in Table 220-11.	

5	**7,650** (line 4) − 3,000 = 5	**4,650** (if 117,000 or less, skip to line 8)	6	**----** − 117,000 = 6	**----**
7	**----** (line 6) × 25% = 7	**----**	8	**4,650** (smaller of line 5 or 117,000) × 35% = 8	**1,628**

9	Total General Lighting and Receptacle Load 3,000 + **----** (line 7) + **1,628** (line 8) = 9	**4,628**

10	Fastened-In-Place Appliances 220-17 *Use the nameplate rating. Do not include electric ranges, clothes dryers, space-heating equipment, or air-conditioning equipment.* water heater / **4,500** dishwasher / **1,200** disposer / **900** compactor / microwave oven / **1,200** / / / If fewer than four units, put total volt-amperes on line 10. If four or more units, multiply total volt-amperes by 75%. **7,800** (volt-amps of four or more) × 75% = 10	**5,850**
11	Clothes Dryers 220-18 *(If present, otherwise skip to line 12.) Use 5,000 watts or the nameplate rating, whichever is larger. (The neutral demand load is 70% for feeders. 220-22)*	**5,000**
12	Ranges, Ovens, Cooktops, and Other Household Cooking Appliances Over 1,750 Watts 220-19 *(If present, otherwise skip to line 13.) Use Table 220-19 and all of the applicable Notes. (The neutral demand load is 70% for feeders. 220-22)*	**8,000**
13	Heating or Air-Conditioning System (Compare the heat and A/C, and omit the smaller.) 220-21 *Include the air handler when using either one. For heat pumps, include the compressor and the maximum amount of electric heat which can be energized while the compressor is running.*	**5,345**
14	Largest Motor (one motor only) 220-14 and 430-24 *Multiply the volt-amperes of the largest motor by 25%.* **900** (volt-amps of largest motor) × 25% = 14	**225**
15	Total Volt-Ampere Demand Load: *Add lines 9 through 14 to find the minimum required volt-amperes.*	**29,048**

Using the standard method, the minimum required volt-amperes for this single unit is 29,048.

NOTE
Load calculations are computed for each individual unit containing a panelboard (sub-panel). Although located in a multi-family dwelling, calculate each unit by one of the methods for one-family dwellings »Article 220«. Performing a load calculation on multiple units, and then dividing by the number of units, will give an incorrect result.

■ Minimum Size Feeder and Overcurrent Protection

Feeder conductors must have an ampacity not less than required to supply the load as computed in Article 220, Parts B, C, and D. The minimum feeder conductor size, before the application of any adjustment or correction factors, must have an allowable ampacity equal to, or greater than, the noncontinuous load plus 125% of the continuous load 》215-2(a)《.

Feeders must be protected against overcurrent in accordance with the Article 240, Part A provisions 》215-3《.

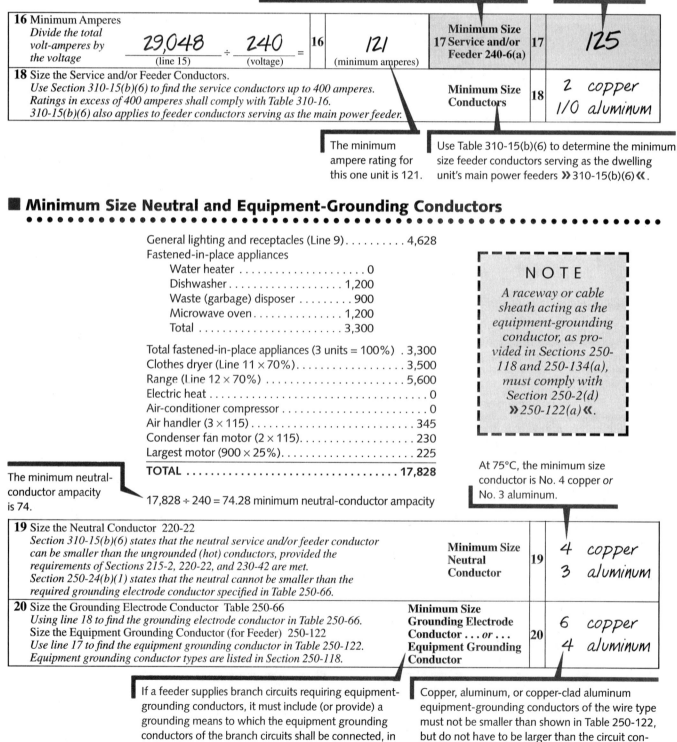

16 Minimum Amperes *Divide the total volt-amperes by the voltage*	$\dfrac{29,048}{\text{(line 15)}} \div \dfrac{240}{\text{(voltage)}} =$	16	*121* (minimum amperes)	**Minimum Size 17 Service and/or Feeder 240-6(a)**	17	*125*
18 Size the Service and/or Feeder Conductors. *Use Section 310-15(b)(6) to find the service conductors up to 400 amperes.* *Ratings in excess of 400 amperes shall comply with Table 310-16.* *310-15(b)(6) also applies to feeder conductors serving as the main power feeder.*				**Minimum Size Conductors**	18	*2 copper* *1/0 aluminum*

The minimum ampere rating for this one unit is 121.

Use Table 310-15(b)(6) to determine the minimum size feeder conductors serving as the dwelling unit's main power feeders 》310-15(b)(6)《.

■ Minimum Size Neutral and Equipment-Grounding Conductors

General lighting and receptacles (Line 9). 4,628
Fastened-in-place appliances
    Water heater 0
    Dishwasher. 1,200
    Waste (garbage) disposer 900
    Microwave oven. 1,200
    Total 3,300

Total fastened-in-place appliances (3 units = 100%) . 3,300
Clothes dryer (Line 11 × 70%). 3,500
Range (Line 12 × 70%) 5,600
Electric heat 0
Air-conditioner compressor 0
Air handler (3 × 115) 345
Condenser fan motor (2 × 115). 230
Largest motor (900 × 25%). 225
TOTAL **17,828**

17,828 ÷ 240 = 74.28 minimum neutral-conductor ampacity

The minimum neutral-conductor ampacity is 74.

> **NOTE**
> *A raceway or cable sheath acting as the equipment-grounding conductor, as provided in Sections 250-118 and 250-134(a), must comply with Section 250-2(d)* 》250-122(a)《.

At 75°C, the minimum size conductor is No. 4 copper *or* No. 3 aluminum.

19 Size the Neutral Conductor 220-22 *Section 310-15(b)(6) states that the neutral service and/or feeder conductor can be smaller than the ungrounded (hot) conductors, provided the requirements of Sections 215-2, 220-22, and 230-42 are met.* *Section 250-24(b)(1) states that the neutral cannot be smaller than the required grounding electrode conductor specified in Table 250-66.*	**Minimum Size Neutral Conductor**	19	*4 copper* *3 aluminum*
20 Size the Grounding Electrode Conductor Table 250-66 *Using line 18 to find the grounding electrode conductor in Table 250-66.* Size the Equipment Grounding Conductor (for Feeder) 250-122 *Use line 17 to find the equipment grounding conductor in Table 250-122.* *Equipment grounding conductor types are listed in Section 250-118.*	**Minimum Size Grounding Electrode Conductor . . . or . . . Equipment Grounding Conductor**	20	*6 copper* *4 aluminum*

If a feeder supplies branch circuits requiring equipment-grounding conductors, it must include (or provide) a grounding means to which the equipment grounding conductors of the branch circuits shall be connected, in accordance with Section 250-134 provisions 》215-6《.

Copper, aluminum, or copper-clad aluminum equipment-grounding conductors of the wire type must not be smaller than shown in Table 250-122, but do not have to be larger than the circuit conductors supplying the equipment 》250-122(a)《.

The following page shows the complete one-family-dwelling load calculation for one multi-family dwelling-unit feeder and remote panelboard (subpanel), according to the standard method.

Standard Method Load Calculation for One-Family Dwellings

1 General Lighting and Receptacle Loads 220-3(a) *Do not include open porches, garages, unused or unfinished spaces not adaptable for future use.*	$3 \times \underline{\hspace{1cm} 1,050 \hspace{1cm}}$ = (sq. ft. using outside dimensions)	**1**	**3,150**
2 Small Appliance Branch Circuits 220-16(a) *At least **two** small appliance branch circuits must be included. 210-11(c)(1)*	$1,500 \times \underline{\hspace{1cm} 2 \hspace{1cm}}$ = (minimum of two)	**2**	**3,000**
3 Laundry Branch Circuit(s) 220-16(b) *At least **one** laundry branch circuit must be included. 210-11(c)(2)*	$1,500 \times \underline{\hspace{1cm} 1 \hspace{1cm}}$ = (minimum of one)	**3**	**1,500**

4 Add lines 1, 2, and 3	**4**	**7,650**	Lines 5 through 8 utilize the demand factors found in Table 220-11.

5 $\underline{\hspace{0.5cm} 7,650 \hspace{0.5cm}}$ $- 3,000 =$ (line 4)	**5**	**4,650** (if 117,000 or less, skip to line 8)	**6** $\underline{\hspace{1cm} ---- \hspace{1cm}} - 117,000 =$ (line 5, if more than 117,000)	**6**	**----**
7 $\underline{\hspace{0.5cm} ---- \hspace{0.5cm}} \times 25\% =$ (line 6)	**7**	**----**	**8** $\underline{\hspace{0.5cm} 4,650 \hspace{0.5cm}} \times 35\% =$ (smaller of line 5 or 117,000)	**8**	**1,628**

9 Total General Lighting and Receptacle Load	$3,000 + \underline{\hspace{0.5cm} ---- \hspace{0.5cm}} + \underline{\hspace{0.5cm} 1,628 \hspace{0.5cm}} =$ (line 7) (line 8)	**9**	**4,628**

10 Fastened-In-Place Appliances 220-17 *Use the nameplate rating. Do not include electric ranges, clothes dryers, space-heating equipment, or air-conditioning equipment.*	water heater / 4,500 dishwasher / 1,200 disposer / 900 compactor / microwave oven / 1,200 / / / /		
If fewer than four units, put total volt-amperes on line 10. If four or more units, multiply total volt-amperes by 75%.	$\underline{\hspace{0.5cm} 7,800 \hspace{0.5cm}} \times 75\% =$ (volt-amps of four or more)	**10**	**5,850**

11 Clothes Dryers 220-18 *(If present, otherwise skip to line 12.) Use 5,000 watts or the nameplate rating, whichever is larger. (The neutral demand load is 70% for feeders. 220-22)*		**11**	**5,000**
12 Ranges, Ovens, Cooktops, and Other Household Cooking Appliances Over 1,750 Watts 220-19 *(If present, otherwise skip to line 13.) Use Table 220-19 and all of the applicable Notes. (The neutral demand load is 70% for feeders. 220-22)*		**12**	**8,000**
13 Heating or Air-Conditioning System (Compare the heat and A/C, and omit the smaller.) 220-21 *Include the air handler when using either one. For heat pumps, include the compressor and the maximum amount of electric heat which can be energized while the compressor is running.*		**13**	**5,345**
14 Largest Motor (one motor only) 220-14 and 430-24 *Multiply the volt-amperes of the largest motor by 25%.*	$\underline{\hspace{0.5cm} 900 \hspace{0.5cm}} \times 25\% =$ (volt-amps of largest motor)	**14**	**225**

15 Total Volt-Ampere Demand Load: *Add lines 9 through 14 to find the minimum required volt-amperes.*		**15**	**29,048**

16 Minimum Amperes *Divide the total volt-amperes by the voltage*	$\underline{\hspace{0.5cm} 29,048 \hspace{0.5cm}} \div \underline{\hspace{0.5cm} 240 \hspace{0.5cm}} =$ (line 15) (voltage)	**16** **121** (minimum amperes)	**Minimum Size 17 Service and/or Feeder 240-6(a)**	**17** **125**
18 Size the Service and/or Feeder Conductors. *Use Section 310-15(b)(6) to find the service conductors up to 400 amperes. Ratings in excess of 400 amperes shall comply with Table 310-16. 310-15(b)(6) also applies to feeder conductors serving as the main power feeder.*		**Minimum Size Conductors**	**18**	**2 copper** **1/0 aluminum**
19 Size the Neutral Conductor 220-22 *Section 310-15(b)(6) states that the neutral service and/or feeder conductor can be smaller than the ungrounded (hot) conductors, provided the requirements of Sections 215-2, 220-22, and 230-42 are met. Section 250-24(b)(1) states that the neutral cannot be smaller than the required grounding electrode conductor specified in Table 250-66.*		**Minimum Size Neutral Conductor**	**19**	**4 copper** **3 aluminum**
20 Size the Grounding Electrode Conductor Table 250-66 *Using line 18 to find the grounding electrode conductor in Table 250-66. Size the Equipment Grounding Conductor (for Feeder) 250-122 Use line 17 to find the equipment grounding conductor in Table 250-122. Equipment grounding conductor types are listed in Section 250-118.*		**Minimum Size Grounding Electrode Conductor . . . or . . . Equipment Grounding Conductor**	**20**	**6 copper** **4 aluminum**

OPTIONAL METHOD: MULTI-FAMILY DWELLINGS

Optional Method Load Calculation for Multi-Family Dwellings

1 General Lighting and Receptacle Loads 220-32(c)(2) *Do not include open porches, garages, unused or unfinished spaces not adaptable for future use.*	$3 \times$ _____ $\times$ _____ $=$ (sq. ft. outside dimensions) (number of units)	**1**	
2 Small Appliance Branch Circuits 220-32(c)(1) *At least **two** small appliance branch circuits must be included. 210-11(c)(1)*	$1{,}500 \times$ _____ $\times$ _____ $=$ (minimum of two) (number of units)	**2**	
3 Laundry Branch Circuit(s) 220-32(c)(1) *At least **one** laundry branch circuit must be included. 210-11(c)(2)*	$1{,}500 \times$ _____ $\times$ _____ $=$ (*minimum of two) (number of units)	**3**	
4 through 11 Appliances and Motors 220-32(c)(3) and (4) *Use the nameplate rating of **ALL** appliances (fastened-in-place, permanently connected, or connected to a specific circuit), ranges, wall-mounted ovens, Counter-mounted cooking units, motors, water heaters, and clothes dryers.* *Convert any nameplate rating given in amperes to volt-amperes by multiplying the amperes by the rated voltage.* ***Do not include any heating or air-conditioning equipment in this section.***	*water heater* ____ / _____ $\times$ _____ $=$ (volt-amperes each) (number)	**4**	
	dishwashers ____ / _____ $\times$ _____ $=$ (volt-amperes each) (number)	**5**	
	disposers ____ / _____ $\times$ _____ $=$ (volt-amperes each) (number)	**6**	
	clothes dryers ____ / _____ $\times$ _____ $=$ (volt-amperes each) (number)	**7**	
	ranges ____ / _____ $\times$ _____ $=$ (volt-amperes each) (number)	**8**	
	microwave ovens ____ / _____ $\times$ _____ $=$ (volt-amperes each) (number)	**9**	
	____ / _____ $\times$ _____ $=$ (volt-amperes each) (number)	**10**	
	____ / _____ $\times$ _____ $=$ (volt-amperes each) (number)	**11**	
12 Heating or Air-Conditioning System (Compare the heat and A/C, and omit the smaller.) 220-32(c)(5) *Include the air handler when using either one.* *For heat pumps, include the compressor and the maximum amount of electric heat which can be energized while the compressor is running.*	_____ $\times$ _____ $=$ (volt-amperes each) (number)	**12**	
13 Total Volt-Ampere Demand Load: *Multiply total VA by Table 220-32 demand factor percent.*	_____ $+$ _____ $=$ (total volt-amperes from lines 1 through 12) (Table 220-32 demand factor)	**13**	
14 House Load *(If present, otherwise skip to line 15)* *Compute in accordance with Article 220, Part B.* *Do not include in Table 220-32 demand factors.*		**14**	

15 Minimum Amperes *Divide the total volt-amperes by the voltage*	_____ $\div$ _____ $=$ (lines 13 and 14) (voltage)	**15** (minimum amperes)	**16** **Minimum Size Service and/or Feeder 240-6(a)**	**16**	
17 Size the Service and/or Feeder Conductors. *Use Section 310-15(b)(6) to find the service conductors up to 400 amperes.* *Ratings in excess of 400 amperes shall comply with Table 310-16.* *310-15(b)(6) also applies to feeder conductors serving as the main power feeder.*			**Minimum Size Conductors**	**17**	
18 Size the Neutral Conductor 220-22 **Note: There is no optional method for calculating the neutral conductor.** *Section 310-15(b)(6) states that the neutral service and/or feeder conductor can be smaller than the ungrounded (hot) conductors, provided the requirements of Sections 215-2, 220-22, and 230-42 are met.* *Section 250-24(b)(1) states that the neutral cannot be smaller than the required grounding electrode conductor specified in Table 250-66.*			**Minimum Size Neutral Conductor**	**18**	
19 Size the Grounding Electrode Conductor Table 250-66 *Using line 18 to find the grounding electrode conductor in Table 250-66.* Size the Equipment Grounding Conductor (for Feeder) 250-122 *Use line 17 to find the equipment grounding conductor in Table 250-122.* *Equipment grounding conductor types are listed in Section 250-118.*			**Minimum Size Grounding Electrode Conductor . . . or . . . Equipment Grounding Conductor**	**19**	

Lines 1 Through 3—General Lighting and Receptacle Load

In the **Optional Method**, Lines 1, 2, and 3 are calculated exactly as described in the **Standard Method**. (See explanations on page 211.)

The feeder or service multi-family dwelling load can be computed in accordance with Table 220-32 (instead of Article 220, Part B) where all of the following conditions are met: (1) no dwelling unit is supplied by more than one feeder; (2) each dwelling unit is equipped with electric cooking equipment; and (3) each dwelling unit is equipped with either electric space heating or air conditioning, or both ≫220-32(a)≪.

Optional Method Load Calculation for Multi-Family Dwellings

1 General Lighting and Receptacle Loads 220-32(c)(2) *Do not include open porches, garages, unused or unfinished spaces not adaptable for future use.*	3 × ___1,050___ (sq. ft. outside dimensions) × ___12___ (number of units) =	**1**	*37,800*
2 Small Appliance Branch Circuits 220-32(c)(1) *At least **two** small appliance branch circuits must be included. 210-11(c)(1)*	1,500 × ___2___ (minimum of two) × ___12___ (number of units) =	**2**	*36,000*
3 Laundry Branch Circuit(s) 220-32(c)(1) *At least **one** laundry branch circuit must be included. 210-11(c)(2)*	1,500 × ___1___ (*minimum of two) × ___12___ (number of units) =	**3**	*18,000*

> **N O T E**
>
> *When the computed load for multi-family dwellings, without electric cooking, in Article 220, Part B exceeds that computed under Part C for the identical load, plus electric cooking (based on 8 kW per unit), use the lesser of the two loads ≫220-32(a) 2. Exception≪.*

This optional load calculation is computed on the same 12-unit multi-family dwelling used for the standard method (located in the first part of this unit).

■ Lines 4 through 11—All Appliances

This Section includes appliances (fastened-in-place, permanently connected, or connected to a specific circuit), ranges, wall-mounted ovens, counter-mounted cooking units, clothes dryers, and water heaters **»220-32(c)(3)«**.

Low-power-factor loads and motors (except air-conditioning motors), are also included **»220-32(c)(4)«**.

Use the exact clothes dryer nameplate rating, if any, even if less than 5,000 volt-amperes. Do not apply the demand factors in Table 220-19 to household cooking equipment (ranges, cooktops, ovens, etc.); Instead, list the nameplate ratings as they appear.

Insert the number of appliances (such as water heaters) having the same rating, which may not necessarily be the number of dwelling units. Here, all dwelling units contain identical equipment, so the number of each type of appliance is the same as the number of dwelling units. In another application, if only half of the units are equipped with laundry hook-ups, the number of units is not the same as the number of a given type of appliance.

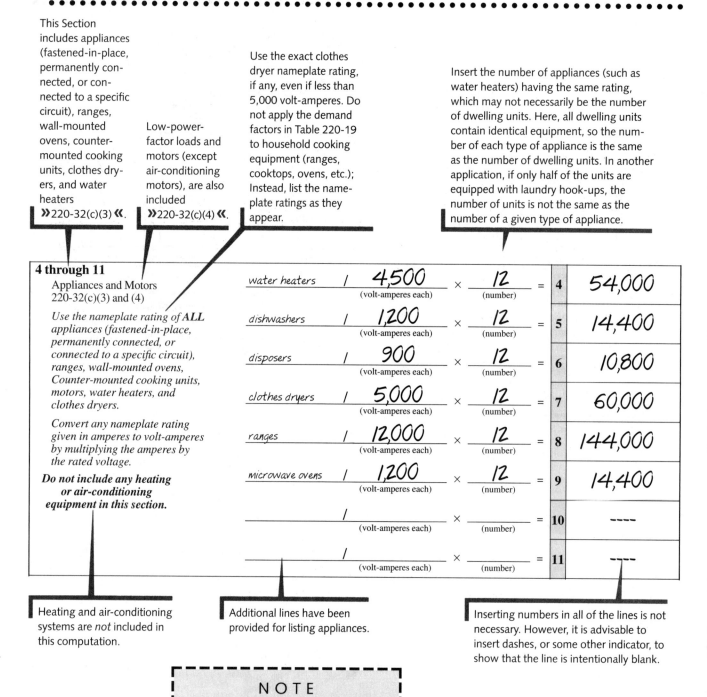

4 through 11

Appliances and Motors
220-32(c)(3) and (4)

*Use the nameplate rating of **ALL** appliances (fastened-in-place, permanently connected, or connected to a specific circuit), ranges, wall-mounted ovens, Counter-mounted cooking units, motors, water heaters, and clothes dryers.*

Convert any nameplate rating given in amperes to volt-amperes by multiplying the amperes by the rated voltage.

Do not include any heating or air-conditioning equipment in this section.

	(volt-amperes each)		(number)			
water heaters	/ 4,500	×	12	=	4	54,000
dishwashers	/ 1,200	×	12	=	5	14,400
disposers	/ 900	×	12	=	6	10,800
clothes dryers	/ 5,000	×	12	=	7	60,000
ranges	/ 12,000	×	12	=	8	144,000
microwave ovens	/ 1,200	×	12	=	9	14,400
	/	×		=	10	----
	/	×		=	11	----

Heating and air-conditioning systems are *not* included in this computation.

Additional lines have been provided for listing appliances.

Inserting numbers in all of the lines is not necessary. However, it is advisable to insert dashes, or some other indicator, to show that the line is intentionally blank.

NOTE

If water-heater elements are interlocked so that all elements cannot be activated simultaneously, the nameplate load is considered the maximum possible load **»220-32(c)(3)«**.

■ Line 12—Heating or Air-Conditioning System

Include only the larger of the heating or air-conditioning loads (except for heat pumps).

Electric heating loads (including associated motors) are computed at 100% of the total connected load. Because the air handler (blower) works with either the heating or the air conditioning, add it to both calculations.

Air-conditioning loads (including associated motors) are calculated at 100% of the total connected load. No derating is allowed. Each unit contains an air-conditioner compressor (14 × 230 = 3,220 VA), condenser fan motor (2 × 115 = 230 VA), and blower motor (3 × 115 = 345 VA). The combined load is 3,795 volt-amperes. Because this is less than the heating load, omit the A/C compressor and the condenser fan motor.

12 Heating or Air-Conditioning System (Compare the heat and A/C, and omit the smaller.) 220-32(c)(5)
Include the air handler when using either one.
For heat pumps, include the compressor and the
maximum amount of electric heat which can be
energized while the compressor is running.

$$\underline{5,345} \times \underline{12} =$$
(volt-amperes each)   (number)

12 **64,140**

> **N O T E**
> *A heat pump with supplementary heat is not considered a*
> *noncoincident load. Add the compressor full-load current*
> *to the maximum amount of heat that can be energized*
> *while the compressor is running. Remember to include*
> *all associated motor(s), i.e., blower, condenser, etc.*

The air handler (blower motor) has a rating of 345 volt-amperes (3 × 115 = 345). Since both loads are energized simultaneously, add the air handler to the 5 kW of electric heat (345 + 5,000 = 5,345 volt-amperes).

■ Line 13—Total Volt-Ampere Demand Load

Adding Lines 1 through 12 results in the volt-ampere load before derating.

Table 220-32 contains demand factors (percentages) for derating three (or more) multi-family dwelling units. Find the number of dwelling units in the first column and follow that row across to the demand factor. The demand factors listed in the second column are percentages. The demand factor for a 12-unit multi-family dwelling is 41 (41%).

Total computed load excluding house loads (if any).

13 Total Volt-Ampere Demand Load:
Multiply total VA by Table 220-32
demand factor percent.

$$\underline{453,540} + \underline{41\%} =$$
(total volt-amperes from lines 1 through 12)   (Table 220-32 demand factor)

13 **185,951**

■ Line 14—House Load

House loads must be computed as stipulated in Article 220, Part B, and must be in addition to the dwelling-unit loads computed in accordance with Table 220-32 »220-32(b)«.

House loads include areas such as: clubhouse, office, hallway and stairway lighting, parking lot lighting, etc.

Insert a line (or dashes) in spaces left empty intentionally.

14 House Load *(If present, otherwise skip to line 15)*
Compute in accordance with Article 220, Part B.
Do not include in Table 220-32 demand factors.

14 ----

■ Lines 15 and 16—Minimum Service and/or Feeder

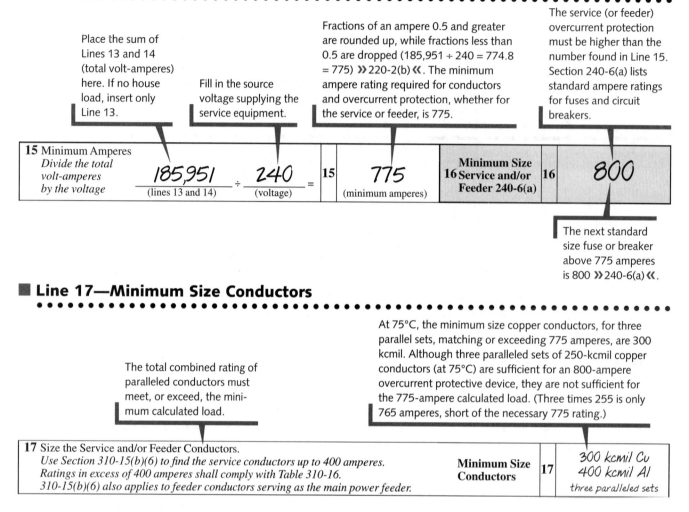

Place the sum of Lines 13 and 14 (total volt-amperes) here. If no house load, insert only Line 13.

Fill in the source voltage supplying the service equipment.

Fractions of an ampere 0.5 and greater are rounded up, while fractions less than 0.5 are dropped (185,951 ÷ 240 = 774.8 = 775) ≫ 220-2(b) ≪. The minimum ampere rating required for conductors and overcurrent protection, whether for the service or feeder, is 775.

The service (or feeder) overcurrent protection must be higher than the number found in Line 15. Section 240-6(a) lists standard ampere ratings for fuses and circuit breakers.

| 15 Minimum Amperes *Divide the total volt-amperes by the voltage* | $\dfrac{185,951}{\text{(lines 13 and 14)}} \div \dfrac{240}{\text{(voltage)}} =$ | 15 | 775 (minimum amperes) | Minimum Size 16 Service and/or Feeder 240-6(a) | 16 | 800 |

The next standard size fuse or breaker above 775 amperes is 800 ≫ 240-6(a) ≪.

■ Line 17—Minimum Size Conductors

The total combined rating of paralleled conductors must meet, or exceed, the minimum calculated load.

At 75°C, the minimum size copper conductors, for three parallel sets, matching or exceeding 775 amperes, are 300 kcmil. Although three paralleled sets of 250-kcmil copper conductors (at 75°C) are sufficient for an 800-ampere overcurrent protective device, they are not sufficient for the 775-ampere calculated load. (Three times 255 is only 765 amperes, short of the necessary 775 rating.)

| 17 Size the Service and/or Feeder Conductors. *Use Section 310-15(b)(6) to find the service conductors up to 400 amperes. Ratings in excess of 400 amperes shall comply with Table 310-16. 310-15(b)(6) also applies to feeder conductors serving as the main power feeder.* | Minimum Size Conductors | 17 | 300 kcmil Cu 400 kcmil Al *three paralleled sets* |

▪ Minimum Size Neutral Conductors

There is no optional method for neutral conductor calculation. This neutral calculation is performed in accordance with Article 220, Part B.

General lighting and receptacles (Line 9)........ 34,080
Fastened-in-place appliances
    Water heaters...................... 0
    Dishwashers (1,200 × 12)....... 14,400
    Waste disposers (7.5 × 120 × 12). 10,800
    Microwave ovens (1,200 × 12) .. 14,400
    Total 39,600

Line 9 on the Multi-family Dwelling Standard Method Load Calculation form.

Lines 11 and 12 from the Standard Method Load Calculation form for Multi-family Dwellings.

Total fastened-in-place appliances (39,600 × 75%) . 29,700
Clothes dryers (Line 11 × 70%)............... 18,900
Ranges (Line 12 × 70%)..................... 18,900
Electric heat 0
Air-conditioner compressors 0
Air handlers (3 × 115 × 12) 4,140
Condenser fan motors (2 × 115 × 12) 2,760
Largest motor (900 × 25%)................... 225
TOTAL 108,705

The minimum computed volt-ampere load is 108,705.

108,705 ÷ 240 = 452.9 = 453 minimum neutral ampacity
453 ÷ 3 = 151 minimum amperes per neutral conductor

At 75°C, the minimum size copper conductor is No. 2/0, with an individual rating of 175 amperes ≫ Table 310-16 ≪.

The minimum neutral ampacity is 453.

The minimum rating for three paralleled sets of neutral conductors is 151 amperes each.

18 Size the Neutral Conductor 220-22 **Note: There is no optional method for calculating the neutral conductor.** *Section 310-15(b)(6) states that the neutral service and/or feeder conductor can be smaller than the ungrounded (hot) conductors, provided the requirements of Sections 215-2, 220-22, and 230-42 are met.* *Section 250-24(b)(1) states that the neutral cannot be smaller than the required grounding electrode conductor specified in Table 250-66.*	**Minimum Size Neutral Conductor** **18**	*2/0 Cu* *4/0 Al* *three paralleled sets*

At 75°C, the minimum size aluminum conductor, if determined by ampacity alone, would be No. 3/0, with a rating of 155 amperes each. However, No. 3/0 would not be acceptable because the neutral cannot be smaller than the required grounding electrode conductor, which in this example is 4/0 ≫ 250-24(b)(1) ≪.

▪ Line 19—Grounding Electrode Conductor

The grounding electrode conductor size is determined by the largest service-entrance conductor (or the equivalent area) if the service-entrance conductors are paralleled. The equivalent area for three parallel sets in this calculation is 900 kcmil copper (3 × 300) or 1,200 kcmil aluminum (3 × 400).

The minimum size grounding electrode conductor is No. 2/0 copper *or* No. 4/0 aluminum ≫ Table 250-66 ≪.

19 Size the Grounding Electrode Conductor Table 250-66 *Using line 18 to find the grounding electrode conductor in Table 250-66.* Size the Equipment Grounding Conductor (for Feeder) 250-122 *Use line 17 to find the equipment grounding conductor in Table 250-122.* *Equipment grounding conductor types are listed in Section 250-118.*	**Minimum Size Grounding Electrode Conductor ... or ... Equipment Grounding Conductor** **19**	*2/0 Cu* *4/0 Al*

The following page shows the complete optional method for a twelve-unit, multi-family dwelling, having all units combined into a single service.

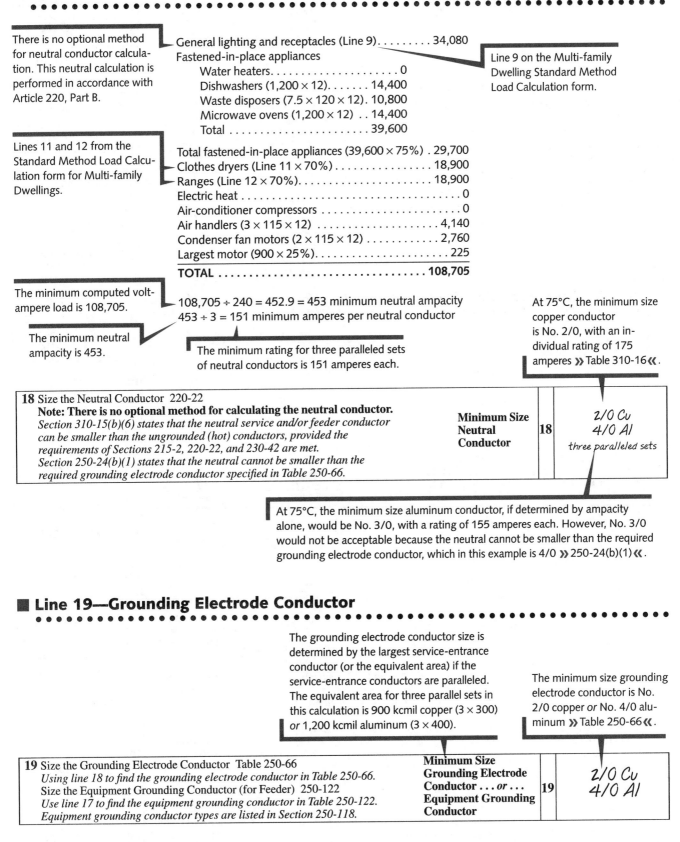

Optional Method Load Calculation for Multi-Family Dwellings

1 General Lighting and Receptacle Loads 220-32(c)(2) *Do not include open porches, garages, unused or* *unfinished spaces not adaptable for future use.*	$3 \times$ __1,050__ (sq. ft. outside dimensions)	$\times$ __12__ (number of units)	$=$ **1**	**37,800**
2 Small Appliance Branch Circuits 220-32(c)(1) *At least **two** small appliance branch circuits* *must be included. 210-11(c)(1)*	$1,500 \times$ __2__ (minimum of two)	$\times$ __12__ (number of units)	$=$ **2**	**36,000**
3 Laundry Branch Circuit(s) 220-32(c)(1) *At least **one** laundry branch circuit* *must be included. 210-11(c)(2)*	$1,500 \times$ __1__ (*minimum of two)	$\times$ __12__ (number of units)	$=$ **3**	**18,000**

4 through 11					
Appliances and Motors 220-32(c)(3) and (4)	water heaters	1 __4,500__ (volt-amperes each)	$\times$ __12__ (number)	$=$ **4**	**54,000**
*Use the nameplate rating of **ALL*** *appliances (fastened-in-place,* *permanently connected, or*	dishwashers	1 __1,200__ (volt-amperes each)	$\times$ __12__ (number)	$=$ **5**	**14,400**
connected to a specific circuit), *ranges, wall-mounted ovens,* *Counter-mounted cooking units,*	disposers	1 __900__ (volt-amperes each)	$\times$ __12__ (number)	$=$ **6**	**10,800**
motors, water heaters, and *clothes dryers.*	clothes dryers	1 __5,000__ (volt-amperes each)	$\times$ __12__ (number)	$=$ **7**	**60,000**
Convert any nameplate rating *given in amperes to volt-amperes* *by multiplying the amperes by* *the rated voltage.*	ranges	1 __12,000__ (volt-amperes each)	$\times$ __12__ (number)	$=$ **8**	**144,000**
Do not include any heating **or air-conditioning** **equipment in this section.**	microwave ovens	1 __1,200__ (volt-amperes each)	$\times$ __12__ (number)	$=$ **9**	**14,400**
		1 _____ (volt-amperes each)	$\times$ _____ (number)	$=$ **10**	**----**
		1 _____ (volt-amperes each)	$\times$ _____ (number)	$=$ **11**	**----**

12 Heating or Air-Conditioning System (Compare the heat and A/C, and omit the smaller.) 220-32(c)(5) *Include the air handler when using either one.* *For heat pumps, include the compressor and the* *maximum amount of electric heat which can be* *energized while the compressor is running.*	__5,345__ $\times$ __12__ (volt-amperes each) (number)	$=$ **12**	**64,140**

13 Total Volt-Ampere Demand Load: *Multiply total VA by Table 220-32* *demand factor percent.*	__453,540__ $+$ __41%__ (total volt-amperes from lines 1 through 12) (Table 220-32 demand factor) $=$ **13**	**185,951**

14 House Load *(If present, otherwise skip to line 15)* *Compute in accordance with Article 220, Part B.* *Do not include in Table 220-32 demand factors.*	**14**	**----**

15 Minimum Amperes *Divide the total* *volt-amperes* *by the voltage*	__185,951__ $\div$ __240__ **15** (lines 13 and 14) (voltage) $=$	**775** (minimum amperes)	**16** Minimum Size Service and/or Feeder 240-6(a)	**16** **800**

17 Size the Service and/or Feeder Conductors. *Use Section 310-15(b)(6) to find the service conductors up to 400 amperes.* *Ratings in excess of 400 amperes shall comply with Table 310-16.* *310-15(b)(6) also applies to feeder conductors serving as the main power feeder.*	**Minimum Size** **Conductors**	**17**	**300 kcmil Cu** **400 kcmil Al** *three paralleled sets*
18 Size the Neutral Conductor 220-22 **Note: There is no optional method for calculating the neutral conductor.** *Section 310-15(b)(6) states that the neutral service and/or feeder conductor* *can be smaller than the ungrounded (hot) conductors, provided the* *requirements of Sections 215-2, 220-22, and 230-42 are met.* *Section 250-24(b)(1) states that the neutral cannot be smaller than the* *required grounding electrode conductor specified in Table 250-66.*	**Minimum Size** **Neutral** **Conductor**	**18**	**2/0 Cu** **4/0 Al** *three paralleled sets*
19 Size the Grounding Electrode Conductor Table 250-66 *Using line 18 to find the grounding electrode conductor in Table 250-66.* Size the Equipment Grounding Conductor (for Feeder) 250-122 *Use line 17 to find the equipment grounding conductor in Table 250-122.* *Equipment grounding conductor types are listed in Section 250-118.*	**Minimum Size** **Grounding Electrode** **Conductor . . . or . . .** **Equipment Grounding** **Conductor**	**19**	**2/0 Cu** **4/0 Al**

SIX-UNIT MULTI-FAMILY DWELLING CALCULATION (OPTIONAL METHOD)

■ Lines 1 Through 12

This optional method calculation is based on six units instead of twelve.

Optional Method Load Calculation for Multi-Family Dwellings

1 General Lighting and Receptacle Loads 220-32(c)(2) *Do not include open porches, garages, unused or unfinished spaces not adaptable for future use.*	$3 \times \underline{1,050}$ (sq. ft. outside dimensions) $\times \underline{6}$ (number of units) =	**1**	*18,900*
2 Small Appliance Branch Circuits 220-32(c)(1) *At least two small appliance branch circuits must be included. 210-11(c)(1)*	$1,500 \times \underline{2}$ (minimum of two) $\times \underline{6}$ (number of units) =	**2**	*18,000*
3 Laundry Branch Circuit(s) 220-32(c)(1) *At least one laundry branch circuit must be included. 210-11(c)(2)*	$1,500 \times \underline{1}$ (*minimum of two) $\times \underline{6}$ (number of units) =	**3**	*9,000*

4 through 11

Appliances and Motors
220-32(c)(3) and (4)

Use the nameplate rating of ALL appliances (fastened-in-place, permanently connected, or connected to a specific circuit), ranges, wall-mounted ovens, Counter-mounted cooking units, motors, water heaters, and clothes dryers.

Convert any nameplate rating given in amperes to volt-amperes by multiplying the amperes by the rated voltage.

Do not include any heating or air-conditioning equipment in this section.

water heaters	/ $\underline{4,500}$ (volt-amperes each)	$\times \underline{6}$ (number) =	**4**	*27,000*
dishwashers	/ $\underline{1,200}$ (volt-amperes each)	$\times \underline{6}$ (number) =	**5**	*7,200*
disposers	/ $\underline{900}$ (volt-amperes each)	$\times \underline{6}$ (number) =	**6**	*5,400*
clothes dryers	/ $\underline{5,000}$ (volt-amperes each)	$\times \underline{6}$ (number) =	**7**	*30,000*
ranges	/ $\underline{12,000}$ (volt-amperes each)	$\times \underline{6}$ (number) =	**8**	*72,000*
microwave ovens	/ $\underline{1,200}$ (volt-amperes each)	$\times \underline{6}$ (number) =	**9**	*7,200*
	/ _____ (volt-amperes each)	$\times$ _____ (number) =	**10**	----
	/ _____ (volt-amperes each)	$\times$ _____ (number) =	**11**	----

12 Heating or Air-Conditioning System (Compare the heat and A/C, and omit the smaller.) 220-32(c)(5) *Include the air handler when using either one. For heat pumps, include the compressor and the maximum amount of electric heat which can be energized while the compressor is running.*	$\underline{5,345}$ (volt-amperes each) $\times \underline{6}$ (number) =	**12**	*32,070*

Since everything is computed at 100%, Lines 1 through 12 are exactly half of that calculated for twelve units.

■ Line 13—Total Volt-Ampere Demand Load

Add Lines 1 through 12 to find the volt-ampere load before derating.

Find the demand factor, located in Table 220-32, and insert here. The demand factor for a six-unit multi-family dwelling is 44 (44%).

13 Total Volt-Ampere Demand Load: *Multiply total VA by Table 220-32 demand factor percent.*	$\underline{226,770}$ (total volt-amperes from lines 1 through 12) $+ \underline{44\%}$ (Table 220-32 demand factor) =	**13**	*99,779*

■ Lines 15 and 16—Minimum Service and/or Feeder

Place the sum of Lines 13 and 14 (total volt-amperes) here. If no house load, insert only Line 13.

Insert the source voltage supplying the service equipment.

The minimum amperage rating required for conductors and overcurrent protection is 416 (99,779 ÷ 240 = 415.75 = 416).

The minimum service (or feeder), determined by the optional method, for these six units is 450 amperes.

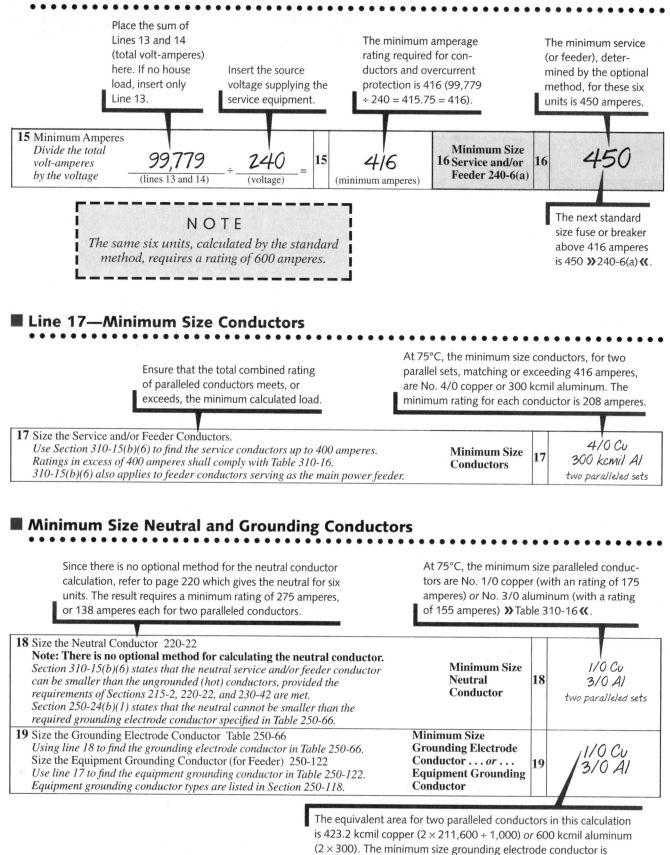

| **15** Minimum Amperes *Divide the total volt-amperes by the voltage* | $\dfrac{99{,}779}{\text{(lines 13 and 14)}} \div \dfrac{240}{\text{(voltage)}} =$ | **15** | *416* (minimum amperes) | **16** Minimum Size Service and/or Feeder 240-6(a) | **16** | *450* |

NOTE
The same six units, calculated by the standard method, requires a rating of 600 amperes.

The next standard size fuse or breaker above 416 amperes is 450 ≫ 240-6(a) ≪.

■ Line 17—Minimum Size Conductors

Ensure that the total combined rating of paralleled conductors meets, or exceeds, the minimum calculated load.

At 75°C, the minimum size conductors, for two parallel sets, matching or exceeding 416 amperes, are No. 4/0 copper or 300 kcmil aluminum. The minimum rating for each conductor is 208 amperes.

| **17** Size the Service and/or Feeder Conductors. *Use Section 310-15(b)(6) to find the service conductors up to 400 amperes. Ratings in excess of 400 amperes shall comply with Table 310-16. 310-15(b)(6) also applies to feeder conductors serving as the main power feeder.* | Minimum Size Conductors | **17** | *4/0 Cu 300 kcmil Al two paralleled sets* |

■ Minimum Size Neutral and Grounding Conductors

Since there is no optional method for the neutral conductor calculation, refer to page 220 which gives the neutral for six units. The result requires a minimum rating of 275 amperes, or 138 amperes each for two paralleled conductors.

At 75°C, the minimum size paralleled conductors are No. 1/0 copper (with an rating of 175 amperes) *or* No. 3/0 aluminum (with a rating of 155 amperes) ≫ Table 310-16 ≪.

| **18** Size the Neutral Conductor 220-22 **Note: There is no optional method for calculating the neutral conductor.** *Section 310-15(b)(6) states that the neutral service and/or feeder conductor can be smaller than the ungrounded (hot) conductors, provided the requirements of Sections 215-2, 220-22, and 230-42 are met. Section 250-24(b)(1) states that the neutral cannot be smaller than the required grounding electrode conductor specified in Table 250-66.* | Minimum Size Neutral Conductor | **18** | *1/0 Cu 3/0 Al two paralleled sets* |
| **19** Size the Grounding Electrode Conductor Table 250-66 *Using line 18 to find the grounding electrode conductor in Table 250-66.* Size the Equipment Grounding Conductor (for Feeder) 250-122 *Use line 17 to find the equipment grounding conductor in Table 250-122. Equipment grounding conductor types are listed in Section 250-118.* | Minimum Size Grounding Electrode Conductor . . . *or* . . . Equipment Grounding Conductor | **19** | *1/0 Cu 3/0 Al* |

The equivalent area for two paralleled conductors in this calculation is 423.2 kcmil copper (2 × 211,600 ÷ 1,000) *or* 600 kcmil aluminum (2 × 300). The minimum size grounding electrode conductor is No. 1/0 copper *or* No. 3/0 aluminum ≫ Table 250-66 ≪.

The following page shows the complete optional method for a six-unit, multi-family dwelling load calculation.

Optional Method Load Calculation for Multi-Family Dwellings

1 General Lighting and Receptacle Loads 220-32(c)(2) *Do not include open porches, garages, unused or* *unfinished spaces not adaptable for future use.*		3 × __1,050__ (sq. ft. outside dimensions)	× __6__ (number of units)	=	**1**	*18,900*
2 Small Appliance Branch Circuits 220-32(c)(1) *At least **two** small appliance branch circuits* *must be included. 210-11(c)(1)*		1,500 × __2__ (minimum of two)	× __6__ (number of units)	=	**2**	*18,000*
3 Laundry Branch Circuit(s) 220-32(c)(1) *At least **one** laundry branch circuit* *must be included. 210-11(c)(2)*		1,500 × __1__ (*minimum of two)	× __6__ (number of units)	=	**3**	*9,000*
4 through 11 Appliances and Motors 220-32(c)(3) and (4) *Use the nameplate rating of **ALL** appliances (fastened-in-place, permanently connected, or connected to a specific circuit), ranges, wall-mounted ovens, Counter-mounted cooking units, motors, water heaters, and clothes dryers.* *Convert any nameplate rating given in amperes to volt-amperes by multiplying the amperes by the rated voltage.* **Do not include any heating or air-conditioning equipment in this section.**	water heaters 1 __4,500__ (volt-amperes each)		× __6__ (number)	=	**4**	*27,000*
	dishwashers 1 __1,200__ (volt-amperes each)		× __6__ (number)	=	**5**	*7,200*
	disposers 1 __900__ (volt-amperes each)		× __6__ (number)	=	**6**	*5,400*
	clothes dryers 1 __5,000__ (volt-amperes each)		× __6__ (number)	=	**7**	*30,000*
	ranges 1 __12,000__ (volt-amperes each)		× __6__ (number)	=	**8**	*72,000*
	microwave ovens 1 __1,200__ (volt-amperes each)		× __6__ (number)	=	**9**	*7,200*
	__1__ _____ (volt-amperes each)		× _____ (number)	=	**10**	----
	__1__ _____ (volt-amperes each)		× _____ (number)	=	**11**	----
12 Heating or Air-Conditioning System (Compare the heat and A/C, and omit the smaller.) 220-32(c)(5) *Include the air handler when using either one.* *For heat pumps, include the compressor and the maximum amount of electric heat which can be energized while the compressor is running.*		__5,345__ (volt-amperes each)	× __6__ (number)	=	**12**	*32,070*
13 Total Volt-Ampere Demand Load: *Multiply total VA by Table 220-32 demand factor percent.*		__226,770__ (total volt-amperes from lines 1 through 12)	+ __44%__ (Table 220-32 demand factor)	=	**13**	*99,779*
14 House Load *(If present, otherwise skip to line 15)* *Compute in accordance with Article 220, Part B.* *Do not include in Table 220-32 demand factors.*					**14**	----

15 Minimum Amperes *Divide the total volt-amperes by the voltage*	__99,779__ ÷ __240__ = **15** (lines 13 and 14) (voltage)	__416__ (minimum amperes)	**Minimum Size 16 Service and/or Feeder 240-6(a)** **16** *450*

17 Size the Service and/or Feeder Conductors. *Use Section 310-15(b)(6) to find the service conductors up to 400 amperes. Ratings in excess of 400 amperes shall comply with Table 310-16. 310-15(b)(6) also applies to feeder conductors serving as the main power feeder.*	**Minimum Size Conductors**	**17**	*4/0 Cu 300 kcmil Al two paralleled sets*
18 Size the Neutral Conductor 220-22 **Note: There is no optional method for calculating the neutral conductor.** *Section 310-15(b)(6) states that the neutral service and/or feeder conductor can be smaller than the ungrounded (hot) conductors, provided the requirements of Sections 215-2, 220-22, and 230-42 are met. Section 250-24(b)(1) states that the neutral cannot be smaller than the required grounding electrode conductor specified in Table 250-66.*	**Minimum Size Neutral Conductor**	**18**	*1/0 Cu 3/0 Al two paralleled sets*
19 Size the Grounding Electrode Conductor Table 250-66 *Using line 18 to find the grounding electrode conductor in Table 250-66.* Size the Equipment Grounding Conductor (for Feeder) 250-122 *Use line 17 to find the equipment grounding conductor in Table 250-122.* *Equipment grounding conductor types are listed in Section 250-118.*	**Minimum Size Grounding Electrode Conductor . . . or . . . Equipment Grounding Conductor**	**19**	*1/0 Cu 3/0 Al*

OPTIONAL LOAD CALCULATION FOR EACH MULTI-FAMILY DWELLING UNIT

■ Optional Method—One Unit

The first three lines of both methods (optional and standard) of one-family dwelling calculations are identical.

Although located in a multi-family dwelling, this unit contains one feeder and one panelboard. Therefore, it is computed in accordance with a one-family dwelling method.

Optional Method Load Calculation for One-Family Dwellings

1	General Lighting and Receptacle Loads 220-30(b)(2) *Do not include open porches, garages, unused or unfinished spaces not adaptable for future use.*	3 × ___1,050___ (sq. ft. using outside dimensions)	= 1	3,150
2	Small Appliance Branch Circuits 220-30(b)(1) *Include at least **two** small appliance branch circuits. 210-11(c)(1)*	1,500 × ___2___ (minimum of two)	= 2	3,000
3	Laundry Branch Circuit(s) 220-30(b)(1) *Include at least **one** laundry branch circuit. 210-11(c)(2)*	1,500 × ___1___ (minimum of one)	= 3	1,500

Under certain conditions, a laundry branch circuit is optional in multi-family dwellings 》210-52(f)《.

■ Line 4—Appliances

This Section includes appliances (fastened-in-place, permanently connected, or connected to a specific circuit), ranges, wall-mounted ovens, counter-mounted cooking units, clothes dryers, and water heaters 》[220-30(b)(3)《.

Low-power-factor loads and motors (except air-conditioning motors) are also found in this Section 》220-30(b)(4)《.

Heating and air-conditioning systems are **not** part of this computation.

Use the exact nameplate rating of the clothes dryer, even if less than 5,000 volt-amperes.

| 4 | Appliances 220-30(b)(3) and (4) *Use the nameplate rating of **ALL** appliances (fastened-in-place, permanently connected, or connected to a specific circuit), ranges, ovens, cook tops, motors, and clothes dryers. Convert any nameplate rating given in amperes to volt-amperes by multiplying the amperes by the rated voltage.* | *Do not include any heating or air-conditioning equipment in this section.* | Total volt-amperes of all appliances LISTED BELOW | 4 | 24,800 |

water heater / 4,500	clothes dryer / 5,500	range / 12,000
dishwasher / 1,200	disposer / 900	microwave oven / 1,200
/	/	/
/	/	/
/	/	/

> **NOTE**
> *Do not apply the demand factors in Table 220-19 to household cooking equipment (ranges, cook tops, and ovens). Instead, list the nameplate ratings as they appear.*

> **NOTE**
> *Do not increase motor loads by 25%* 》220-30(b)(4)《.

■ Line 5—Applying Demand Factors to Lines 1 through 4

This demand factor is found in the last two lines of Table 220-30.

5 Apply Section 220-30(b) demand factor to the total of lines 1 through 4.

$$\underline{\hspace{1cm}32{,}450\hspace{1cm}} - 10{,}000 = \underline{\hspace{1cm}22{,}450\hspace{1cm}} \times 40\% = \underline{\hspace{1cm}8{,}980\hspace{1cm}} + 10{,}000 = \boxed{5 \quad 18{,}980}$$

(total of lines 1 through 4)

The newly calculated load includes everything except the heating and air-conditioning systems. (3,152 + 3,000 + 1,500 + 24,800)

Because the first 10,000 volt-amperes are calculated at 100%, 10,000 is subtracted here.

The remainder of the load is calculated at 40%.

The result on Line 5 will be part of Line 7's computation.

■ Line 6—Heating and/or Air-Conditioning Systems

The air-conditioner compressor (14 × 230 = 3,220 VA), condenser fan motor (2 × 115 = 230 VA), and blower motor (3 × 115 = 345VA) have a combined load of 3,795 volt-amperes. Calculate the air-conditioning system(s) at 100%.

The electric heat (5,000) added to the blower motor (3 × 115 = 345) is 5,345 volt-amperes. The heating system is multiplied by 65% and the result placed on Line 6c.

6 Heating and/or Air-Conditioning System 220-30(c)
*Use the nameplate rating(s) in volt-amperes for all applicable systems in lines **a** through **e**.*

a) Air-conditioning and cooling system(s), including heat pump compressors and supplemental heating unless the controller prevents the compressor and supplemental heating from operating at the same time:
$$\underline{\hspace{1cm}3{,}795\hspace{1cm}} \times 100\% = \boxed{\text{a)} \quad 3{,}795}$$

b) Electric thermal storage and other heating systems where the usual load is expected to be continuous at full nameplate value: *Systems qualifying under this selection shall not be figured under any other selection in 220-30(c).*
$$\underline{\hspace{1cm}} \times 100\% = \boxed{\text{b)}}$$

c) Central electric space heating equipment, including integral supplemental heating in heat pumps where the controller prevents the compressor and supplemental heating from operating at the same time:
$$\underline{\hspace{1cm}5{,}345\hspace{1cm}} \times 65\% = \boxed{\text{c)} \quad 3{,}474}$$

d) Electric space heating equipment, if less than four separately controlled units:
$$\underline{\hspace{1cm}} \times 65\% = \boxed{\text{d)}}$$

e) Electric space heating equipment, if four or more separately controlled units:
$$\underline{\hspace{1cm}} \times 40\% = \boxed{\text{e)}}$$

■ Line 7—Total Volt-Ampere Demand Load

The largest volt-ampere rating of the heating and/or air-conditioning system(s), after application of demand factors.

The volt-ampere demand load from Line 5.

The total volt-ampere load calculated by the optional method. (For comparison, this one-family dwelling, calculated by the standard method, is 29,048 volt-amperes.)

7 Total Volt-Ampere Demand Load:
$$\underline{\hspace{1cm}3{,}795\hspace{1cm}} + \underline{\hspace{1cm}18{,}980\hspace{1cm}} = \boxed{7 \quad 22{,}775}$$
(largest VA rating from line 6a through 6e) (line 5)

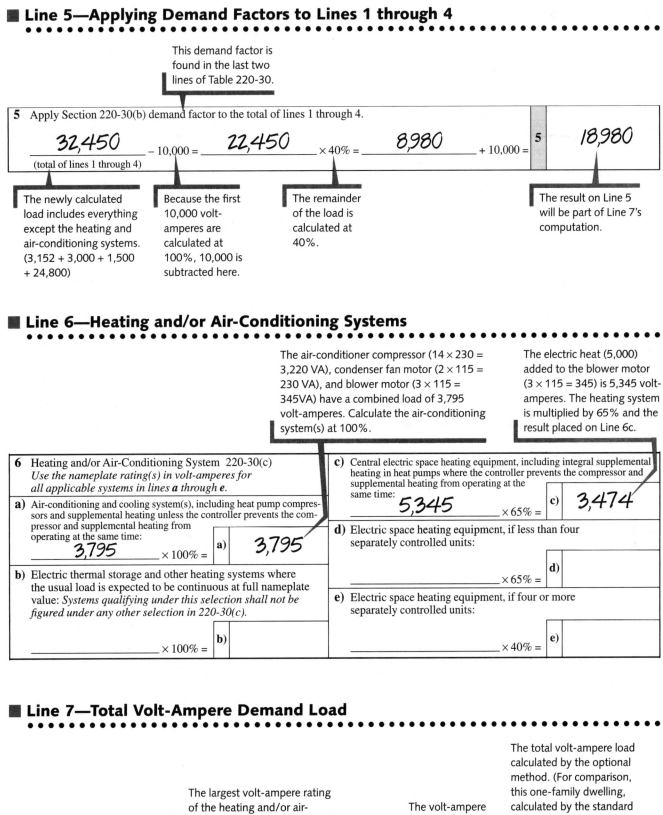

■ Lines 8 And 9—Minimum Feeder

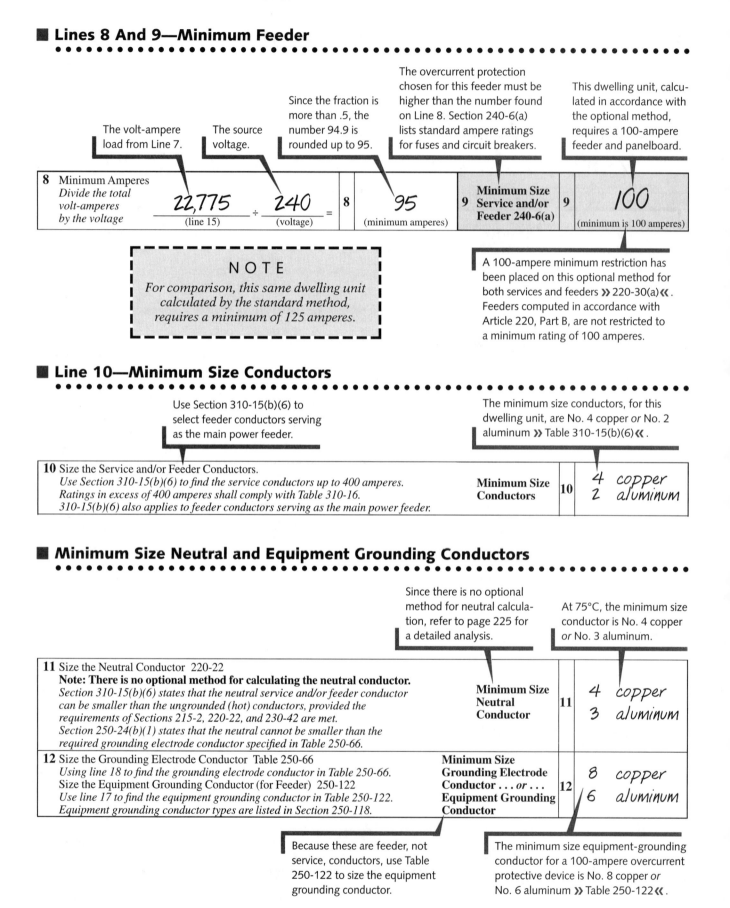

The volt-ampere load from Line 7.

The source voltage.

Since the fraction is more than .5, the number 94.9 is rounded up to 95.

The overcurrent protection chosen for this feeder must be higher than the number found on Line 8. Section 240-6(a) lists standard ampere ratings for fuses and circuit breakers.

This dwelling unit, calculated in accordance with the optional method, requires a 100-ampere feeder and panelboard.

| 8 | Minimum Amperes *Divide the total volt-amperes by the voltage* | $\dfrac{22{,}775}{\text{(line 15)}} \div \dfrac{240}{\text{(voltage)}} =$ | 8 | *95* (minimum amperes) | 9 | **Minimum Size Service and/or Feeder 240-6(a)** | 9 | *100* (minimum is 100 amperes) |

NOTE

For comparison, this same dwelling unit calculated by the standard method, requires a minimum of 125 amperes.

A 100-ampere minimum restriction has been placed on this optional method for both services and feeders ≫ 220-30(a) ≪. Feeders computed in accordance with Article 220, Part B, are not restricted to a minimum rating of 100 amperes.

■ Line 10—Minimum Size Conductors

Use Section 310-15(b)(6) to select feeder conductors serving as the main power feeder.

The minimum size conductors, for this dwelling unit, are No. 4 copper *or* No. 2 aluminum ≫ Table 310-15(b)(6) ≪.

| 10 | Size the Service and/or Feeder Conductors. *Use Section 310-15(b)(6) to find the service conductors up to 400 amperes. Ratings in excess of 400 amperes shall comply with Table 310-16. 310-15(b)(6) also applies to feeder conductors serving as the main power feeder.* | **Minimum Size Conductors** | 10 | *4 copper* *2 aluminum* |

■ Minimum Size Neutral and Equipment Grounding Conductors

Since there is no optional method for neutral calculation, refer to page 225 for a detailed analysis.

At 75°C, the minimum size conductor is No. 4 copper *or* No. 3 aluminum.

| 11 | Size the Neutral Conductor 220-22 **Note: There is no optional method for calculating the neutral conductor.** *Section 310-15(b)(6) states that the neutral service and/or feeder conductor can be smaller than the ungrounded (hot) conductors, provided the requirements of Sections 215-2, 220-22, and 230-42 are met. Section 250-24(b)(1) states that the neutral cannot be smaller than the required grounding electrode conductor specified in Table 250-66.* | **Minimum Size Neutral Conductor** | 11 | *4 copper* *3 aluminum* |
| 12 | Size the Grounding Electrode Conductor Table 250-66 *Using line 18 to find the grounding electrode conductor in Table 250-66. Size the Equipment Grounding Conductor (for Feeder) 250-122 Use line 17 to find the equipment grounding conductor in Table 250-122. Equipment grounding conductor types are listed in Section 250-118.* | **Minimum Size Grounding Electrode Conductor . . . or . . . Equipment Grounding Conductor** | 12 | *8 copper* *6 aluminum* |

Because these are feeder, not service, conductors, use Table 250-122 to size the equipment grounding conductor.

The minimum size equipment-grounding conductor for a 100-ampere overcurrent protective device is No. 8 copper *or* No. 6 aluminum ≫ Table 250-122 ≪.

The following page shows the complete optional method one-family dwelling load calculation for one multi-family dwelling-unit feeder and remote panelboard (subpanel).

Optional Method Load Calculation for One-Family Dwellings

1	General Lighting and Receptacle Loads 220-30(b)(2) *Do not include open porches, garages, unused or unfinished spaces not adaptable for future use.*	$3 \times$ ___**1,050**___ = *(sq. ft. using outside dimensions)*	**1** **3,150**
2	Small Appliance Branch Circuits 220-30(b)(1) *Include at least **two** small appliance branch circuits. 210-11(c)(1)*	$1,500 \times$ ___**2**___ = *(minimum of two)*	**2** **3,000**
3	Laundry Branch Circuit(s) 220-30(b)(1) *Include at least **one** laundry branch circuit. 210-11(c)(2)*	$1,500 \times$ ___**1**___ = *(minimum of one)*	**3** **1,500**

4 Appliances 220-30(b)(3) and (4)
*Use the nameplate rating of **ALL** appliances (fastened-in-place, permanently connected, or connected to a specific circuit), ranges, ovens, cook tops, motors, and clothes dryers.*
Convert any nameplate rating given in amperes to volt-amperes by multiplying the amperes by the rated voltage.

Do not include any heating or air-conditioning equipment in this section.

Total volt-amperes of all appliances LISTED BELOW

4 **24,800**

water heater / **4,500**	clothes dryer / **5,500**	range / **12,000**
dishwasher / **1,200**	disposer / **900**	microwave oven / **1,200**
/	/	/
/	/	/
/	/	/

5 Apply Section 220-30(b) demand factor to the total of lines 1 through 4.

___**32,450**___ $- 10,000 =$ ___**22,450**___ $\times 40\% =$ ___**8,980**___ $+ 10,000 =$
(total of lines 1 through 4)

5 **18,980**

6 Heating and/or Air-Conditioning System 220-30(c)
*Use the nameplate rating(s) in volt-amperes for all applicable systems in lines **a** through **e**.*

a) Air-conditioning and cooling system(s), including heat pump compressors and supplemental heating unless the controller prevents the compressor and supplemental heating from operating at the same time:
___**3,795**___ $\times 100\% =$ **a)** **3,795**

b) Electric thermal storage and other heating systems where the usual load is expected to be continuous at full nameplate value: *Systems qualifying under this selection shall not be figured under any other selection in 220-30(c).*

_____ $\times 100\% =$ **b)**

c) Central electric space heating equipment, including integral supplemental heating in heat pumps where the controller prevents the compressor and supplemental heating from operating at the same time:
___**5,345**___ $\times 65\% =$ **c)** **3,474**

d) Electric space heating equipment, if less than four separately controlled units:
_____ $\times 65\% =$ **d)**

e) Electric space heating equipment, if four or more separately controlled units:
_____ $\times 40\% =$ **e)**

7 Total Volt-Ampere Demand Load:
___**3,795**___ $+$ ___**18,980**___ $=$ **7** **22,775**
(largest VA rating from line 6a through 6e) *(line 5)*

8 Minimum Amperes *Divide the total volt-amperes by the voltage*
___**22,775**___ $\div$ ___**240**___ = **8** **95**
(line 15) *(voltage)* *(minimum amperes)*

9 **Minimum Size Service and/or Feeder 240-6(a)** **9** **100** *(minimum is 100 amperes)*

10 Size the Service and/or Feeder Conductors.
Use Section 310-15(b)(6) to find the service conductors up to 400 amperes. Ratings in excess of 400 amperes shall comply with Table 310-16. 310-15(b)(6) also applies to feeder conductors serving as the main power feeder.
Minimum Size Conductors **10** **4 copper** **2 aluminum**

11 Size the Neutral Conductor 220-22
Note: There is no optional method for calculating the neutral conductor.
Section 310-15(b)(6) states that the neutral service and/or feeder conductor can be smaller than the ungrounded (hot) conductors, provided the requirements of Sections 215-2, 220-22, and 230-42 are met. Section 250-24(b)(1) states that the neutral cannot be smaller than the required grounding electrode conductor specified in Table 250-66.
Minimum Size Neutral Conductor **11** **4 copper** **3 aluminum**

12 Size the Grounding Electrode Conductor Table 250-66
Using line 18 to find the grounding electrode conductor in Table 250-66.
Size the Equipment Grounding Conductor (for Feeder) 250-122
Use line 17 to find the equipment grounding conductor in Table 250-122. Equipment grounding conductor types are listed in Section 250-118.
Minimum Size Grounding Electrode Conductor ... or ... Equipment Grounding Conductor **12** **8 copper** **6 aluminum**

Summary

• •

- Multi-family dwelling calculations are a progression of the one-family dwelling calculations (standard and optional).
- Some of the information needed for each individual dwelling unit includes: total square-foot area, the number of small-appliance branch circuits and laundry branch-circuits (if any), appliance ratings, dryer ratings (if any), cooking equipment ratings, and heating and air-conditioning loads.
- Identifying the number of dwelling units supplied by each service and/or disconnecting means is essential to the performance of multi-family load calculations.
- In certain instances, the laundry branch circuit may be omitted from both the dwelling unit and the calculation.
- Laundry (if installed) and small appliance branch circuits are calculated using a rating of at least 1,500 volt-amperes each.
- Table 220-11 demand factors are applied to general lighting, small appliance, and laundry loads.
- The total volt-ampere rating of four (or more) fastened-in-place appliances is multiplied by 75% in the standard method calculation.

- Demand factors for household electric clothes dryers are found in Table 220-18.
- Table 220-19 lists the demand loads for household cooking appliances (over 1¾-kW rating).
- The heating and air-conditioning loads are compared, and the larger of the two is used.
- For heat pumps, include the compressor and the maximum amount of electric heat which can be energized while the compressor is running.
- Two multi-family load calculation methods are provided in Article 220: **Standard** and **Optional**.
- The neutral conductor can be smaller than the ungrounded (hot) conductors, provided the requirements of Sections 215-2, 220-22, and 230-42 are met.
- Table 250-66 is used to size grounding electrode conductors.
- Remote panelboards (sub-panels) require equipment grounding conductors, which can be one, or more, or a combination of the types listed in Section 250-118.
- Equipment grounding conductors are sized in accordance with Table 250-122.

Unit 11 Competency Test

• •

NEC® Reference   Answer

_____ _____ 1. One unit in a multi-family dwelling has a 125-ampere, 240-volt panelboard, fed with 75°C conductors. What is the minimum size aluminum equipment-grounding conductor required?

_____ _____ 2. What is the dryer demand load (in kW) for a 25-unit multi-family dwelling with a 4.5-kW clothes dryer in each unit? (The service is 120/240 volt, single-phase.)

_____ _____ 3. A four-unit apartment has the following ranges: a 15 kW, a 14 kW, a 10 kW, and a 9 kW. What is the kW demand load added to the service by these ranges?

_____ _____ 4. An 18-unit multi-family dwelling contains the following appliances in each unit: a 1-kVA, 115-volt dishwasher; a ⅓-HP, 115-volt waste (garbage) disposer; and a 4,500-watt, 230-volt water heater. Using the standard method, what is the service neutral load contribution, in volt-amperes, for these appliances? (Assume water heater watt rating equivalent to volt-ampere.)

_____ _____ 5. A multi-family dwelling's service-entrance conductors consist of four paralleled sets of No. 3/0 THHN copper conductors. What size copper grounding electrode conductor is required?

_____ _____ 6. A 5.4-kVA, 240-volt clothes dryer will contribute _____ amperes to the neutral load, when calculating the service by the standard method.

_____ _____ 7. What is the minimum kW service demand load for twenty 6.5-kW ranges in a multi-family dwelling?

8. An apartment has 3,000 watts allocated for general lighting and receptacles. How many 15-ampere circuits are required? (The apartment contains laundry facilities.)

9. Two paralleled sets of 500-kcmil copper service-entrance conductors supply a multi-family dwelling disconnecting means. What is the minimum size aluminum grounding electrode conductor required?

10. A 50-unit apartment building has a 4.5-kW clothes dryer in each apartment. By using the optional method, what is the service demand load?

Questions 11 through 18 are based on a 120/240 volt, 3-wire, single-phase, eight-unit multi-family dwelling. The eight identical units contain the following:

Floor area . . . . 1,200 sq. ft.	Electric heat . . . . 4.2 kW, 240 volt
Range . . . . 8¾ kW, 240 volt	Air handler (blower motor) . . . . ¼ HP, 115 volt
Water heater . . . . 4 kW, 240 volt	Air-conditioner compressor . . . . 3 HP, 230 volt
Dishwasher . . . . 1 kW, 120 volt	Condenser fan motor . . . . ⅙ HP, 115 volt
Clothes dryer . . . . 5 kW, 240 volt	

(Assume water heater, clothes dryer, range, and electric heat kW ratings equivalent to kVA.)

Using Article 220, Part B (the standard method):

11. What is the service demand load for the clothes dryers?

12. What is the service demand load for the cooking appliances?

13. Of the heating or air-conditioning (and associated motors), which is omitted?

14. What is the minimum rating (in amperes) for the service overcurrent device?

15. A parallel run (2 sets) of service-entrance conductors will be installed in two raceways. What is the minimum size THWN copper ungrounded conductors that can be used?

16. What is the minimum rating (in amperes) for the neutral conductors?

17. What is the minimum size THWN copper neutral (grounded) conductors that can be installed?

18. What is the minimum size copper grounding electrode conductor?

Questions 19 through 22 are based on a 120/240 volt, 3-wire, single-phase, six-unit multi-family dwelling. The six identical units contain the following:

Floor area . . . . 1,350 sq. ft.	Clothes dryer . . . . 4 kW, 240 volt
Range . . . . 10.6 kW, 240 volt	Electric heat (2 banks @ 3 kW each) . . 6 kW, 240 volt
Water heater . . . . 4.2 kW, 240 volt	Air handler (blower motor) . . . . ⅓ HP, 115 volt
Dishwasher . . . . 1.2 kW, 120 volt	Air-conditioner compressor . . . . 3 HP, 230 volt
Garbage disposer . . . . ⅓ HP, 115 volt	Condenser fan motor . . . . ⅙ HP, 115 volt

(Assume water heater, clothes dryer, range, and electric heat kW ratings equivalent to kVA.)

Using the Optional Method:

19. What is minimum rating (in amperes) for the service overcurrent device?

20. What is the minimum size THWN copper ungrounded conductors that can be installed? (Do not parallel the conductors.)

21. What is the minimum size THWN copper neutral (grounded) conductor that can be installed?

22. What is the minimum acceptable size copper grounding electrode conductor?

General Provisions

Objectives

• •

After studying this unit, the student will:

✔ know that commercial occupancy receptacle placement differs from one-family and multi-family dwellings.

✔ understand that commercial bathrooms do not require receptacle outlets.

✔ be aware that each commercial building (or occupancy) accessible to pedestrians must have at least one sign outlet.

✔ be familiar with Article 430 motor provisions.

✔ know that air-conditioning and refrigeration provisions are found in Article 440.

✔ be able to determine receptacle volt-ampere ratings for single, duplex, quad, etc.

✔ know the maximum number of receptacles permitted on 15- and 20-ampere branch circuits.

✔ be aware that at least two receptacle outlets must be readily accessible in guest rooms of hotels, motels, and similar occupancies.

✔ thoroughly understand showcase and show window provisions.

✔ be able to determine an occupancy's general lighting load based on square-foot area.

✔ be familiar with fluorescent, HID, recessed, and track-lighting provisions.

✔ understand that while certain lighting fixtures can be used as raceways, the number of branch circuits permitted therein is limited.

✔ know that a metal pole supporting a lighting fixture(s) requires a handhole, unless an exception is met.

Introduction

• •

The broad topic of commercial wiring encompasses many types of structures; office buildings, restaurants, stores, schools, and warehouses, to name a few. In fact, most of the occupancies found in Table 220-3(a) are indeed commercial. Many of the electrical requirements that apply to one-family and multi-family dwellings also apply to commercial and industrial locations. Because this overlap of requirements is common, it is especially important to begin with a solid understanding of not only some, but all of, the preceding material. Of particular importance are Units 3 through 5 of Section 1, which contain fundamental principles applicable to all types of occupancies. In addition, Unit 9 (Services, Feeders, and Electric Equipment) explains one-family dwelling provisions which extend into the realm of commercial wiring systems. Unit 12 (General Provisions) likewise covers overlapping topics, such as: branch circuits, receptacles, and lighting. Because it would be very difficult, if not impossible, to cover such a large field as commercial wiring in the framework of this text, only a small, but representative, portion is presented. The next unit explains non-dwelling (inclusive of commercial) load calculations, which are helpful in determining service and feeder loads. Complete subsequent units, and portions of others, apply to commercial wiring as well.

The inception of most commercial electrical projects includes multi-faceted plans (blueprints) and detailed specifications. As a reference, illustrations for a retail store's electrical and lighting floor plans are provided. Various general provisions (branch circuits, receptacles, lighting, etc.) and numerous specific provisions (required receptacle outlets, show windows, showcases, etc.) are discussed. Proper receptacle placement is demonstrated through the use of a hotel/motel guest-room floor plan, including furniture placement. Receptacle and lighting branch circuit load calculations are also part of this unit.

While this unit is by no means a complete reference for commercial electrical wiring, taken together with the extensive information presented throughout this text, it does provide a substantial foundation toward understanding commercial electrical wiring systems.

BRANCH CIRCUITS

■ Required Sign Outlet

Each commercial building and occupancy open to pedestrians must have at least one accessible outlet at every entrance to each tenant space for sign, or outline lighting systems, use ≫ 600-5(a)≪ .

The outlet(s) must be supplied by a branch circuit rated 20 amperes or more and that supplies no other load≫ ≫ 600-5(a)≪ .

Each sign and outline lighting system, or feeder or branch circuit supplying a sign or outline lighting system, must be controlled by an externally operable switch (or circuit breaker) that opens all ungrounded conductors, unless: (1) the sign is an exit directional sign located within a building; or (2) the sign is cord-connected with an attachment plug ≫ 600-6≪ .

Entry

20 ampere, 120 volt, sign outlet
(located above ceiling)

1'-6"   6'-4"   1'-0" 1'-0"   6'-4"   1'-6"

Sign and outline lighting outlets must be computed at a minimum of 1,200 volt-amperes per required branch circuit specified in Section 600-5(a) ≫ 220-3(b)(6)≪ .

> **NOTE**
> *Switches, flashers, and similar devices controlling transformers and electronic power supplies must either be rated for controlling inductive loads, or have a current rating not less than twice the transformer's current rating ≫ 600-6(b)≪ .*

The disconnecting means must be within sight of the sign or outline lighting system that it controls. If out of the line of sight from any section that may be energized, the disconnecting means must be capable of locking in the open position ≫ 600-6(a)≪ .

Signs or outline lighting systems operated by external electronic or electromechanical controllers can have a disconnecting means located within sight of the controller or within the controller enclosure. The disconnecting means must disconnect both the sign (or outline lighting system) and the controller from all ungrounded supply conductors. It must be designed so that no pole operates independently and it must be capable of locking in the open position ≫ 600-6(a)≪ .

> **NOTE**
> *Service hallways or corridors are not considered accessible to pedestrians ≫ 600-5(a)≪ .*

■ Commercial (Retail) Occupancy

● ●

No receptacle wall placement provisions exist for commercial occupancies. The "6 foot rule," which is the maximum distance to a receptacle (measured horizontally along the wall), applies only to dwelling units ▶▶ 210-52(a) ◀◀.

Receptacles are not required in non-dwelling bathrooms. If installed however, they must be GFCI protected ▶▶ 210-8(b)(1) ◀◀.

A unit load at least equal to that shown in Table 220-3(a) for a specified occupancy constitutes the minimum lighting load for each square foot of floor area. Floor area (for each floor) is computed using the outside dimensions of the building, dwelling unit, or other area involved. The minimum general lighting load for a store is 3 volt-amperes per square foot ▶▶ 220-3(a) ◀◀. The minimum general lighting load for this retail space is 3,726 volt-amperes. (27 × 46 = 1,242 × 3 = 3,726)

NOTE

Refer to Section 210-2 for a list of other Articles (and Sections) pertaining to branch circuit requirements. The provisions for branch circuits supplying equipment in that list amend or supplement Article 210 provisions and apply to the referenced branch circuits ▶▶ *210-2* ◀◀.

In a building having more than one nominal voltage system, each accessible ungrounded (hot) conductor of a multiwire branch circuit, must be identified by phase and system. The identification method can be separate color coding, marking, tape, tagging, or other approved means, and must be permanently posted at each branch circuit panelboard ▶▶ 210-4(d) ◀◀.

Branch circuits recognized by Article 220 must be rated in accordance with the maximum permitted ampere rating (or setting) of the overcurrent device. Circuits, other than individual branch circuits, must be rated 15, 20, 30, 40, and 50 amperes. If higher ampacity conductors are used for any reason, the ampere rating (or setting) of the specified overcurrent device determines the circuit rating ▶▶ 210-3 ◀◀.

Branch circuit conductors must have an ampacity not less than the maximum load served. In a branch circuit supplying continuous loads, or any combination of continuous and noncontinuous loads, the minimum branch circuit conductor size (before the application of any adjustment or correction factors) must have an allowable ampacity equal to, or greater, than the noncontinuous load plus 125% of the continuous load ▶▶ 210-19(a) ◀◀.

Equipment and branch circuit conductors must be protected by overcurrent protective devices rated in compliance with 210-20(a) through (d) ▶▶ 210-20 ◀◀.

Outlet devices must have an ampere rating not less than the load served, and must comply with Article 220-21(a) and (b) ▶▶ 210-21 ◀◀.

The requirements for multiple outlet or receptacle are summarized in Table 210-24 ▶▶ 210-24 ◀◀.

The minimum number of branch circuits is determined from the total computed load and the size (or rating) of the circuits used. In all installations, the number of circuits must be sufficient to supply the load served. In no case can the load on any circuit exceed the maximum specified by Section 220-4 ▶▶ 210-11(a) ◀◀.

Article 210 covers branch circuits with the exception of branch circuits supplying only motor loads. These are covered in Article 430. Articles 210 and 430 provisions apply to branch circuits which have combination loads ▶▶ 210-1 ◀◀.

■ Showcases

NOTE
Other equipment must not be electrically connected to showcases »410-29(d) «.

In cord-connected showcases, the secondary circuit(s) of each electric-discharge lighting ballast is limited to one showcase »410-29(e) «.

Receptacles, connectors, and attachment plugs must be of a listed grounding type, rated 15 or 20 amperes »410-29(b) «.

Individual moveable showcases can be connected by flexible cord to permanently installed receptacles. Groups of not more than six such showcases can be coupled together (by flexible cord and separable locking-type connectors) provided one of the group is connected (by flexible cord) to a permanently installed receptacle. The installation must comply with Section 410-29(a) through (e) »410-29 «.

Unless used in wiring of chain-supported lighting fixtures or as supply cords for portable lamps and other merchandise being displayed or exhibited, flexible cords used in show cases and show windows must be: Type S, SE, SEO, SEOO, SJ, SJE, SJEO, SJEOO, SJO, SJOO, SJT, SJTO, SJTOO, SO, SOO, ST, STO, or STOO »400-11 «.

Flexible cords must be secured to showcase undersides so that: (1) wiring will not be exposed to mechanical damage; (2) cases cannot separate more than 2 inches, nor can the first case extend more than 12 inches from the supply receptacle; and (3) the free lead at the end of the showcase group has a female fitting which does not extend beyond the case »410-29(c) «.

Flexible cord must be hard-service type, have conductors not smaller than the branch-circuit overcurrent device, and have an equipment grounding conductor »410-29(a) «.

■ Motor Loads

Motor load outlets must be computed in accordance with Sections 430-22 and 430-24 requirements »220-3(b)(3) «.

If a circuit supplies only motor-operated loads, apply Article 430 »220-4(a) «.

Branch circuit conductors supplying a single motor, used in a continuous duty application, must have an ampacity of not less than 125% of the motor's full load current rating as determined by Section 430-6(a)(1) »430-22(a) «.

NOTE
Conductors used for a motor in a short-time, intermittent, periodic, or varying duty application must have an ampacity of not less than the percentage of the motor nameplate current rating shown in Table 430-22(b), unless the authority having jurisdiction grants special permission for lower ampacity conductors »430-22(b) «.

NOTE
Any motor application is considered continuous duty unless the nature of the apparatus driven is such that the motor will not operate continuously with load under any condition of use »Note under Table 430-22(b) «.

■ Motor-Operated and Combination Loads

For circuits supplying loads consisting of motor-operated utilization equipment (fastened in place with a motor larger than ⅛ HP) in combination with other loads, the total computed load must be based on 125% of the largest motor load plus the sum of the other loads ≫220-4(a)≪.

Because devices such as room air-conditioners, household refrigerators and freezers, drinking water coolers, and beverage dispensers are considered appliances, Article 422 provisions also apply ≫440-3(c)≪.

For cord- and plug-connected appliances, an accessible, separable connector or an accessible plug and receptacle can serve as the disconnecting means. Where the separable connector or plug and receptacle are not accessible, cord- and plug-connected appliances must be provided with a disconnecting means in accordance with Section 422-31 ≫422-32(a)≪.

The rating of a receptacle or a separable connector must not be less than the rating of any connected appliance ≫422-32(c)≪.

Maximum = 50% of Branch Circuit Rating

Maximum = 80% of Branch Circuit Rating

To Panel

To Panel

Receptacle outlet for cord- and plug-connected, non fastened-in-place utilization equipment.

Combination branch circuit, consisting of two duplex receptacles

Dedicated branch circuit, consisting of only one receptacle.

For cord-connected equipment such as room air-conditioners, household refrigerators (and freezers), drinking water coolers, and beverage dispensers, a separable connector or an attachment plug and receptacle can serve as the disconnecting means ≫440-13≪.

If supplied by a 20-ampere circuit, this drinking fountain's maximum rating is 10-amperes (20 × 50%). If supplied by a 15-ampere circuit the maximum rating is 7.5 (15 × 50%).

If supplied by a 20-ampere circuit, the maximum rating for this drinking fountain cooler is 16 amperes (20 × 80%). If supplied by a 15-ampere circuit the maximum rating is 12 amperes (15 × 80%).

NOTE

A 15- or 20-ampere branch circuit can supply lighting units, other utilization equipment, or a combination of both. The rating of any one cord- and plug-connected utilization equipment must not exceed 80% of the branch circuit ampere rating. The total rating of utilization equipment fastened in place (other than lighting fixtures) cannot exceed 50% of the branch circuit ampere rating in the event that lighting units, cord- and plug-connected utilization equipment (not fastened in place), or both, are also supplied ≫210-23(a)≪.

■ Air-Conditioning and/or Refrigeration Equipment

Article 440, Part B provisions require a means capable of disconnecting air-conditioning and refrigerating equipment (including motor-compressors and controllers) from the circuit conductors >> 440-11 <<.

The provisions of Article 440, Part C specify devices intended to protect the branch circuit conductors, control apparatus, and motors in circuits supplying hermetic refrigerant motor-compressors against overcurrent due to short circuits and grounds. These provisions amend and/or supplement those in Article 240 >> 440-21 <<.

Article 310 and Article 440, Part D, provisions specify conductor ampacities required to carry the motor current (without overheating) under the conditions specified, except as modified in Section 440-6(a), *Exception No. 1* >> 440-31 <<.

Article 440, Part F specifies devices intended to protect the motor-compressor against excessive heating due to motor overload and failure to start. See Sections 240-3(g) and 430-31 >> 440-51 <<.

Motor load outlets must be computed in accordance with Article 440 requirements >> 220-3(b)(3) <<.

Where a circuit supplies only air-conditioning equipment, refrigeration equipment, or both, apply Article 440 >> 220-4(a) <<.

Disconnecting means must be located within sight from, and readily accessible from, the air-conditioning or refrigerating equipment. The disconnecting means can be installed on or within the equipment >> 440-14 <<.

For a hermetic refrigerant motor-compressor, the rated-load current marked on the nameplate of the equipment in which the motor-compressor is employed must be used in determining the rating or ampacity of the disconnecting means, the branch circuit conductors, the controller, the branch-circuit short-circuit and ground-fault protection, and the separate motor overload protection. Where no rated-load current is shown on the equipment nameplate, use the rated-load current shown on the compressor nameplate. For disconnecting means and controllers, see also Sections 440-12 and 440-41 >> 440-6(a) <<.

Article 440 provisions apply to electric motor-driven air-conditioning and refrigerating equipment, and to their branch circuits and controllers. It provides the special considerations necessary for circuits supplying hermetic refrigerant motor-compressors and for any air-conditioning or refrigerating equipment supplied from a branch circuit which also supplies a hermetic refrigeration motor-compressor >> 440-1 <<.

RECEPTACLES

■ Maximum Number of Receptacles on a Branch Circuit

Ten receptacles are permitted on a 15-ampere overcurrent protective device (15 ÷ 1.5 = 10).

Receptacles are calculated at 180 volt-amperes per strap >> 220-3(b)(9) <<. Therefore, when the source is 120 volts, the rating of each receptacle is 1.5 amperes (180 ÷ 120 = 1.5).

Thirteen receptacles are permitted on a 20-ampere overcurrent protective device (20 ÷ 1.5 = 13.3 or 13).

A load having the maximum level of current sustained for three hours or more is referred to as a continuous load. A general purpose receptacle is not a continuous load >> Article 100 <<.

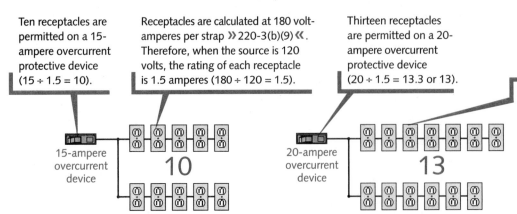

■ Non-Dwelling Receptacle Placement

Dwelling Occupancy

6 Feet | 12 Feet | 6 Feet

Dwelling unit receptacles must be installed so that no point along the floor line in any wall space is more than 6 feet, measured horizontally, from a receptacle outlet in that space ≫210-52(a)(1)≪.

Receptacle outlets in hotel, motel, and similar occupancy guest-rooms must be installed in accordance with Section 210-52 ≫210-60(a)≪.

Non-dwelling Occupancy

24 Feet
No Receptacle Outlet Required

Although no provisions pertain to non-dwelling general receptacle placement (except hotel and motel guest rooms), certain requirements exist for specific occupancies and/or areas. Examples include, but are not limited to: show windows ≫210-62≪; rooftop, attic, and crawl space receptacle outlets for servicing heating, air-conditioning, and refrigeration equipment ≫210-63≪; and health care patient bed location receptacles ≫517-18(b)≪.

Generally speaking, there are no receptacle placement provisions for non-dwelling occupancies.

■ Bathroom Receptacles

Non-dwelling bathrooms do not require receptacles. A bathroom receptacle is required only in dwelling units and hotel or motel guest rooms ≫210-52(d)≪ and 210-60(a)≪.

Restroom

Where installed, 125-volt, single-phase, 15- and 20-ampere bathroom receptacle must have ground-fault, circuit-interrupter protection for personnel ≫210-8(b)≪.

NOTE
Ground-fault, circuit-interrupters are designed to trip when a 4 through 6 milliampere (0.004 through 0.006 ampere) difference between the ungrounded (hot) conductor and grounded conductor occurs.

■ Fixed Multioutlet Assemblies

Where appliances are unlikely to be used simultaneously, each 5 feet or fraction thereof (of separate and continuous length) shall be considered one outlet of not less than 180 volt-amperes ≫220-3(b)(8)(a)≪.

Because fixed multioutlet assemblies used in dwelling units or hotel and motel guest rooms are included in the general lighting load calculation, no additional load calculation is required ≫220-3(b)(8)≪.

Receptacles *unlikely* to be used simultaneously

5 feet=
180 VA

Where appliances are likely to be used simultaneously, each 1 foot or fraction thereof shall be considered one outlet of not less than 180 volt-amperes ≫220-3(b)(8)(b)≪.

It is not stipulated that each receptacle be rated 180 volt-amperes. This illustration shows one receptacle per foot; therefore, each receptacle is rated 180 volt-amperes. Multioutlet assemblies having two receptacles per foot (6 inches on center), have a rating of 180 volt-amperes for two receptacles.

Receptacles *likely* to be used simultaneously

1 foot= 180 VA | 1 foot= 180 VA | 1 foot= 180 VA | 1 foot= 180 VA | 1 foot= 180 VA

■ Receptacle Volt-Ampere Rating

Except as covered in one-, two-, and multi-family dwellings, and in hotel and motel guest rooms, receptacle outlets must be computed at not less than 180 volt-amperes for each single, or multiple, receptacle on one strap »220-3(b)(9)«.

one strap

180
Volt-Amperes

one strap

180
Volt-Amperes

one strap

180
Volt-Amperes

A single piece of equipment (consisting of four or more receptacles) must be computed at not less than 90 volt-amperes per receptacle »220-3(b)(9)«.

90 VA per receptacle

360
Volt-Amperes

■ Provisions for Hotel and Motel Guest Rooms

When performing service (or feeder) calculations, use Table 220-11 to derate hotel and motel (include apartment houses without cooking equipment) guest room lighting and receptacle loads. Do not apply Table 220-11 demand factors to areas where all lighting will probably be used at one time. Examples include, but are not limited to: office, lobby, restaurant, hallway, parking area, etc.

Ground-fault circuit-interrupter (GFCI) protection is required for every 125-volt bathroom receptacle »210-8(a)(1)«. A bathroom is an area including a basin (lavatory or sink) with one or more of the following: toilet, tub, or shower »Article 100«. As used here, area does not necessarily mean a single room.

Locate at least one wall receptacle within 36 inches of the outside edge of each basin (lavatory or sink). The receptacle outlet must be located on a wall adjacent to the basin »210-52(d)«.

Hallways of less than 10 feet in length do not require a receptacle »210-52(h)«.

At least one wall switch-controlled lighting outlet or receptacle must be installed in hotel, motel, or similar occupancy guest rooms »210-70(b)«.

Receptacles must be installed so that no point along the floor line in any wall space is more than 6 feet, measured horizontally, from an outlet in that space »210-52(a)(1)«. (See Section 210-60(b) provisions allowing a greater distance due to permanent fixture layout.)

Guest rooms in hotels, motels, and similar occupancies must have receptacles outlets installed in accordance with Section 210-52 (dwelling unit receptacle provisions) »210-60(a)«. (See Units 6 and 7, in this text, for dwelling unit receptacle requirements.)

Section 410-4(d) contains bathtub and shower area lighting provisions. (See illustrated explanation in Unit 7 of this text.)

Receptacles installed behind the bed must either be positioned so that the bed does not contact any installed attachment plug, or the receptacle must include a suitable guard »210-60(b)«.

The total number of receptacle outlets must comply with the minimum number of receptacles provision of Section 210-52(a). These receptacle outlets can be located conveniently for permanent furniture layout »210-60(b)«. This receptacle could be located between the beds, even though the distance to the receptacle behind the desk exceeds 12 feet.

Bathroom

Hallway

Closet

Hotel or Motel Guest Room

At least two receptacle outlets must be readily accessible »210-60(b)«.

Branch circuit conductors supplying more than one receptacle for cord- and plug-connected portable loads must have an ampacity at least equal to the rating of the branch circuit »210-19(b)«.

NOTE

Hotels, motels, and similar occupancies having guest room kitchen facilities must meet Sections 210-52(b) and (c) provisions.

NOTE

It is not required that occupants in hotel/motel guest rooms (intended for transient use) have ready access to all overcurrent devices protecting conductors supplying their room »240-24(b)«.

CAUTION <<<<<<<<<<<<<<<<

In dwelling units and guest rooms of hotels and motels, overcurrent devices (other than supplementary overcurrent protection) must not be located in bathrooms as defined in Article 100 »240-24(e)«.

■ Show Windows

A **show window** is any window designed or used for the display of goods or advertising material, whether it is fully or partly enclosed, entirely open at the rear, and whether or not it has a platform raised higher than the street floor level »Article 100 «.

Unless used in wiring of chain-supported lighting fixtures, or as supply cords for portable lamps and other merchandise being displayed or exhibited, flexible cords used in show cases and show windows must be Type S, SE, SEO, SEOO, SJ, SJE, SJEO, SJEOO, SJO, SJOO, SJT, SJTO, SJTOO, SO, SOO, ST, STO, or STOO »400-11 «.

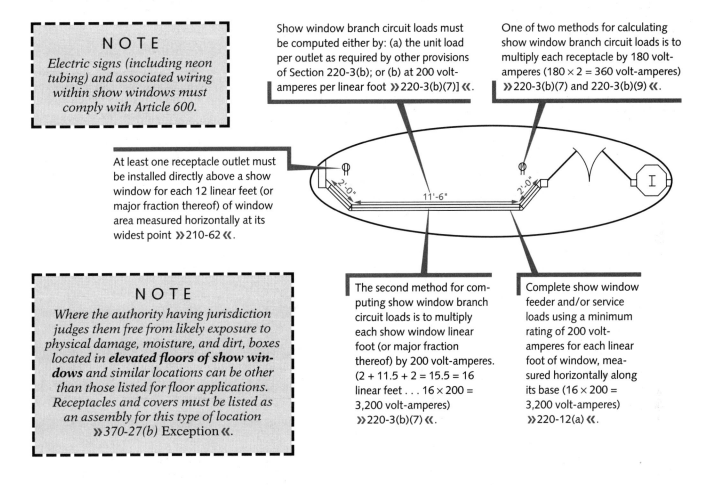

> **N O T E**
> *Electric signs (including neon tubing) and associated wiring within show windows must comply with Article 600.*

Show window branch circuit loads must be computed either by: (a) the unit load per outlet as required by other provisions of Section 220-3(b); or (b) at 200 volt-amperes per linear foot »220-3(b)(7)] «.

One of two methods for calculating show window branch circuit loads is to multiply each receptacle by 180 volt-amperes (180 × 2 = 360 volt-amperes) »220-3(b)(7) and 220-3(b)(9) «.

At least one receptacle outlet must be installed directly above a show window for each 12 linear feet (or major fraction thereof) of window area measured horizontally at its widest point »210-62 «.

> **N O T E**
> *Where the authority having jurisdiction judges them free from likely exposure to physical damage, moisture, and dirt, boxes located in **elevated floors of show windows** and similar locations can be other than those listed for floor applications. Receptacles and covers must be listed as an assembly for this type of location »370-27(b) Exception «.*

The second method for computing show window branch circuit loads is to multiply each show window linear foot (or major fraction thereof) by 200 volt-amperes. (2 + 11.5 + 2 = 15.5 = 16 linear feet . . . 16 × 200 = 3,200 volt-amperes) »220-3(b)(7) «.

Complete show window feeder and/or service loads using a minimum rating of 200 volt-amperes for each linear foot of window, measured horizontally along its base (16 × 200 = 3,200 volt-amperes) »220-12(a) «.

LIGHTING

■ Lampholder Installations

Lampholder

Screw-shell type lampholders must be installed only for use as lampholders »410-47 «.

If the supply circuit has a grounded conductor, it must be connected to the screw shell »410-47 «.

> **N O T E**
> *Only weatherproof type lampholders can be installed in wet or damp locations »410-49 «.*

■ Reflective Ceiling/Lighting Plan

For circuits of over 250 volts to ground, the electrical continuity of metal raceways and cables with metal sheaths, containing any conductor other than service conductors, must be ensured by one or more of the methods specified for services in Section 250-94(1) through (4) ≫250-97 ≪.

Circuit breakers used as switches in 120-volt and 277-volt fluorescent lighting circuits must be listed and must be marked "SWD" ≫240-83(d) ≪. All switches and circuit breakers used as switches must be located so that they are operable from a readily accessible place. The center of the operating-handle grip of the switch or circuit breaker, when in its highest position, must not be more than 6 feet, 7 inches, above the floor, or working platform ≫380-8(a) ≪.

NOTE

Suspended ceiling system framing members used to support fixtures must be securely fastened to each other and to the building structure at appropriate intervals. Fixtures must be securely fastened to the ceiling framing member by mechanical means such as bolts, screws, or rivets. Listed clips identified for use with the type of ceiling framing member(s) and fixture(s) are also permitted ≫410-16(c) ≪.

Branch circuits recognized by Article 210 must be rated in accordance with the maximum permitted overcurrent device ampere rating (or setting) ≫210-3 ≪.

Branch circuits for lighting and appliances (including motor-operated appliances) must supply the loads in accordance with Section 220-3 ≫210-11 ≪.

Branch circuit conductor ampacity must meet or exceed the maximum load being served. Where a branch circuit supplies continuous loads, or any combination of continuous and non-continuous loads, the minimum **branch-circuit conductor** size, before the application of any adjustment or correction factors, must have an allowable ampacity equal to or greater than the noncontinuous load plus 125% of the continuous load ≫210-19(a) ≪.

Where a branch circuit supplies continuous loads, or any combination of continuous and noncontinuous loads, the rating of the **overcurrent device** must not be less than the noncontinuous load plus 125% of the continuous load ≫210-20(a) ≪.

A 15- or 20-ampere branch circuit can supply lighting units or other utilization equipment, or a combination of both ≫210-23(a) ≪.

Wall switch-controlled lighting outlets are not required in commercial occupancies, except for attic and underfloor spaces containing equipment which requires servicing ≫210-70(c) ≪.

A unit load meeting Table 220-3(a) specifications for listed occupancies constitutes the minimum lighting load for each square foot of floor area. The area for each floor must be computed using the outside dimensions of the building, dwelling unit, or other area involved ≫220-3(a) ≪.

Supplementary overcurrent protection used for lighting fixtures, appliances, and other equipment (including internal circuits and components) must not be used as a substitute for the branch circuit protection specified in Article 210. Ready accessibility is not required for supplementary overcurrent devices ≫240-10 ≪.

A 30-, 40- or 50-ampere branch circuit can supply fixed lighting units with heavy-duty lampholders in other than dwelling unit(s) ≫210-23(b) and (c) ≪.

Lighting outlet loads must not be supplied by branch circuits larger than 50 amperes ≫210-23(d) ≪.

■ Lighting Fixture Schedule

Conductors of circuits rated no more than 600 volts, nominal, ac circuits, and dc circuits can occupy the same equipment wiring enclosure, cable, or raceway. The conductors' insulation rating must equal or exceed the maximum circuit voltage applied to any conductor within the enclosure, cable, or raceway ≫300-3(c)(1) ≪.

For lighting units that have ballasts, transformers, or auto-transformers, the supply circuit's computed load is based on the total ampere ratings of such units; not on the lamps' total wattage ≫220-4(b) ≪. Obtain the ampere rating from the manufacturer or from the information listed on the ballast. If the fixture contains more than one ballast, add the ampere ratings. Multiply the total ampere rating by the voltage to find the fixture's volt-ampere rating.

LIGHTING FIXTURE SCHEDULE

Type	Manufacturer	Volts	Catalog Number	Comments
A	Lighthouse	277	P622-232U277	2' × 2' lay-in fluorescent deep cell parabolic
B	Lighthouse	277	P622-232U277-EM	emergency light
C	Lighthouse	120	10R-CM70MH120	10" recessed, 70W metal halide
D	Lighthouse	277	RT24–432EB277	2' × 4' lay-in fluorescent
E	Lighthouse	120	8001BK/HT308	8' lighting track with 5 heads
F	Lighthouse	120	8001BK/HT308	12' lighting track with 7 heads
G	Lighthouse	120	SL48-32EB120	4' Single lamp fluorescent strip
H	Lighthouse	120	SL24-20EB120	2' Single lamp fluorescent strip
X	Lighthouse	120	EX7B1LEDR	Exit light with battery

LAMP SCHEDULE

Lamps per Fixture	Watts per Lamp	Lamp Size
2	32	F32-T8-U
2	32	F32-T8-U
1	70	MH70/C/U/M
4	32	F32-T8
5	75	75W-PAR30FL
7	75	75W-PAR30FL
1	32	F32-T8
1	20	F20-T8
Lamps are included in the fixture.		

The minimum number of branch circuits is determined from the total computed load, and the size (or rating) of the circuits used. In all installations, the number of circuits must be sufficient to supply the load served. In no case shall the load on any circuit exceed the maximum specified by Section 220-4 ≫210-11(a) ≪.

The load calculated on a volt-ampere per square foot basis, must be adequately served by the wiring system up to and including the branch circuit panelboard(s). This load must be evenly proportioned among multioutlet branch circuits within the panelboard(s). Branch circuits and overcurrent devices must only be installed to serve the connected load ≫210-11(b) ≪.

> **N O T E**
> *The total load on any branch circuit conductor must not exceed 80% of its rating, after the application of any adjustment or correction factors, where the load will continue for three hours or more, i.e. a continuous load ≫210-19(a) ≪. Most commercial lighting is considered continuous.*

> **N O T E**
> *The total load on any panelboard overcurrent device must not exceed 80% of its rating where, in normal operation, the load will continue for three hours or more ≫384-16(d) ≪.*

■ Fluorescent Lighting Fixtures

Suspended-ceiling system framing members used to support fixtures must be securely fastened to each other and to the building structure at appropriate intervals. Fixtures must be fastened securely to the ceiling framing member by mechanical means such as bolts, screws, or rivets. Listed clips identified for use with the type of ceiling framing member(s) and fixture(s) are also permitted ≫410-16(c)≪.

Fixtures and equipment are considered grounded if mechanically connected to an equipment grounding conductor per Section 250-118 and sized in accordance with Section 250-122 ≫410-21≪.

Branch circuits recognized by Article 210 can be used in multiwire circuits. A multiwire branch circuit can be thought of as multiple circuits. All conductors must originate from the same panelboard ≫210-4(a)≪. Unless an exception is met, multiwire branch circuits can only supply line-to-neutral loads ≫210-4(c)≪.

Wiring on, or within, fixtures must be neatly arranged and must not be exposed to physical damage. Avoid excess wiring. Arrange conductors so they are not subjected to temperatures greater than those for which they are rated ≫410-22≪.

A 3-phase, 4-wire, wye-connected power system supplying nonlinear loads may necessitate a design allowing for the possibility of high harmonic neutral currents ≫210-4(a) FPN≪.

Fixtures must be equipped with shades (or guards), constructed, or installed so that combustible material will not be subjected to temperatures exceeding 90°C ≫410-5≪.

Securely support all fixtures, lampholders, and receptacles ≫410-15(a)≪.

Exposed metal parts must be: (1) grounded or insulated from ground and other conducting surfaces; or (2) inaccessible to unqualified personnel. Grounding is not required for lamp tie wires, mounting screws, clips, and decorative bands on glass spaced at least 1½ inches from lamp terminals ≫410-18(a)≪.

Raceway fittings supporting a lighting fixture(s) must be capable of supporting the combined weight of the fixture assembly and lamp(s) ≫410-16(f)≪.

> **NOTE**
>
> *Install fixtures so that the connections between fixture conductors and circuit conductors can be inspected without having to disconnect any of the wiring (unless the fixtures are connected by attachment plugs and receptacles) ≫410-16(b)≪.*

> **NOTE**
>
> ***Luminaire***, *the international term for "lighting fixture" is defined as a complete lighting unit consisting of lamp(s), parts designed to distribute the light, parts positioning and protecting the lamp(s), and parts connecting the lamp(s) to the power supply ≫410-1 FPN≪.*

■ Show-Window and Track Lighting

Lighting track is a manufactured assembly designed to support and energize lighting fixtures which can be readily repositioned along the track. Its length may be altered by the addition or subtraction of track sections »410-100«.

Track system ends must be insulated and capped »410-105(a)«.

The load connected to a lighting track must not exceed the track's rating. The rating of the branch circuit which supplies lighting track cannot exceed the track rating »410-101(b)«.

Fittings identified for use on lighting track must be designed specifically for the track on which they are installed. They must be securely fastened to the track, maintain polarization and grounding, and be designed to be suspend directly from the track »410-101(d)«.

Lighting track must be permanently installed as well as permanently connected to a branch circuit. Install only lighting track fittings on the lighting track. Fittings equipped with general-purpose receptacles cannot be used on lighting track »410-102(a)«.

Lighting track must be securely mounted so that each fastening suitably supports the maximum weight of fixtures that can be installed. Unless identified for greater support intervals, a single section 4 feet or less in length must have two supports. If installed in a continuous row, each individual section of not more than 4 feet in length must have one additional support »410-104«.

When computing feeder and/or service track lighting in other than dwelling units or hotel/motel guest rooms, include an additional load of 150 volt-amperes for every 2 feet of lighting track or fraction thereof »220-12(b)«.

No externally wired fixture, except chain-supported, can be used in show windows »410-7«.

Type E (typical)

Type F

When computing feeder and/or service show-window lighting, include a unit load of not less than 200 volt-amperes per linear foot of show window, measured horizontally along the base »220-12(a)«. (2 + 11.5 + 2 = 15.5 = 16 × 200 = 3,200 volt-amperes)

Lighting track must be grounded in accordance with Article 250. Track sections must be securely coupled to maintain continuity of the circuitry, polarization, and grounding throughout »410-105(b)«.

Heavy-duty lighting track is identified for use exceeding 20 amperes. Each fitting attached to a heavy-duty lighting track must have individual overcurrent protection »410-103«.

NOTE
Do not install lighting track in the following locations:
1. *Where physical damage is likely*
2. *In wet or damp locations*
3. *Where subject to corrosive vapors*
4. *In storage battery rooms*
5. *In hazardous (classified) locations*
6. *Where concealed*
7. *Where extended through walls or partitions*
8. *Less than 5 feet above the finished floor except where protected from physical damage or where the track operates at less than 30 volts rms open-circuit voltage*
9. *Within the zone measured 3 feet horizontally and 8 feet vertically from the top of a bathtub rim* »410-101(c)«.

■ Recessed Fixtures

A fixture can be recessed in fire-resistant material in a building of fire-resistant construction, subject to temperatures between 90°C (194°F) and 150°C (302°F), provided the fixture is plainly marked as listed for that service 》410-65(b)《.

Do not install thermal insulation above a recessed fixture or within 3 inches of the fixture's enclosure (wiring compartment, or ballast) unless it is identified for contact with insulation, Type IC 》410-66(b)《.

High-intensity discharge (HID) fixtures supply power to HID lamps. HID lamps produce light from gaseous discharge arc tubes. HID lamps include: mercury lamps, metal halide lamps; and high-pressure sodium lamps.

A recessed remote ballast for a high-intensity discharge fixture must have integral thermal protection and be identified as such 》410-73(f)(4)《.

Recessed high-intensity fixtures designed for installations in wall or ceiling cavities must have thermal protection and be identified as such 》410-73(f)(1)《. Thermal protection is not required in a recessed high-intensity fixture whose design, construction, and thermal performance characteristics are equivalent to a thermally protected fixture, and are identified as inherently protected 》410-73(f)(2)《.

Install fixtures so that adjacent combustible material will not be subjected to temperatures greater than 90°C (194°F) 》410-65(a)《.

Recessed HID (Typical)

Fixtures installed in recessed cavities (walls or ceilings) must comply with Sections 410-65 through 410-72 》410-64《.

A recessed fixture, not identified for contact with insulation, must have all recessed parts spaced at least ½ inch from combustible materials. Support points and the trim finishing off the ceiling or wall opening can contact combustible materials 》410-66(a)(1)《.

A recessed fixture identified for contact with insulation, Type IC, can contact combustible materials at recessed parts, points of support, and portions passing through or finishing off the structural opening 》410-66(a)(2)《.

A recessed high-intensity discharge fixture identified for use and installed in poured concrete does not require thermal protection 》410-73(f)(3)《.

NOTE
Install fixtures with exposed ballasts or transformers so that such ballasts or transformers will not contact combustible material 》410-76(a)《.

NOTE
An outlet supplying recessed lighting fixture(s) must be computed based on the maximum volt-ampere rating of the equipment and lamps for which the fixture(s) is rated 》220-3(b)(4)《.

■ Exit Signs and Emergency Lighting

A ballast in a fluorescent exit fixture shall not have thermal protection »410-73(e)(3)«.

> **NOTE**
>
> *Article 700 provisions apply to the electrical safety of emergency systems installation, operation, and maintenance. These systems consist of circuits and equipment intended to supply, distribute, and control electricity for facility illumination, power, or both, whenever the normal electrical service is interrupted. Emergency systems are legally mandated by municipal, state, federal, or other codes, or by any governmental agency having jurisdiction. The purpose of these systems is to automatically supply illumination, power, or both, to designated areas and equipment in the event of failure of the normal supply, or in the event of accidental damage to systems elements intended to supply, distribute, and control power and illumination essential for safety to human life* »700-1«.

Suspended ceiling systems framing members supporting fixtures must be securely fastened to each other and to the building structure at appropriate intervals. Fixtures must be securely fastened to the ceiling framing member by mechanical means such as bolts, screws, or rivets. Listed clips identified for use with the type of ceiling framing member(s) and fixture(s) are also permitted »410-16(c)«.

The ballast of a fluorescent fixture installed indoors must have integral thermal protection. Replacement ballasts must also have similar thermal protection »410-73(e)(1)«.

A fluorescent fixture ballast used for egress lighting, energized only during an emergency, shall not have thermal protection »410-73(e)(4)«.

An exit directional sign located within a building does not require a disconnecting means »600-6 *Exception No. 1*«.

■ Adjustable Fixtures

Two conduits can support an enclosure containing a device under the conditions found in Section 370-23(f): (1) The box must not exceed 100 cubic inches; (2) The box has either threaded entries or hubs identified for the purpose; (3) The box must be supported by two or more conduits threaded wrenchtight into the enclosure or hubs; (4) Each conduit must be secured within 18 inches of the enclosure provided all entries are on the same side.

Equipment grounding conductors that are part of flexible cords, or used with fixture wires in accordance with Section 240-4, cannot be smaller than No. 18 copper and not smaller than the circuit conductors »250-122(e)«.

A wet location receptacle, where the product plugged in is unattended while in use, requires an enclosure that is weatherproof at all times (plug inserted or not) »410-57(b)(1)«.

Fixtures that require adjusting or aiming after installation do not require an attachment plug or cord connector, provided the exposed cord is of the hard or extra-hard usage type and is not longer than that required for maximum adjustment. The cord must not be subject to strain or physical damage »410-30(b)«.

Fixtures Used as Raceways

Fixtures designed for end-to-end connection, thereby forming a continuous assembly, or fixtures connected by recognized wiring methods, are permitted to contain the supply conductors of a 2-wire or a multiwire branch circuit supplying the fixtures »410-31 *Exception No. 2* «.

One additional 2-wire branch circuit separately supplying one or more of the connected fixtures described in Exception No. 2 can be carried through the fixtures »410-31 Exception No. 3 «.

Branch-circuit conductors, within 3 inches of a ballast (within the ballast compartment), must have an insulation temperature rating no lower than 90°C (194°F), such as Types RHH, THW, THHN, THHW, FEP, FEPB, SA, and XHHW »410-31 «.

Fixtures must not be used as a raceway for circuit conductors unless listed for such use »410-31 *Exception No. 1* «.

NOTE

Electric-discharge lighting fixtures, supported independent of the outlet box, must be connected to the branch circuit through metal raceway, nonmetallic raceway, Type MC cable, Type AC cable, Type MI able, nonmetallic sheathed cable, or by flexible cord as permitted in Section 410-30(b) or (c) »410-14(a) «. (Some of these wiring methods are acceptable only in certain occupancies.)

In a completed installation, a cover must be provided for each outlet box, unless covered by means of a fixture canopy, lampholder, receptacle, or similar device »410-12 «.

Branch circuit conductors must have an ampacity at least equal to the maximum load being served. In a branch circuit supplying continuous loads, the minimum branch-circuit conductor size (before the application of any adjustment or correction factors) must have an allowable ampacity equal to or greater than 125% of the continuous load »210-19(a) «.

The overcurrent device, on a branch circuit supplying continuous loads, must be rated at least 125% of the load »210-20(a) «.

Auxiliary equipment for electric-discharge lamps must be enclosed in noncombustible cases and be treated as heat sources »410-54(a) «.

Hinged Metal Poles, 20 Feet or Less, Supporting Lighting Fixtures

A metal pole 20 feet or less in height above grade that has a hinged base does not require a handhole »410-15(b)(1) *Exception No. 2* «.

If equipped with a hinged base, the pole and base must be bonded together »410-15(b)(4) «.

No Handhole Required

Hinged Base

Grounding Terminal Required

20 Feet Maximum

A hinged-base pole longer than 20 feet, supporting a lighting fixture(s), requires a handhole.

A metal pole must have an accessible grounding terminal within the base »410-15(b)(3)(b) «.

Metal raceways or other equipment grounding conductors must be bonded to the pole by means of an equipment grounding conductor recognized by Section 250-118 and sized in accordance with Section 250-122 »410-15(b)(5) «.

■ Cord Connected Lampholders and Fixtures

The inlet of a metal lampholder attached to a flexible cord, must be equipped with an insulating bushing that, if threaded, is not smaller than nominal ⅜-inch pipe size. The cord hole must be of an appropriate size, and all burrs/fins must be removed providing a smooth bearing surface for the cord. Bushings having holes ⁹⁄₃₂-inch in diameter are permitted when using plain pendant cord and holes ¹³⁄₃₂-inch in diameter with reinforced cord **»** 410-30(a) **《** .

Electric-discharge lighting fixtures provided with mogul-base, screw-shell lampholders can be connected to branch circuits of 50 amperes or less by cords which comply with Section 240-4. While receptacles and attachment plugs can be of a lower ampere rating than the branch circuit, they cannot be less than 125% of the fixture load current **»** 410-30(c)(2) **《** .

Cord-connected fixtures must be located directly below the outlet box or busway.

A listed fixture or a listed assembly can be cord connected if:
a. the fixture is located directly below the outlet box or busway, and
b. the flexible cord must be:
  1. Visible over its entire length outside the fixture
  2. Not subject to strain or physical damage
  3. Terminated in a grounding-type attachment plug cap, busway plug or have a fixture assembly with a strain relief and canopy **»** 410-30(c)(1) **《** .

For circuits which supply lighting units with ballasts, transformers, or auto-transformers, the computed load must be based on the total ampere rating of such units and not on the total lamp(s) wattage **»** 220-4(b) **《** .

A load expected to continue at maximum current for 3 hours or more is a continuous load **»** Article 100 **《** .

Electric-discharge lighting fixtures having a flanged surface inlet can be supplied by cord pendants equipped with cord connectors. Inlets and connectors can be of a lower ampere rating than the branch circuit but not less than 125% of the fixture load current **»** 410-30(c)(3) **《** .

■ Portable Lamps

Portable lamps must be wired with flexible cord, recognized by Section 400-4, and a polarized or grounding type attachment plug **»** 410-42(a) **《** .

If used with Edison-base lampholders, the grounded conductor must be identified and attached to the screw shell as well as the attachment plug's identified blade **»** 410-42(a) **《** .

In addition to the provisions of Section 410-42(a), portable handlamps must comply with Section 410-42(b)(1) through (5):
1. Metal shell, paper-lined lampholders must not be used.
2. Handlamps must be equipped with a handle of molded composition or other insulating material.
3. Handlamps must be equipped with a substantial guard attached to the lampholder or handle.
4. Metallic guards must be grounded by means of an equipment grounding conductor run with circuit conductors within the power-supply cord.
5. Portable handlamps supplied through an isolating transformer, with an ungrounded secondary of not over 50 volts, do not require grounding.

■ Metal Poles, 8 Feet or Less, Supporting Lighting Fixtures

No handhole is required in a pole no more than 8 feet in height (above grade) where the supply wiring method continues without splice or pull point and where the pole interior and splices are accessible by removing the fixture �410-15(b)(1) Exception No. 1 ◀◀.

Removable Fixture

Splices must remain accessible

The wiring method must continue without splice to the lighting fixture termination point.

8 Feet Maximum

A pole longer than 8 feet, supporting a lighting fixture(s), requires a handhole.

No Handhole Required

No Grounding Terminal Required

No grounding terminal is required in a pole 8 feet (or less) in height above grade where the supply wiring method continues without splice or pull point, and where the pole interior and splices are accessible by fixture removal ➤➤410-15(b)(3) Exception ◀◀.

> **NOTE**
>
> *A receptacle installed in a pole, supporting a lighting fixture, may require GFCI protection (See Section 210-8). A wet-location receptacle where the product will be attended while in use (e.g. portable tools, etc.) must have a weatherproof enclosure when the attachment plug is removed* ➤➤410-57(b)(2) ◀◀.

■ Required Handhole in a Metal Pole

Metal pole more than 8 feet tall above grade

Metal raceways or other equipment grounding conductors must be bonded to the pole by means of an equipment grounding conductor recognized by Section 250-118 and sized in accordance with Section 250-122 ➤➤410-15(b)(5) ◀◀.

2" x 4" Minimum

Grounding Terminal Required

Handhole Required

Table 300-5

Support conductors in vertical metal poles used as raceway as provided in Section 300-19 ➤➤410-15(b)(6) ◀◀.

A metal pole must have a handhole not less than 2 inches by 4 inches with a raintight cover to provide access to the supply terminations within the pole or pole base ➤➤410-15(b)(1) ◀◀.

A metal pole must be provided with a handhole accessible grounding terminal ➤➤410-15(b)(3)(a) ◀◀.

Direct-buried cable, conduit, or other raceways must meet Table 300-5 minimum cover requirements ➤➤300-5(a) ◀◀.

> **NOTE**
>
> *Where conductor size is adjusted to compensate for voltage drop, equipment grounding conductors must be adjusted proportionately according to circular mil area* ➤➤250-122(b) ◀◀.

■ Supply Conductors Not Installed Within the Pole

Where cable or raceway risers are not installed within the pole, a threaded fitting (or nipple) must be brazed (or welded) to the pole, opposite the hand-hole for the supply connection **»**410-15(b)(2) **«**.

Installed raceways subject to physical damage may require additional protection.

Raceways, fittings, boxes, etc. must be securely fastened in place **»**300-11(a) **«**. Follow specific raceway support provisions for the type of raceway used.

See Unit 5 for raceway provisions.

Brazed or Welded

Handhole

Conduit Strap

Summary

- Receptacle placement provisions for one-family and multi-family dwellings do not apply to commercial occupancies.
- Although a bathroom receptacle is not required in a commercial occupancy, it must be GFCI protected if installed.
- Each commercial building/occupancy open to pedestrians must be provided with at least one 20-ampere outlet in an accessible location at each entrance to every tenant space for sign or outline lighting systems use.
- One similarity in calculating branch circuit conductors for continuous loads and motor loads is that each is multiplied by 125%.
- Article 440 contains air-conditioning and refrigeration equipment requirements.
- Commercial receptacle(s) are computed at not less than 180 volt-amperes per strap.
- No more than 10 receptacles are permitted on a 15-ampere branch circuit, and 13 on a 20-ampere branch circuit.

- At least 2 receptacle outlets must be readily accessible in guest rooms of hotels, motels, and similar occupancies.
- Table 220-3(a) gives the volt-ampere (per square foot) lighting load for listed occupancies.
- Compute feeder and/or service track lighting at 150 volt-amperes for every 2 feet of lighting track or fraction thereof.
- Calculate feeder and/or service show-window lighting at not less than 200 volt-amperes per linear foot of show window.
- Certain lighting fixtures, such as end-to-end connected fluorescent lights, have a limited use as raceways.
- A handhole is required in a metal pole taller than 8 feet, above grade, unless the pole is no more than 20 feet with a hinged base.

Unit 12 Competency Test

NEC® Reference   Answer

_____ _____ 1. In computing the load of lighting units which employ an autotransformer, the load shall be based on _____.

_____ _____ 2. _____ fixtures used in a show window can be externally wired.

_____ _____ 3. Fixtures shall be wired with conductors having insulation suitable for the _____ to which the conductors will be subjected.

NEC® Reference   Answer

_____ _____   4. Systems of illumination utilizing fluorescent lamps, high intensity discharge lamps, or neon tubing is the definition for _____.

_____ _____   5. The international term for a lighting fixture is _____ and is defined as a complete lighting unit consisting of lamp(s) together with the parts designed to distribute the light, to position and protect the lamps, and to connect the lamps to the power supply.

_____ _____   6. The total rating of utilization equipment fastened in place, other than lighting fixtures, shall not exceed _____ % of the branch-circuit ampere rating where lighting units, cord- and plug-connected utilization equipment not fastened in place, or both, are also supplied.

_____ _____   7. A metal pole supporting a lighting fixture must have a handhole not less than _____ with a raintight cover to provide access to the supply terminations within the pole or pole base.

_____ _____   8. Branch circuits that supply neon tubing installations shall not be rated in excess of _____ amperes.

_____ _____   9. Multioutlet assembly shall not be permitted:

       I. to be run through hoistways

      II. to be run within dry partitions

     III. to be run through dry partitions

     a) II only   b) III only   c) I and II only   d) I, II, and III

_____ _____   10. Where portable lamps are used with Edison-base lampholders, the grounded conductor shall be identified and attached to the _____ of the attachment plug.

_____ _____   11. When computing track lighting loads for a commercial service, an additional load of _____ volt-amperes shall be included for every _____ feet of lighting track or fraction thereof.

_____ _____   12. The terminal of an electric-discharge lamp shall be considered as a live part where any lamp terminal is connected to a circuit of over _____.

_____ _____   13. _____ fixtures surface mounted over concealed outlet, pull, or junction boxes shall be installed with suitable openings in back of the fixture to provide access to the boxes.

_____ _____   14. When computing show-window lighting loads for a commercial service, a load of not less than _____ shall be included for each linear foot of show window, measured horizontally along its base.

_____ _____   15. Twenty-four feet (3 eight-foot sections) of track lighting must be installed in a continuous row in a retail store. What is the minimum number of supports required?

_____ _____   16. No handhole shall be required in a pole, supporting a lighting fixture, that is _____ or less in height above grade where the supply wiring method continues without splice or pull point, and where the interior of the pole and any splices are accessible by removing the fixture.

_____ _____   17. A hermetic refrigerant motor-compressor's branch-circuit short-circuit and ground-fault protective device shall be capable of carrying _____ percent of the motor's starting current.

     a) 100%   b) 125%   c) 150%   d) 175%

NEC® Reference   Answer

18. Heavy-duty lighting track is lighting track identified for use exceeding _____.

19. At least _____ receptacle outlet(s) shall be readily accessible in hotel and motel guest rooms.

20. Flexible cords shall be secured to the undersides of showcases so that a separation between cases not in excess of _____, nor more than _____ between the first case and the supply receptacle, will be ensured.

21. Branch circuits that supply sign and outline lighting systems containing incandescent and fluorescent forms of illumination shall be rated not to exceed _____.

22. No handhole shall be required in a metal pole, supporting a lighting fixture, that is _____ or less in height above grade that is provided with a hinged base.

23. A single piece of equipment consisting of a multiple receptacle comprised of four or more receptacles shall be computed at not less than _____ per receptacle.

24. Lighting fixtures shall not be used as a raceway for circuit conductors unless _____ for use as a raceway.

25. At least one receptacle outlet shall be installed above a show window for each _____ linear feet or major fraction thereof of show window area measured horizontally at its maximum width.

26. On hermetic refrigerant motor-compressors, the value of branch-circuit selection current will always be _____ the marked rated-load current.

    I. less than

    II. equal to

    III. greater than

    a) I only   b) I or II   c) II or III   d) III only

27. Electric-discharge lighting fixtures provided with mogul-base, screw-shell lampholders shall be permitted to be connected to branch circuits of _____ amperes or less by cords complying with Section 240-4.

28. The load for the required sign branch circuit shall be computed at a minimum of _____.

29. A(n) _____ branch circuit consists of two or more ungrounded conductors having a potential difference between them, and a grounded conductor having equal potential difference between it and each ungrounded conductor of the circuit and that is connected to the neutral or grounded conductor of the system.

30. Lighting track shall be securely mounted so that each fastening will be suitable for supporting the maximum weight of fixtures that can be installed. Unless identified for supports at greater intervals, a single section _____ or shorter in length shall have _____ supports.

31. Switches, flashers, and similar devices controlling sign transformers and electronic power supplies shall be rated for controlling inductive loads or have a current rating not less than _____ of the transformer's current rating.

    a) 80%   b) 100%   c) 125%   d) 200%

NEC® Reference   Answer

—————— —————— 32. Conductors of circuits rated 600 volts, nominal, or less, ac circuits, and dc circuits shall be permitted to occupy the same enclosure if:

      a) all conductors within the enclosure have an insulation rating equal to at least 600 volts.

      b) all conductors operating at different voltage levels within the enclosure are separated by a permanent barrier.

      c) all conductors have an insulation rating equal to at least the maximum circuit voltage applied to any conductor within the enclosure.

      d) each conductor has an insulation rating equal to at least the maximum circuit voltage applied to each individual conductor within the enclosure.

—————— —————— 33. Branch circuit conductors within _____ of a ballast shall have an insulation temperature rating not lower than 90°C, unless supplying a fixture listed and identified as suitable for a different insulation temperature.

—————— —————— 34. The screw shell of all sign lampholders supplied by a grounded circuit shall be connected to the _____ conductor.

—————— —————— 35. Where connected to a branch circuit having a rating in excess of _____, lampholders shall be of the heavy-duty type.

—————— —————— 36. A recessed lighting fixture that is not identified for contact with insulation shall have all recessed parts spaced at least _____ from combustible materials.

Non-Dwelling Load Calculations

Objectives

After studying this unit, the student will:

- ✔ have a good understanding of the elements required to perform a non-dwelling load calculation.
- ✔ know that receptacle outlets are counted, unlike dwelling-unit load calculations.
- ✔ understand how to apply Table 220-13 demand factor to receptacle loads in excess of 10 kVA.
- ✔ be able to compute the receptacle load for banks and office buildings where the actual number of receptacle outlets is unknown.
- ✔ be familiar with volt-ampere unit loads for different types of occupancies, and even for different areas within certain occupancies.
- ✔ know when and how to apply Table 220-11 demand factors.

- ✔ be aware that track lighting is computed in addition to the general lighting load.
- ✔ know when to include a sign and/or outline lighting outlet in a load calculation.
- ✔ understand the method for calculating show window lighting loads.
- ✔ be able to determine whether a load is continuous or noncontinuous.
- ✔ understand that continuous loads require the inclusion of an additional 25% volt-ampere rating in the load calculation.
- ✔ know that the load calculation rating represents only a minimum rating.

Introduction

Unit 13 contains a non-dwelling load calculation form, along with a detailed explanation of each line. Dwelling calculations are covered in Unit 8 (one-family) and Unit 11 (multi-family) of this text. Unlike dwelling unit load calculations, receptacle outlets are counted. Receptacle outlets (if known) and fixed multioutlet assemblies (if any) are entered into the calculation, and if the load is great enough, a demand factor is applied. As with dwelling-unit load calculations, general lighting is computed using outside dimensions. Other items, such as a sign outlet (where required) and show window(s) (if present) are part of the non-dwelling load calculation. Continuous Loads (Line 10) is a very important computation. All continuous load ratings must be increased by 25%. While Kitchen Equipment (Line 11) is not included in every type of calculation, be aware that kitchen equipment is not limited to restaurants. For instance, kitchen equipment could be a portion of a load calculation for a school. Line 14 (All Other Loads) is a catchall for any load not included otherwise in the calculation.

The load calculation form has little room for listing individual items, such as motors, equipment, etc. Depending upon the size of the occupancy, the calculation could contain hundreds, if not thousands, of individual items. Space is also limited for certain calculations. When the need arises, simply attach additional sheets of paper containing extra items and/or calculations. Some procedure(s) will not apply to certain load calculations. It is recommended that a line not be left completely empty. Some predetermined marking should fill the space; i.e., a dashed line, the letters NA, etc. One of the procedures (Line 4) is applicable only in banks and office buildings where the actual number of receptacle outlets is unknown.

The load calculation form results (overcurrent protection and conductors) represent only a *minimum* requirement. No consideration is given for the addition of future electrical loads. The size service and/or feeder is not restricted to the form's calculated size. For example, an electrician might install a 200-ampere service in an occupancy where the load calculation only required a 125-ampere rating.

Because certain cities, states, etc. require that some (if not all) electrical installations be designed by a licensed Electrical Engineer, caution is advised. Check with local authorities to determine these, as well as other, requirements.

NON-DWELLING LOAD CALCULATIONS

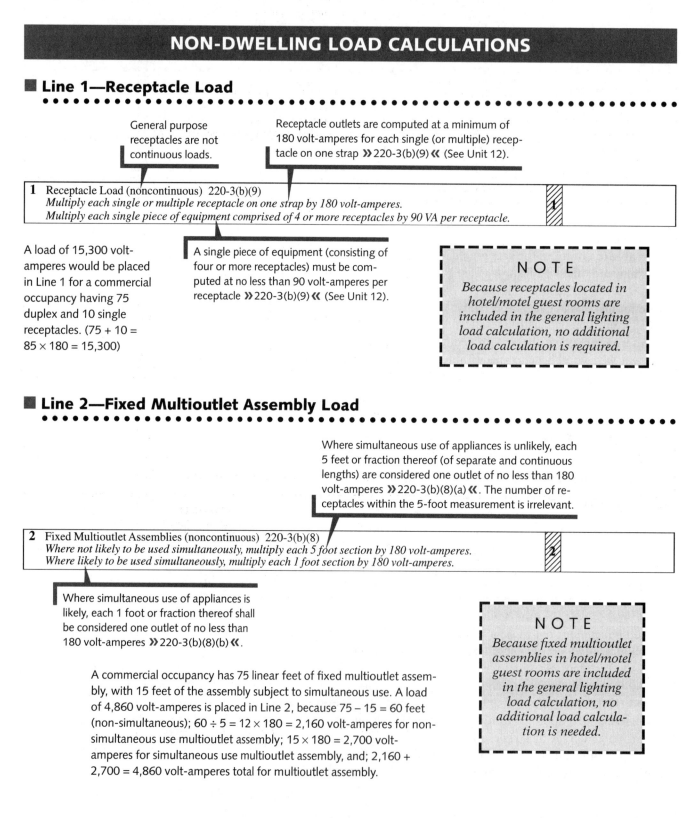

■ Line 1—Receptacle Load

General purpose receptacles are not continuous loads.

Receptacle outlets are computed at a minimum of 180 volt-amperes for each single (or multiple) receptacle on one strap ≫ 220-3(b)(9) ≪ (See Unit 12).

1 Receptacle Load (noncontinuous) 220-3(b)(9)
Multiply each single or multiple receptacle on one strap by 180 volt-amperes.
Multiply each single piece of equipment comprised of 4 or more receptacles by 90 VA per receptacle.

A load of 15,300 volt-amperes would be placed in Line 1 for a commercial occupancy having 75 duplex and 10 single receptacles. (75 + 10 = 85 × 180 = 15,300)

A single piece of equipment (consisting of four or more receptacles) must be computed at no less than 90 volt-amperes per receptacle ≫ 220-3(b)(9) ≪ (See Unit 12).

NOTE
Because receptacles located in hotel/motel guest rooms are included in the general lighting load calculation, no additional load calculation is required.

■ Line 2—Fixed Multioutlet Assembly Load

Where simultaneous use of appliances is unlikely, each 5 feet or fraction thereof (of separate and continuous lengths) are considered one outlet of no less than 180 volt-amperes ≫ 220-3(b)(8)(a) ≪. The number of receptacles within the 5-foot measurement is irrelevant.

2 Fixed Multioutlet Assemblies (noncontinuous) 220-3(b)(8)
Where not likely to be used simultaneously, multiply each 5 foot section by 180 volt-amperes.
Where likely to be used simultaneously, multiply each 1 foot section by 180 volt-amperes.

Where simultaneous use of appliances is likely, each 1 foot or fraction thereof shall be considered one outlet of no less than 180 volt-amperes ≫ 220-3(b)(8)(b) ≪.

NOTE
Because fixed multioutlet assemblies in hotel/motel guest rooms are included in the general lighting load calculation, no additional load calculation is needed.

A commercial occupancy has 75 linear feet of fixed multioutlet assembly, with 15 feet of the assembly subject to simultaneous use. A load of 4,860 volt-amperes is placed in Line 2, because 75 – 15 = 60 feet (non-simultaneous); 60 ÷ 5 = 12 × 180 = 2,160 volt-amperes for non-simultaneous use multioutlet assembly; 15 × 180 = 2,700 volt-amperes for simultaneous use multioutlet assembly, and; 2,160 + 2,700 = 4,860 volt-amperes total for multioutlet assembly.

Non-dwelling Feeder/Service Load Calculation

1 Receptacle Load (noncontinuous) 220-3(b)(9) *Multiply each single or multiple receptacle on one strap by 180 volt-amperes.* *Multiply each single piece of equipment comprised of 4 or more receptacles by 90 VA per receptacle.*	**1**
2 Fixed Multioutlet Assemblies (noncontinuous) 220-3(b)(8) *Where not likely to be used simultaneously, multiply each 5 foot section by 180 volt-amperes.* *Where likely to be used simultaneously, multiply each 1 foot section by 180 volt-amperes.*	**2**
3 Receptacle Load Demand Factor (for nondwelling receptacles) 220-13 *If the receptacle load is more than 10,000 volt-amperes, apply the demand factor from Table 220-13.* *Add lines 1 and 2. Multiply the first 10 kVA or less by 100%. Then, multiply the remainder by 50%.*	**3**
4 Unknown Receptacle Load (Banks and Office buildings only) *Where the actual number of general purpose receptacle outlets are unknown, include 1 volt-ampere per sq. ft. Table 220-3(a) footnote* [b] $1 \times \underline{\hspace{2cm}} = $ (sq. ft. outside dimensions)	**4**
5 General Lighting Load Table 220-3(a) *Multiply the volt-ampere unit load (for the type of occupancy) by the square foot outside dimensions.* $\underline{\hspace{1cm}} \times \underline{\hspace{1cm}} = $ (VA unit load) (sq. ft. outside dimensions)	**5**
6 Lighting Load Demand Factors 220-11 . . . *Apply Table 220-11 demand factors to certain portions of hospitals, hotels, motels, apartment houses (without provisions for cooking), and storage warehouses. Do not include areas in hospitals, hotels, and motels where the entire lighting will be used at one time.*	**6**
7 Track Lighting (in addition to general lighting) 220-12(b) *Include 150 volt-amperes for every 2 feet, or fraction thereof, for lighting track.* $\underline{\hspace{1cm}} \div 2 \underline{\hspace{1cm}} \times 150 = $ (total linear feet)	**7**
8 Sign and/or Outline Lighting Outlet (where required) 220-3(b)(6) *Each commercial building (or occupancy) accessible to pedestrians must have at least one outlet per tenant space entrance. 600-5(a) Each outlet must be at least 1,200 volt-amperes.*	**8**
9 Show Window Lighting 220-12(a) *Include at least 200 volt-amperes for each linear foot, measured horizontally along the show window's base.* $\underline{\hspace{1cm}} \times 200 = $ (total linear feet of show window)	**9**
10 Continuous Loads 215-2(a), 215-3, and 230-42(a) . . . *Multiply the continuous load volt-amperes (listed above) by 25%. (General purpose receptacles are not considered continuous.)* $\underline{\hspace{1cm}} \times 25\% = $ (total continuous load volt-amperes)	**10**
11 Kitchen Equipment 220-20 *Multiply three or more pieces of equipment by Table 220-20 demand factor (percent).* Use Table 220-19 for household cooking equipment used in instructional programs. Table 220-19 Note 5	**11**
12 Noncoincident Loads 220-21 . . . *The smaller of two (or more) noncoincident loads can be omitted, as long as they will never be energized simultaneously (such as certain portions of heating and A/C systems). Calculate fixed electric space heating loads at 100% of the total connected load. 220-15*	**12**
13 Motor Loads 220-4(a), 430-24, 430-25, 430-26, and Article 440 *Motor-driven air-conditioning and refrigeration equipment is found in Article 440. Multiply the largest motor (one motor only) by 25% and add to load.*	**13**
14 All Other Loads . . . *Add all other noncontinuous loads into the calculation at 100%. Multiply all other continuous loads (operating for 3 hours or more) by 125%.*	**14**
15 Total Volt-Ampere Demand Load: *Add Lines 3 through 14 to find the minimum required volt-amperes.*	**15**

16 Minimum Amperes *Divide the total volt-amperes by the voltage* $\underline{\hspace{1cm}} \div \underline{\hspace{1cm}} = $ (line 15) (voltage) (minimum amperes)	**16**	**Minimum Size** **17 Service and/or** **Feeder 240-6(a)**	**17**
18 Size the Service and/or Feeder Conductors. Tables 310-16 through 310-19 *Use the tables along with Section 310-15(b)(1) through (7) to determine conductor size. If the overcurrent device is rated more than 800 amperes, the conductor ampacity must be equal to, or greater than, the rating of the overcurrent device. 240-3(c)*		**Minimum Size** **Conductors**	**18**
19 Size the Neutral Conductor 220-22 *The neutral service and/or feeder conductor can be smaller than the ungrounded (hot) conductors, but not smaller than the maximum unbalanced load determined by Article 220. Section 250-24(b)(1) states that the neutral cannot be smaller than the required grounding electrode conductor specified in Table 250-66. A further demand factor is permitted for any neutral load over 200 amperes.*		**Minimum Size** **Neutral** **Conductor**	**19**
20 Size the Grounding Electrode Conductor (for Service) 250-66 *Using line 18 to find the grounding electrode conductor in Table 250-66.* Size the Equipment Grounding Conductor (for Feeder) 250-122 *Use line 17 to find the equipment grounding conductor in Table 250-122. Equipment grounding conductor types are listed in Section 250-118.*		**Minimum Size** **Grounding Electrode** **Conductor . . . or . . .** **Equipment Grounding** **Conductor**	**20**

■ Line 3—Receptacle Load Demand Factor

Receptacle loads in other than dwelling units are computed at no more than 180 volt-amperes per outlet in accordance with Section 220-3(b)(9) and fixed multioutlet assemblies are computed per Section 220-3(b)(8). Both can be added to the lighting load and made subject to the demand factors given in Table 220-11, or they can be made subject to Table 220-13 demand factors ❯❯ 220-13 ❮❮ .

If the total receptacle load is no more than 10,000 volt-amperes, insert the number directly in Line 3.

| 3 | Receptacle Load Demand Factor (for nondwelling receptacles) 220-13 *If the receptacle load is more than 10,000 volt-amperes, apply the demand factor from Table 220-13. Add lines 1 and 2. Multiply the first 10 kVA or less by 100%. Then, multiply the remainder by 50%.* | **3** | |

A commercial occupancy has a receptacle load of 15,300 volt-amperes and a fixed multioutlet assembly load of 4,860. The total receptacle load is 20,160 volt-amperes. First, subtract 10,000 from the total receptacle load (20,160 − 10,000 = 10,160). Next, multiply the remainder by 50% (10,160 × 50% = 5,080). Finally, add the result back to the original 10,000 volt-amperes (5,080 + 10,000 = 15,080). The receptacle load, after demand, is then entered into Line 3.

NOTE
General purpose receptacles are not considered continuous loads.

If the total receptacle load is more than 10,000 volt-amperes, subtract 10,000 from the total, and multiply the remainder by 50%. Finally, add that number to the original 10,000 and place the total in Line 3.

■ Line 4—Unknown Receptacle Load

In addition, a unit load of 1 volt-ampere per square foot must be included for general purpose receptacle outlets where the actual number is unknown ❯❯ Table 220-3(a) footnote [b] ❮❮ .

If the actual number of general purpose receptacle outlets in a bank or office building is known, Line 4 would not contain a number.

Line 4 would contain a figure only if the occupancy is a bank or office building.

| 4 | Unknown Receptacle Load (Banks and Office buildings only) *Where the actual number of general purpose receptacle outlets are unknown, include 1 volt-ampere per sq. ft. Table 220-3(a) footnote* [b] | $1 \times \underline{\hspace{3cm}}$ (sq. ft. outside dimensions) = | **4** | |

It is not known how many receptacles are in an 8,500-square-foot bank. A load of 8,500 volt-amperes should be placed in Line 4 (8,500 × 1 = 8,500). Since the actual number of receptacle outlets is unknown, Line 3 would not contain a number.

CAUTION ❮❮❮❮❮❮❮❮❮❮❮❮❮❮❮
Line 4 applies only to bank and office building occupancies.

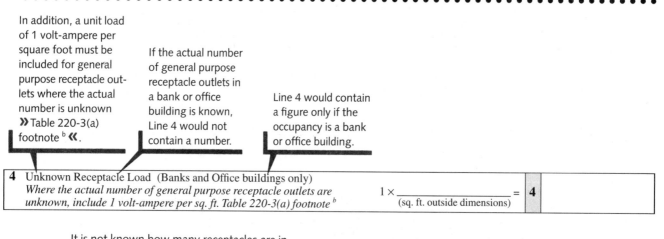

■ Line 5—General Lighting Load

A unit load which meets or exceeds that specified in Table 220-3(a) for occupancies listed therein constitutes the minimum lighting load for each square foot of floor area **»220-3(a) «**.

Find the correct volt-ampere unit load located across from the occupancy type, and insert into the calculation **»** Table 220-3(a) **«**.

Each floor's area must be computed using the building's (or area's) outside dimensions **»220-3(a) «**.

| 5 | General Lighting Load Table 220-3(a)
Multiply the volt-ampere unit load (for the type of occupancy) by the square foot outside dimensions. | $\dfrac{}{\text{(VA unit load)}} \times \dfrac{}{\text{(sq. ft. outside dimensions)}} =$ | 5 | |

An 8,500-square-foot bank has a general lighting load of 29,750 volt-amperes (8,500 × 3.5 = 29,750). The receptacle load, whether known or unknown, is not used here.

A 30,000-square-foot store has a general lighting load of 90,000 volt-amperes (30,000 × 3 = 90,000).

> **N O T E**
>
> *In all Table 220-3(a) occupancies (except one-family dwellings and individual dwelling units of two-family and multifamily dwellings), specific areas can be separately multiplied by different volt-ampere unit loads. For example: assembly halls and auditoriums have a unit load of 1; halls, corridors, closets, and stairways have a unit load of 0.5; and storage spaces have a unit load of 0.25 volt-amperes per square foot.*

■ Line 6—Lighting Load Demand Factors

> **N O T E**
>
> *Do not use Table 220-11 to determine the total number of branch circuits* **»220-11 «**.

Demand factors apply to certain areas in hospitals, hotels, motels, apartment houses (without cooking provisions), and warehouses.

| 6 | Lighting Load Demand Factors 220-11 . . . *Apply Table 220-11 demand factors to certain portions of hospitals, hotels, motels, apartment houses (without provisions for cooking), and storage warehouses. Do not include areas in hospitals, hotels, and motels where the entire lighting will be used at one time.* | 6 | |

Do not include areas in hospitals, hotels, and motels where all lighting is subject to simultaneous use. Primarily, such lighting is considered continuous loads. Continuous loads must not be derated by Table 220-11, but instead, are increased by 25% (Line 10).

After demand, the general lighting load for hospital patients' rooms, where the room dimensions total 100,000 square feet is 30,000 volt-amperes. First, multiply 100,000 by the hospital volt-ampere unit load (2), found in Table 220-3(a) (100,000 × 2 = 200,000). Next, multiply the first 50,000 by 40% (50,000 × 40% = 20,000). Then, multiply the remainder by 20% (100,000 − 50,000 = 50,000 × 20% = 10,000). Finally, add the two figures (20,000 + 10,000 = 30,000).

■ Line 7—Track Lighting

Do not consider track lighting in dwelling unit (or hotel/motel guest room) service and/or feeder load calculations **»220-12(b) «**.

Track lighting is calculated in addition to the occupancy general lighting load, found in Table 220-3(a).

A store having 80 feet of lighting track has an additional lighting load of 6,000 volt-amperes. First, divide the total length of lighting track by two (80 ÷ 2 = 40). Then, multiply the result by 150 volt-amperes (40 × 150 = 6,000).

| 7 | Track Lighting (in addition to general lighting) 220-12(b)
Include 150 volt-amperes for every 2 feet, or fraction thereof, for lighting track. | $\dfrac{}{\text{(total linear feet)}} \div 2 \underline{\hspace{2cm}} \times 150 =$ | 7 | |

Multiply each 2 foot section, or fraction thereof, by 150 volt-amperes **»220-12(b) «**.

■ Line 8—Sign and Outline Lighting

Each commercial building (or occupancy) open to pedestrians must be provided with at least one outlet in a location accessible to each tenant space entrance for sign or outline lighting system use ≫ 220-3(b)(6) ≪.

NOTE

Because service hallways or corridors are not considered accessible to pedestrians, no circuit is required ≫ 220-3(b)(6) ≪.

| 8 | Sign and/or Outline Lighting Outlet (where required) 220-3(b)(6)
Each commercial building (or occupancy) accessible to pedestrians must have at least one outlet per tenant space entrance. 600-5(a) Each outlet must be at least 1,200 volt-amperes. | 8 | |

CAUTION ≪≪≪≪≪≪≪≪≪≪≪

If the rating is more than 1,200 volt-amperes, use the actual rating of the sign and/or outline lighting.

Each sign and outline lighting outlet is computed at a minimum of 1,200 volt-amperes ≫ 220-3(b)(6) ≪.

■ Line 9—Show Window Lighting

If an occupancy has show window(s), include at least 200 volt-amperes for each linear foot, measured horizontally along the show window's base ≫ 220-12(a) ≪.

NOTE

Calculate show window branch circuits in accordance with Section 220-3(b)(6).

A store having 75 linear feet of show window space has a feeder/service load of 15,000 volt-amperes (75 × 200 = 15,000).

| 9 | Show Window Lighting 220-12(a)
Include at least 200 volt-amperes for each linear foot, measured horizontally along the show window's base. | $\underline{\hspace{4cm}} \times 200 =$
(total linear feet of show window) | 9 | |

■ Line 10—Continuous Loads

A load where the maximum current is expected to continue for 3 hours or more is a continuous load ≫ Article 100 ≪.

Feeder conductors and overcurrent protection must be sized at 100% of noncontinuous loads *plus* 125% of continuous loads ≫ 215-2(a) and 215-3 ≪. In other words, an additional 25% of the volt-ampere rating must be added to the continuous loads. This form simplifies that computation. Add all of the continuous loads (listed above Line 10) and multiply by 25%. This rule also applies to service entrance conductors ≫ 230-42(a) ≪.

| 10 | Continuous Loads 215-2(a), 215-3, and 230-42(a) . . . *Multiply the continuous load volt-amperes (listed above) by 25%. (General purpose receptacles are not considered continuous.)* | $\underline{\hspace{4cm}} \times 25\% =$
(total continuous load volt-amperes) | 10 | |

General purpose receptacles and multioutlet assemblies are not included in this calculation.

Total all of the *continuous* loads listed in Lines 5 through 9. These lines are not automatically considered continuous. Only if the load is expected to continue for 3 hours or more is it a continuous load. Examples may include, but are not limited to: general lighting, track lighting, show window lighting, signs and outline lighting.

■ Line 11—Kitchen Equipment

Compute the load for commercial electric cooking equipment, dishwasher booster heaters, water heaters, and other kitchen equipment in accordance with Table 220-20. Apply the demand factors to all equipment that: (1) has thermostatic control, or; (2) is intermittently used as kitchen equipment ≫220-20≪.

Table 220-20 is relatively simple. No derating is allowed for only 1 or 2 pieces of equipment. The demand factor for 3 pieces of equipment is 90%; for 4 pieces, 80%; and for 5 pieces, 70%. The demand factor for more than 5 pieces of equipment is 65%.

> **NOTE**
> *Table 220-20 does not apply to space-heating, ventilating, or air-conditioning equipment.*

| 11 Kitchen Equipment 220-20 *Multiply three or more pieces of equipment by Table 220-20 demand factor (percent).* Use Table 220-19 for household cooking equipment used in instructional programs. Table 220-19 Note 5 | **11** | |

> **CAUTION** ≪≪≪≪≪≪≪≪≪≪≪≪
>
> *The feeder or service demand load can never be smaller than the combined rating of the two largest kitchen equipment loads ≫220-20≪.*

For household cooking appliances rated over 1¾ kW and used in instructional programs, refer to Table 220-19 ≫Table 220-19 Note 5≪. For example, a feeder supplying twenty 12-kW household electric ranges, in a school home-economics classroom, requires a minimum rating of only 35 kW ≫Table 220-19, Column A≪.

■ Line 12—Noncoincident Loads

> **NOTE**
> *Where reduced loading of the conductors results from units operating on duty-cycle, intermittently, or from all units not operating at one time, the authority having jurisdiction may grant permission for feeder conductors to have an ampacity less than 100 percent, provided the conductors have an ampacity for the load so determined ≫220-15 Exception≪.*

Where it is unlikely that multiple noncoincident loads will be used simultaneously, use only the largest load(s) that will be used at one time, in computing the total feeder/service load ≫220-21≪.

| 12 Noncoincident Loads 220-21 . . . *The smaller of two (or more) noncoincident loads can be omitted, as long as they will never be energized simultaneously (such as certain portions of heating and A/C systems). Calculate fixed electric space heating loads at 100% of the total connected load. 220-15* | **12** | |

While fixed electric space heating loads are computed at 100% of the total connected load, in no case shall a feeder or service load current rating be less than the rating of the largest branch circuit supplied ≫220-15≪.

■ Line 13—Motor Loads

The ampacity of the conductors supplying multi-motor and combination-load equipment must not be less than the minimum circuit ampacity marked on the equipment per Section 430-7(d). If the individual equipment nameplates are visible (per Section 430-7(d)(2), but the equipment is not factory-wired, use Section 430-24 to determine conductor ampacity >> 430-25 «.

Where conductor heating is reduced as a result of motors operating on duty-cycle, intermittently, or from all motors not operating at one time, the authority having jurisdiction may grant permission for feeder conductors to have an ampacity less than specified in Section 430-24, provided the conductors have sufficient ampacity for the number of motors supplied and the nature of the their loads and duties >> 430-26 «.

13 Motor Loads 220-4(a), 430-24, 430-25, 430-26, and Article 440 *Motor-driven air-conditioning and refrigeration equipment is found in Article 440.* *Multiply the largest motor (one motor only) by 25% and add to load.*	**13**

Conductors supplying several motors, or a motor(s) and other load(s), must have an ampacity at least equal to the sum of the full-load current rating of all the motors (as determined by Section 430-6(a)(1), *plus* 25 percent of the highest-rated motor in the group >> 430-24 «.

Where one or more of the motors of the group are used for short-time, intermittent, periodic, or varying duty, the ampere rating of such motors used in the summation shall be determined in accordance with Section 430-22(b). For the highest-rated motor, the greater of either the ampere rating from 430-22(b) or the largest continuous-duty motor full-load current multiplied by 1.25 must be used in the summation >> 430-24 *Exception No.1* «.

Where interlocked circuitry prevents operation of selected motors or other loads at the same time, the conductor ampacity can be based on the summation of the currents of the motors and other loads being operated at the same time resulting in the highest total current >> 430-24 *Exception No.3* «.

■ Line 14—All Other Loads

This line is a catchall for loads not falling into one of the previous categories.

Calculate all noncontinuous loads (not yet input) at 100% of their volt-ampere rating.

14 All Other Loads . . . *Add all other noncontinuous loads into the calculation at 100%.* *Multiply all other continuous loads (operating for 3 hours or more) by 125%.*	**14**

Total all continuous loads (not yet calculated) and multiply by 125%. Add the result of the continuous loads to that for noncontinuous loads and put the total on Line14.

■ Line 15—Total Volt-Ampere Demand Load

Add all volt-ampere loads listed in Lines 3 through 14 and place the result in Line 15.

15 Total Volt-Ampere Demand Load: *Add Lines 3 through 14 to find the minimum required volt-amperes.*	**15**

> ### NOTE
> *Not all lines will contain volt-ampere loads. Be sure to add only the lines containing loads. For example, if the actual number of receptacles in a bank are known, Line 4 will not contain a figure.*

■ Lines 16 and 17—Minimum Size Service and/or Feeder

• •

Place the total volt-ampere amount found in Line 15 here.

Write down the source voltage that supplies the feeder/service.

Fractions of an ampere 0.5 and higher are rounded up, while fractions less than 0.5 are dropped ≫ 220-2(b) ≪.

The service (or feeder) overcurrent protection must be higher than the number found in Line 16. Refer to Section 240-6(a) for a list of fuse and circuit breaker standard ampere ratings.

| 16 Minimum Amperes *Divide the total volt-amperes by the voltage* _____ ÷ _____ = | |16| | Minimum Size 17 Service and/or Feeder 240-6(a) | |17| | |
(line 15)   (voltage)   (minimum amperes)

Divide the volt-amperes by the voltage to find the amperage.

Three-phase voltage is found by multiplying the voltage (single-phase) by 1.732. For example, the source voltage for 208-volt, three-phase is 360 (208 × 1.732).

The result on Line 16 is the minimum amperage rating required for the service and/or feeder being calculated.

■ Line 18—Minimum Size Conductors

• •

No. 1/0 and larger conductors can be connected in parallel (electrically joined at both ends to form a single conductor) ≫ 310-4 ≪.

The number and type of paralleled conductor sets is a design consideration, not necessarily a *Code* issue. Exercise extreme care when designing a paralleled conductor installation without violating provisions, such as 110-14(a), 110-14(c), 240-3(c), 250-66, 250-122(f), 300-20(a), 310-4, etc.

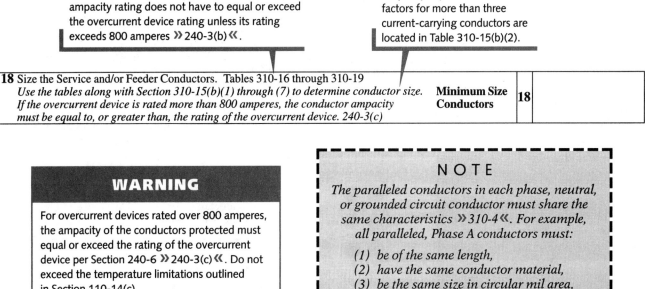

Using Tables 310-16 through 310-19, choose a conductor size which equals (or exceeds) Line 16's minimum ampacity rating. The conductor's ampacity rating does not have to equal or exceed the overcurrent device rating unless its rating exceeds 800 amperes ≫ 240-3(b) ≪.

Ampacity adjustment (correction) factors for more than three current-carrying conductors are located in Table 310-15(b)(2).

| 18 Size the Service and/or Feeder Conductors. Tables 310-16 through 310-19 *Use the tables along with Section 310-15(b)(1) through (7) to determine conductor size. If the overcurrent device is rated more than 800 amperes, the conductor ampacity must be equal to, or greater than, the rating of the overcurrent device. 240-3(c)* | Minimum Size Conductors | 18 | |

NOTE

The paralleled conductors in each phase, neutral, or grounded circuit conductor must share the same characteristics ≫ 310-4 ≪. For example, all paralleled, Phase A conductors must:

(1) be of the same length,
(2) have the same conductor material,
(3) be the same size in circular mil area,
(4) have the same insulation type,
(5) be terminated in the same manner.

■ Line 19—Neutral Conductor

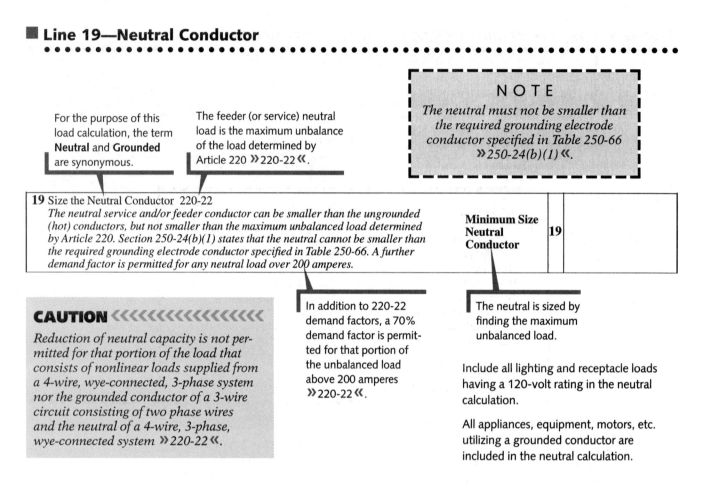

For the purpose of this load calculation, the term **Neutral** and **Grounded** are synonymous.

The feeder (or service) neutral load is the maximum unbalance of the load determined by Article 220 »220-22«.

NOTE

The neutral must not be smaller than the required grounding electrode conductor specified in Table 250-66 »250-24(b)(1)«.

19 Size the Neutral Conductor 220-22
The neutral service and/or feeder conductor can be smaller than the ungrounded (hot) conductors, but not smaller than the maximum unbalanced load determined by Article 220. Section 250-24(b)(1) states that the neutral cannot be smaller than the required grounding electrode conductor specified in Table 250-66. A further demand factor is permitted for any neutral load over 200 amperes.

Minimum Size Neutral Conductor	**19**	

CAUTION «««««««««««««««

Reduction of neutral capacity is not permitted for that portion of the load that consists of nonlinear loads supplied from a 4-wire, wye-connected, 3-phase system nor the grounded conductor of a 3-wire circuit consisting of two phase wires and the neutral of a 4-wire, 3-phase, wye-connected system »220-22«.

In addition to 220-22 demand factors, a 70% demand factor is permitted for that portion of the unbalanced load above 200 amperes »220-22«.

The neutral is sized by finding the maximum unbalanced load.

Include all lighting and receptacle loads having a 120-volt rating in the neutral calculation.

All appliances, equipment, motors, etc. utilizing a grounded conductor are included in the neutral calculation.

■ Line 20—Grounding Conductor

Grounding electrode conductor size is based on the largest service-entrance conductor or equivalent area for parallel conductors »Table 250-66«. (The minimum size service-entrance conductors were determined on Line 18.)

While Table 250-66 is used for service installations, it is also used in other wiring applications, such as separately derived systems »(250-30[a][2])«, and connections at separate buildings or structures »(250-32)«.

Where the grounding electrode conductor connects to made electrode, and that conductor portion is the sole grounding electrode connection, the maximum size required is No. 6 copper or No. 4 aluminum »250-66(a)«.

For a grounding electrode conductor connected to a concrete-encased electrode, serving as the only grounding electrode connection, the maximum size required is No. 4 copper »250-66(b)«.

If the grounding electrode conductor is connected to a ground ring, and that conductor portion is the sole connection to the grounding electrode, the maximum size required is the size used for the ground ring »250-66(c)«.

20 Size the Grounding Electrode Conductor (for Service) 250-66
Using line 18 to find the grounding electrode conductor in Table 250-66.
Size the Equipment Grounding Conductor (for Feeder) 250-122
Use line 17 to find the equipment grounding conductor in Table 250-122. Equipment grounding conductor types are listed in Section 250-118.

Minimum Size Grounding Electrode Conductor . . . or . . . Equipment Grounding Conductor	**20**	

While the equipment bonding jumper on the load side of the service overcurrent devices must be sized, as a minimum, in accordance with Table 250-122 sizes, they do not have to be larger than the circuit conductors supplying the equipment »250-102(d)«.

The equipment grounding conductor size is based on the rating (or setting) of the circuit's automatic overcurrent device ahead of equipment, conduit, etc. »Table 250-122«.

SAMPLE LOAD CALCULATION—STORE

Load Calculation Information

Load calculation data for a small retail store.

Calculate the floor area from the outside dimensions. In this example, the square-foot area is already provided.

Store's sq. ft. area . . . . 1,200 sq. ft.
Receptacles 10
Lighting track . . . . 80 feet
Two sign circuits . . . . 14.5 amperes, 120 volt (each)
Show window . . . . 32 feet
Water heater . . . . 2 kW, 120 volt
Electric heat . . . . 15 kW, 480 volt, 3-phase
Air handler (blower motor) . . . . 1.8 amperes, 480 volt, 3-phase
Air-conditioner compressor . . . . 8.5 amperes, 480 volt, 3-phase
Condenser fan motor . . . . 0.8 amperes, 480 volt, 3-phase

Assume water heater and electric heat kW ratings equivalent to kVA.

The electric supply consists of a 480 volt, 3-phase, 4-wire service.

A three-phase transformer steps down the voltage from 480 to 208Y/120 volts.

Lines 1 and 2—Store

The different style of box on Line 1 (unlike Lines 3 through 14) indicates that the load will become part (if not all) of Line 3. Line 1 is not included in Line 15's total volt-ampere demand load.

Since there are 10 receptacles, the calculated load is 1,800 volt-amperes (10 × 180).

1	Receptacle Load (noncontinuous) 220-3(b)(9) *Multiply each single or multiple receptacle on one strap by 180 volt-amperes.* *Multiply each single piece of equipment comprised of 4 or more receptacles by 90 VA per receptacle.*	1	*1,800*
2	Fixed Multioutlet Assemblies (noncontinuous) 220-3(b)(8) *Where not likely to be used simultaneously, multiply each 5 foot section by 180 volt-amperes.* *Where likely to be used simultaneously, multiply each 1 foot section by 180 volt-amperes.*	2	----

As with Line 1, Line 2 has a different style box from Lines 3 through 14 because the load (if any) becomes part of Line 3. Line 2 is not included in Line 15's total volt-ampere demand load.

Because this store has no fixed multioutlet assembly, a dashed line occupies Line 2.

■ Lines 3 and 4—Store

Line 3 is the first of twelve lines whose values are combined to determine the total volt-ampere demand load.

Because there is no multi-outlet assembly, and the receptacle load is less than 10 kVA, the 1,800 volt-ampere load from Line 1 is inserted in Line 3.

| 3 | Receptacle Load Demand Factor (for nondwelling receptacles) 220-13
If the receptacle load is more than 10,000 volt-amperes, apply the demand factor from Table 220-13.
Add lines 1 and 2. Multiply the first 10 kVA or less by 100%. Then, multiply the remainder by 50%. | | **3** | *1,800* |
| 4 | Unknown Receptacle Load (Banks and Office buildings only)
Where the actual number of general purpose receptacle outlets are
unknown, include 1 volt-ampere per sq. ft. Table 220-3(a) footnote [b] | $1 \times \dfrac{\text{----}}{\text{(sq. ft. outside dimensions)}} =$ | **4** | ---- |

Use Line 4 only when the actual number of receptacles, in banks and office buildings, is unknown.

A dashed line has been drawn in Line 4 showing that it has intentionally been left blank.

■ Lines 5 and 6—Store

The volt-ampere unit load for a store is 3 » Table 220-3(a) « .

Since the store's square foot area is provided, no calculation is required.

Store lighting loads are continuous. The result of Line 5 becomes part of the continuous volt-ampere load calculation (Line 10).

| 5 | General Lighting Load Table 220-3(a)
Multiply the volt-ampere unit load (for the type of
occupancy) by the square foot outside dimensions. | $\dfrac{3}{\text{(VA unit load)}} \times \dfrac{1,200}{\text{(sq. ft. outside dimensions)}} =$ | **5** | *3,600* |
| 6 | Lighting Load Demand Factors 220-11 . . . *Apply Table 220-11 demand factors to certain portions of*
hospitals, hotels, motels, apartment houses (without provisions for cooking), and storage warehouses.
Do not include areas in hospitals, hotels, and motels where the entire lighting will be used at one time. | | **6** | ---- |

Use Line 6 only for certain potions of hospitals, hotels, motels, storage warehouses, and apartment houses without provisions for cooking.

The marking in Line 6 indicates that it has intentionally been left blank.

■ Lines 7 and 8—Store

Except for dwelling units or hotel/motel guest rooms, track lighting loads are additional to Table 220-3(a)'s general lighting loads » 220-12(b) « .

This store contains 80 feet of lighting track.

Base the calculation on every 2 feet, or fraction thereof, of track.

A store's track lighting load is continuous. Line 7's result becomes part of the continuous volt-ampere load calculation (Line 10).

| 7 | Track Lighting (in addition to general lighting) 220-12(b)
Include 150 volt-amperes for every 2 feet,
or fraction thereof, for lighting track. | $\dfrac{80}{\text{(total linear feet)}} \div 2 \quad \underline{40} \times 150 =$ | **7** | *6,000* |
| 8 | Sign and/or Outline Lighting Outlet (where required) 220-3(b)(6)
Each commercial building (or occupancy) accessible to pedestrians must have at least one
outlet per tenant space entrance. 600-5(a) Each outlet must be at least 1,200 volt-amperes. | | **8** | *3,480* |

At least one outlet, having a minimum rating of 1,200 volt-amperes, is required.

This store's actual sign load is 3,420 volt-amperes (2 × 14.5 × 120). This store has two sign circuits.

Since a store's sign load is continuous, the result of Line 8 becomes part of the continuous volt-ampere load calculation (Line 10).

■ Line 9—Store

Not all commercial occupancies have show windows.

The store in this example has 32 linear feet of show window.

Because show window load is continuous, the result of Line 9 becomes part of the continuous volt-ampere load calculation (Line 10).

9 Show Window Lighting 220-12(a) *Include at least 200 volt-amperes for each linear foot, measured horizontally along the show window's base.*	$\underline{32}$ (total linear feet of show window) × 200 =	9	6,400

■ Lines 10 and 11—Store

Lines 5, 7, 8, and 9 represent the store's continuous loads.

3,600 + 6,000 + 3,480 + 6,400 = 19,480

An additional 25% volt-ampere load has been added because of the continuous loads.

10 Continuous Loads 215-2(a), 215-3, and 230-42(a) . . . *Multiply the continuous load volt-amperes (listed above) by 25%. (General purpose receptacles are not considered continuous.)*	$\underline{19,480}$ (total continuous load volt-amperes) × 25% =	10	4,870
11 Kitchen Equipment 220-20 *Multiply three or more pieces of equipment by Table 220-20 demand factor (percent).* Use Table 220-19 for household cooking equipment used in instructional programs. Table 220-19 Note 5		11	– – – –

The marking in Line 11 indicates a lack of kitchen equipment.

■ Lines 12 and 13—Store

Because the combined electric heat and air handler load is larger than the combined compressor, condenser fan motor, and air handler, omit the compressor and condenser fan motor load.

12 Noncoincident Loads 220-21 . . . *The smaller of two (or more) noncoincident loads can be omitted, as long as they will never be energized simultaneously (such as certain portions of heating and A/C systems). Calculate fixed electric space heating loads at 100% of the total connected load. 220-15*	12	16,456
13 Motor Loads 220-4(a), 430-24, 430-25, 430-26, and Article 440 *Motor-driven air-conditioning and refrigeration equipment is found in Article 440. Multiply the largest motor (one motor only) by 25% and add to load.*	13	374

Find the largest motor. In this example that would be the air-handler motor (1.8 × 480 × 1.732 = 1,496.45). Increase the largest motor by 25% (1,496.45 × 25% = 374).

■ Lines 14 and 15—Store

The only load not previously calculated is the water heater. Under normal conditions, a water heater is a noncontinuous load. Add noncontinuous loads to the calculation at 100%.

14 All Other Loads . . . *Add all other noncontinuous loads into the calculation at 100%. Multiply all other continuous loads (operating for 3 hours or more) by 125%.*	14	2,000
15 Total Volt-Ampere Demand Load: *Add Lines 3 through 14 to find the minimum required volt-amperes.*	15	44,980

Add Lines 3 through 14 to find the minimum volt-ampere load (Lines 4, 6, and 11 are blank)

■ Lines 16 and 17—Store

Line 15's minimum volt-ampere load

The total voltage for a 480-volt, three-phase system is 831 (480 × 1.732).

The minimum ampacity for a 480-volt, three-phase service is 54 amperes.

The next-higher standard size fuse (or breaker) above 54 amperes is 60 ≫240-6(a)≪.

| **16** Minimum Amperes *Divide the total volt-amperes by the voltage* | $\dfrac{44,980}{\text{(line 15)}} \div \dfrac{831}{\text{(voltage)}} =$ | **16** | 54 (minimum amperes) | **Minimum Size** **17 Service and/or** **Feeder 240-6(a)** | **17** | 60 |

■ Line 18—Store

The minimum 60/75°C copper service conductor is No. 6 ≫Table 310-16≪.

| **18** Size the Service and/or Feeder Conductors. Tables 310-16 through 310-19 *Use the tables along with Section 310-15(b)(1) through (7) to determine conductor size. If the overcurrent device is rated more than 800 amperes, the conductor ampacity must be equal to, or greater than, the rating of the overcurrent device. 240-3(c)* | **Minimum Size** **Conductors** | **18** | No. 6 Cu No. 4 Al |

The minimum 60/75°C aluminum service conductor is No. 4 ≫Table 310-16≪.

■ Line 19—Store

Although there will be little (if any) load on the neutral, a neutral conductor is required. Where an ac system (operating at less than 1,000 volts) is grounded at any point, the grounded conductor(s) must run to each service disconnecting means, and must bond to each disconnecting means enclosure ≫250-24(b)≪.

While the grounded conductor must not be smaller than the required grounding electrode conductor specified in Table 250-66, it does not have to be larger than the largest ungrounded service-entrance phase conductor ≫250-24(b)(1)≪.

Since the minimum size copper grounding electrode conductor is No. 8, the neutral must also be No. 8.

| **19** Size the Neutral Conductor 220-22 *The neutral service and/or feeder conductor can be smaller than the ungrounded (hot) conductors, but not smaller than the maximum unbalanced load determined by Article 220. Section 250-24(b)(1) states that the neutral cannot be smaller than the required grounding electrode conductor specified in Table 250-66. A further demand factor is permitted for any neutral load over 200 amperes.* | **Minimum Size** **Neutral** **Conductor** | **19** | No. 8 Cu No. 6 Al |

Because the minimum size aluminum grounding electrode conductor is No. 6, the neutral must also be No. 6.

■ Line 20—Store

The grounding electrode conductor's size is based on the service-entrance conductor ≫Table 250-66≪.

The minimum size copper grounding conductor is No. 8.

| **20** Size the Grounding Electrode Conductor (for Service) 250-66 *Using line 18 to find the grounding electrode conductor in Table 250-66.* Size the Equipment Grounding Conductor (for Feeder) 250-122 *Use line 17 to find the equipment grounding conductor in Table 250-122. Equipment grounding conductor types are listed in Section 250-118.* | **Minimum Size** **Grounding Electrode** **Conductor . . . or . . .** **Equipment Grounding** **Conductor** | **20** | No. 8 Cu No. 6 Al |

The minimum size aluminum grounding conductor is No. 6.

The following page shows the store's completed load calculation.

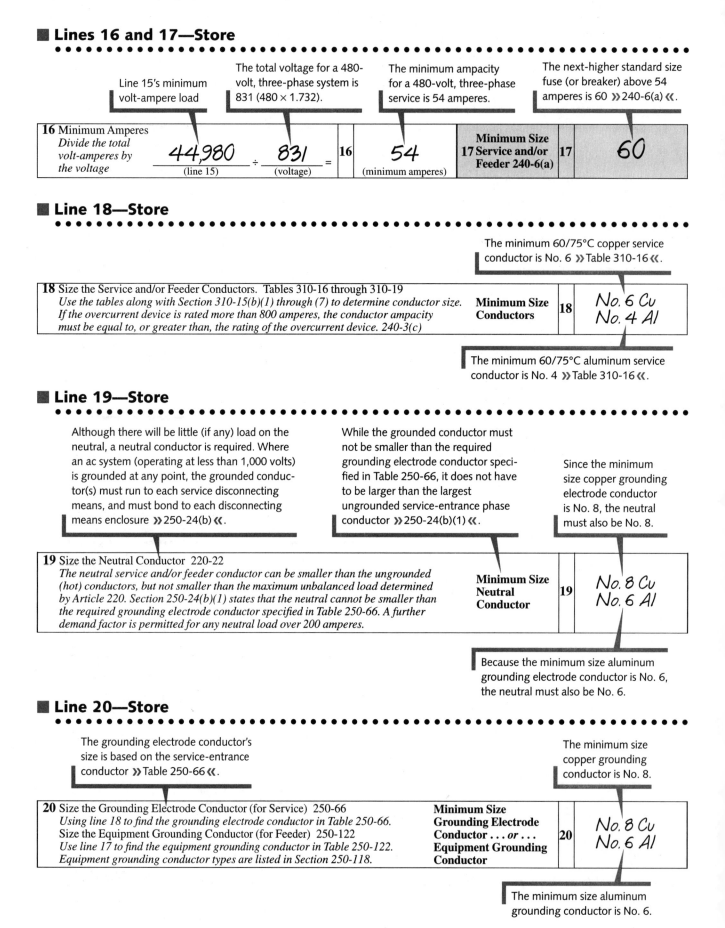

Non-dwelling Feeder/Service Load Calculation

1	Receptacle Load (noncontinuous) 220-3(b)(9) *Multiply each single or multiple receptacle on one strap by 180 volt-amperes.* *Multiply each single piece of equipment comprised of 4 or more receptacles by 90 VA per receptacle.*		1	1,800
2	Fixed Multioutlet Assemblies (noncontinuous) 220-3(b)(8) *Where not likely to be used simultaneously, multiply each 5 foot section by 180 volt-amperes.* *Where likely to be used simultaneously, multiply each 1 foot section by 180 volt-amperes.*		2	----
3	Receptacle Load Demand Factor (for nondwelling receptacles) 220-13 *If the receptacle load is more than 10,000 volt-amperes, apply the demand factor from Table 220-13.* *Add lines 1 and 2. Multiply the first 10 kVA or less by 100%. Then, multiply the remainder by 50%.*		3	1,800
4	Unknown Receptacle Load (Banks and Office buildings only) *Where the actual number of general purpose receptacle outlets are* *unknown, include 1 volt-ampere per sq. ft. Table 220-3(a) footnote [b]*	$1 \times \dfrac{----}{\text{(sq. ft. outside dimensions)}} =$	4	----
5	General Lighting Load Table 220-3(a) *Multiply the volt-ampere unit load (for the type of* *occupancy) by the square foot outside dimensions.*	$\dfrac{3}{\text{(VA unit load)}} \times \dfrac{1,200}{\text{(sq. ft. outside dimensions)}} =$	5	3,600
6	Lighting Load Demand Factors 220-11 . . . *Apply Table 220-11 demand factors to certain portions of hospitals, hotels, motels, apartment houses (without provisions for cooking), and storage warehouses. Do not include areas in hospitals, hotels, and motels where the entire lighting will be used at one time.*		6	----
7	Track Lighting (in addition to general lighting) 220-12(b) *Include 150 volt-amperes for every 2 feet,* *or fraction thereof, for lighting track.*	$\dfrac{80}{\text{(total linear feet)}} \div 2 \quad 40 \times 150 =$	7	6,000
8	Sign and/or Outline Lighting Outlet (where required) 220-3(b)(6) *Each commercial building (or occupancy) accessible to pedestrians must have at least one outlet per tenant space entrance. 600-5(a) Each outlet must be at least 1,200 volt-amperes.*		8	3,480
9	Show Window Lighting 220-12(a) *Include at least 200 volt-amperes for each linear foot,* *measured horizontally along the show window's base.*	$\dfrac{32}{\text{(total linear feet of show window)}} \times 200 =$	9	6,400
10	Continuous Loads 215-2(a), 215-3, and 230-42(a) . . . *Multiply the continuous load volt-amperes (listed above) by 25%. (General purpose receptacles are not considered continuous.)*	$\dfrac{19,480}{\text{(total continuous load volt-amperes)}} \times 25\% =$	10	4,870
11	Kitchen Equipment 220-20 *Multiply three or more pieces of equipment by Table 220-20 demand factor (percent).* Use Table 220-19 for household cooking equipment used in instructional programs. Table 220-19 Note 5		11	----
12	Noncoincident Loads 220-21 . . . *The smaller of two (or more) noncoincident loads can be omitted, as long as they will never be energized simultaneously (such as certain portions of heating and A/C systems). Calculate fixed electric space heating loads at 100% of the total connected load. 220-15*		12	16,456
13	Motor Loads 220-4(a), 430-24, 430-25, 430-26, and Article 440 *Motor-driven air-conditioning and refrigeration equipment is found in Article 440.* *Multiply the largest motor (one motor only) by 25% and add to load.*		13	374
14	All Other Loads . . . *Add all other noncontinuous loads into the calculation at 100%.* *Multiply all other continuous loads (operating for 3 hours or more) by 125%.*		14	2,000
15	Total Volt-Ampere Demand Load: *Add Lines 3 through 14 to find the minimum required volt-amperes.*		15	44,980

16 Minimum Amperes *Divide the total* *volt-amperes by* *the voltage*	$\dfrac{44,980}{\text{(line 15)}} \div \dfrac{831}{\text{(voltage)}} =$ **16**	$\dfrac{54}{\text{(minimum amperes)}}$	**Minimum Size** **17 Service and/or** **Feeder 240-6(a)** **17**	60

18	Size the Service and/or Feeder Conductors. Tables 310-16 through 310-19 *Use the tables along with Section 310-15(b)(1) through (7) to determine conductor size.* *If the overcurrent device is rated more than 800 amperes, the conductor ampacity* *must be equal to, or greater than, the rating of the overcurrent device. 240-3(c)*	**Minimum Size** **Conductors** **18**	No. 6 Cu No. 4 Al
19	Size the Neutral Conductor 220-22 *The neutral service and/or feeder conductor can be smaller than the ungrounded (hot) conductors, but not smaller than the maximum unbalanced load determined by Article 220. Section 250-24(b)(1) states that the neutral cannot be smaller than the required grounding electrode conductor specified in Table 250-66. A further demand factor is permitted for any neutral load over 200 amperes.*	**Minimum Size** **Neutral** **Conductor** **19**	No. 8 Cu No. 6 Al
20	Size the Grounding Electrode Conductor (for Service) 250-66 *Using line 18 to find the grounding electrode conductor in Table 250-66.* Size the Equipment Grounding Conductor (for Feeder) 250-122 *Use line 17 to find the equipment grounding conductor in Table 250-122.* *Equipment grounding conductor types are listed in Section 250-118.*	**Minimum Size** **Grounding Electrode** **Conductor . . . or . . .** **Equipment Grounding** **Conductor** **20**	No. 8 Cu No. 6 Al

SAMPLE LOAD CALCULATION—BANK

■ Load Calculation Information

Load calculation data for a bank

Bank's outside dimensions . . . . 50' × 100'
Actual inside connected lighting load . . . 16,200 volt-amperes
Receptacles . . . . unknown
Sign . . . . 9 amperes, 120 volt
Show window . . . . none
Parking lot lighting . . . . 57 amperes, 120 volt
Water heater . . . . 4 kW, 208 volt
Electric heating unit #1 . . . . 6 kW, 208 volt, 3-phase
Electric heating unit #2 . . . . 10 kW, 208 volt, 3-phase
Air handler (blower motor) #1 . . . . 4 amperes, 208 volt, 3-phase
Air handler (blower motor) #2 . . . . 4 amperes, 208 volt, 3-phase
Air-conditioner compressor #1 . . . . 22 amperes, 208 volt, 3-phase
Air-conditioner compressor #2 . . . . 22 amperes, 208 volt, 3-phase
Condenser fan motor #1 . . . . 2.5 amperes, 208 volt, 3-phase
Condenser fan motor #2 . . . . 2.5 amperes, 208 volt, 3-phase

Assume water heater and electric heat kW ratings equivalent to kVA.

The electric supply consists of a 208 volt, 3-phase, 4-wire service.

> **NOTE**
> *Because the actual number of receptacles is unknown, Line 1 does not contain a number.*

> **NOTE**
> *Line 2 will not contain a number since the bank has no fixed multi-outlet assemblies,*

■ Lines 3 and 4—Bank

> **NOTE**
> *Use Sections 220-3(b)(8) and (9) for banks (or office buildings) having a known number of receptacles.*

Since Lines 1 and 2 are blank, Line 3 also does not contain a number. A dashed line has been drawn in Line 3 showing that it has been left blank intentionally.

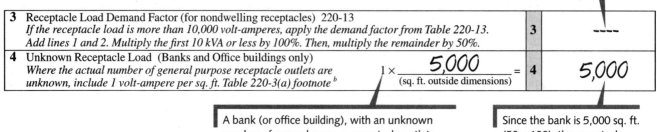

| 3 | Receptacle Load Demand Factor (for nondwelling receptacles) 220-13
If the receptacle load is more than 10,000 volt-amperes, apply the demand factor from Table 220-13.
Add lines 1 and 2. Multiply the first 10 kVA or less by 100%. Then, multiply the remainder by 50%. | | **3** | - - - - |
| 4 | Unknown Receptacle Load (Banks and Office buildings only)
Where the actual number of general purpose receptacle outlets are
unknown, include 1 volt-ampere per sq. ft. Table 220-3(a) footnote [b] | $1 \times \dfrac{5,000}{\text{(sq. ft. outside dimensions)}} =$ | **4** | *5,000* |

A bank (or office building), with an unknown number of general purpose receptacle outlets, requires a rating of 1 volt-ampere per square foot.

Since the bank is 5,000 sq. ft. (50 × 100), the receptacle load is 5,000 volt-amperes.

■ Lines 5 and 6—Bank

> **NOTE**
> *This bank has no lighting track. The dashed line drawn shows that it has intentionally been left blank.*

The volt-ampere unit load for a bank is 3.5 ≫ Table 220-3(a) ≪.

This bank, with outside dimensions of 50' × 100', has a total calculated area of 5,000 sq. ft.

A bank's lighting load is continuous. The result of Line 5 becomes part of the continuous volt-ampere load calculation (Line 10).

5	General Lighting Load Table 220-3(a) *Multiply the volt-ampere unit load (for the type of occupancy) by the square foot outside dimensions.*	$\underset{\text{(VA unit load)}}{3.5} \times \underset{\text{(sq. ft. outside dimensions)}}{5{,}000} =$	5	*17,500*
6	Lighting Load Demand Factors 220-11 . . . *Apply Table 220-11 demand factors to certain portions of hospitals, hotels, motels, apartment houses (without provisions for cooking), and storage warehouses. Do not include areas in hospitals, hotels, and motels where the entire lighting will be used at one time.*		6	----

> **NOTE**
> *Although the actual connected lighting load, excluding exterior lighting, is 16,200 volt-amperes, Table 220-3(a) requires a minimum lighting load of 17,500 volt-amperes (3.5 × 50 × 100). The greater value of 17,500 must be used in this calculation. Use the actual connected lighting load (if any) only when it is greater than the value computed from Table 220-3(a). Actual versus computed lighting loads are both continuous and must be increased by 25%. This function is performed in Line 10 as a separate computation. Another example can be found in Appendix D, Example No. D3.*

Because Line 6 does not apply to this occupancy, a dashed line has been drawn showing that it has been left blank intentionally.

■ Line 8—Bank

> **NOTE**
> *Because this bank has no show window, a show window lighting load is not required.*

Although the bank's actual sign load is 1,080 volt-amperes (9 × 120), each sign outlet must have a minimum rating of 1,200 volt-amperes. Use the actual sign and/or outline load(s) only if the rating is greater than the required minimum.

8	Sign and/or Outline Lighting Outlet (where required) 220-3(b)(6) *Each commercial building (or occupancy) accessible to pedestrians must have at least one outlet per tenant space entrance. 600-5(a) Each outlet must be at least 1,200 volt-amperes.*	8	*1,200*

This bank's sign load is continuous. Therefore, the rating is part of the continuous volt-ampere load calculation (Line 10).

■ Line 10—Bank

This store's continuous loads are found on Lines 5 and 8.

17,500 + 1,200 = 18,700

An additional 25% volt-ampere load is added to the computation because of the continuous loads.

> **NOTE**
> *Line 11 contains a dashed line because there is no kitchen equipment.*

10	Continuous Loads 215-2(a), 215-3, and 230-42(a) . . . *Multiply the continuous load volt-amperes (listed above) by 25%. (General purpose receptacles are not considered continuous.)*	$\underset{\text{(total continuous load volt-amperes)}}{18{,}700} \times 25\% =$	10	*4,675*

■ Lines 12 and 13—Bank

Each air handler load is 1,441 volt-amperes ($4 \times 208 \times 1.732$). Each condenser-fan motor load is 901 volt-amperes ($2.5 \times 208 \times 1.732$). Each compressor motor load is 7,926 volt-amperes ($22 \times 208 \times 1.732$).

The total combined compressor, condenser-fan motor, and air handler load is 20,536 volt-amperes ($7,926 + 7,926 + 901 + 901 + 1,441 + 1,441$).

The total combined electric heat and air handler load is 18,882 ($6,000 + 10,000 + 1,441 + 1,441$).

Because the combined compressor, condenser-fan motor, and air handler load is larger than the combined electric heat and air handler, omit the electric heat and air handler load.

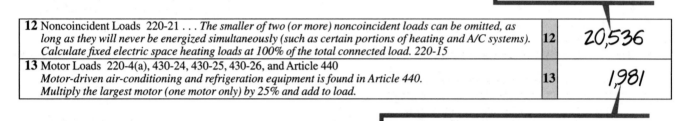

12 Noncoincident Loads 220-21 . . . *The smaller of two (or more) noncoincident loads can be omitted, as long as they will never be energized simultaneously (such as certain portions of heating and A/C systems). Calculate fixed electric space heating loads at 100% of the total connected load. 220-15*	**12**	*20,536*
13 Motor Loads 220-4(a), 430-24, 430-25, 430-26, and Article 440 *Motor-driven air-conditioning and refrigeration equipment is found in Article 440. Multiply the largest motor (one motor only) by 25% and add to load.*	**13**	*1,981*

Find the largest motor. In this example, that is the compressor motor ($22 \times 208 \times 1.732 = 7,925.63$). Increase the largest motor by 25% ($7,925.63 \times 25\% = 1,981$).

■ Lines 14 and 15—Bank

At this point, two loads have not been calculated: one continuous and one noncontinuous load. Because parking-lot lighting is continuous, multiply by 125% ($57 \times 120 \times 125\% = 8,550$ volt-amperes). The water heater is a noncontinuous load. These two loads have a combined total rating of 12,550 volt-amperes ($8,550 + 4,000$).

14 All Other Loads . . . *Add all other noncontinuous loads into the calculation at 100%. Multiply all other continuous loads (operating for 3 hours or more) by 125%.*	**14**	*12,550*
15 Total Volt-Ampere Demand Load: *Add Lines 3 through 14 to find the minimum required volt-amperes.*	**15**	*63,442*

Add Lines 3 through 14 to determine the minimum volt-ampere load (Lines 3, 6, 7, 9, and 11 are blank).

■ Lines 16 and 17—Bank

Line 15's minimum volt-ampere load

The total voltage for a 208 volt, three-phase system is 360 (208×1.732).

The minimum ampacity for a 208 volt, three-phase service is 176 amperes.

The next higher standard size fuse (or breaker) above 176 amperes is 200 ≫240-6(a)≪.

16 Minimum Amperes *Divide the total volt-amperes by the voltage*	*63,442* (line 15)	÷	*360* (voltage)	=	**16**	*176* (minimum amperes)	**Minimum Size 17 Service and/or Feeder 240-6(a)**	**17**	*200*

■ Line 18—Bank

The minimum 75°C copper service conductor is No. 3/0 ≫ Table 310-16 ≪.

18 Size the Service and/or Feeder Conductors. Tables 310-16 through 310-19 *Use the tables along with Section 310-15(b)(1) through (7) to determine conductor size.* *If the overcurrent device is rated more than 800 amperes, the conductor ampacity* *must be equal to, or greater than, the rating of the overcurrent device. 240-3(c)*	**Minimum Size** **Conductors**	**18** *3/0 Cu* *4/0 Al*

The minimum 75°C aluminum service conductor is No. 4/0 ≫ Table 310-16 ≪.

■ Line 19—Bank

The neutral conductor computation requires only loads having a neutral or grounded conductor termination. In this example, include Lines 4, 5, 8, 10, and part of Line 14. Line 14's neutral load calculation is result 8,550 volt-amperes (57 × 120 × 125%). The total neutral load is 36,925 volt-amperes (5,000 + 17,500 + 1,200 + 4,675 + 8,550). The minimum required neutral rating is 103 amperes (36,925 ÷ 360).

The minimum 75°C copper neutral conductor is No. 2 ≫ Table 310-16 ≪.

19 Size the Neutral Conductor 220-22 *The neutral service and/or feeder conductor can be smaller than the ungrounded* *(hot) conductors, but not smaller than the maximum unbalanced load determined* *by Article 220. Section 250-24(b)(1) states that the neutral cannot be smaller than* *the required grounding electrode conductor specified in Table 250-66. A further* *demand factor is permitted for any neutral load over 200 amperes.*	**Minimum Size** **Neutral** **Conductor**	**19** *No. 2 Cu* *1/0 Al*

The minimum 75°C aluminum neutral conductor is No. 1/0 ≫ Table 310-16 ≪.

> ### N O T E
> *While the grounded conductor must not be smaller than the required grounding electrode conductor (specified in Table 250-66), it does not have to be larger than the largest ungrounded service-entrance phase conductor ≫250-24(b)(1)≪. This provision does not apply to this example, because the minimum size neutral is larger than the required grounding electrode conductor.*

■ Line 20—Bank

The grounding electrode conductor's size is based on the service-entrance conductor ≫ Table 250-66 ≪.

The minimum size copper grounding conductor is No. 4.

20 Size the Grounding Electrode Conductor (for Service) 250-66 *Using line 18 to find the grounding electrode conductor in Table 250-66.* Size the Equipment Grounding Conductor (for Feeder) 250-122 *Use line 17 to find the equipment grounding conductor in Table 250-122.* *Equipment grounding conductor types are listed in Section 250-118.*	**Minimum Size** **Grounding Electrode** **Conductor . . . or . . .** **Equipment Grounding** **Conductor**	**20** *No. 4 Cu* *No. 2 Al*

The minimum size aluminum grounding conductor is No. 2.

The following page shows the bank's completed load calculation.

Non-dwelling Feeder/Service Load Calculation

1 Receptacle Load (noncontinuous) 220-3(b)(9) *Multiply each single or multiple receptacle on one strap by 180 volt-amperes* *Multiply each single piece of equipment comprised of 4 or more receptacles by 90 VA per receptacle.*	**1**	----
2 Fixed Multioutlet Assemblies (noncontinuous) 220-3(b)(8) *Where not likely to be used simultaneously, multiply each 5 foot section by 180 volt-amperes.* *Where likely to be used simultaneously, multiply each 1 foot section by 180 volt-amperes.*	**2**	----
3 Receptacle Load Demand Factor (for nondwelling receptacles) 220-13 *If the receptacle load is more than 10,000 volt-amperes, apply the demand factor from Table 220-13.* *Add lines 1 and 2. Multiply the first 10 kVA or less by 100%. Then, multiply the remainder by 50%.*	**3**	----
4 Unknown Receptacle Load (Banks and Office buildings only) *Where the actual number of general purpose receptacle outlets are unknown, include 1 volt-ampere per sq. ft. Table 220-3(a) footnote* [b] $1 \times \underline{5{,}000}$ (sq. ft. outside dimensions) =	**4**	5,000
5 General Lighting Load Table 220-3(a) *Multiply the volt-ampere unit load (for the type of occupancy) by the square foot outside dimensions.* $\underline{3.5}$ (VA unit load) $\times \underline{5{,}000}$ (sq. ft. outside dimensions) =	**5**	17,500
6 Lighting Load Demand Factors 220-11 . . . *Apply Table 220-11 demand factors to certain portions of hospitals, hotels, motels, apartment houses (without provisions for cooking), and storage warehouses. Do not include areas in hospitals, hotels, and motels where the entire lighting will be used at one time.*	**6**	----
7 Track Lighting (in addition to general lighting) 220-12(b) *Include 150 volt-amperes for every 2 feet, or fraction thereof, for lighting track.* $\underline{----}$ (total linear feet) $\div 2 \underline{----} \times 150 =$	**7**	----
8 Sign and/or Outline Lighting Outlet (where required) 220-3(b)(6) *Each commercial building (or occupancy) accessible to pedestrians must have at least one outlet per tenant space entrance. 600-5(a) Each outlet must be at least 1,200 volt-amperes.*	**8**	1,200
9 Show Window Lighting 220-12(a) *Include at least 200 volt-amperes for each linear foot, measured horizontally along the show window's base.* $\underline{----}$ (total linear feet of show window) $\times 200 =$	**9**	----
10 Continuous Loads 215-2(a), 215-3, and 230-42(a) . . . *Multiply the continuous load volt-amperes (listed above) by 25%. (General purpose receptacles are not considered continuous.)* $\underline{18{,}700}$ (total continuous load volt-amperes) $\times 25\% =$	**10**	4,675
11 Kitchen Equipment 220-20 *Multiply three or more pieces of equipment by Table 220-20 demand factor (percent).* Use Table 220-19 for household cooking equipment used in instructional programs. Table 220-19 Note 5	**11**	----
12 Noncoincident Loads 220-21 . . . *The smaller of two (or more) noncoincident loads can be omitted, as long as they will never be energized simultaneously (such as certain portions of heating and A/C systems). Calculate fixed electric space heating loads at 100% of the total connected load. 220-15*	**12**	20,536
13 Motor Loads 220-4(a), 430-24, 430-25, 430-26, and Article 440 *Motor-driven air-conditioning and refrigeration equipment is found in Article 440. Multiply the largest motor (one motor only) by 25% and add to load.*	**13**	1,981
14 All Other Loads . . . *Add all other noncontinuous loads into the calculation at 100%. Multiply all other continuous loads (operating for 3 hours or more) by 125%.*	**14**	12,550
15 Total Volt-Ampere Demand Load: *Add Lines 3 through 14 to find the minimum required volt-amperes.*	**15**	63,442

16 Minimum Amperes *Divide the total volt-amperes by the voltage* $\underline{63{,}442}$ (line 15) $\div \underline{360}$ (voltage) =	**16** 176 (minimum amperes)	**Minimum Size** **17** Service and/or Feeder 240-6(a)	**17** 200

18 Size the Service and/or Feeder Conductors. Tables 310-16 through 310-19 *Use the tables along with Section 310-15(b)(1) through (7) to determine conductor size. If the overcurrent device is rated more than 800 amperes, the conductor ampacity must be equal to, or greater than, the rating of the overcurrent device. 240-3(c)*	**Minimum Size Conductors**	**18** 3/0 Cu 4/0 Al
19 Size the Neutral Conductor 220-22 *The neutral service and/or feeder conductor can be smaller than the ungrounded (hot) conductors, but not smaller than the maximum unbalanced load determined by Article 220. Section 250-24(b)(1) states that the neutral cannot be smaller than the required grounding electrode conductor specified in Table 250-66. A further demand factor is permitted for any neutral load over 200 amperes.*	**Minimum Size Neutral Conductor**	**19** No. 2 Cu 1/0 Al
20 Size the Grounding Electrode Conductor (for Service) 250-66 *Using line 18 to find the grounding electrode conductor in Table 250-66.* Size the Equipment Grounding Conductor (for Feeder) 250-122 *Use line 17 to find the equipment grounding conductor in Table 250-122. Equipment grounding conductor types are listed in Section 250-118.*	**Minimum Size Grounding Electrode Conductor . . . or . . . Equipment Grounding Conductor**	**20** No. 4 Cu No. 2 Al

Summary

- Article 220 contains load calculation procedures.
- Unlike dwelling unit load calculations, receptacle outlets must be included.
- Receptacle loads, including fixed multioutlet assemblies, can be derated if the load is greater than 10,000 volt-amperes.
- If the actual number of receptacle outlets in a bank or office building is unknown, a load of one volt-ampere per square foot must be included.
- Various occupancy types have different volt-ampere unit loads per Table 220-3(a).
- One similarity between dwelling unit and non-dwelling unit load calculations is that a volt-ampere unit load must be multiplied by the square footage using outside dimensions.

- Table 220-11 contains lighting load demand factors for certain portions of hospitals, hotels, motels, storage warehouses, and apartment houses without provisions for cooking.
- Track lighting loads must be included in addition to the general lighting load.
- Some occupancies require a sign or outline lighting outlet.
- Show window minimum volt-ampere lighting loads are based on a linear foot measurement.
- Continuous load calculations must be increased by an additional 25% volt-ampere rating.
- The volt-ampere rating, calculated in accordance with Article 220, represents only the minimum rating.

Unit 13 Competency Test

NEC® Reference **Answer**

_____ _____ 1. What is the demand load contribution for a commercial laundry with ten 5-kVA clothes dryers and ten 1.5-kVA washing machines?

_____ _____ 2. What is the volt-ampere unit load per square foot for a beauty salon?

_____ _____ 3. Outlets for heavy-duty lampholders shall be computed at a minimum of _____ volt-amperes.

_____ _____ 4. What is the minimum general lighting load for an 11,750-square-foot night club with an actual connected lighting load of 22,800 volt-amperes? (For this question, do not apply the continuous load demand factor.)

_____ _____ 5. A restaurant contains the following commercial kitchen equipment: two 12-kW ovens, one 10-kW grill, two 8-kW deep fryers, one 1.2-kW disposer, one 1.5-kW dishwasher, one 10-kW booster heater, and one 4.5-kW water heater. What is the feeder demand load (in kW) for this equipment?

_____ _____ 6. For track lighting in a church, a load of not less than _____ must be included for every _____ of lighting track or fraction thereof.

_____ _____ 7. What is the receptacle demand load for a 9,600-square-foot bank with fifty-two receptacles?

_____ _____ 8. Where it is unlikely that two or more _____ loads will be in use simultaneously, it shall be permissible to use only the largest load(s) that will be used at one time in computing the total load of a feeder or service.

_____ _____ 9. What is the receptacle load (after demand factors) for an office building with 250 receptacle outlets?

_____ _____ 10. What is the minimum general lighting load for an 8,000-square-foot assembly hall? (Do not apply the continuous load demand factor.)

_____ _____ 11. A manufacturing plant has 320 feet of multioutlet assemblies, of which 60 feet will contain equipment subject to simultaneous use. What is the volt-ampere demand load contribution for these multioutlet assemblies?

_____ _____ 12. A school home-economics class has twelve 8-kW household ranges. The kitchen equipment is supplied by a 230-volt, single-phase panelboard. What is the minimum size THW copper conductors required for this panelboard?

_____ _____ 13. A motel has fifty guest rooms, each room measuring 18' x 30'. Each room contains 8 duplex receptacles, of which one is GFCI-protected. Exterior hallways have a total area of 3,000 square feet. What is the lighting and receptacle load (after demand factors) for this motel?

_____ _____ 14. Fixed electric space heating loads shall be computed at _____ percent of the total connected load; however, in no case shall a feeder or service load current rating be less than the rating of the largest branch circuit supplied.

_____ _____ 15. An outlet supplying recessed lighting fixture(s) shall be computed based on _____.

_____ _____ 16. For circuits supplying loads consisting of motor-operated utilization equipment that is fastened in place and that has a motor larger than ⅛ hp in combination with other loads, the total computed load shall be based on _____.

_____ _____ 17. In a commercial storage garage, duplex receptacle outlets shall be computed at not less than _____.

_____ _____ 18. A restaurant contains the following commercial kitchen equipment: one 12-kW range, one 2.8-kW mixer, one 4.2-kW deep fryer, two 0.8-kW soup wells, and one 6-kW water heater. What is the feeder demand load (in kW) for this equipment?

_____ _____ 19. A hospital has 150 patient's rooms, each room measuring 15' × 25'. What is the lighting load (after demand factors) for these rooms?

_____ _____ 20. For circuits supplying lighting units that have ballasts, transformers, or autotransformers, the computed load shall be based on _____.

Questions 21 through 30 are based on an office containing the following:

Office's total square foot area. . . . . 19,000 sq. ft.
Receptacle outlets 200
Multioutlet assemblies . . . . 80 feet
Multioutlet assemblies . . . . 25 feet
   (simultaneously used)
Sign . . . . 15 amperes, 120 volt
Track lighting . . . . 48 feet
Water heater . . . . 4 kW, 208 volt
Four electric heating units . . . . 10 kW, 208 volt (each)
Four air handlers (blower motors) . . . . . . 4 amperes, 208 volt, 3-phase (each)
Four air-conditioner compressors. . . . . . . 24.6 amperes, 208 volt, 3-phase (each)
Eight condenser fan motors. . . . . 2.2 amperes, 208 volt, 3-phase (each)

Assume water heater and electric heat kW ratings equivalent to kVA.

The electric supply is a 208-volt, 3-phase, 4-wire service.

_____ _____ 21. What is the total receptacle (including multioutlet assembly) volt-ampere, service load after demand factors are applied?

_____ _____ 22. What is the general lighting load for this 19,000-square-foot office? (Do not apply the continuous load demand factor.)

NEC® Reference   **Answer**

_____ _____ 23. Without applying the continuous load demand factor, what track lighting load (if any) must be contributed to this office calculation?

_____ _____ 24. What volt-ampere load must be included for the sign? (For this question, do not apply the continuous load demand factor.)

_____ _____ 25. Which load(s), if any, can be omitted?

_____ _____ 26. What is the volt-ampere load contribution for the heating and/or air-conditioning system(s)? (Do not include 25% of the largest motor.)

_____ _____ 27. What is the minimum size service overcurrent protective device required for this office?

_____ _____ 28. A parallel run (2 sets) of service-entrance conductors will be installed in two raceways. What is the minimum size THWN copper ungrounded conductors that can be used?

_____ _____ 29. What is the minimum size THWN copper neutral (grounded) conductors that can be installed?

_____ _____ 30. What is the minimum size copper grounding electrode conductor?

Services, Feeders, and Equipment

Objectives

• •

After studying this unit, the student will:

✔ know the required minimum vertical wiring clearances for each electrical installation.

✔ be aware that other associated equipment, above or below the electrical installation, cannot extend more than 6 inches beyond the front of that electrical equipment.

✔ be familiar with the dedicated space above panelboards, switchboards, and motor control centers.

✔ be aware that a space equal to the equipment's width and depth, above panelboards, switchboards, and motor control centers, must be clear of foreign systems unless protection is provided.

✔ know that working space clearances vary depending on the conditions and voltages.

✔ understand the conditions which require two equipment working space entrances.

✔ know where to connect grounded and grounding conductors on both the service supply side and load side.

✔ understand the conditions which require ground-fault protection of equipment.

✔ have a good understanding of transformer and generator provisions.

✔ know that, while grounding and bonding connection points must be made in the same place, this can be either the panel or the transformer.

✔ have an extensive understanding of busway provisions.

Introduction

• •

Unit 14 contains electrical provisions relative to commercial type occupancies. Before beginning this unit, a review of Unit 9 (Services and Electrical Equipment) is recommended. Some of the same provisions which apply to one-family dwelling likewise apply to commercial installations. Vertical wiring clearances, found in Articles 225 and 230, apply to all electrical feeder and service installations. Also applicable to both dwelling and commercial areas is Section 110-26, which contains minimum working space provisions (600 volts or less) above, below, and in front of electrical equipment. Some requirements, such as for equipment rated 1,200 amperes or more, over 6 feet wide, and containing devices (overcurrent, switching, or control), are not included in one-family dwellings. This unit contains separately derived system provisions which pertain to certain battery, solar photovoltaic, generator, transformer, and converter winding-system installations. Due to space limitations, only transformer illustrations are used to explain these provisions. In addition to the separately derived system provisions, transformers have their own article (450) which covers all transformer installations. Following the transformer material is an illustration briefly touching on generator provisions. Although generators have their own article (445), other generator provisions are found throughout of the *Code*. Generators and their associated wiring and equipment must also adhere to applicable provisions of Articles 695, 700, 701, 702, and 705. Busway provisions close this unit. Busways can also be found in multiple occupancy types, i.e., multifamily, commercial, and industrial.

Since state and local jurisdictions may modify *National Electrical Code®* provisions, it is expedient to obtain a copy of local rules and regulations.

CLEARANCES AND WORKING SPACE

■ Vertical Clearances

A 10-foot minimum vertical clearance is required from any platform or projection, from which overhead conductors (of 150 volts, or less, to ground) might be reached ≫230-24(b) and 225-18≪.

A minimum 12-foot vertical clearance is required over commercial areas (not subject to truck traffic) provided conductor voltage does not exceed 300 volts to ground ≫230-24(b) and 225-18≪.

The minimum vertical clearance increases from 12 feet to 15 feet for conductors whose voltage exceeds 300 volts to ground ≫230-24(b) and 225-18≪.

A drip loop's lowest point has a minimum vertical clearance of 10 feet ≫230-24(b)≪.

The point of attachment of the service-drop conductors to a building (or other structure) must meet Section 230-24 minimum clearance specifications. This point of attachment must always be at least 10 feet above finished grade ≫230-26≪.

Public streets, alleys, roads, parking areas subject to truck traffic, driveways on non-residential property, and other land traversed by vehicles (such as cultivated, grazing, forest, and orchard) require a minimum 18-foot vertical clearance ≫230-24(b) and 225-18≪.

18 Feet

■ Maximum Depth of Associated Equipment

The working space depth, in the direction of access to live parts, must not be less than indicated by Table 110-26(a) ≫110-26(a)(1)≪.

All electrical equipment must be constantly surrounded by sufficient access and work space permitting ready and safe equipment operation and maintenance. Electrical apparatus enclosures, controlled by lock and key, are considered accessible to qualified persons ≫110-26≪.

Wireways are troughs (sheet-metal or nonmetallic) with hinged or removable covers used to house, as well as protect electric wires and cables. A wireway must be installed as a complete system before conductors are laid in place ≫362-1 and 362-14≪. Article 362 outlines wireway provisions.

Other associated equipment, located above or below electrical equipment, cannot extend more than **6 inches** beyond the front of that electrical equipment ≫110-26(a)(3)≪.

Side View

Panelboard

6 inches maximum

Trough

More than 6 inches

Transformer

Top View

Trough

Panelboard

6 inches maximum

Transformer

More than 6 inches

NOTE

Surrounding space (equal to the equipment's width and depth) must be clear of foreign systems, unless protection is provided against damage from condensation, leaks, or breaks in such foreign systems. This zone extends from the top of the equipment to the structural ceiling ≫110-26(f)(1)(b)≪. A dropped, suspended, or similar ceiling, that does not strengthen the building, is not a structural ceiling ≫110-26(f)(1)(d)≪. Sprinkler protection can occupy dedicated space, provided the piping complies with Section 110-26 ≫110-26(f)(1)(c)≪.

■ Dedicated Equipment Space

Space equal to the equipment's width and depth must be clear of foreign systems **unless protection is provided** to avoid damage from condensation, leaks, or breaks in such foreign systems. This zone extends from the top of the electrical equipment to the structural ceiling ➤➤110-26(f)(1)(b)◀◀.

The space equal to the equipment's width and depth extending from the floor to a height of 6 feet above the equipment or to the structural ceiling (whichever is lower) is dedicated to the electrical installation. No piping, ducts, or foreign equipment shall be located in this zone ➤➤110-26(f)(1)(a)◀◀.

Section 110-26(f) applies only to motor control centers and equipment within the scope of Article 384.

Section 110-26(f)(1)(b) pertains to piping, ducts, etc. which could damage electrical equipment due to condensation, leaks, or breaks.

Foreign systems may enter the zone if the electrical equipment is protected from condensation, leaks, or breaks in such foreign systems ➤➤110-26(f)(1)(b)◀◀.

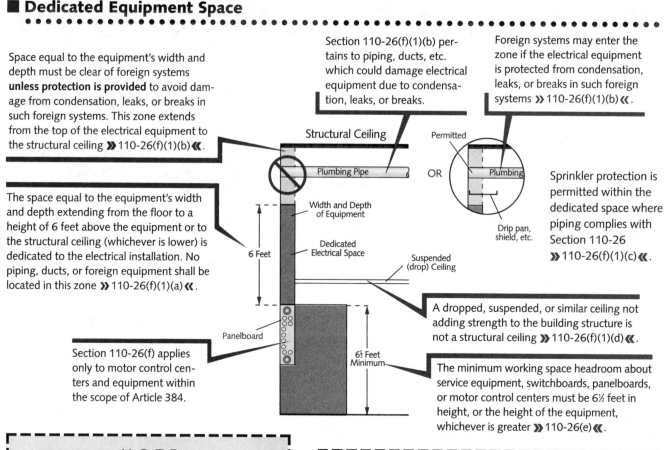

Sprinkler protection is permitted within the dedicated space where piping complies with Section 110-26 ➤➤110-26(f)(1)(c)◀◀.

A dropped, suspended, or similar ceiling not adding strength to the building structure is not a structural ceiling ➤➤110-26(f)(1)(d)◀◀.

The minimum working space headroom about service equipment, switchboards, panelboards, or motor control centers must be 6½ feet in height, or the height of the equipment, whichever is greater ➤➤110-26(e)◀◀.

> **N O T E**
> *Equipment isolated from the foreign equipment by height, physical enclosures, or covers affording adequate mechanical protection from vehicular traffic or accidental unauthorized personnel contact; or, that complies with 110-26(f)(1)(b), shall be permitted in areas that do not have the dedicated space described in Section 110-26(a)* ➤➤*110-26(f)(1)(a) Exception*◀◀*.*

> **N O T E**
> *Outdoor electrical equipment must be installed in suitable enclosures and must be protected from accidental contact by unauthorized personnel, or by vehicular traffic, or by accidental piping system leakage. The working clearance space includes the zone described in Section 110-26(a). No architectural appurtenance or other equipment can occupy this zone* ➤➤*110-26(f)(2)*◀◀*.*

■ Working Space Access and Entrances

One access means is permitted where the location provides a continuous and unobstructed exit ➤➤110-26(c) Exception No. 1◀◀.

When normally enclosed live parts are exposed for inspection or servicing, any surrounding work space, in a passageway or general open area, must be suitably guarded ➤➤110-26(b)◀◀.

> **N O T E**
> *Equipment operating over 600 volts, nominal, must comply with Article 110, part C.*

Equipment rated 1,200 amperes (or more) and over 6 feet wide containing overcurrent devices, switching devices, or control devices, requires an entrance at least 24 inches wide and 6½ feet high at each end of the working space ➤➤110-26(c)◀◀.

If the work space specified by Section 110-26(a) is doubled, only one entrance to the working space is required. The edge of the entrance nearest the equipment must meet the minimum clear distance given in Table 110-26(a) ➤➤110-26(c) Exception No. 2◀◀.

> **WARNING**
> Do not use working space required by Section 110-26 for storage ➤➤110-26(b)◀◀.

■ Working Space

• •

Condition 1—Exposed live parts on one side with no live or grounded parts on the other side of the working space, or exposed live parts on both sides effectively guarded by suitable wood or other insulating materials. Insulated wire/busbars operating at 300 volts or less to ground, are not considered live parts » Table 110-26(a) «.

Sufficient access and working space must be continuously maintained about all electric equipment permitting ready, safe equipment operation and maintenance. Enclosures housing electrical apparatus, controlled by lock and key, are considered accessible to qualified persons » 110-26 «.

The working space depth in the direction of access to live parts must meet Table 110- 26(a) specifications. If exposed, measure the distance from the live part(s). Where enclosed, distance measurement begins from the enclosure front (or opening) » 110-26(a)(1) «.

0 to 600 volts

Non-grounded Surface

←3'→

Working space is not required behind or beside assemblies, such as dead-front switchboards or motor control centers, having no renewable or adjustable parts (such as fuses or switches) on the back or sides, provided all connections are accessible from other locations » 110-26(a)(1) *Exception No. 1* «.

Where rear access is required to work on de-energized parts at the back of enclosed equipment, a minimum working space of 30 inches horizontally is required » 110-26(a)(1) *Exception No. 1* «.

CAUTION «««««««

Working space does not apply only to panelboards and switchboards . . . Working space for equipment operating at 600 volts, nominal, or less to ground and likely to require examination, adjustment, servicing, or maintenance while energized must comply with Section 110-26(a)(1) through (3) dimensions, or as required/permitted elsewhere in the Code » 110-26(a) «.

Condition 2—Exposed live parts on one side with grounded parts on the other side. Concrete, brick, or tile walls qualify as grounded » Table 110-26(a) «.

0 to 150 volts

Grounded Surface

←3'→

151 to 600 volts

Grounded Surface

←3½'→

N O T E

By special permission, smaller spaces are allowed where all uninsulated parts are no greater than 30 volts rms, 42 peak, or 60 volts dc » 110-26(a)(1) Exception No. 2 «.

Condition 3—The work space having exposed live parts on both sides (not guarded as provided in Condition 1) with the operator between » Table 110-26(a) «.

0 to 150 volts

←3'→

151to 600 volts

←4'→

N O T E

Equipment operating over 600 volts, nominal, must comply with Article 110, part C.

In replacing equipment within existing buildings, Condition 2 working clearance is permitted between dead-front switchboards, and panelboards, or motor control centers located across the aisle from each other where: (1) conditions of maintenance and supervision ensure that written procedures have been adopted to prohibit equipment on both sides of the aisle from being open at the same time, and; (2) authorized qualified personnel will service the installation » 110-26(a)(1) *Exception No. 3* «.

■ Guarding and Warning

Except as required (or permitted) elsewhere by the *NEC®*, live parts of electric equipment operating at 50 volts or more must be guarded against accidental contact by approved enclosures or by any of the following means listed in Section 110-27(a):

(1) By location in a room, vault, or similar enclosure accessible only to qualified persons.

(2) By suitable permanent, substantial partitions (or screens) arranged so that only qualified persons have access to the space within reach of live parts. Any openings in such partitions (barriers) must be sized and located so that accidental human contact with the live parts and/or conducting objects is unlikely.

(3) By locations on a suitable balcony, gallery, or elevated platform whose arrangement excludes unqualified persons.

(4) By elevations of 8 feet or more above the floor or other working surface.

Where potential for physical damage to electric equipment exists, enclosures (or guards) must be of sufficient strength and placement to prevent such damage ≫110-27(b) ≪.

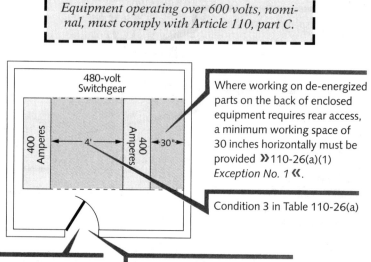

NOTE

Equipment operating over 600 volts, nominal, must comply with Article 110, part C.

Where working on de-energized parts on the back of enclosed equipment requires rear access, a minimum working space of 30 inches horizontally must be provided ≫110-26(a)(1) *Exception No. 1* ≪.

Condition 3 in Table 110-26(a)

At least one entrance of sufficient area must be provided giving access to the electrical equipment's working space ≫110-26(c) ≪.

Entrances to rooms and other guarded locations containing exposed live parts must be marked with conspicuous warning signs forbidding entrance of unqualified persons ≫110-27(c) ≪.

SWITCHBOARDS AND PANELBOARDS

■ Phase Arrangement

The phase arrangement on 3-phase busses are A, B, C from front to back, top to bottom, or left to right, as viewed from the front of the switchboard or panelboard. The B phase must be that phase having the higher voltage to ground on 3-phase, 4-wire, delta-connected systems. Other clearly marked busbar arrangements are permitted for existing installation additions ≫384-(3)(f) ≪.

Equipment sharing the same single section or multisectional switchboard or panelboard as the meter (on 3-phase, 4-wire, delta-connected systems) is allowed to have the same phase configuration as the metering equipment ≫384-3(f) *Exception No. 1* ≪.

Top to Bottom

Front to Back

Left to Right

NOTE

A switchboard or panelboard supplied from a 4-wire, delta-connected system where the midpoint of one phase winding is grounded, requires the phase busbar or conductor having the higher voltage to ground to be durably and permanently marked by an outer orange finish, or by other effective means ≫384-3(e) ≪.

■ Switchboards

A switchboard is a large single panel, frame, or assembly of panels with front and/or rear mounted, switches, protective devices (overcurrent, etc.), buses, and usually instruments. Switchboards are generally accessible from both the front and the rear and are not intended for cabinet installation ≫Article 100 ≪.

Each switchboard or panelboard used as service equipment must be provided with a main bonding jumper sized according to Section 250-28(d) or the equivalent placed within the panelboard or one of the switchboard sections for the grounded service conductor connection (on its supply side) to the switchboard or panelboard frame. All switchboard sections must be bonded together using an equipment-grounding conductor sized per Table 250-122 ≫384-3(c) ≪.

Switchboards having exposed live parts must be: (1) located in permanently dry locations; (2) under competent supervision, and (3) accessible only to qualified persons. Switchboards must be located to minimize the probability of damage from equipment or processes ≫384-5 ≪.

Anything other than a totally enclosed switchboard, requires a space not less than 3 feet between the top of the switchboard and any combustible ceiling, unless a noncombustible shield separates them ≫384-8(a) ≪.

Conduit or other raceway entering a switchboard, floor-standing panelboard, or similar enclosure from the bottom, requires sufficient space to permit installation of conductors in the enclosure. The wiring space must meet Table 384-10 specifications where the conduit or raceway enters or leaves the enclosure below the busbars, their supports, or other obstructions. The conduit or raceway, including end fittings, must not rise more than 3 inches above the bottom of the enclosure ≫384-10 ≪.

Position switchboards to reduce the probability of communicating fire to adjacent combustible materials to a minimum. Where installed over a combustible floor, suitable protection must be provided ≫384-7 ≪.

> ### NOTE
> *A bushing, or terminal fitting, with an integral bushed opening must be used at the end of a conduit or other raceway terminating underground where the conductors or cables emerge as a direct burial wiring method. A seal incorporating the same physical protection characteristics can be used in lieu of a bushing ≫300-5(h) ≪.*

■ Power Panelboard

> ### NOTE
> *A power panelboard, used as service equipment, with multiple disconnecting means per Section 230-71, does not require an individual overcurrent protective device ≫384-16(b) Exception ≪.*

All panelboards must have a rating not less than the minimum feeder capacity required for the load computed according to Article 220 ≫384-13 ≪.

Panelboards must be durably marked by the manufacturer with the voltage, current rating, and the number of phases for which they are designed. The manufacturer's name or trademark must also be visible after installation, without disturbing the interior parts or wiring ≫384-13 ≪.

Panelboards classifications are either: (1) lighting and appliance branch-circuit; or (2) power ≫384-14 ≪.

A 3-pole circuit breaker counts as three overcurrent devices when determining the maximum number allowed ≫384-15 ≪.

The purpose (or use) of a panelboard's circuits (and circuit modifications) must be legibly identified on a circuit directory affixed to the panel door(s) face or inside ≫384-13 ≪. The marking must durably withstand surrounding environmental conditions ≫110-22 ≪.

No more than 10% of a power panelboard's overcurrent devices protect lighting and appliance branch circuits ≫384-14(b) ≪.

In addition to Section 384-13 requirements, a power panelboard whose supply conductors include a neutral, having more than 10% of its overcurrent devices protecting branch circuits rated 30 amperes (or less), must be protected on the supply side by an overcurrent protective device with a rating not greater than that of the panelboard ≫384-16(b) ≪.

■ Lighting and Appliance Branch-Circuit Panelboards

Lighting and appliance branch circuit panel-boards physically inhibit the installation of excess overcurrent devices. These panelboards accommodate only that number for which the panelboard was designed, rated and approved »384-15«.

A lighting and appliance branch circuit connects to the panelboard neutral, and has overcurrent protection of 30 amperes or less in one or more conductors »384-14(a)«.

A 2-pole circuit breaker counts as two overcurrent devices for Section 384-15 purposes.

Individual protection for a lighting and appliance panelboard is not required if the feeder has overcurrent protection not exceeding the panelboard's rating »384-16(a) *Exception No. 1*«.

A grounding connection must not be made to any grounded circuit conductor on the load side of the service disconnecting means, except as otherwise permitted in Article 250 »250-24(a)(5)«.

More than 10% of a lighting and appliance branch-circuit panel-board's overcurrent devices protect lighting and appliance branch circuits »384-14(a)«.

41   42

A maximum of 42 overcurrent devices (excluding those provided for in the mains) can be installed in any one cabinet/cutout box of a lighting and appliance branch circuit panelboard »384-15«.

┌─────────────────────────────┐
│   N O T E   │
│ *Panelboards used as service equip-* │
│ *ment in pre-existing single residen-* │
│ *tial occupancy installations, do* │
│ *not require individual protection* │
│ *for lighting and appliance branch-* │
│ *circuit panelboards »384-16(a)* │
│ *Exception No. 2«.* │
└─────────────────────────────┘

┌─────────────────────────────┐
│   N O T E   │
│ *Each lighting and appliance branch circuit* │
│ *panelboard must be individually protected* │
│ *on the supply side by no more than two* │
│ *main circuit breakers (or two sets of fuses)* │
│ *having a combined rating not greater than* │
│ *that of the panelboard »384-16(a)«.* │
│ *(See the two exceptions to this main rule.)* │
└─────────────────────────────┘

■ Grounding—Supply Side of Service

> **NOTE**
>
> *The interior metal water piping system must be bonded to either the service equipment enclosure, the grounded conductor at the service, the grounding electrode conductor (if of sufficient size), or the grounding electrode(s) used. The bonding jumper, sized according to Table 250-66, must be installed per Section 250-64(a), (b), and (e). The bonding jumper's point of attachment must be accessible* » 250-104(a) «.

Section 250-92(a) requires that the noncurrent-carrying metal equipment parts indicated below must be effectively bonded together:

(1) Service raceways, cable trays, cablebus framework, or service cable armor/sheath except as Section 250-84 permits.

(2) All service enclosures containing service conductors, including meter fittings, boxes, or the like, interposed in the service raceway or armor.

(3) Any metallic raceway or armor enclosing a grounding electrode conductor as specified in Section 250-64(b). Bonding applies at each end and to all intervening raceways, boxes, and enclosures between the service equipment and the grounding electrode.

Ungrounded (hot) Conductors

Size the grounding electrode conductor according to Section 250-66 for the derived phase conductors » 250-30(a)(2) «.

The connection between the grounded circuit conductor (neutral) and the equipment grounding conductor at the service is the main bonding jumper » Article 100 «. If a screw serves as the main bonding jumper, it must be identified with a green finish which remains visible after installation » 250-28(b) «.

Main Bonding Jumper

Grounding Electrode Conductor

Neutral and Equipment Grounding Terminal Bar

Grounded (Neutral) Conductor

Two to six circuit breakers (or sets of fuses) are permitted as the overcurrent device providing overload protection. The sum of the circuit breakers (or fuses) ratings can exceed the service conductor's ampacity, provided the calculated load (per Article 220) does not exceed the ampacity of the service conductors » 230-90(a) *Exception No. 3* «.

Each service-disconnecting means permitted by Section 230-2, or for each set of service-entrance conductors permitted by Section 230-40 Exception Nos. 1 or 3, must consist of no more than six switches/ circuit breakers mounted within a single enclosure, in a group of separate enclosures, or in (or on) a switchboard. The maximum number of disconnects per service, grouped in any one location, is six » 230-71(a) «.

The bonding jumper must meet the size requirements shown in Table 250-66 for grounding electrode conductors. If the service-entrance phase conductors are larger than 1,100 kcmil copper (or 1,750 kcmil aluminum), the bonding jumper must have an area not less than 12½ percent of the largest phase conductor's area. If the phase conductors and the bonding jumper are of different materials (copper or aluminum), the minimum size of the bonding jumper is based on the assumed use of the same material and with an ampacity equal to that of the installed phase conductors. Where the service-entrance conductors are paralleled in multiple raceways or cables, the equipment bonding jumper, where routed with the raceway or cables, must also run in parallel. The bonding jumper size for each raceway or cable must be based on the size of the service-entrance conductors in each raceway or cable » 250-102(c) «.

A grounded circuit conductor can ground noncurrent-carrying metal equipment parts, raceways, and other enclosures at any of the following locations listed in Section 250-142(a):

(1) On the supply side, or within the enclosure, of the ac service-disconnecting means

(2) On the supply side, or within the enclosure, of the main-disconnecting means for separate buildings as provided in Section 250-32(b)

(3) On the supply side, or within the enclosure, of the main-disconnecting means or overcurrent devices of a separately derived system where Section 250-30(a)(1) permits.

■ Grounding—Load Side of Service

Equipment bonding jumpers (on the load side of the service overcurrent devices) must be: (1) sized, as a minimum, in accordance with Table 250-122; but (2) are not required to be larger than the circuit conductors supplying the equipment; and (3) cannot be smaller than No. 14 copper ≫ 250-102(d) ≪ .

While equipment grounding conductors of wire type copper, aluminum, or copper-clad aluminum must not be less than shown in Table 250-122, they are not required to be larger than the circuit conductors supplying the equipment. A raceway, cable armor, or sheath used as the equipment grounding conductor (as provided in Section 250-134(a) and 250-118), must comply with Section 250-2(d) ≫ 250-122(a) ≪ .

Do *not* make a grounding connection to any grounded circuit conductor on the load side of the service disconnecting means except as otherwise allowed in Article 250 ≫ 250-24(a)(5) ≪ .

Equipment grounding conductors can be bare, covered, or insulated. Individually covered (or insulated) equipment grounding conductors must have a continuous outer finish, either green or green with yellow stripe(s), unless otherwise permitted by Article 250, Part F ≫ 250-119 ≪ .

Ungrounded (hot) Conductors

Grounding Conductor

Bonding Jumper **Not** Installed

Equipment Grounding Terminal Bar

Neutral Terminal Bar

Grounded (Neutral) Conductor

Unless grounded by connection to the grounded circuit conductor (permitted by Sections 250-32, 250-140, and 250-142), noncurrent-carrying metal parts of equipment, raceways, and other enclosures, must be grounded (where required) by one of the following methods: (a) by any of the equipment grounding conductors permitted by Section 250-118; or (b) by an equipment grounding conductor contained within the same raceway, cable, or otherwise run with the circuit conductors ≫ 250-134 ≪ .

Individual overcurrent protection for a lighting and appliance panelboard is not required if the panelboard feeder has overcurrent protection which does not exceed the panelboard rating ≫ 384-16(a) *Exception No. 1* ≪ .

Except as permitted in Sections 250-30(a)(1) and 250-32(b), a grounded circuit conductor cannot be used for grounding noncurrent-carrying metal equipment parts on the load side of the service disconnecting means, or on the load side of a separately derived system disconnecting means, or the overcurrent device for a separately derived system not having a main disconnecting means, unless an exception has been met ≫ 250-142(b) ≪ .

Metal panelboard cabinets and frames must physically contact each other and be grounded. If the panelboard is used with nonmetallic raceway (or cable), or if separate grounding conductors are provided, secure a terminal bar for the grounding conductors within the cabinet. The terminal bar must be bonded to the metal cabinet and panelboard frame. Otherwise, connect it to the grounding conductor which runs with the conductors feeding the panelboard ≫ 384-20 ≪ .

NOTE

Paralleled conductors within multiple raceways or cables (permitted in Section 310-4) require that the equipment grounding conductors (where used) also run in parallel within each raceway or cable. Each parallel equipment grounding conductor must be sized on the basis of the ampere rating of the overcurrent device protecting the circuit conductors in the raceway or cable per Table 250-122, unless ground-fault protection of equipment is implemented ≫ 250-122(f) ≪ .

Equipment grounding conductor and equipment bonding jumper sizes (on the load side of the service) are determined by the size of the overcurrent device protecting the equipment.

Where conductor sizes are adjusted to compensate for voltage drop, equipment grounding conductors (if required) must be proportionate, according to circular mil area ≫ 250-122(b) ≪ .

▪ Ground-Fault Protection of Equipment

Ground-fault protection of equipment must be provided for solidly grounded wye electrical systems of more than 150 volts to ground, but not exceeding 600 volts phase-to-phase for each service disconnect rated 1,000 amperes or more. **Solidly Grounded** means that the grounded conductor is grounded without implementing any resistor or impedance device 》 230-95 and 240-13 《.

Ground-fault protection systems, when activated, must cause the service disconnect to open all ungrounded conductors of the faulted circuit. The maximum ground-fault setting is 1,200 amperes 》 230-95(a) 《.

One second is the maximum time delay for ground-fault currents of 3,000 amperes or more 》 250-95(a) 《.

The service disconnect rating is considered to be the rating of the largest fuse that can be installed, or the highest continuous current trip setting for which the actual overcurrent device installed in a circuit breaker is rated (or can be adjusted) 》 230-95 《.

480Y/277 Volt Supply

1,000 Amperes or More

Ground-Fault Protection of Equipment is a system intended to provide protection of equipment from damaging line-to-ground fault currents by causing the disconnecting means to open all ungrounded conductors of the faulted circuit. This protection is provided at current levels less than those required to protect conductors from damage through the operation of a supply circuit overcurrent device 》 Article 100 《.

> **N O T E**
>
> *The ground-fault protection provisions of Section 230-95 do not apply to a service disconnect for a continuous industrial process where a nonorderly shutdown introduces additional or increased hazards* 》 *230-95 Exception No. 1 and 240-13(1)* 《.

CAUTION ≪≪≪≪≪≪≪≪≪≪≪≪≪

Bonding for circuits over 250 volts to ground must comply with Section 250-97.

If a switch and fuse combination is used, the fuses employed must be capable of interrupting any current higher than the interrupting capacity of the switch when the ground-fault protective system will not cause the switch to open 》 250-95(b) 《.

Ground-fault protection that opens the service disconnect will afford no protection from faults on the line side of the protective element. It serves only to limit damage to conductors and equipment only on the load side of the protective element in the event of an arcing ground fault 》 230-95(c) FPN No. 1 《.

This added protective apparatus at the service equipment may necessitate a review of the overall wiring system for proper selective overcurrent protection coordination. Additional installations of ground-fault protective equipment may be needed on feeders and branch circuits requiring maximum continuity of electrical service 》 230-95(c) FPN No. 2 《.

Where ground-fault protection is provided for the service disconnect, and another supply system is interconnected by a transfer device, a method may be needed to ensure proper sensing by the ground-fault protection equipment 》 230-95(c) FPN No. 3 《.

> **N O T E**
>
> *Section 230-95 ground-fault protection provisions do not apply to fire pumps* 》 *230-95 Exception No. 2 and 240-13(3)* 《.

> **N O T E**
>
> *Services exceeding 600 volts, nominal, must comply with Article 230, Part H.*

SEPARATELY DERIVED SYSTEMS

■ Transformers

The intended meaning of **transformer** is an individual transformer, single- or polyphase, identified by a single nameplate, unless otherwise indicated by Article 450 ≫450-2≪.

Overcurrent protection for transformers over 600 volts, nominal, must be provided in accordance with Table 450-3(a) ≫450-3(a)≪.

Each transformer must have a nameplate giving the manufacturer's name; rated kilovolt-amperes; frequency; primary and secondary voltage; impedance of transformers (25 kVA and larger); required clearances for transformers with ventilating openings; and the amount and kind of insulating liquid (where used). In addition, the nameplate of each dry-type transformer must include the insulation system's temperature class ≫450-11≪.

Transformers with ventilation openings must be installed so that such openings are not blocked by walls or other obstructions. The required clearances must be legibly marked on the transformer ≫450-9≪.

Dry-type transformers rated 600 volts, nominal, or less, and not exceeding 50 kVA, are permitted in hollow spaces of buildings which are not permanently closed in, provided they meet Section 450-9 ventilation requirements and Section 450-21(a) separation from combustible material requirements. Transformers so installed do not have to be readily accessible ≫450-13(b)≪.

Exposed noncurrent-carrying metal parts of transformer installations (fences, guards, etc.), must be grounded, where required, under the conditions and in a manner specified for electric equipment and other exposed metal parts in Article 250 ≫450-10≪.

It is not required that dry-type transformers rated 600 volts, nominal, or less, located in the open on walls, columns, or structures, be readily accessible ≫450-13(a)≪.

Article 450 covers the installation of all transformers except:

(1) Current transformers.

(2) Dry-type transformers that constitute a component of other apparatus and comply with these requirements.

(3) Transformers that are an integral part of X-ray, high-frequency, or electrostatic-coating apparatus.

(4) Transformers used with Class 2 and Class 3 circuits in compliance with Article 725.

(5) Sign and outline lighting transformers that comply with Article 600.

(6) Transformers for electric-discharge lighting per Article 410.

(7) Power-limited fire alarm circuit transformers which comply with Article 760, Part C.

(8) Transformers used for research, development, or testing, where effective safeguards protect persons from contacting energized parts ≫450-1 *Exception Nos. 1 through 8*≪.

Article 450 covers the installation of transformers dedicated to supplying power to a fire pump installation as modified by Article 695 ≫450-1≪.

Article 450 covers the installation of transformers in hazardous (classified) locations as modified by Articles 501 through 504 ≫450-1≪.

Autotransformers are covered in Sections 450-4 and 5.

Appropriate provisions must be made to minimize the possibility of damage to transformers from external causes where the transformers are exposed to such dangers ≫450-8(a)≪.

Use primary fuses to protect indoor, or otherwise enclosed, voltage transformers ≫450-3(c)≪.

Article 450, Part B covers specific provisions applicable to a variety of transformers.

Article 450, Part C covers transformer vaults.

> **N O T E**
>
> *All transformers and their vaults must be readily accessible to qualified personnel for inspection and maintenance, or must otherwise meet Section 450-13(a) or (b) requirements* ≫450-13≪.

■ Transformer—Primary Only Protection
9 Amperes or More—600 Volts or Less

Where 125% of this current does not correspond to a standard fuse or nonadjustable circuit breaker rating, the next higher rating described in Section 240-6 is permitted » Table 450-3(b) Note 1 «.

No secondary overcurrent protection is required for certain transformers 600 volts, nominal, or less, with currents of at least 9 amperes and a maximum primary overcurrent protection of 125%. (The absence of secondary protection applies only to certain panelboards. See Section 384-16[e] *Exception*.)

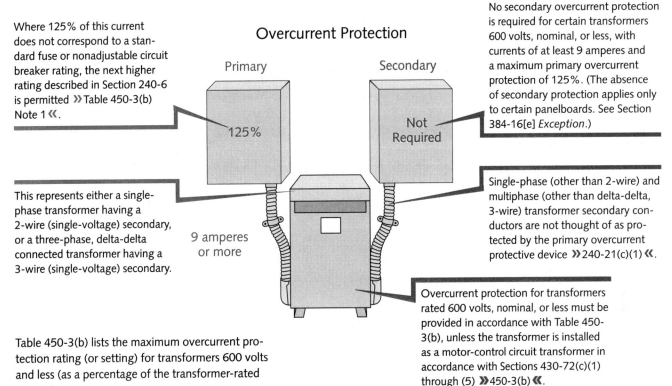

Overcurrent Protection

Primary — 125%

Secondary — Not Required

9 amperes or more

This represents either a single-phase transformer having a 2-wire (single-voltage) secondary, or a three-phase, delta-delta connected transformer having a 3-wire (single-voltage) secondary.

Single-phase (other than 2-wire) and multiphase (other than delta-delta, 3-wire) transformer secondary conductors are not thought of as protected by the primary overcurrent protective device » 240-21(c)(1) «.

Overcurrent protection for transformers rated 600 volts, nominal, or less must be provided in accordance with Table 450-3(b), unless the transformer is installed as a motor-control circuit transformer in accordance with Sections 430-72(c)(1) through (5) » 450-3(b) «.

Table 450-3(b) lists the maximum overcurrent protection rating (or setting) for transformers 600 volts and less (as a percentage of the transformer-rated current).

Overcurrent protection for transformers 600 volts or less having a rating less than 9 amperes must be provided in accordance with Table 450-3(b) » 450-3(b) «.

Conductors supplied by the secondary side of a single-phase transformer having a 2-wire (single-voltage) secondary, or a three-phase, delta-delta connected transformer having a 3-wire (single-voltage) secondary, can be protected by overcurrent protection situated on the primary (supply) side of the transformer, provided this protection is in accordance with Section 450-3 and does not exceed the value determined when multiplying the secondary conductor ampacity by the secondary to primary transformer voltage ratio » 240-21(c)(1) «.

WARNING

Although Table 450-3(b) does not require secondary overcurrent protection if the primary overcurrent protection is limited to 125%, Section 384-16(e) requires secondary overcurrent for most panelboards. Where a panelboard is supplied through a transformer, locate the overcurrent protection in 384-16(a), (b), and (c) on the transformer's secondary side » 384-16(e) «. A panelboard supplied by the secondary side of a transformer is considered as protected by the overcurrent protection provided on the transformer's primary side, where that protection complies with Section 240-21(c)(1) » 384-16(e) *Exception* «.

■ Transformer—Primary and Secondary Protection 9 Amperes or More—600 Volts or Less

Overcurrent Protection

Primary

Secondary

250%

125%

9 amperes or more

Unlike the note allowing the next standard rating above 125 percent, the **maximum** rating or setting is 250 percent.

A transformer equipped with coordinated thermal overload protection (by the manufacturer) and arranged to interrupt the primary current can have primary overcurrent protection rated (or set) at a current value not more than six times the rated current of transformers having not more than 6% impedance; and not more than four times the rated current of transformers having more than 6%, but not more than 10%, impedance »Table 450-3(b) Note 3 «.

Overcurrent protection for transformers rated 600 volts, nominal, or less must be provided according to Table 450-3(b), unless the transformer is installed as a motor-control circuit transformer per Sections 430-72(c)(1) through (5) »450-3(b) «.

Where 125% of this current does not correspond to a standard fuse or nonadjustable circuit breaker rating, the next higher rating described in Section 240-6 is permitted »Table 450-3(b) Note 1 «.

If secondary overcurrent protection is required, these devices are permitted to consist of no more than six circuit breakers (or six sets of fuses) grouped in one location. Where multiple overcurrent devices are implemented, the total of all the device ratings must not exceed the single overcurrent device allowed value. If both breakers and fuses constitute the overcurrent device, the total of the device ratings must not exceed that allowed for fuses »Table 450-3(b) Note 2 «.

Grounding and Bonding—Terminating in the Panel

NOTE

Transformers, which are only one type of separately derived system, will be used to illustrate and explain Section 250-30.

A bonding jumper complying with Section 250-28(a) through (d), and sized for the derived phase conductors, must connect the equipment grounding conductors of the separately derived system to the grounded conductor. Except as allowed by Section 250-24(a)(4), this connection is made at any point on the separately derived system from the source to the first system disconnecting means (or overcurrent devices) **»**250-30(a)(1) **«**.

To Grounding Electrode

Grounding Electrode Conductor

Neutral and Equipment Grounding Terminal Bar

The bonding jumper point of connection must be the same as the grounding electrode conductor per Section 250-30(a)(2) requirements **»**250-30(a)(1) **«**.

Bonding Jumper Installed

For vertically operated circuit-breaker handles, the "up" position of the handle must be the "on" position **»**240-81 **«**.

Grounded (Neutral) Conductor

A grounded separately derived ac system must comply with Section 250-30(a)(1) through (4), unless it is a high-impedance grounded neutral system **»**250-30(a) **«**.

Transformer

A bonding jumper at both the source and the first disconnecting means is permitted, if doing so does not establish a parallel path for the grounded circuit conductor. While grounded conductors used in this manner must not be smaller than the size specified for the bonding jumper, they do not have to be larger than the ungrounded conductor(s). In applying this exception, connection through the earth does not qualify as providing a parallel path **»**250-30(a)(1) *Exception No. 1* **«**.

Size the equipment grounding conductor in accordance with Section 250-122. The equipment grounding conductor must be of the types listed in Section 250-118.

Equipment Grounding Conductor

Equipment Ground

Bonding Jumper **Not** Installed

H1 H2 H3 X1 X2 X3 X0

Premises wiring deriving power from systems having no direct electrical connection (such as generators, transformers, or converter windings) to supply conductors originating in another system, and any system having direct electrical connection (such as a battery, or solar photovoltaic system) is a **separately derived system**. Grounded circuit conductors connecting systems do not constitute a direct electrical connection **»**Article 100 **«**.

A grounded circuit conductor can be used to ground noncurrent-carrying metal parts of equipment, raceways, and other enclosures on the supply side, or within the enclosure of, the main disconnecting means (or overcurrent devices) of a separately derived system as permitted by Section 250-30(a)(1) **»**250-142(a)(3) **«**.

Grounding (and bonding) terminations made at the system disconnecting means or overcurrent device, requires that the transformer's grounded (neutral) bus be insulated from the transformer enclosure. If the neutral bus is bonded to the enclosure and the raceway to the panelboard is metallic, then part of the neutral current flows on the raceway. Section 250-6(a) prohibits objectionable current flow over grounding paths or grounding conductors.

■ Grounding and Bonding—Terminating in the Transformer

In all other respects, grounding methods must comply with requirements prescribed through other applicable parts of the *Code* ≫250-30(a)(4)≪.

The equipment of an ungrounded separately derived system must be grounded as specified in Section 250-30(b)(1) through (3).

While a 3-phase separately derived system is used for this illustration, the same rules apply to single-phase systems.

If the grounding (and bonding) terminations are made at the transformer, the system disconnecting means (or overcurrent device) grounded (neutral) terminal bar must be insulated from both the panelboard enclosure and the equipment grounding conductor. If the neutral terminal bar is bonded to the enclosure as well as the equipment grounding conductor, part of the neutral current will flow through the raceway (if metallic) and the equipment grounding conductor. Section 250-6(a) strictly prohibits such a condition.

Except as permitted in Section 250-30(a)(1), a grounded circuit conductor cannot be used for grounding noncurrent-carrying metal equipment parts on the load side of a separately derived system disconnecting means (or the overcurrent devices) for a separately derived system having no main disconnecting means ≫250-142(b)≪.

Neutral Terminal Bar

Equipment Grounding Terminal Bar

Bonding Jumper **Not** Installed

Grounding Conductor

Grounded (Neutral) Conductor

The bonding jumper connection can be made at any point on the separately derived system, i.e., in either the panel or transformer ≫250-30(a)(1)≪. Equipment Grounding Conductor

Transformer

The bonding jumper and the grounding electrode conductor must share the same point of connection as required in Section 250-30(a)(2) ≫250-30(a)(1)≪.

To Grounding Electrode

Equipment Ground

Bonding Jumper Installed

Grounding Electrode Conductor

Use a bonding jumper which complies with Sections 250-28(a) through (d) to connect the equipment grounding conductors of the separately derived system to the grounded conductor. The bonding jumper size, determined by the derived phase conductors, must not be smaller than Table 250-66 grounding electrode conductor sizes ≫250-30(a)(1)≪.

A system supplying power to a Class 1, Class 2, or Class 3 circuit, having the power derived from a transformer rated 1,000 volt-amperes or less, must have a bonding jumper sized no smaller than the derived phase conductors. The smallest size allowed is No. 14 copper or No. 12 aluminum ≫250-30(a)(1) *Exception No. 2*≪.

A system that supplies a Class 1, 2, or 3 circuit and is derived from a transformer rated no more than 1,000 volt-amperes does not require a grounding electrode conductor, provided the system grounded conductor is bonded to the transformer frame (or enclosure) by means of a jumper sized in accordance with Section 250-30(a)(1), Exception No. 2, and the transformer frame (or enclosure) is grounded by one of the means specified in Section 250-134 ≫250-30(a)(2) *Exception*≪.

The grounding electrode conductor installation must comply with Section 250-64.

If the main bonding jumper (specified in Section 250-28) is a wire or busbar, installed from the neutral bar (or bus) to the equipment grounding terminal bar (or bus), the grounding electrode conductor can be connected to the same equipment grounding terminal bar (or bus) as the main bonding jumper ≫250-24(a)(4)≪.

■ Grounding Electrode and Grounding Electrode Conductor

The grounding electrode must be as near as practicable to, and preferably in the same area as, the grounding electrode conductor connection to the system. The grounding electrode must be the nearest one of the following:

(a) An effectively grounded metal member of the structure.

(b) An effectively grounded metal water pipe within 5 feet of the entrance point into the building (in industrial and commercial buildings where maintenance and supervision conditions ensure that only qualified persons will service the installation and the entire length of the interior metal water pipe is exposed, the connection is permitted at any point along the water pipe system. Except as permitted by Section 250-24(a)(4), this connection can be at any point on the separately derived system from the source to the first system disconnecting means (or overcurrent devices) ≫ 250-30(a)(3) *Exception* ≪.

(c) Other electrodes as specified in Sections 250-50 and 250-52 where (a) and (b) above are not available ≫ 250-30(a)(3) ≪.

Grounding electrode conductor metal enclosures must be electrically continuous from the point of attachment to cabinets (or equipment) to the grounding electrode, and must be securely fastened to the ground clamp (or fitting). Metal enclosures that are not physically continuous from cabinet (or equipment) to the grounding electrode must be made electrically continuous by bonding *each end* to the grounding conductor ≫ 250-64(e) ≪.

A grounding electrode conductor that is sized according to Section 250-66 for the derived phase conductors must connect the grounding conductor of the derived system to the grounding electrode per Section 250-30(a)(3) specifications. Except as permitted by Section 250-24(a)(3) or (a)(4), this connection can be made at any point on the separately derived system where the bonding jumper is installed ≫ 250-30(a)(2) ≪.

The connection of a grounding electrode conductor to a grounding electrode must be accessible ≫ 250-68(a) ≪.

The separately derived system's grounded conductor must be bonded to the nearest available point of interior metal water piping in the area served by the separately derived system. This connection can be made at the same point on the separately derived system as the grounding electrode conductor connection. Size the bonding jumper per Table 250-66 ≫ 250-104(a)(4) ≪.

Building Steel

Water Pipe

E.M.T.

Grounding Electrode Conductor

Bonding Locknut

Bonding Jumper Installed

Transformer

Bonding Jumper **Not** Installed

Equipment requiring grounding and having a nonconductive coating (such as paint, lacquer, and enamel) must have the coating removed from threads and other contact surfaces to ensure good electrical continuity (unless connected by fittings that are designed so as to make the removal unnecessary) ≫ 250-12 ≪.

Where a separately derived system originates in listed equipment also used as service equipment, the service equipment's grounding electrode can serve as the separately derived system's grounding electrode, provided it is sufficiently sized for the separately derived system. Where the equipment ground bus (internal to the service equipment) is not smaller than the required grounding electrode conductor, the grounding electrode connection for the separately derived system can be made to that bus ≫ 250-30(a)(3) *Exception for (a), (b), and (c)* ≪.

■ Generators

The conductor ampacity from the generator terminals to the first overcurrent device must not be less than 115% of the generator's nameplate current rating, unless the design and operation prevent overloading »445-5 «.

An alternate ac power source (such as an on-site generator) is not a separately derived system if the neutral is solidly interconnected to a service-supplied system neutral »250-20(d) FPN No. 1 «.

Generators, together with associated wiring and equipment, must also comply with the applicable provisions of Articles 695, 700, 701, 702, and 705 »445-1 «.

The type of generator used must be suitable for the installed location. The motor requirements in Section 430-14 must also be met. Generators installed in hazardous (classified) locations as described in Articles 500 through 503, and in 505, and in other locations as described in Articles 510 through 517, 518, 520, 525, 530, 665, and 695 must also comply with the applicable provisions of those articles »445-2 «.

Depending on the generator type, overcurrent protection must meet Section 445-4 specifications.

Generators must have a disconnect means by which the generator, all protective devices, and control apparatus can be entirely disconnected from the circuits supplied by the generator, except where:
(1) The generator's driving means can be readily shut down; and
(2) The generator is not arranged to operate in parallel with another generator or other source of voltage »445-10 «.

Where wires pass through an enclosure, conduit box, or barrier opening, a bushing must protect the conductors from any sharp edges. The bushing surfaces must be smooth and well-rounded where it may contact the conductors. If oils, grease, or other contaminants are present, the bushing must be made of a material not harmfully affected »445-8 «.

Each generator must have a nameplate identifying the manufacturer; the rated frequency; power factor; number of phases (if of alternating current); the rating in kilowatts or kilovolt amperes; the normal volts and amperes corresponding to the rating; rated revolutions per minute; insulation system class and rated ambient temperature (or rated temperature rise); and time rating »445-3 «.

For systems not separately derived and not required to be grounded as specified in Section 250-30, see Section 445-5 to determine the minimum size of conductors that must carry fault current »250-20(d) FPN No. 2 «.

Live generator parts operating at more than 50 volts to ground must not be exposed to accidental contact by unqualified persons »445-6 «.

Where necessary for the attendant's safety, the requirements of Section 430-133 apply »445-7 «.

Generator terminal housings must comply with Section 430-12 »445-9 «.

Article 700 contains emergency system installation provisions pertaining to generators and associated equipment.

While legally required standby system requirements for generators and associated equipment are located in Article 701, optional standby system provisions are found in Article 702.

BUSWAYS

■ General Installation Provisions

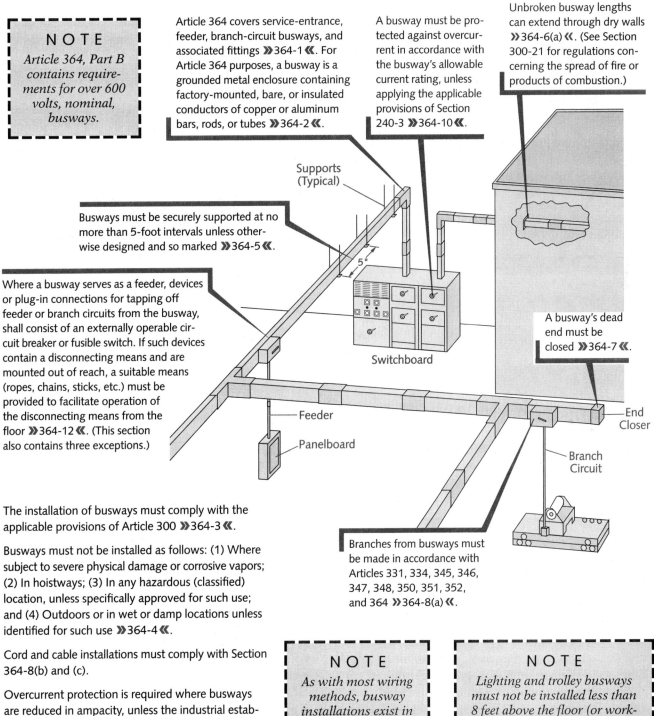

NOTE

Article 364, Part B contains requirements for over 600 volts, nominal, busways.

Article 364 covers service-entrance, feeder, branch-circuit busways, and associated fittings »364-1«. For Article 364 purposes, a busway is a grounded metal enclosure containing factory-mounted, bare, or insulated conductors of copper or aluminum bars, rods, or tubes »364-2«.

A busway must be protected against overcurrent in accordance with the busway's allowable current rating, unless applying the applicable provisions of Section 240-3 »364-10«.

Unbroken busway lengths can extend through dry walls »364-6(a)«. (See Section 300-21 for regulations concerning the spread of fire or products of combustion.)

Supports (Typical)

Busways must be securely supported at no more than 5-foot intervals unless otherwise designed and so marked »364-5«.

5"

Where a busway serves as a feeder, devices or plug-in connections for tapping off feeder or branch circuits from the busway, shall consist of an externally operable circuit breaker or fusible switch. If such devices contain a disconnecting means and are mounted out of reach, a suitable means (ropes, chains, sticks, etc.) must be provided to facilitate operation of the disconnecting means from the floor »364-12«. (This section also contains three exceptions.)

Switchboard

A busway's dead end must be closed »364-7«.

Feeder

Panelboard

End Closer

Branch Circuit

The installation of busways must comply with the applicable provisions of Article 300 »364-3«.

Busways must not be installed as follows: (1) Where subject to severe physical damage or corrosive vapors; (2) In hoistways; (3) In any hazardous (classified) location, unless specifically approved for such use; and (4) Outdoors or in wet or damp locations unless identified for such use »364-4«.

Cord and cable installations must comply with Section 364-8(b) and (c).

Overcurrent protection is required where busways are reduced in ampacity, unless the industrial establishment exception is met »364-11«.

A busway used as a branch circuit must be protected against overcurrent in accordance with Section 210-20. Where so used, the circuit must comply with the applicable requirements of Articles 210, 430, and 440 »364-13«.

Branches from busways must be made in accordance with Articles 331, 334, 345, 346, 347, 348, 350, 351, 352, and 364 »364-8(a)«.

NOTE

As with most wiring methods, busway installations exist in a variety of occupancies (commercial, industrial, etc.).

NOTE

Lighting and trolley busways must not be installed less than 8 feet above the floor (or working platform) unless provided with a cover and identified for the purpose »364-4(b)«.

■ Vertical Installation Provisions

> **NOTE**
>
> *Implement electrical installations in hollow spaces or vertical shafts so that the possible spread of fire, or products of combustion, will not be substantially increased. Openings around electrical penetrations through fire-resistant-rated walls, partitions, floors, or ceilings must be firestopped by approved methods to maintain the fire resistance rating* **》300-21《**.

Within 12 inches of Floor Opening

4 inch high curb

Busway installations must comply with the applicable provisions of Article 300 **》364-3《**.

Busways and busway taps must be protected against overcurrent in accordance with Sections 364-10 through 364-13 **》240-21(e)《**.

Section 364-6(b) requires that floor penetrations comply with (1) and (2):
(1) Vertical busways can extend through dry floors if totally enclosed (unventilated) where passing through, and for a minimum distance of 6 feet above the floor, providing adequate physical damage protection.
(2) In other than industrial establishments, where a vertical riser penetrates multiple dry floors, a minimum 4-inch high curb must be installed around all busway riser floor openings to prevent liquid penetration. The curb must be installed within 12 inches of the floor opening. Locate electrical equipment so that it is not likely to be damaged by curb retained liquids.

Summary

- Sections 225-18 and 230-24(b) contain minimum vertical wiring clearance provisions.
- Associated equipment, located above or below electrical equipment, cannot extend more than 6 inches beyond the electrical equipment's front.
- A space equal to the width and depth of the equipment (panelboards, switchboards, and motor control centers), extending from the floor to a height of 6 feet above the equipment (or to the structural ceiling, whichever is lower) must be dedicated to the electrical installation.
- The space equal to the width and depth of the equipment above panelboards, switchboards, and motor control centers must be clear of foreign systems unless protection is provided to avoid damage from condensation, leaks, or breaks.
- Working space clearances vary, depending on conditions and voltage.
- For equipment rated 1,200 amperes or more, over 6 feet wide, and containing overcurrent devices,

switching devices, or control devices, there must be an entrance at each end of the working space.
- More than 10% of a lighting and appliance branch-circuit panelboard's overcurrent devices protect lighting and appliance branch circuits.
- On the supply side of the service, the grounded circuit conductor and grounding conductors are connected together.
- A grounding connection must not be made to any grounded circuit conductor on the load side of the service disconnecting means except as otherwise allowed by Article 250.
- Transformer provisions are located in Article 450.
- Section 250-30 contains separately derived system grounding provisions.
- An alternate ac power source (such as an on-site generator) is not a separately derived system if the neutral is solidly interconnected to a service-supplied neutral.
- Article 364 covers service-entrance, feeder, and branch-circuit busways and associated fittings.

Unit 14 Competency Test

• •

NEC® Reference   Answer

_____ _____   1. A separately derived system's grounding electrode conductor connection can be made at any point from the source of the separately derived system to the _____ system disconnecting means or overcurrent device.

_____ _____   2. Not more than _____ overcurrent devices (other than those provided for in the mains) of a lighting and appliance branch-circuit panelboard shall be installed in any one cabinet or cutout box.

_____ _____   3. Internal parts of electrical equipment, including busbars, wiring terminals, insulators, and other surfaces, shall not be _____ by foreign materials such as paint, plaster, cleaners, abrasives, or corrosive residues.

_____ _____   4. An alternate ac power source such as an on-site generator is not a separately derived system if the _____ is solidly interconnected to a service-supplied system _____.

_____ _____   5. For equipment rated 1,200 amperes or more and over 6 feet wide that contains overcurrent devices, switching devices, or control devices, there shall be _____ entrance(s).

_____ _____   6. A transformer, less than 600 volts, equipped with coordinated thermal overload protection by the manufacturer and arranged to interrupt the primary current, shall be permitted to have primary overcurrent protection rated or set at a current value that is not more than _____ the rated current of the transformer for a transformer having 8% impedance.

_____ _____   7. The space equal to the width and depth of switchboards, panelboards, and motor control centers shall be kept clear of foreign systems _____ to avoid damage from condensation, leaks, or breaks in such foreign systems.

_____ _____   8. Where wires pass through an opening in an enclosure, conduit box, or barrier, a _____ shall be used to protect the conductors from the edges of an opening having sharp edges within generators.

_____ _____   9. On a switchboard or panelboard supplied from a 4-wire, delta-connected system where the midpoint of one phase winding is grounded, that phase busbar or conductor having the higher voltage to ground shall be durably and permanently marked by an outer finish that is _____ in color or by other effective means.

_____ _____   10. Ground-fault protection of equipment shall be provided for solidly grounded wye electrical systems of more than 150 volts to ground, but not exceeding 600 volts phase-to-phase for each individual device used as a building or structure main disconnecting means rated _____ or more.

_____ _____   11. The _____ conductor of the separately derived system shall be bonded to the nearest available point of the interior metal water piping system in the area served by the separately derived system.

_____ _____   12. Busways shall be securely supported at intervals not exceeding _____ unless otherwise designed and marked.

_____ _____   13. What is the minimum vertical clearance for 208Y/120 volt, 3-phase service-drop conductors in a commercial area subject to minimal truck traffic?

_____ _____   14. Covered, shielded, fenced, enclosed, or otherwise protected by means of suitable covers, casings, barriers, rails, screens, mats, or platforms to remove the likelihood of approach or contact by persons or objects to a point of danger is the definition of _____.

NEC® Reference   Answer

_____   _____   15. Where the overcurrent device is rated over _____ amperes, the ampacity of the conductors it protects shall be equal to or greater than the rating of the overcurrent device as defined in Section 240-6.

_____   _____   16. The enclosure for a panelboard shall have the top and bottom wire-bending space sized in accordance with _____ for the largest conductor entering or leaving the enclosure.

_____   _____   17. A(n) _____ is a premises wiring system whose power is derived from a battery, a solar photovoltaic system, or from a generator, transformer, or converter windings, and that has no direct electrical connection, including a solidly connected grounded circuit conductor, to supply conductors originating in another system.

_____   _____   18. Dry-type transformers 600 volts, nominal, or less and not exceeding _____ shall be permitted in hollow spaces of buildings not permanently closed in by structure.

_____   _____   19. Where rear access is required to work on de-energized parts on the back of enclosed equipment, a minimum working space of _____ horizontally shall be provided.

_____   _____   20. The ampacity of the conductors from the generator terminals to the first overcurrent device shall not be less than _____ percent of the nameplate current rating of the generator.

_____   _____   21. Where exposed to the weather, raceways enclosing service-entrance conductors shall be _____.

_____   _____   22. For industrial establishments only, omission of overcurrent protection shall be permitted at points where busways are reduced in ampacity, provided that the length of the busway having the smaller ampacity does not exceed _____ and has an ampacity at least equal to one-third the rating or setting of the overcurrent device next back on the line, and provided that such busway is free from contact with combustible material.

_____   _____   23. Circuit breakers shall be marked with their ampere rating in a manner that will be _____ after installation.

_____   _____   24. The minimum headroom of working spaces about service equipment, switchboards, panelboards, or motor control centers shall be _____.

_____   _____   25. The load on any overcurrent device located in a panelboard shall not exceed _____ percent of its rating where, in normal operation, the load will continue for _____ hours or more.

_____   _____   26. A 480Y/277 volt switchboard, having exposed live parts on one side and a concrete wall on the other side, requires a minimum working space depth of _____ inches.

_____   _____   27. For systems that are not separately derived and are not required to be grounded as specified in Section 250-30, see _____ for minimum size of conductors that must carry fault current.

_____   _____   28. In other than industrial establishments, where a vertical riser penetrates two or more dry floors, a minimum _____ curb shall be installed around all floor openings for riser busways to prevent liquids from entering the opening.

_____   _____   29. _____ are generally accessible from the rear as well as from the front and are not intended to be installed in cabinets.

NEC® Reference	Answer	
_____	_____	30. Generators shall be equipped with a _____ by means of which the generator and all protective devices and control apparatus are able to be disconnected entirely from the circuits supplied by the generator.
_____	_____	31. The space equal to the width and depth of switchboards, panelboards, and motor control centers and extending from the floor to a height of _____ above the equipment or to the structural ceiling, whichever is lower, shall be dedicated to the electrical installation.
_____	_____	32. For other than a totally enclosed switchboard, a space not less than _____ shall be provided between the top of the switchboard and any combustible ceiling, unless a noncombustible shield in provided between the switchboard and the ceiling.
_____	_____	33. Where the service-entrance phase conductors are larger than 1,100 kcmil copper or 1,750 kcmil aluminum, the main bonding jumper shall have an area that is not less than _____ of the area of the largest phase conductor.
_____	_____	34. Surrounded by a case, housing, fence, or walls that prevent persons from accidentally contacting energized parts is the definition of _____.
_____	_____	35. Busways shall be permitted to be extended vertically through dry floors if totally enclosed (unventilated) where passing through and for a minimum distance of _____ above the floor to provide adequate protection from physical damage.

Hazardous (Classified) Locations

Objectives

• •

After studying this unit, the student will:

✔ understand what constitutes Class I, II, and III locations, as well as the parameters for Divisions 1 and 2 within each class.

✔ have a good understanding of provisions pertaining to commercial garages, repair and storage facilities.

✔ understand that Article 513 applies to buildings/structures, or parts thereof, inside which aircraft are housed, stored, serviced, repaired, or altered.

✔ understand the distinction between Division 1's normal operating conditions, and Division 2's abnormal operations.

✔ know that Articles 500 through 504 cover requirements for electrical and electronic equipment/wiring, of all voltages, within hazardous (classified) locations with potential fire or explosion hazards due to the presence of flammable gases (or vapors), flammable liquids, combustible dust, or ignitible fibers (or flyings).

✔ know that a gasoline dispensing and service station involves the transfer of gasoline, or other volatile flammable liquids or liquefied flammable gases, to the fuel tanks of self-propelled vehicles or to approved containers.

Introduction

• •

Hazardous (classified) locations are divided into certain categories and sub-categories, otherwise known as classes and divisions. Articles 500 through 504 cover the requirements for Class I, Division 1 and 2; Class II, Divisions 1 and 2; and Class III, Divisions 1 and 2 located electrical and electronic equipment/ wiring, of all voltages, where potential fire or explosion hazards exist due to the presence of flammable gases (or vapors), flammable liquids, combustible dust, or ignitible fibers (or flyings). Locations containing potentially sufficient quantities of air-borne gases (or vapors) to cause explosion or ignition are categorized as Class I. Locations hazardous due to the presence of combustible dust are designated Class II. While Class III locations are hazardous because of the presence of easily ignitible fibers (or flyings) which, however, are not likely to be suspended in the air in quantities sufficient to produce ignitible mixtures. Under typical, i.e. normal, operating conditions, Division 1 locations contain ignitible or combustible elements (gases, dust, fibers, etc.) in quantities sufficient to produce explosive (or ignitable) mixtures. Most locations where air-borne combustible or ignitible elements (gases, dust, fibers, etc.) are not ordinarily present in quantities sufficient to produce explosive (or ignitable) mixtures, are classified Division 2. Within these articles, many fine print notes reference other standards for more information on specific applications. Similarly, this test refers the reader, in this instance, to the *Code* itself for Article 504 (*Intrinsically Safe Systems*) and Article 505 (*Class I, Zone 0, 1, and 2 Locations*). The fact that, for various reasons, this text could not address these topics is in no way intended to diminish their significance. Articles 511 through 516 cover occupancies, or parts thereof, that are or may be hazardous because of atmospheric concentrations of flammable liquids, gases, or vapors; or, because of deposits or accumulations of readily ignitable materials.

OVERVIEW

■ Class Categorizations

Locations containing potentially sufficient quantities of air-borne gases (or vapors) to cause explosion or ignition are categorized as Class I locations. Class I locations include those specified in Section 500-7(a) and (b) ≫ 500-7 ≪ .

The Class I temperature marking, specified in Section 500-5(d), must not exceed the existing gas or vapor ignition temperature ≫ 500-5(e) ≪ . (See FPN for the location of additional literature regarding gas/vapor ignition temperatures.)

In most cases, locations where combustible or ignitible elements (gases, dust, fibers, etc.) are not ordinarily present in the air in quantities sufficient to produce explosive (or ignitable) mixtures, are Division 2 locations.

Generally, under normal operating conditions, Division 1 locations contain ignitible or combustible elements (gases, dust, fibers, etc.) in quantities sufficient to produce explosive (or ignitable) mixtures.

Class I group classifications are covered in Section 500-5(a).

Equipment approved for Class I locations are permitted in Class I, Class II, and Class III locations. Equipment approved for Class II locations are permitted in Class II and Class III locations, but not in Class I locations. Equipment approved for Class III locations are not permitted in either Class I or Class II locations.

The temperature marking for Class II locations, per Section 500-5(d), must be less than the ignition temperature of the existing specified dust. The temperature rating of any organic dusts, which may dehydrate or carbonize, must be the lower of the ignition temperature or 165°C (329°F). Table 500-5(f) shows the assumed ignition temperature for approved equipment prior to this requirement ≫ 500-5(f) ≪ .

Locations which are hazardous due to the presence of combustible dust are designated Class II locations. Class II locations include those specified in Section 500-8(a) and (b) ≫ 500-8 ≪ .

Class II group classifications are covered in Section 500-5(b).

Locations that are hazardous because of the presence of easily ignitible fibers (or flyings), but in which such fibers (or flyings) are not likely to be suspended in the air in quantities sufficient to produce ignitible mixtures, are designated Class III. Class III locations include those specified in Section 500-9(a) and (b) ≫ 500-9 ≪ .

> **NOTE**
> *Within a single class and group, equipment approved for Division 1 use, is permitted in a Division 2 location* ≫ 500-5(c) ≪ .

There are no groups within Class III locations.

Threaded Conduit

> **NOTE**
> *Threaded conduit installa-tions must also comply with Chapter 3's requirements, in particular, Article 346 for* Rigid Metal Conduit *and Article 345 for* Intermediate Metal Conduit.

Equipment having metric threaded entries must be so identified, or must provide listed adapters appropriate for connection to conduit or NPT-threaded fittings. Listed metric-threaded cable fittings are also acceptable ≫ 500-3(d) ≪.

Threaded conduit within hazardous (classified) locations must be threaded using an NPT standard conduit cutting die which renders a ¾-inch taper per foot ≫ 500-3(d) ≪.

NPT Standard Conduit Cutting Dies

Threaded conduit must be wrenchtight to prevent sparking should fault current flow through the conduit system; and, to ensure explosionproof/dust-ignitionproof integrity where necessary ≫ 500-3(d) ≪.

General

Scope—Articles 500 through 504 cover the require-ments for electrical and electronic equipment/wiring for all voltages in Class I, Divisions 1 and 2; Class II, Divisions 1 and 2; and Class III, Divisions 1 and 2 locations where fire or explosion hazards may exist due to presence of flammable gases (or vapors), flammable liquids, combustible dust, or ignitible fibers (or flyings) ≫ 500-1 ≪.

What constitutes a hazardous (classified) location, and the require-ments that apply, are determined contingent upon: (1) the proper-ties of flammable vapors, liquids, gases, or combustible dusts/fibers that are present; and (2) the likelihood that a flammable/com-bustible concentration of such substances can occur ≫ 500-3(a) ≪.

Each room, section or area within a hazardous (classified) location, is considered individually to de-termine the appropriate classifi-cation ≫ 500-3(a) ≪. (Here no specific classifications are assigned for each area; they are shown for illustration purposes only.)

Area #6
Area #5
Area #4
Area #3
Area #2
Area #1

> **NOTE**
> *Proper documentation must be maintained on each hazardous (classi-fied) location. Such documentation must be available to those autho-rized to design, install, inspect, maintain, or operate electrical equip-ment at that particular location* ≫ *500-3(b)* ≪.

Article 505 details the electrical/electronic requirements for all voltages in Class, Zone 0, Zone 1, and Zone 2 hazardous (classi-fied) locations where fire or explosion hazards may exist due to flammable gases (or vapors), or flammable liquids ≫ 500-1 FPN ≪.

Except as modified in Articles 500 through 504, all other applicable *Code* rules apply to electrical equipment/wiring installed in hazardous (classified) locations ≫ 500-2 ≪.

Locations using/handling only spontaneously ignitable materials are not classified ≫ 500-3(a) ≪.

Creative design and implementation of hazardous (classified) electrical systems may facilitate the placement of equipment in less hazardous, or even nonhazardous areas, greatly re-ducing the risk as well as the need for special equipment ≫ 500-3(a) FPN ≪.

Article 500-3(c) FPN No's 2 through 5 list numerous additional hazardous (classified) location information sources.

When making determinations regarding classification, ven-tilation, protection against static electricity, lightning, etc., the authority having jurisdiction must be familiar with the recorded industrial experience, together with the standards of the National Fire Protection Association (NFPA), the Amer-ican Petroleum Institute (API), and the Instrument Society of America (ISA) ≫ 500-3(c) FPN No. 1 ≪.

Fiber-optic cable assemblies, containing potentially current-carrying conductors, must be installed in accordances with Articles 500 through 502, as applicable ≫ 500-3(e) ≪.

Section 500-4(a) through (i) covers protection technique provi-sions.

Articles 510 through 517 cover: garages, aircraft hangars, gasoline dispensing/service stations, bulk storage plants, spray applications, dipping/coating processes, and health care facilities ≫ 500-6 ≪. (Health care facilities are covered in Unit 16 of this text.)

■ Class I Group Classifications

Hazardous (classified) locations divide atmosphere identifications for Class I, Divisions 1 and 2, locations into Groups A, B, C, and D ≫ 500-5(a)(1) through (4) ≪.

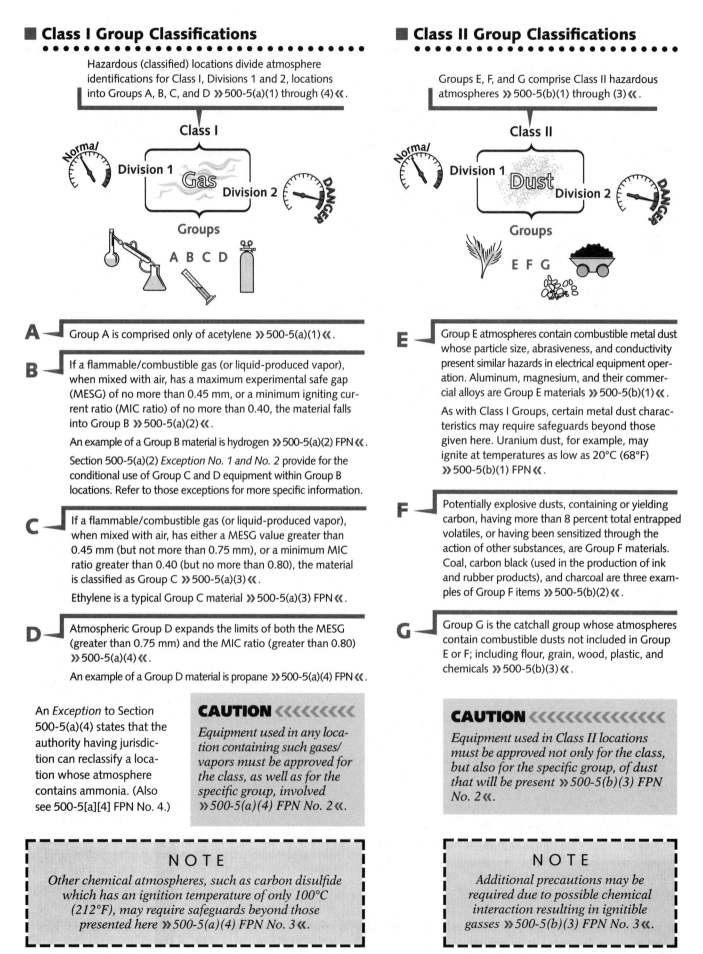

Class I

Normal/ DANGER

Division 1 **Gas** Division 2

Groups

A B C D

A Group A is comprised only of acetylene ≫ 500-5(a)(1) ≪.

B If a flammable/combustible gas (or liquid-produced vapor), when mixed with air, has a maximum experimental safe gap (MESG) of no more than 0.45 mm, or a minimum igniting current ratio (MIC ratio) of no more than 0.40, the material falls into Group B ≫ 500-5(a)(2) ≪.

An example of a Group B material is hydrogen ≫ 500-5(a)(2) FPN ≪.

Section 500-5(a)(2) *Exception No. 1 and No. 2* provide for the conditional use of Group C and D equipment within Group B locations. Refer to those exceptions for more specific information.

C If a flammable/combustible gas (or liquid-produced vapor), when mixed with air, has either a MESG value greater than 0.45 mm (but not more than 0.75 mm), or a minimum MIC ratio greater than 0.40 (but no more than 0.80), the material is classified as Group C ≫ 500-5(a)(3) ≪.

Ethylene is a typical Group C material ≫ 500-5(a)(3) FPN ≪.

D Atmospheric Group D expands the limits of both the MESG (greater than 0.75 mm) and the MIC ratio (greater than 0.80) ≫ 500-5(a)(4) ≪.

An example of a Group D material is propane ≫ 500-5(a)(4) FPN ≪.

An *Exception* to Section 500-5(a)(4) states that the authority having jurisdiction can reclassify a location whose atmosphere contains ammonia. (Also see 500-5[a][4] FPN No. 4.)

> **CAUTION** ≪≪≪≪≪≪≪
>
> *Equipment used in any location containing such gases/vapors must be approved for the class, as well as for the specific group, involved ≫ 500-5(a)(4) FPN No. 2 ≪.*

> **N O T E**
>
> *Other chemical atmospheres, such as carbon disulfide which has an ignition temperature of only 100°C (212°F), may require safeguards beyond those presented here ≫ 500-5(a)(4) FPN No. 3 ≪.*

■ Class II Group Classifications

Groups E, F, and G comprise Class II hazardous atmospheres ≫ 500-5(b)(1) through (3) ≪.

Class II

Normal/ DANGER

Division 1 **Dust** Division 2

Groups

E F G

E Group E atmospheres contain combustible metal dust whose particle size, abrasiveness, and conductivity present similar hazards in electrical equipment operation. Aluminum, magnesium, and their commercial alloys are Group E materials ≫ 500-5(b)(1) ≪.

As with Class I Groups, certain metal dust characteristics may require safeguards beyond those given here. Uranium dust, for example, may ignite at temperatures as low as 20°C (68°F) ≫ 500-5(b)(1) FPN ≪.

F Potentially explosive dusts, containing or yielding carbon, having more than 8 percent total entrapped volatiles, or having been sensitized through the action of other substances, are Group F materials. Coal, carbon black (used in the production of ink and rubber products), and charcoal are three examples of Group F items ≫ 500-5(b)(2) ≪.

G Group G is the catchall group whose atmospheres contain combustible dusts not included in Group E or F; including flour, grain, wood, plastic, and chemicals ≫ 500-5(b)(3) ≪.

> **CAUTION** ≪≪≪≪≪≪≪≪≪≪≪≪
>
> *Equipment used in Class II locations must be approved not only for the class, but also for the specific group, of dust that will be present ≫ 500-5(b)(3) FPN No. 2 ≪.*

> **N O T E**
>
> *Additional precautions may be required due to possible chemical interaction resulting in ignitible gasses ≫ 500-5(b)(3) FPN No. 3 ≪.*

■ Equipment

External Atmosphere Remains Uneffected

Articles 500 through 504 require equipment construction and installation that will ensure safe performance under conditions of proper use and maintenance ≫ 500-5 ≪.

Equipment must be approved not only for the class of location but also for the explosive, combustible, or ignitible properties of the specific gas, vapor, dust, fiber, or flyings that are present. In addition, Class I equipment must not have any exposed surface operating at temperatures in excess of the ignition temperature of the specific gas/vapor. Class II equipment must not have an external temperature higher than that specified in Section 500-5(f). Class III equipment must not exceed the maximum surface temperature specified in Section 503-1 ≫ 500-5(c) ≪.

Internal Explosion

Cooled Gas

Burning Gases

Atmosphere Unthreatened by Cooled Gases

Hot Gas Cool when Escaping Through the Threads

Internal Explosive Force

Five Threads Minimum

Burning Gases

At a minimum, equipment installed must be approved for all applicable levels of Class, Division, and Group, giving careful consideration to normal operating and ignition temperatures. Equipment exceeding the specified requirements is permitted. However, the reverse is not true; i.e., Division 1 approved equipment is permitted in a Division 2 location of the same class and group; but Division 2 approved equipment cannot be used in a Division 1 location ≫ 500-5(c) ≪.

Inspection authorities and users are advised to exercise extraordinary care with regard to installation and maintenance ≫ 500-5 FPN No. 1 ≪.

See Section 500-5 FPN No. 2 for low ambient condition precautions.

Any equipment preventing the entrance of flammable/combustible fluids by means of a single compression seal, diaphragm, or tube must be approved for a Class I, Division 2 location ≫ 500-5(c) ≪. Equipment installed in a Class I, Division 1 location must be suitable for a Division 1 location ≫ 500-5(c) *Exception* ≪.

Where Articles 501 through 503 specifically permit, general-purpose equipment or equipment in general-purpose enclosures can be installed in Division 2 locations if, under normal operating conditions, the equipment does not constitute a source of ignition ≫ 500-5(c) ≪.

Attention must be given to the possibility of simultaneously occurring flammable gases and combustible dust when determining the safe operating temperature of installed electrical equipment ≫ 500-5(c) ≪.

The characteristics of various atmospheric mixtures of gases, vapors, and dusts depend on the actual materials involved ≫ 500-5(c) FPN ≪.

Approved equipment must display markings of the class, group, and operating temperature or temperature range referenced to a 40°C ambient, unless provided otherwise in Section 500-5(d) *Exception No.'s 1 through 5* ≫ 500-5(d) ≪.

Table 500-5(d) lists the identification numbers acceptable on equipment nameplates as well as markings for temperature range, when given ≫ 500-5(d) ≪.

The maximum safe operating temperature must be marked on equipment approved for Class I and Class II, having been determined by simultaneous exposure to the combined conditions of both classes ≫ 500-5(d) ≪.

Inconsistent relationships, explosion properties and ignition temperatures are independent requirements ≫ 500-5(d) FPN ≪.

■ Protection Techniques

Approved protection techniques for electrical and electronic equipment in hazardous (classified) locations include: Explosionproof Apparatus; Dust Ignitionproof; Dusttight; Purged and Pressurized; Intrinsically Safe Systems; Nonincendive (circuit and equipment); Oil Immersion; Hermetically Sealed; and other such techniques employing equipment specifically listed for a given location ≫500-4≪.

Explosionproof Enclosure for Current Interrupting Device

Approved explosionproof apparatus are required for Class I, Division 1 and 2 locations ≫500-4(a)≪.

Explosionproof Junction Box

Dusttight designated equipment is adequate for Class II Division 2 and Class III locations for which it is approved ≫500-4(c)≪.

Dusttight Enclosure

Equipment of the approved Dust Ignitionproof variety is acceptable for Class II, Division 1 and 2 locations ≫500-4(b)≪.

Dust Ignitionproof Enclosure for Current Interrupting Devices

Dustproof: Constructed or protected so that dust does not interfere with successful operation ≫Article 100≪.

Dusttight: Constructed to prevent dust from entering the enclosed case, under stipulated test conditions ≫Article 100 and 500-4(c) FPN No. 1≪.

Explosionproof Apparatus: is encased by constructive design to withstand a potential internal explosion of a specified gas/vapor; effectively preventing the ignition of surrounding gas/vapors by internal sparks, flashes, or explosions; and, operating at such an external temperature that any surrounding flammable atmosphere will not be ignited as a result ≫Article 100≪.

Explosionproof Panelboard

The Purged and Pressurized protection technique is permitted for equipment in any hazardous (classified) location for which it is approved ≫500-4(d)≪.

Hazardous (classified) locations may be limited (by reduction of hazard), or eliminated altogether, by adequate positive-pressure ventilation from a clean air source, provided effective measures are provided in the event of ventilation failure ≫500-4(d) FPN≪.

Intrinsically safe apparatus and wiring is permitted in the specified hazardous (classified) location for which it is approved, provided the installation meets Article 504 requirements. Articles 501 through 503, and 510 through 516, are not applicable to such installations, unless required by Article 504 ≫500-4(e)≪.

■ Protection Techniques (*continued*)

Explosionproof
Incandescent
Light Fixture

Explosionproof
Attachment Plug
and Receptacle

Explosionproof
Fluorescent
Light Fixture

Explosionproof
Outlet Box

Explosionproof
Vertical Seal

Explosionproof
Vertical Seal with
Drain Fitting

Explosionproof Flexible Coupling

Nonincendive Circuit: A circuit, other than field wiring, in which any arc or thermal effect produced under the equipment's intended operating conditions, is incapable, under specified test conditions, of igniting the flammable gas-, vapor-, or dust-air mixture ≫ Article 100 and 500-4(f)(1) ≪ .

Nonincendive Field Wiring: Wiring that enters or leaves an equipment enclosure and, under the equipment's normal operating conditions, is not capable, due to arcing or thermal effects, of igniting the flammable gas-, vapor-, or dust-air mixture. Normal operation includes opening, shorting, or grounding the field wiring ≫ Article 100 ≪ .

Under normal operating conditions, nonincendive equipment's electrical or electronic circuitry is incapable of igniting a given flammable substance due to arcing or thermal means ≫ 500-4(f)(2) ≪ .

A nonincendive component has a contacting mechanism capable of making and breaking an incendive circuit; which mechanism's construction renders the component incapable of igniting the specified flammable gas- or vapor-air mixture. A nonincendive component's housing does not exclude the

flammable atmosphere nor does it contain an explosion. This protection technique is allowed for current-interrupting contacts in those Class I, Divisions 2 locations for which the component is approved ≫ 500-4(f)(2)(a) ≪ .

Nonincendive protection provides, under normal operating conditions, that any arcing or thermal effects are incapable of igniting whatever flammable/combustible substances are present. Classification-approved nonincendive techniques are permitted for equipment found in Class I, Division 2; Class II, Division 2; and Class III locations ≫ 500-4(f) ≪ .

The oil-immersion protective technique is valid for current-interrupting contacts in Class I, Division 2 locations as described in Section 501-6(b)(1)(b) ≫ 500-4(g) ≪ .

Current-limiting contacts in Class I, Division 2 locations, can be hermetically sealed against the intrusion of an external atmosphere via a fusion seal, i.e., soldering brazing, welding, or the fusion of glass to metal ≫ 500-4(h) ≪ .

Other protection techniques inherent to equipment listed for use in hazardous (classified) locations are allowed ≫ 500-4(i) ≪ .

CLASS I LOCATIONS

■ Class I, Division 1 Locations

• •

Class I, Division 1 classifications usually includes the following locations:
(1) Where volatile flammable liquids or liquefied flammable gases are transferred from one container to another
(2) Spray booth interiors and areas in the vicinity of spraying/painting operations where flammable solvents are used
(3) Locations containing open tanks/vats of volatile flammable liquids
(4) Drying rooms or components for the evaporation of flammable solvents
(5) Locations containing fat and oil extraction equipment using volatile flammable solvents
(6) Portions of cleaning and dyeing plants where flammable liquids are used
(7) Gas generator rooms and other portions of gas manufacturing plants where flammable gas may escape
(8) Inadequately ventilated pump rooms for flammable gas or for volatile flammable liquids
(9) Refrigerator/freezer interiors in which volatile flammable materials are stored in open, lightly stoppered, or easily ruptured containers
(10) All other locations where ignitable concentrations of flammable vapors/gases are likely to occur in the course of normal operation »500-7(a) FPN No. 1 «.

Class I, Division 1 locations contain ignitible concentrations of flammable gases/vapors which: (1) can exist during normal operation; (2) may often exist due to repair, maintenance, or leakage; and/or (3) might be released because operational failures cause simultaneous breakdowns of electrical equipment, capable of direct ignition »500-7(a) «.

Locations containing potentially sufficient quantities of air-borne gases (or vapors) to cause explosion or ignition, are categorized as Class I locations »500-7 «.

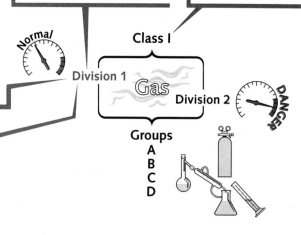

In some Division 1 locations, ignitible concentrations of flammable gasses/vapors may be present continuously or for extended periods of time. Examples include the following:
(1) The inside of inadequately vented enclosures containing instruments normally venting flammable gases/vapors within the enclosure
(2) The inside of vented tanks containing volatile flammable liquids
(3) The area between the inner and outer roof sections of a floating roof tank containing volatile flammable fluids
(4) Inadequately ventilated areas within spraying/coating operations using volatile flammable fluids
(5) The interior of an exhaust duct, used to vent ignitible concentrations of gases/vapors.
Avoiding the installation of electrical components in these areas altogether is strongly recommended. If relocation is unfeasible, use electric equipment or instrumentation approved specifically for the application, or consisting of what Article 504 describes as intrinsically safe systems »500-7(a) FPN No. 2 «.

In addition to the specific requirements found within Article 500, general *Code* rules apply to the electric wiring and equipment in Class I locations »501-1 «.

In Class I, Division 1 locations, transformers and capacitors, containing burnable liquid must be installed only in approved vaults compliant with Sections 450-41 through 450-48, in addition to all the following stipulations: (a) the vault cannot have any form of communicating opening into the Division 1 location; (b) ample ventilation must continually remove flammable gases/vapors; (c) all vents/ducts lead to safe locations, outside of buildings; and (d) vent ducts and openings must be of sufficient area to relieve the vault's internal explosive pressures, having all portions of vent ducts (within the building)

constructed of reinforced concrete »501-2(a)(1) «. Subsequently, transformers and capacitors not containing burnable liquids must either be installed in 501-2(a)(1) compliant vaults, or be approved for Class I locations »501-2(a)(2) «.

Approved enclosures must be provided for Class I, Division 1 located meters, instruments and relays. This includes kilowatt-hour meters, instrument transformers, resistors, rectifiers, and thermionic tubes. Class I, Division 1 approved enclosures include explosionproof as well as purged and pressurized enclosures »501-3(a) «.

Equipment, wiring methods, etc. within Class I, Division 1 locations are covered in the following: Section 501-4(a), *wiring method*; Section 501-5(a), (c), (d), and (f), *conduit/ cable seals*; Section 501-6(a), *switches, circuit breakers, motor controllers, and fuses*; Section 501-7(a), *control transformers and resistors*; Section 501-8(a), *motors, generators, and other rotating electric machinery*; Section 501-9(a), *lighting fixtures*; Section 501-10(a), *utilization equipment*; Section 501-11, *flexible cords*; Section 501-12, *receptacles and attachment plugs*; Section 501-13, *conductor insulation*; Section 501-14(a), *signaling, alarm, remote-control, and communications systems*; Section 501-15, *live parts*; Section 501-17(a), *surge protection*; and, Section 501-18, *multiwire branch circuits*.

Bonding and grounding must comply with Section 501-16(a) and (b) provisions.

Class I, Division 2 Locations

Class I, Division 2 locations: (1) are areas where volatile flammable liquids/gases are handled, processed, or used, but the potentially volatile substance (liquid, gas; vapor) is normally enclosed (containers or systems) and is subject to escape only during accidental rupture (or breakdown) of such enclosures or in case of abnormal conditions; (2) are equipped with positive mechanical ventilation which normally prevents the buildup of gas/vapor concentrations to an ignition level; and (3) adjoin a Class I, Division 1 location which might communicate ignitible gaseous concentrations unless prevented by adequate positive-pressure ventilation from a clean air source, with adequate safeguards against ventilation failure »500-7(b)«.

Class I, Division 2 classification usually includes locations where volatile flammable liquids/gases, or vapors are used; but, in the judgement of the authority having jurisdiction, would become hazardous only in case of an accident or unusual operating condition. The quantity of flammable material accidentally escaping, adequacy of ventilating equipment, total area involved, and the industry/business record with respect to explosions (or fires) are all factors that merit consideration in determining each location's classification and extent »500-7(b) FPN No. 1«.

Piping without valves, checks, meters, and similar devices would not ordinarily introduce a hazard even though used for flammable liquids or gases. Depending on factors such as quantity and size of containers and ventilation, locations used for the sealed-container storage of flammable liquids or liquefied or compressed gases may be considered either hazardous (classified) or unclassified locations »500-7(b) FPN No. 2«.

Unless modified by this article, Section 505-10 approved equipment, specifically listed for use in Class I, Zone 0, 1, or 2 locations, is acceptable in Class I, Division 2 applications for the same gas when appropriately rated for the temperature »501-1 and *Exception*«.

In Class I, Division 2 locations, transformers and capacitors must comply with Sections 450-21 through 450-27 »501-2(b)«.

Detailed requirements for Class I, Division 2 meters, instruments and relays are given in Section 501-3(b)(1) through (6).

Switches, circuit breakers, make-and-break pushbutton contacts, relays, alarm bells and horns, must have Class I, Division 1 rated enclosures per Section 501-3(a) »501-3(b)(1)«. General-purpose enclosures are permitted if the current-interrupting contacts are: (1) immersed in oil; (2) enclosed within a hermetically sealed chamber; (3) nonincendive circuits; or (4) part of a listed nonincendive component. »501-3(b)(1) *Exception*«.

Resistors, resistance devices, thermionic tubes, rectifiers, and similar equipment used in association with meters, instruments and relays must comply with Section 501-3(a) »501-3(b)(2)«. For particular instances where general-purpose enclosures are permitted, see Section 501-3(b)(2) *Exception*.

Transformer windings, impedance coils, solenoids, etc., without sliding/make-or-break contacts must have enclosures, which may be of the general-purpose-type »501-3(b)(3)«.

A single general-purpose enclosure is allowed for any assembly whose individual components qualify for general-purpose enclosures according to 501-3(b)(1) through (3). The exterior of an assembly enclosure containing resistors and similar equipment must be clearly and permanently marked with the maximum obtainable surface temperature of any internal component. Approved equipment can be marked with Table 500-5(d) identification numbers specifying its suitable temperature range »501-3(b)(4)«.

Where general-purpose enclosures are allowed in 501-3(b)(1)–(4), they may contain overcurrent-protective fuses for instrument circuits (normally exempt from overloading), provided each fuse is preceded by a Section 501-3(b)(1) compliant switch »501-3(b)(5)«.

Flexible cord, attachment plug, and receptacle connections are acceptable for process control instruments, provided all of the following conditions are met:

(1) A Section 501-3(b)(1) compliant switch, rather than the attachment plug, provides for current interruption; and

(2) Maximum current is 3 amperes at 120 volts, nominal; and

(3) The power-supply cord is no more than 3 feet in length, is approved for extra-hard usage (where exposed) or for hard usage (if protected by location), and is supplied through a locking and grounding type attachment plug and receptacle; and

(4) Only necessary receptacles are provided; and

(5) A receptacle label warns against unplugging under load »501-3(b)(6)«.

■ Class I, Division 2 Locations (*continued*)

Class I, Division 2 location equipment, wiring methods, etc., are covered in the following: Section 501-4(b), *wiring method*; Section 501-5(b), (c), (e), and (f), *conduit/cable seals*; Section 501-6(b), *switches, circuit breakers, motor controllers, and fuses*; Section 501-7(b), *control transformers and resistors*; Section 501-8(b), *motors, generators, and other rotating electric machinery*; Section 501-9(b), *lighting fixtures*; Section 501-10(b), *utilization equipment*; Section 501-11, *flexible cords*; Section 501-12, *receptacles and attachment plugs*; Section 501-13, *conductor insulation*; Section 501-14(b), *signaling, alarm, remote-control, and communications systems*; Section 501-15, *live parts*; and, Section 501-17(b), *surge protection*.

Bonding and grounding must comply with Section 501-16(a) and (b) provisions.

■ Wiring Methods—Class I, Division 1

Threaded joints must fully engage at least five threads ≫501-4(a)(1)≪.

1
2
3
4
5

NOTE

An equipment grounding conductor is required for the dual purpose of providing raceway system electrical continuity, as well as grounding of noncurrent-carrying metal parts ≫501-4(a)(1) Exception No. 1≪.

Avoid tensile stress at Type MI cable termination fittings through proper installation and support methods ≫501-4(a)(1)≪.

Rigid nonmetallic conduit (complying with Article 347) can be used when first surrounded by at least 2 inches of concrete, followed by a minimum cover of 24 inches (top of conduit to grade). In some cases, Section 511-4, Exception; 514-8, Exception No. 2; and Section 515-5(a) permit the omission of the concrete encasement ≫501-4(a)(1) Exception No. 1≪.

Threaded metal conduit (either rigid or steel intermediated) must comprise the last 24 inches of a rigid nonmetallic underground conduit. This measurement ends at either the point of emergence, or the point of aboveground raceway connection ≫501-4(a)(1) Exception No. 1≪.

Section 501-4(a)(1) Exception No.'s 2 and 3 respectively introduce the use of Type MC and Type ITC cable in certain industrial establishments. Refer to these **exceptions** for more information. Sections 334-3 and 334-4 outline Type MC cable restrictions.

Fixed wiring in Class I, Division 1 locations must consist of threaded metal conduit, either rigid metal or steel intermediate metal, or Type MI cable with location-approved termination fittings ≫501-4(a)(1)≪.

Connections requiring flexibility, such as those at motor terminals, can employ Class I listed flexible fittings ≫501-4(a)(2)≪. Flexible cord, installed per Section 501-11, is also acceptable ≫501-4(a)(2) Exception≪.

Use only threaded, explosion proof boxes, fittings, and joints for conduit connections or cable terminations ≫501-4(a)(1)≪.

■ Wiring Methods—Class I, Division 2

Explosionproof boxes, fittings, and joints are not mandatory unless required by Sections 501-3(b)(1), 501-6(b)(1), and 501-14(b)(1) »501-4(b)«.

Nonincendive field wiring is permitted using any of the methods suitable for wiring in ordinary locations »501-4(b) *Exception*«.

In Class I, Division 2 locations, use threaded rigid metal conduit, threaded steel intermediate metal conduit, enclosed gasketed busways, enclosed gasketed wireways, Type PLTC cable (per Article 725), or Type ITC cable in cable trays/raceways, supported by messenger wire, or directly buried where the cable is listed for this use; Type MI, MC, MV, or TC cable with approved termination fittings must be the wiring method employed. Type ITC, PLTC, MI, MC, MV, or TC cable which can be placed in cable tray systems, must be installed in a manner to avoid tensile stress at the termination fitting »501-4(b)«.

Where limited flexibility is necessary (at motor terminals), one of the following must be used: flexible metal fittings, flexible metal conduit with approved fittings, liquidtight flexible metal conduit with approved fittings, liquidtight flexible nonmetallic conduit with approved fittings, or flexible cord approved for extra-hard usage and provided with approved bushed fittings. Flexible cord must include an additional conductor for grounding »501-4(b)«.

Type MC cable

See Section 501-16(b) for grounding requirements where flexible conduit is used »501-4(b) FPN«.

■ Gasoline Dispensing and Service Stations

A *gasoline dispensing and service station* is a location where gasoline, or other volatile flammable liquids or liquefied flammable gases are transferred to the fuel tanks (including auxiliary fuel tanks) of self-propelled vehicles or to approved containers »514-1«.

GAS STATION

Wiring and equipment above Class I locations defined in Section 514-2 must comply with Sections 511-6 and 511-7 »514-4«.

All electrical equipment/ wiring within Class I locations defined in Section 514-2 must comply with the applicable provisions of Article 501 »514-3«.

Areas used as lubritoriums, service rooms, repair rooms, offices, salesrooms, compressor rooms, and similar locations must comply with Articles 510 and 511 with respect to electric wiring and equipment »514-1«.

If the authority having jurisdiction can satisfactorily determine that flammable liquids having a flash point below 38°C (100°F), such as gasoline, will not be handled, such location does not require a hazardous classification »514-1«.

A Class I location does not extend beyond any unpierced wall, roof, or other solid partition »514-2«.

All metal raceways, the metal armor or metallic sheath on cables, and all noncurrent-carrying metal parts of fixed/portable electrical equipment, regardless of voltage, must be grounded according to Article 250. Grounding in Class I locations must comply with Section 501-16 »514-16«.

> ┌─────────────────────────────────┐
> │   N O T E   │
> │ *Table 514-2 must be used to delineate* │
> │ *and classify service stations as well as* │
> │ *any other location storing, handling, or* │
> │ *dispensing Class I liquids* »514-2«. │
> └─────────────────────────────────┘

■ Conduit and Cable Seals—Class I, Divisions 1 and 2

Seals in conduit and cable systems, within Class I, Divisions 1 and 2, must comply with Section 501-5(a) through (f) ≫ 501-5 ≪.

Explosionproof Vertical Seal

Explosionproof Horizontal or Vertical Seal

In a seal, the conductors' cross-sectional area must not exceed 25% of the cross-sectional area of a rigid metal conduit of the same trade size, unless it is specifically approved for a higher percentage of fill ≫ 501-5(c)(6) ≪.

Explosionproof Vertical Seal with Drain Fitting

If a probability exists that liquid or other condensed vapor may be trapped within control equipment enclosures, or at any point in the raceway system, install approved means to prevent accumulation or to permit periodic draining ≫ 501-5(f)(1) ≪. See Section 501-5(f)(2) for motor and generator draining provisions. Seal drainage provisions for canned pump, process or service connection, etc., are found in Section 501-5(f)(3).

Conduit/cable system seals, minimize the passage of gasses/vapors and prevent the passage of flames from one portion of the electrical installation to another ≫ 501-5 FPN ≪.

Smaller Plug

Plug

Fiber Dam   Sealing Compound   Fiber Dam

In a completed seal, the sealing compound's minimum thickness must at least equal the trade size of the sealing fitting and must be in no case less than ⅝ inch ≫ 501-5(c)(3) ≪.

Plug

Fiber Dam   Sealing Compound

Sealing compound must: (1) be approved; (2) provide a seal against passage of gas/vapors through the seal fitting; (3) not be affected by surrounding atmosphere or liquids; and (4) have a melting point of 93°C (200°F), or higher ≫ 501-5(c)(2) ≪. Sealing compound must be of a type approved for the conditions and use ≫ 501-5 ≪.

Type MI cable termination fittings must contain sealing compound which excludes moisture/fluids from the cable insulation ≫ 501-5 ≪.

Class I, Division 1 conduit seals must be installed within 18 inches from the enclosure. Only explosionproof unions, couplings, reducers, elbows, capped elbows, and conduit bodies similar to L, T, and Cross types no larger than the trade size of the conduit are permitted between the sealing fitting and the explosionproof enclosure ≫ 501-5(a)(1) ≪.

CAUTION ≪≪≪

Sealing fittings must be accessible ≫ 501-5(c)(1) ≪.

Install a conduit seal in each conduit run leaving a Class I, Division 1 (or Division 2) location. The conduit seal is permitted on either side, and within 10 feet of such location's boundary, and must be designated and installed to minimize the amount of gas/vapor within the Division 1 (or Division 2) portion of the conduit from being communicated to the conduit beyond the seal ≫ 501-5(a)(4) and 501-5(b)(2) ≪.

In Class I, Division 1 locations, conduit seals must be placed in each conduit entry into an explosionproof enclosure where either: (a) the enclosure contains apparatus, such as switches, circuit breakers, fuses, relays, or resistors, that may produce arcs, sparks, or high temperatures that are an ignition source in normal operation; or (b) the entry is 2-inch size or larger and the enclosure contains terminals, splices, or taps ≫ 501-5(a)(1) ≪.

Class I, Division 2 connections to required explosionproof enclosures must have a conduit seal in accordance with 501-5(a)(1)(a) and (a)(3). The entire conduit run or nipple between the seal and such enclosure must comply with Section 501-4(a) ≫ 501-5(b)(1) ≪.

Splices/taps must not be made inside fittings intended only for sealing with compound, nor can other fittings which contain splices/taps be filled with compound ≫ 501-5(c)(4) ≪.

Class I, Division 1 location cable seals must comply with Section 501-5(d)(1) through (3).

Class I, Division 2 location cable seals must comply with Section 501-5(e)(1) through (4).

Location Boundary

10 Feet Maximum

Non-hazardous Area

Hazardous Area

Except for explosionproof reducers at the conduit seal, there must be no union, coupling, box, or fitting between the conduit seal and the point at which the conduit leaves the Division 1 or 2 location ≫ 501-5(a)(4) and 501-5(b)(2) ≪.

OR

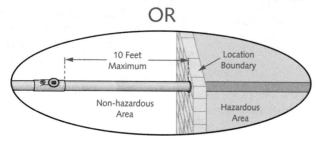

10 Feet Maximum

Location Boundary

Non-hazardous Area

Hazardous Area

NOTE

Study Section 501-5 carefully, because many exceptions and fine print notes are contained therein.

■ Commercial Garages, Repair, and Storage

Article 511 occupancies include those used for service and repair of self-propelled vehicles (including, but not limited to, passenger automobiles, buses, trucks, and tractors) which use volatile flammable liquids for fuel or power »511-1«.

For pendants, flexible cord suitable for the type of service and approved for hard usage must be used »511-6(b)«.

Areas adjoining defined locations, in which flammable vapor release is unlikely (such as stock rooms, switchboard rooms, and similar locations), are not classified if: (1) mechanically ventilated at a rate of four (or more) air changes per hour; or (2) effectively separated by walls or partitions »511-3(c)«. Adjacent areas having ventilation, air pressure differentials, or physical spacing such that, in the opinion of the authority enforcing the NEC®, no ignition hazard exists, are unclassified »511-3(d)«.

The entire area up to a level of 18 inches above each floor is a Class 1, Division 2 location »511-3(a)«. The Division 2 classification may not be required if the enforcing agency determines that mechanical ventilation provides at least four air changes per hour »511-3(a) Exception«.

18 Inches

Class I, Division 2

Ventilated Pit

A pit or depression, in which six air changes per hour are exhausted at the pit's floor level, can be judged by the enforcing agency to be a Class I, Division 2 location »511-3(b)«.

CAUTION ‹‹‹‹‹‹‹‹‹‹‹‹‹‹‹‹‹‹

All 125-volt, single-phase, 15- and 20-ampere receptacles in areas where electrical diagnostic equipment, electrical hand tools, or portable lighting equipment are used must have ground-fault circuit-interrupter protection for personnel »511-10«.

Any area where flammable fuel is transferred to vehicle fuel tanks must conform to Article 514 »511-2«.

Article 514 requirements apply for fuel dispensing units (other than liquid petroleum gas, which is prohibited) located within buildings »511-3(e)«.

Special provisions apply to portable lighting equipment. See Section 511-3(f).

Within Class I locations as defined in Section 511-3, wiring and equipment must conform to applicable Article 501 provisions. Raceways embedded in a masonry wall, or buried beneath a floor, are considered to be within the Class I location above the floor, if any connections or extensions lead into (or through) such areas »511-4«. (See exception pertaining to rigid nonmetallic conduit.)

Approved seals conforming to Section 501-5 provisions must be provided. Section 501-5(b)(2) applies not only to horizontal but also to vertical boundaries of the defined Class I locations »511-5«.

Equipment less than 12 feet above the floor level which may produce arcs, sparks, or hot metal particles (such as cutouts, switches, charging panels, generators, motors, or other equipment; and excluding receptacles, lamps, and lampholders) having make-and-break or sliding contacts, must be of the totally enclosed type or constructed to prevent the escape of sparks or hot metal particles »511-7(a)«.

Fixed lighting lamps and lampholders that are located over lanes through which vehicles are commonly driven or that may otherwise be exposed to physical damage must be located at least 12 feet above floor level, unless totally enclosed or constructed to prevent the escape of sparks or hot metal particles »511-7(b)«.

12 Feet

Unventilated Pit

Class I, Division 2

Class I, Division 1

Any pit or depression below floor level, lacking adequate ventilation, is considered a Class I, Division 1 location which extends up to said floor level »511-3(b)«.

■ Commercial Garages, Repair, and Storage (*continued*)

All fixed wiring above Class I locations must be placed within metal raceways, rigid nonmetallic conduit, electrical nonmetallic tubing, flexible metal conduit, liquidtight flexible metal conduit, or liquidtight flexible nonmetallic conduit; or must be Type MC, MI, manufactured wiring systems; or PLTC cable in accordance with Article 725; or Type TC cable »511-6(a)«.

Class I, Division 2

Class I, Division 1

NOTE

Fixed position attachment plug receptacles must either be located above the level of any Class I defined location, or be approved for the location »511-6(d)«.

Any circuit, above a Class I location, supplying portables (or pendants) which includes a grounded conductor (per Article 200), must have receptacles, attachment plugs, connectors, and similar devices of the grounding type; and the flexible cord's grounded conductor must either be connected to the screw shell of any lampholder or to the grounded terminal of any utilization equipment supplied. Grounding conductor continuity (between the fixed wiring system and the noncurrent-carrying metal portions of pendant fixtures, portable lamps, and portable utilization equipment) must be maintained via approved means »511-6(c)«.

Battery chargers, their control equipment, and batteries being charged, must not be located within areas classified in Section 511-3 »511-8«.

Electric vehicle charging equipment and wiring are covered in Section 511-9, as well as Article 625.

All metal raceways, the metal armor or metallic sheath on cables, and all noncurrent-carrying metal parts of fixed/portable electrical equipment, regardless of voltage, must be grounded according to Article 250. Grounding in Class I locations must comply with Section 501-16 »511-16«.

■ Parking Garages

For further information, see *Standard for Parking Structures*, NFPA 88A-1995, and *Standard for Repair Garages*, NFPA 88B-1997 »511-2 FPN«.

Garages used for parking or storage, where no repair work is done except exchange of parts and routine maintenance not requiring use of electrical equipment, open flame, welding, or volatile flammable liquids, are **not classified** »511-2«.

■ Aircraft Hangars

Article 513 applies to buildings/structures, or parts thereof inside which aircraft (containing Class I liquids or Class II liquids whose temperatures exceed their flash points) are housed, stored, serviced, repaired, or altered. It does not apply to locations used exclusively for aircraft that do not now, nor have ever, contained fuel ≫513-1≪.

The area within 5 feet horizontally from aircraft power plants or fuel tanks is a Class I, Division 2 location extending upward from the floor to a level 5 feet above the upper surface of wings and engine enclosures ≫513-3(c)≪.

Class I, Division 1

Class I, Division 2

18 Inches Above Floor

Any pit or depression below the hangar floor level is a Class I, Division 1 location which extends up to floor level ≫513-3(a)≪.

The hangar's entire area, including any adjacent and communicating areas not suitably separated from the hangar, is designated as a Class I, Division 2 location up to a level 18 inches above the floor ≫513-3(b)≪.

No Rear Engine

Class I, Division 2

> **NOTE**
>
> *In locations other than those described in Section 513-3, equipment less than 10 feet above aircraft wings and enclosures, which may produce arcs, sparks, or particles of hot metal (such as lamps and lampholders for fixed lighting, cutouts, switches, receptacles, charging panels, generators, motors, or other equipment having make-and-break or sliding contacts), must be of the totally enclosed type or by construction must prevent the escape of sparks or hot metal particles ≫513-6(a)≪.*

Mobile Equipment: Equipment having electric components designed for movement via mechanical aids, or having wheels to facilitate movement by person(s) or powered devices ≫513-2≪.

Portable Equipment: Equipment with electric components which can be moved by a single person without using mechanical aids ≫513-2≪.

All wiring and equipment intended for installation, installed or operated within any of the Section 513-3 defined Class I locations, must comply with the applicable provisions of Article 501. All wiring installed in (or under) the hangar floor must comply with Class I, Division 1 requirements. Adequate drainage must be provided for such wiring located within vaults, pits, or ducts ≫513-4(a)≪.

Attachment plugs and receptacles in Class I locations must either be approved for Class I locations, or designed so that they cannot be energized during connection/disconnection ≫513-4≪.

All fixed wiring in a hangar, not within a Class I location as defined in Section 513-3, must be: (1) installed in metal raceways; or (2) Type MI, TC, or MC cable ≫513-5(a)≪.

Portable utilization equipment and lamps require flexible cord suitable for the type of service and approved for extra-hard usage. Such cords must include a separate equipment grounding conductor ≫513-5(c)≪.

Adjacent areas in which flammable liquids or vapors are not likely to be released (stock rooms, electrical control rooms, etc.) are not classified, provided such areas are adequately ventilated and effectively separated from the hangar itself by walls or partitions ≫513-3(d)≪.

■ Aircraft Hangars (*continued*)

Portable utilization equipment which can be, or is, used within a hangar must be of a type suitable for use in Class I, Division 2 locations ≫513-6(d)≪.

Electric wiring, outlets, and equipment (including lamps) on, or attached to, stanchions (stationary or mobile), rostrums, or docks must comply with Section 513-7 provisions.

Approved seals must be provided per Section 501-5. Sealing requirements specified in Sections 501-5(a)(4) and (b)(2) apply both to horizontal and vertical Class I location boundaries. Raceways embedded within a concrete floor, or buried beneath the floor, are considered to be within the Class I location above the floor ≫513-8≪.

Aircraft battery chargers and their control equipment are covered in Section 513-10.

Section 513-11 requirements apply to aircraft energizers (stationary and mobile).

Mobile servicing equipment (such as vacuum cleaners, air compressors, air movers, etc.) must comply with Section 513-12 provisions.

All metal raceways, the metal armor or metallic sheath on cables, and all noncurrent-carrying metal parts of fixed/portable electrical equipment, regardless of voltage, must be grounded per Article 250. Grounding in Class I locations must comply with Section 501-16 ≫513-16≪.

■ Dispensing Devices

Within 18 inches horizontally, extending in all directions from grade to the dispenser enclosure, or that portion of the dispenser enclosure containing liquid handling components, is a Class I, Division 2 location ≫Table 514-2≪.

The space inside the dispenser enclosure is classified as covered in *Power Operated Dispensing Devices for Petroleum Products*, ANSI/UL 87-1995 ≫Table 514-2≪.

Each circuit leading to (or through) dispensing equipment, including remote pumping system equipment, must have a clearly identified, readily accessible switch (or other acceptable means), located away from the dispensing devices, to simultaneously disconnect all circuit conductors (including the grounded conductor, if any) form the supply source ≫514-5(a)≪.

From grade level up to a height of 18 inches within 20 feet horizontally of any enclosure edge, is a Class I, Division 2 location ≫Table 514-2≪.

All or part of any pit, box, or space below grade level within a Division 1 or 2 classified location, is designated Class I, Division 1 ≫Table 514-2≪.

An approved seal must be provided in each conduit run entering/leaving a dispenser or any cavities (or enclosures) in direct communication therewith. The sealing fitting must be the first fitting after the conduit emerges from the earth or concrete ≫514-7(a)≪.

Underground wiring must be installed either in threaded rigid metal conduit or threaded steel intermediate metal conduit. Any portion of electrical wiring or equipment, below the surface of a Class I, Division 1 or 2 location (as defined in Table 514-2), qualifies as a Class I, Division 1 location, and extends at least to the point of emergence above grade ≫514-8≪. (See exceptions for Type MI cable and rigid nonmetallic conduit.)

> **N O T E**
> *Each dispensing device must have a means to remove all external voltage sources, including feedback, during periods of maintenance/service ≫514-6≪.*

Unattended self-service station emergency controls as specified in Section 514-5(a) must be installed at a location acceptable to the authority having jurisdiction, such location being more than 20, but less than 100, feet from dispensers. Additional emergency controls must either be installed on each group of dispensers, or on the outdoor equipment controlling the dispensers. Emergency controls must completely shut off power to all of the station's dispensing equipment. Controls must be manually reset only in a manner approved by the authority having jurisdiction ≫514-5(c)≪.

Attended self-service station emergency controls, as specified in Section 514-5(a), must be installed at a location acceptable to the authority having jurisdiction and not more than 100 feet from dispensers ≫514-5(b)≪.

Additional seals must be provided per Section 501-5. Sections 501-5(a)(4) and (b)(2) apply to horizontal as well as vertical boundaries of the Class I locations ≫514-7(b)≪.

■ Bulk Storage Plants

A **bulk storage plant** is that portion of a property where flammable liquids are received (by tank vessel, pipelines, tank car, or tank vehicle), stored or blended in bulk for the purpose of distribution by vessel, pipeline, tank car, tank vehicle, portable tank, or container »515-1«.

All fixed wiring above Class I locations must be contained within metal raceways or PVC Schedule 80 rigid nonmetallic conduit (or equivalent) or be Type MI, TC, or MC cable. Fixed equipment which may produce arcs, sparks, or particles of hot metal (such as lamps and lampholders for fixed lighting, cutouts, switches, receptacles, motors, or other equipment having make-and-break or sliding contacts), must either be of the totally enclosed type or otherwise be constructed to prevent the escape of sparks or hot metal particles. Portable lamps or other utilization equipment, and associated flexible cords, must comply with Article 501's requirements for the class of location above which they are connected or used »515-4«.

Apply Table 515-2 where Class I liquids are stored, handled, or dispensed. Table 515-2 is also used to delineate and classify bulk storage plants. The Class I location does not extend beyond any floor, wall, roof, or other solid partition that has no communicating openings »515-2«.

All electric wiring and equipment within the Class I locations, defined in Section 515-2, must comply with the applicable provisions of Article 501 »515-3«.

Sealing requirements in Section 501-5(a)(4) and (b)(2) apply to horizontal as well as vertical boundaries of the defined Class I locations. Raceways buried under a defined Class I area are considered to be within a Class I, Division 1 location »515-6«.

Underground wiring must be enclosed in either threaded rigid metal conduit or threaded steel intermediate metal conduit; if buried under at least 2 feet of cover, rigid nonmetallic conduit (or an approved cable) can be used. If rigid nonmetallic conduit is used, threaded metal conduit (either rigid or steel intermediate) must be used for the last 2 feet of the conduit run to emergence or to the point of connection to the above ground raceway. Cable, if used, must also be enclosed in threaded metal conduit (either rigid or steel intermediate) from the point of lowest buried cable to the point of aboveground raceway connection »515-5(a)«.

Conductor insulation must comply with Section 501-13 »515-5(b)«.

If rigid nonmetallic conduit or cable with a nonmetallic sheath is used, include an equipment grounding conductor to provide electrical continuity of the raceway system and grounding of noncurrent-carrying metal parts »515-5(c)«.

Where gasoline, or other volatile flammable liquids or liquefied flammable gases, are dispensed at bulk stations, the appropriate provisions of Article 514 apply »515-7«.

All metal raceways, the metal armor or metallic sheath on cables, and all noncurrent-carrying metal parts of fixed/portable electrical equipment, regardless of voltage, must be grounded according to Article 250. Grounding in Class I locations must comply with Section 501-16 »515-16«.

■ Spray Application, Dipping, and Coating Process

Classification is based on dangerous quantities of flammable vapors, combustible mists, residues, dusts, or deposits »516-2«.

The following spaces are considered Class I or Class II, Division 1 locations, as applicable:

(1) Spray booth/room interiors except as specifically provided in Section 516-3(d).

(2) Exhaust duct interiors.

(3) Any area in the direct path of spray operations.

(4) For dipping and coating operations, all space within a 5-foot radius of the vapor source extending from the surface to the floor. The vapor source is liquid exposed during the process; which includes draining, and any dipped or coated object from which measurable vapor concentrations exceed 25 percent of the lower flammable limit within 1 foot, in any direction, from the object.

(5) Pits within 25 feet horizontally of the vapor source. Pits extending beyond 25 feet must have a vapor stop, without which the entire pit is designated Class I, Division 1.

(6) The interior of any enclosed coating/dipping process »516-2(a)«.

■ Spray Application, Dipping, and Coating Process (*continued*)

Article 516 covers the regular (or frequent) spray application of flammable liquids, combustible liquids, and combustible powders, as well as the application of flammable/combustible liquids at temperatures above their flashpoint, by dipping, coating, or other means »516-1«.

A space 3 feet above the floor, extending 20 feet horizontally in all directions from the Class I, Division 1 location, is designated Class I, Division 2 »516-2(b)«.

Class I, Division 2
(3 feet above floor,
extending 20 feet horizontally)

Class I, Division 1

Class I, Division 2
(outside area)

Class I, Division 1
(inside area)

Class I, Division 2
(3 feet above floor,
extending 20 feet horizontally)

All space within a 5-foot radius of the vapor source, extending from the surface to the floor, is designated Class I, Division 1 »516-2(a)(4)«.

A 3-foot space surrounding the dip-tank and drain-board's Class I, Division 1 location, is a Class I, Division 2 location »516-2(b)«.

```
N O T E
Wiring and equipment above Class I
and Class II locations must comply
with Section 516-7 specifications.
```

The following spaces are considered Class I or Class II, Division 2 locations, as applicable:

(1) All spaces outside of open sparing, but within 20 feet horizontally and 10 feet vertically of the Class I, Division 1 location as defined in Section 516-2(a), and not separated by partitions. (See *NEC®* illustration "Figure 516-2(b)(1).")

(2) If spray application operations are conducted within a closed-top, open-face, or open-front booth/room, any electrical wiring or utilization equipment located outside of the booth/room, but within the Division 2 boundaries shown in Figure 516-2(b)(2), must be suitable for Class I, Division 2 or Class II Division 2 locations, whichever is applicable. Open-face or open-front applications must comply with either Section 516-2(b)(2)(a) or 516-2-(b)(2)(b) requirements, depending upon the exhaust ventilation system setup.

(3) In open top spray booth operations, the space 3 feet vertically above the booth, and within 3 feet of all other booth openings, qualifies as Class I or Class II, Division 2.

(4) For enclosed spray booth/room operations, the space within 3 feet in all directions from any opening is considered Class I or Class II, Division 2 as shown in Figure 516-2(b)(4).

(5) The 3-foot space surrounding the dip-tank and drain-board's Class I, Division 1 location, as defined in Section 516-2(a)(4) and as shown in Figure 516-2(a)(5).

(6) For dip tanks and drain boards, the space 3 feet above the floor, extending 20 feet horizontally in all directions from the Class I, Division 1 location, unless an exception is met »516-2(b)«.

The space adjacent to an enclosed dipping (or coating) process/apparatus is unclassified »516-2(c)«.

Adjacent locations, cut off from the defined Class I or Class II locations by tight partitions without communicating openings, where release of flammable vapors or combustible powders is unlikely, are unclassified »516-2(d)«.

Locations using drying, curing, or fusion apparatus having (1) positive mechanical ventilation adequate to prevent accumulation of flammable vapor concentrations, and (2) effective interlocks to de-energize all electrical equipment (other than equipment approved for Class I locations) in case of inoperative ventilating equipment, are unclassified where the authority having jurisdiction deems appropriate »516-2(e)«.

Wiring and equipment (including illumination) provisions within Class I locations are covered in Section 516-3(a) through (e).

While fixed electrostatic spraying and detearing equipment requirements are located in Section 516-4(a) through (j), electrostatic hand-spraying equipment provisions are found in Section 516-5(a) through (e).

Section 516-6 pertains to combustible dry powder application processes.

All metal raceways, the metal armor or metallic sheath on cables, and all noncurrent-carrying metal parts of fixed/portable electrical equipment, regardless of voltage, must be grounded according to Article 250. Grounding in Class I locations must comply with Section 501-16 »516-16«.

CLASS II LOCATIONS

Class II, Division 1 Locations

A Class II, Division 1 location is an area: (1) In which combustible dust is airborne under normal operating conditions in quantities sufficient to produce explosive or ignitible mixtures; or (2) Where mechanical failure or abnormal operation of machinery/equipment might not only produce such explosive or ignitible mixtures, but also provide an ignition source through simultaneous failure of electric equipment, operation of protection devices, or by other causes; or (3) In which combustible dusts of an electrically conductive nature may exist in hazardous quantities »500-8(a)«.

Class II locations are hazardous because of the presence of combustible dust »500-8«.

Class II

Division 1

Dust

Division 2

Groups E F G

Normal

DANGER

Noncombustible dusts include dust produced in the handling and processing of grain and grain products, pulverized sugar and cocoa, dried egg and milk powders, pulverized spices, starch and pastes, potato and wood-flour, oil meal from beans and seed, dried hay, and other organic materials. Only Group E dusts are considered electrically conductive for classification purposes. Because dusts containing magnesium or aluminum are particularly hazardous, extreme precautions are necessary to avoid ignition and explosion »500-8(a) FPN«.

In addition to the requirements found in Article 500, wiring and equipment in Class II locations must meet Article 502's provisions »502-1«.

Dust-ignitionproof, as used by Article 502, means enclosed in a manner which excludes dust; and, where installed and protected in accordance with this *Code*, which does not permit arcs, sparks, or heat otherwise generated or liberated inside the enclosure to ignite exterior accumulations or atmospheric suspensions of a specified dust on or near the enclosure »502-1«.

Equipment in Class II locations must be able to function at full rating without developing surface temperatures high enough to excessively dehydrate or gradually carbonize any organic dust deposits that may occur »502-1«. Carbonized/excessively dry dust is highly susceptible to spontaneous ignition »502-1 FPN«.

The type of equipment and wiring defined as explosionproof (in Article 100) is not required and therefore not acceptable for Class II locations unless approved for such use »502-1«.

All locations containing Class II, Group E dusts in hazardous quantities are Division 1 »502-1«.

In Class II, Division 1 locations, transformers and capacitors, containing burnable liquid, must be installed only in approved vaults compliant with Sections 450-41 through 450-48, in addition to all the following stipulations: (a) Openings (doors, etc.) communicating with the Division 1 locations must have self-closing fire doors on both sides of the wall, carefully fitted and having suitable seals (such as weather stripping) to minimize the entrance of dust into the vault; (b) Vent openings and ducts must connect directly with outside air; and (c) Suitable pressure-relief openings to the outside air must be provided »502-2(a)(1)«. Subsequently, transformers and capacitors not containing burnable liquids must either be installed in vaults complying with Sections 450-41 through 450-48, or be approved as a complete assembly, including terminal connections for Class II locations »502-2(a)(2)«.

No transformer or capacitor can be installed in a location where dust from magnesium, aluminum bronze powders, or other similarly hazardous metals may be present »502-2(a)(3)«.

Equipment, wiring methods, etc. within Class II, Division 1 locations are covered in the following: Section 502-4(a), *wiring methods*; Section 502-5, *conduit/cable seals*; Section 502-6(a), *switches, circuit breakers, motor controllers, and fuses*; Section 502-7(a), *control transformers and resistors*; Section 502-8(a), *motors, generators, and other rotating electric machinery*; Section 502-9(a), *ventilating piping*; Section 502-10(a), *utilization equipment*; Section 502-11(a), *lighting fixtures*; Section 502-12, *flexible cords*; Section 502-13(a), *receptacles and attachment plugs*; Section 502-14(a), *signaling, alarm, remote-control, communications systems, meters, instruments, and relays*; Section 502-15, *live parts*; Section 502-17, *surge protection*; and, Section 502-18, *multiwire branch circuits*.

Bonding and grounding must comply with Section 502-16(a) and (b) provisions.

■ Class II, Division 2 Locations

A Class II, Division 2 location is an area: (1) where airborne combustible dust is not normally present in quantities sufficient to produce explosive or ignitible mixtures, and any dust accumulations typically do not interfere with normal operation of electrical equipment or other apparatus, but combustible dust may become airborne as a result of infrequent malfunctioning of processing/handling equipment; and (2) where combustible dust accumulations on, in, or near electrical equipment may be sufficient to interfere with safe heat dissipation from said equipment, or may be ignited by abnormal operation or failure of electrical equipment »500-8(b) «.

The quantity of combustible dust that may be present, and the adequacy of dust removal systems, are factors that merit consideration in determining classification. Such factors may even result in an unclassified area »500-8(b) FPN No. 1 «.

Where the handling of substances such as seed produces low quantities of dust, the amount of dust deposited may not warrant classification »500-8(b) FPN No. 2 «.

In Class II, Division 2 locations, transformers and capacitors containing burnable liquid must be installed within vaults compliant with Sections 450-41 through 450-48 »502-2(b)(1) «.

Askarel-containing transformers, rated in excess of 25 kVA, must have: (a) pressure-relief vents; (b) a means for absorbing any gases generated by arcing inside the case, or chimney/flue connected pressure-relief vents must carry such gases outside the building; and (c) at least 6 inches of airspace between the transformer case and any adjacent combustible material »502-2(b)(2) «.

Dry-type transformers, operating at not over 600 volts nominal, must either be installed in vaults or must have their windings and terminal connections enclosed in tight metal housings without ventilating or other openings »502-2(b)(3) «.

Class II, Division 2 location equipment, wiring methods, etc., are covered in the following: Section 502-4(b), *wiring methods*; Section 502-5, *conduit/cable seals*; Section 502-6(b), *switches, circuit breakers, motor controllers, and fuses*; Section 502-7(b), *control transformers and resistors*; Section 502-8(b), *motors, generators, and other rotating electric machinery*; Section 502-9(b), *ventilating piping*; Section 502-10(b), *utilization equipment*; Section 502-11(b), *lighting fixtures*; Section 502-12, *flexible cords*; Section 502-13(b), *receptacles and attachment plugs*; Section 502-14(b), *signaling, alarm, remote-control, communications systems, meters, instruments, and relays*; Section 502-15, *live parts*; and Section 502-17, *surge protection*.

Bonding and grounding must comply with Section 502-16(a) and (b) provisions.

■ Wiring Methods—Class II, Divisions 1 and 2

Class II locations are hazardous because of the presence of combustible dust »500-8 «.

TOP SEED Grain Company

Approved wiring methods in Class II, Division 1 locations include: threaded rigid metal conduit, threaded steel intermediate metal conduit, or type MI cable with termination fittings approved for the location. Type MI cable must be installed and supported in a manner which avoids tensile stress at the termination fittings »502-4(a) «. In industrial establishments with limited public access, where maintenance and supervision conditions ensure that only qualified persons will service the installation, Class II, Division 1 listed Type MC cable with a gas/vaportight continuous corrugated aluminum sheath, an overall jacket of suitable polymeric material, separate grounding conductors (per Section 250-122), and termination fittings listed for the application are permitted »502-4(a) *Exception* «.

Class II, Division 1 fittings and boxes must: (1) be provided with threaded bosses for conduit connections or cable terminations; (2) have close-fitting covers; and (3) have no openings (such as attachment screw holes) through which dust can enter or sparks or burning metal might escape. Fittings and boxes containing taps, joints, or terminal connections, or that are located where dusts of a combustible, electrically conductive nature exist, must be approved for Class II locations »502-4(a)(1) «.

Use only dusttight flexible connectors, liquidtight flexible metal conduit with approved fittings, liquidtight flexible nonmetallic conduit with approved fittings, or flexible cord approved for extra-hard usage and provided with bushed fittings in Class II, Divisions 1 and 2 locations necessitating flexible connections. Any flexible cords used must comply with Section 502-12. If flexible connections are subject to oil or other corrosive conditions, the conductor's insulation must either be of a type approved for the condition, or must be protected by means of a suitable sheath »502-4(a)(2) «. See Section 502-16(b) for flexible conduit grounding requirements »502-4(a)(2) FPN «.

Sealing fittings must be accessible »502-5 «.

Unless an exception is met, approved wiring methods in Class II, Division 2 locations include: rigid metal conduit, intermediate metal conduit, electrical metallic tubing, dusttight wireways, Type MC or MI cable with approved termination fittings, Type PLTC in cable trays, Type ITC in cable trays, or Type MC, MI, or TC cable installed in cable trays (ladder, ventilated trough, or ventilated channel) in a single layer, with a space at least equal to the larger cable diameter between the two adjacent cables »502-4(b) «. Exception No. 1 and Exception No. 2 pertain to nonincendive field wiring and Type MC cable respectively.

Class II, Division 2 wireways, fittings, and boxes containing taps, joints, or terminal connections must not only be designed to minimize dust infiltration, but also must: (1) have telescoping or close-fitting covers (or other effective means) to prevent the escape of sparks or burning material; (2) have no post-installation openings (such as holes for attachment screws) through which sparks or burning material can escape, or through which adjacent combustible material might be ignited »502-4(b)(1) «.

In Class II, Division 1 and 2 locations, where a raceway provides communication between an enclosure that is required to be dust-ignitionproof and one that is not, suitable means must prevent the entrance of dust into the dust-ignitionproof enclosure via the raceway. The following means are permitted: (1) a permanent and effective seal; (2) a horizontal raceway at least 10 feet long; or (3) a vertical raceway not less than 5 feet long which extends downward from the dust-ignitionproof enclosure »502-5 «.

Seals are not required where a raceway provides communication between an enclosure that is required to be dust-ignitionproof and an enclosure in an unclassified location »502-5 «.

In Class II, Division 1 and 2 locations, explosionproof seals are not required »502-5 «.

CLASS III LOCATIONS

■ Class III, Division 1 Locations

A Class III, Division 1 location is an area where easily ignitible fibers or materials producing combustible flyings are handled, manufactured, or used ❯❯ 500-9(a) ❮❮ .

Such locations usually include some parts of textile mills (rayon, cotton, etc.); cotton gins and cotton seed mills; flax-processing plants; clothing manufacturing facilities; woodworking plants; and establishments/industries involving similar hazardous processes or conditions ❯❯ 500-9(a) FPN No. 1 ❮❮ .

Easily ignitible fibers and flyings include rayon, cotton (including cotton linters and cotton waste), sisal or henequen, istle, jute, hemp, tow, cocoa fiber, oakum, baled waste kapok, Spanish moss, excelsior, and other materials of similar nature ❯❯ 500-9(a) FPN No. 2 ❮❮ .

Class III locations are hazardous because of the presence of easily ignitible fibers or flyings, which are not likely to be airborne in quantities sufficient to produce ignitible mixtures ❯❯ 500-9 ❮❮ .

In addition to Article 500 requirements, wiring and equipment in locations designated Class III must meet Article 503's provisions ❯❯ 503-1 ❮❮ .

Class III located equipment must be able to function at full rating without developing surface temperatures high enough to cause excessive dehydration or gradual carbonization of accumulated fibers or flyings. Carbonized or excessively dried organic material is highly susceptible to spontaneous ignition. During operation, the maximum surface temperature must not exceed 165°C (329°F) for equipment not subject to overloading, and 120°C (248°F) for equipment that may be overloaded (such as motors or power transformers) ❯❯ 503-1 ❮❮ .

Transformers and capacitors in Class III, Division 1 locations, must comply with Section 502-2(b) ❯❯ 503-2 ❮❮ .

Equipment, wiring methods, etc. within Class III, Division 1 locations are covered in the following: Section 503-3(a), *wiring methods*; Section 503-4, *switches, circuit breakers, motor controllers, and fuses*; Section 503-5, *control transformers and resistors*; Section 503-6, *motors, generators, and other rotating machinery*; Section 503-7, *ventilating piping*; Section 503-8, *utilization equipment*; Section 503-9, *lighting fixtures*; Section 503-10, *flexible cords*; Section 503-11, *receptacles and attachment plugs*; Section 503-12, *signaling, alarm, remote-control, and local loudspeaker intercommunications systems*; Section 503-13, *electric cranes, hoists, and similar equipment*; Section 503-14, *storage battery charging equipment*; and, Section 503-15, *live parts*.

Bonding and grounding must comply with Section 503-16(a) and (b) provisions.

■ Class III, Division 2 Locations

In Class III, Division 2 locations, transformers and capacitors must comply with Section 502-2(b) ❯❯ 503-2 ❮❮ .

Equipment, wiring methods, etc. within Class III, Division 2 locations are covered in the following: Section 503-3(b), *wiring methods*; Section 503-4, *switches, circuit breakers, motor controllers, and fuses*; Section 503-5, *control transformers and resistors*; Section 503-6, *motors, generators, and other rotating machinery*; Section 503-7, *ventilating piping*; Section 503-8, *utilization equipment*; Section 503-9, *lighting fixtures*; Section 503-10, *flexible cords*; Section 503-11, *receptacles and attachment plugs*; Section 503-12, *signaling, alarm, remote-control, and local loudspeaker intercommunications systems*; Section 503-13, *electric cranes, hoists, and similar equipment*; Section 503-14, *storage battery charging equipment*; and, Section 503-15, *live parts*.

Bonding and grounding must comply with Section 503-16(a) and (b) provisions.

A Class III, Division 2 location is an area in which easily ignitible fibers are stored or handled other than in the manufacturing process ❯❯ 500-9(b) ❮❮ .

■ Wiring Methods—Class III, Division 1

Such locations usually include some parts of textile mills (rayon, cotton, etc.); cotton gins and cotton seed mills; flax-processing plants; clothing manufacturing facilities; woodworking plants; and establishments/industries involving similar hazardous processes or conditions ≫500-9(a) FPN No. 1 ≪.

Woodworking Plant

Textile Production Equipment

Approved wiring methods for Class III, Division 1 locations include: rigid metal conduit, rigid nonmetallic conduit, intermediate metal conduit, electrical tubing, dusttight wireways, and Type MC or MI cable with approved fittings ≫503-3(a) ≪.

All Class III, Division 1 boxes and fittings must be dusttight ≫503-3(a)(1) ≪.

Use only dusttight flexible connectors, liquidtight flexible metal conduit with approved fittings, liquidtight flexible nonmetallic conduit with approved fittings, or flexible cord per Section 503-10 where flexible connections are necessary ≫503-3(a)(2) ≪. For grounding requirements where flexible conduit is used, see Section 503-16(b).

Class III, Division 1 and 2 switches, circuit breakers, motor controllers, and fuses, (including pushbuttons, relays, and similar devices), must have dusttight enclosures ≫503-4 ≪.

Class III, Division 1 and 2 flexible cords must comply with the following: (1) be of a type approved for extra-hard usage; (2) contain, in addition to the conductors of the circuit, a Section 400-23 compliant grounding conductor; (3) be connected to terminals or supply conductors in an approved manner; (4) be supported by clamps or other suitable means so that there will be no tension on the terminal connections; and (5) be provided with suitable means to prevent the entrance of fibers or flyings where the cord enters boxes or fittings ≫503-10 ≪.

Class III, Division 1 and 2 receptacles and attachment plugs must: (1) be of the grounding type; (2) be designed to minimize the accumulation (or entry) of fibers or flyings; and (3) prevent the escape of sparks or molten particles ≫503-11 ≪. If, in the judgement of the authority having jurisdiction, only moderate accumulations of lint or flyings are likely to collect in the vicinity of a receptacle, and where such receptacle is readily accessible for routine cleaning, general-purpose grounding-type receptacles (mounted to minimize the entry of fibers or flyings) are permitted ≫503-11 *Exception* ≪.

Live parts must not be exposed except as provided in Section 503-13 ≫503-15 ≪.

■ Wiring Methods—Class III, Division 2

Unless an exception is met, Class III, Division 2 location wiring methods must comply with Section 503-3(a) **》**503-3(b)**《** .

In storage sections, compartments, or areas containing no machinery, open wiring on insulators is permitted if installed per Article 320, but only on the condition that Section 320-14 required protection is provided where conductors are not run in roof spaces, and are well removed from sources of physical damage **》**503-3(b) *Exception***《** .

All Class III, Division 2 boxes and fittings must be dusttight **》**503-3(a)(1)**《** .

Class III, Division 1 and 2 pendant lighting fixtures must be suspended by stems of threaded rigid metal conduit, threaded intermediate metal conduit, threaded metal tubing of equivalent thickness, or by chains with approved fittings. Stems longer than 12 inches, must be permanently and effectively braced against lateral displacement at a level no more than 12 inches above the lower end of the stem, or flexibility in the form of an approved fitting or a flexible connector must be provided no more than 12 inches from the point of attachment to the supporting box or fitting **》**503-9(c)**《** .

Class III, Division 1 and 2 portable lighting equipment must have both handles and substantial guards. Lampholders must be of the unswitched type, incapable of receiving attachment plugs. There must be no exposed current-carrying metal parts, and all exposed noncurrent-carrying metal parts must be grounded. In all other respects, portable lighting equipment must comply with Section 503-9(a) **》**503-9(d)**《** .

Class III, Division 1 and 2 luminaires for fixed lighting must provide enclosures for lamps and lampholders designed to minimize entrance of fibers and flyings and to prevent the escape of sparks, burning material, or hot metal. Each fixture must be clearly marked to show the maximum permitted lamp wattage without exceeding an exposed surface temperature of 165°C (329°F) under normal use conditions **》**503-9(a)**《** .

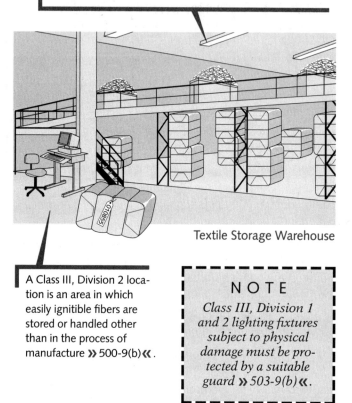

Textile Storage Warehouse

A Class III, Division 2 location is an area in which easily ignitible fibers are stored or handled other than in the process of manufacture **》**500-9(b)**《** .

> ### N O T E
> *Class III, Division 1 and 2 lighting fixtures subject to physical damage must be protected by a suitable guard* **》***503-9(b)***《**.

Summary

- Hazardous locations are divided into three classifications (Class I, II, and III); each class is subdivided into two categories (Division 1 and 2).
- While Article 500 covers general requirements for all classes, Articles 501, 502, and 503 provide specific provisions for each class individually.
- Locations containing potentially sufficient quantities of air-borne gases (or vapors) to cause explosion or ignition are categorized as Class I locations.
- Class II locations are hazardous due to the presence of combustible dust.
- Class III locations are hazardous because of the presence of easily ignitible fibers or flyings, which are not likely to be airborne in quantities sufficient to produce ignitible mixtures.

- Under typical operating conditions, Division 1 locations contain ignitible or combustible elements (gases, dust, fibers, etc.) in quantities sufficient to produce explosive (or ignitable) mixtures.
- Locations where ordinarily present quantities of combustible or ignitible elements (gases, dust, fibers, etc.) in the air are insufficient to produce explosive (or ignitable) mixtures, are classified as Division 2.
- Article 511 occupancies include those used for service and repair of self-propelled vehicles (including, but not limited to, passenger automobiles, buses, trucks, and tractors) which use volatile flammable liquids for fuel or power.

- Article 513 applies to buildings/structures, or parts thereof, inside which aircraft (containing Class I or II liquids whose temperatures exceed their flash points) are housed, stored, serviced, repaired, or altered.
- Gasoline dispensing and service station requirements are found in Article 514.

- Article 515 provisions address bulk storage plants.
- The regular (or frequent) spray application of flammable liquids, combustible liquids, and combustible powders, as well as the application of flammable/combustible liquids at temperatures above their flashpoint, by dipping, coating, or other means, are covered by Article 516.

Unit 15 Competency Test

• •

NEC® Reference   Answer

_____   _____   1. In aircraft hangars, any pit or depression below the level of the hangar floor shall be classified as a _____ location that shall extend up to said floor level.

_____   _____   2. Equipment, regardless of the classification of the location in which it is installed, that depends on a single compression seal, diaphragm, or tube to prevent flammable or combustible fluids from entering the equipment, shall be approved for a _____ location.

_____   _____   3. A _____ location is a location in which volatile flammable liquids or flammable gases are handles, process, or used, but in which the liquids, vapors, or gases will normally be confined within closed containers or closed systems from which they can escape only in case of accident rupture or breakdown of such containers or systems, or in case of abnormal operation of equipment.

_____   _____   4. All hazardous (classified) location threaded conduit shall be threaded with an NPT standard conduit cutting die that provides _____ taper per foot.

_____   _____   5. In bulk storage plants, aboveground tank vents shall be classified as _____ within 5 feet of the open end, extending in all directions.

_____   _____   6. In Class I, Division 1 locations, threaded joints shall be made up with at least _____ threads fully engaged.

_____   _____   7. A(n) _____ enclosure is constructed or protected so that dust will not interfere with its successful operation.

_____   _____   8. In Class II, Division 1 locations, fittings and boxes shall be provided with _____ for connection to conduit or cable terminations, shall have close-fitting covers, and shall have no openings through which dust might enter or through which sparks or burning material might escape.

_____   _____   9. A 1-inch threaded rigid metal conduit is terminating into a 2-inch sealing fitting. The minimum thickness of the sealing compound shall be _____.

_____   _____   10. In aircraft hangars, the area within _____ horizontally from aircraft power plants or aircraft fuel tanks shall be classified as a Class I, Division 2 location that shall extend upward from the floor to a level _____ above the upper surface of wings and of the engine enclosures.

_____   _____   11. _____ is defined as enclosed in a manner that will exclude dusts and, where installed and protected in accordance with the *NEC®*, will not permit arcs, sparks, or heat otherwise generated or liberated inside of the enclosure to cause ignition of exterior accumulations or atmospheric suspensions of a specified dust on or in the vicinity of the enclosure.

_____   _____   12. Class I locations are those in which _____ are or may be present in the air in quantities sufficient to produce explosive or ignitible mixtures.

_____   _____   13. In commercial garages, lamps and lampholders for fixed lighting that is located over lanes through which vehicles are commonly driven shall be located no less than _____ above the floor level.

_____   _____ 14. A _____ location is a location in which easily ignitible fibers or materials producing combustible flyings are handled, manufactured, or used.

_____   _____ 15. _____, _____, _____, _____, and _____ are five acceptable hazardous (classified) location protection techniques for electrical and electronic equipment.

_____   _____ 16. A(n) _____ is defined as a circuit, other than field wiring, in which any arc or thermal effect produced under intended operating conditions of the equipment, is not capable, under specified test conditions, of igniting the flammable gas-, vapor-, or dust-air mixture.

_____   _____ 17. In each conduit run passing from a Class I, Division 2 location into an unclassified area, the sealing fitting must be located within _____ feet of the boundary.

_____   _____ 18. For dipping and coating operations, all space within a _____ radial distance from the vapor sources extending from these surfaces to the floor, shall be considered a Class I, Division 1 location.

_____   _____ 19. In Class III, Division 2 locations, transformers and capacitors shall comply with Section _____.

_____   _____ 20. An outdoor gasoline dispensing device is considered to be a Class I, Division 2 location up to _____ above grade level within _____ horizontally of any edge of the enclosure.

_____   _____ 21. In bulk storage plants, the space between _____ and _____ from the open end of an aboveground tank vent, extending in all directions, is a Class I, Division 2 location.

_____   _____ 22. A(n) _____ enclosure is constructed so that dust will not enter the enclosing case under specified test conditions.

_____   _____ 23. For dip tanks and drain boards, the space _____ above the floor and extending _____ horizontally in all directions from the Class I, Division 1 location, shall be considered a Class I, Division 2 location.

_____   _____ 24. Aircraft energizers shall be designed and mounted so that all electric equipment and fixed wiring will be at least _____ above floor level.

_____   _____ 25. A _____ location is a location in which combustible dust is in the air under normal operating conditions in quantities sufficient to produce explosive or ignitible mixtures.

_____   _____ 26. For each commercial garage floor, the entire area up to a level of _____ above the floor shall be considered to be a Class I, Division 2 location.

_____   _____ 27. All Class I, Division 1 boxes, fittings, and joints shall be threaded for connection to conduit or cable terminations and shall be _____.

_____   _____ 28. A ½-inch threaded steel intermediate metal conduit is terminating into a ½-inch sealing fitting. The minimum thickness of the sealing compound shall be _____.

_____   _____ 29. The entire area of an aircraft hangar, including any adjacent and communicating areas not suitably cut off from the hangar, shall be classified as a Class I, Division 2 location up to a level _____ above the floor.

_____   _____ 30. For unattended self-service stations, emergency controls shall be more than _____ but less than _____ from the dispensers.

_____   _____ 31. In Class I, Division 2 locations, resistors shall be provided with enclosures; and the assembly shall be approved for Class I locations, unless resistance is nonvariable and maximum operating temperature, in degrees Celsius, will not exceed _____ of the ignition temperature of the gas or vapor involved, or has been tested and found incapable of igniting the gas or vapor.

Health Care

Objectives

● ●

After studying this unit, the student will:

✔ have an adequate understanding of health care facility terminology.

✔ be introduced to general, as well as specific, hospital, nursing home, and limited care facility requirements.

✔ know the criteria for judging an ambulatory health care facility.

✔ know what area constitutes the patient vicinity.

✔ be familiar with grounding and bonding requirements.

✔ have a good understanding of patient care areas (general care, critical care, and wet locations).

✔ be familiar with general and critical care area branch circuit and receptacle requirements.

✔ be introduced to the receptacle/receptacle cover tamper resistant requirements for pediatric areas.

✔ understand the two types of systems (equipment and emergency), as well as the two types of branches (life safety and critical) required by hospitals.

✔ know the required nursing home and limited care facility branches (life safety and critical).

✔ have a thorough understanding of hazardous (classified) anesthetizing location requirements.

✔ be familiar with X-ray installation provisions, including both fixed (stationary) and mobile equipment.

Introduction

● ●

Health care facilities, by definition, are buildings, or portions thereof, which contain, at least in part, occupancies such as hospitals, nursing homes, limited care, supervisory care, clinics, medical/dental offices, and ambulatory care, whether permanent or movable. Electrical installation criteria and wiring methods for health care facilities, are far more specialized, than for a standard commercial or industrial project. Article 517 of the *NEC*® governs electrical installations within health care facilities. Applicable requirements of *NEC*® Chapters 1 through 4 must be met, unless specifically modified by Article 517. The provisions of Article 517, Parts B and C, apply not only to single-function buildings, but are also intended for individual application to their respective forms of occupancy within a multifunction structure (e.g., Section 517-10 requirements must be met by a doctor's examining room situated within a limited care facility). As with any specialized endeavor, a working knowledge of the pertinent terminology is highly recommended. Because most of these terms apply only to health care facilities, Section 517-3 provides a comprehensive list of the most commonly used words and phrases. Article 517, Part B *(Wiring and Protection)* covers patient care areas throughout the spectrum of health care facilities. *Essential Electrical Systems* (Part C) outlines electrical system requirements for specific occupancies, i.e., hospitals, nursing homes, etc. *Inhalation Anesthetizing Location* provisions are presented in Part D of Article 517. Part E covers *X-Ray Installations* of all varieties: fixed, stationary, portable, and mobile, as well as transportable. In no way is Part E to be construed as specifying safeguards against the useful beam or stray X-ray radiation. As for *Communications, Signaling Systems, Data Systems, Fire Alarm Systems, and Systems Less than 120 Volts, Nominal,* see Part F for particular provisions. And finally, Article 517, Part G, applies to the installation of *Isolated Power Systems* which are comprised of an isolating transformer (or equivalent), a line isolation monitor, and its ungrounded circuit conductors. All in all, Article 517, together with the explanations found in this Unit, offers a reasonably complete picture of health care facility electrical system installations.

GENERAL

◼ Nursing Home and Limited Care Facility

Nursing Home: A building (or part thereof) used for 24-hour lodging, boarding, and nursing care, of four (or more) persons who, because of mental or physical incapacity, may be unable to safely provide for their own needs without assistance. **Nursing home,** whenever used in the *NEC®*, includes nursing and convalescent homes, skilled nursing facilities, intermediate care facilities, and infirmaries of homes for the aged ≫ 517-3 ≪.

Nursing homes and limited care facilities are included in the term **health care facilities**.

NOTE
Nursing homes and limited care facilities providing inpatient hospital care must comply with the requirements of Part C, Sections 517-30 through 517-35 ≫ *517-40(b)* ≪.

Limited Care Facility: A building (or part thereof) used for 24-hour housing of four or more persons incapable of self-preservation because of age; physical limitation, whether due to accident or illness; or mental limitations, such as retardation/developmental disability, mental illness, or chemical dependency ≫ 517-3 ≪.

Part C, Sections 517-40(c) through 517-44 requirements apply to nursing homes and limited care facilities ≫ 517-40(a) ≪.

Part C, Sections 517-40(c) through 517-44 requirements do not apply to freestanding buildings used as nursing homes and limited care facilities, provided that:
(a) Admitting and discharge policies preclude providing care to any patient (or resident) who may require electrical life-support.
(b) No surgical treatment requiring general anesthesia is offered.
(c) Automatic battery-operated system(s) or equipment is provided, effective for a minimum of 1½ hours, and otherwise in accordance with Section 700-12. The system must capably supply lighting for exit lights, exit corridors, stairways, nursing stations, medical preparation areas, boiler rooms, and communications areas, as well as power to operate all alarm systems ≫ 517-40(a) *Exception* ≪.

Nursing homes and limited care facilities that are contiguous with, or adjacent to, a hospital, can have essential electrical systems supplied by the hospital ≫ 517-40(c) ≪.

■ Hospital

●●

Hospital: A building (or part thereof) used for 24-hour medical, psychiatric, obstetrical, or surgical care, of four or more inpatients. **Hospital,** *wherever used in the Code, includes general hospitals, mental hospitals, tuberculosis hospitals, children's hospitals, and any such other facilities providing inpatient care* ≫517-3≪.

Health Care Facilities: Buildings (or portions of buildings) that contain, but are not limited to, occupancies such as hospitals; nursing homes; limited care; supervisory care; clinics; medical and dental offices; and ambulatory care, whether permanent or portable ≫517-3≪.

> N O T E
> *See Section 517-3 for definitions pertaining to health care facilities.*

Article 517, Part B, applies to patient care areas of all health care facilities ≫517-10(a)≪.

Part C, Sections 517-30 through 517-35 requirements apply to hospitals requiring an essential electrical system ≫517-30(a)≪.

Article 517 specifies the installation criteria and wiring methods that minimize electrical hazards by maintaining adequately low-potential differences between exposed conductive surfaces that are likely to become energized and are subject to patient contact ≫517-11≪. (See 517-11 FPN for additional explanation.)

Wiring methods must comply with the applicable requirements of Chapters 1 through 4, unless modified in Article 517 ≫517-12≪.

Psychiatric Hospital: A building used exclusively for 24-hour psychiatric care, with at least four inpatients ≫517-3≪.

While Article 517 provisions apply to health care facilities, it does not apply to veterinary facilities ≫517-1 FPN No. 1≪.

Refer to the appropriate health care facility documents for performance, maintenance, and testing criteria information ≫517-1 FPN No. 2≪.

Parts B and C requirements apply not only to single-function buildings, but also to their respective forms of occupancy within a multi-function building (e.g., a doctor's examining room, located within a limited care facility, must meet Section 517-10 provisions) ≫517-2≪.

Systems used for communications, signaling, data, fire alarm, and other systems less than 120 volts, nominal, must comply with Article 517, Part F (517-80 through 82).

Patient Bed Location: The location of an inpatient sleeping bed; or the bed (or procedure table) used in an area for critical patient care ≫517-3≪.

Patient Care Area: Any portion of a health care facility wherein patients are examined or treated. Patient care areas of health care facilities are classified as *general care areas* or *critical care areas,* either of which may be a wet location ≫517-3≪.

> N O T E
> *Article 517, Part B does not apply to the following: (1) Business offices, corridors, waiting rooms, and the like in clinics, medical and dental offices, and outpatient facilities; and (2) areas of nursing homes and limited care facilities, wired according to Chapters 1 through 4 of the NEC®, where these areas are used exclusively as patient sleeping rooms* ≫517-10(b)≪.

■ Patient Vicinity

In an area where patient care normally occurs, the **patient vicinity** is the space having surfaces subject to contact by the patient or by an attendant who may touch the patient. In a patient room, this typically encloses a space within the room no less than 6 feet beyond the bed's perimeter (in its normal location), and extending vertically no less than 7½ feet from the floor ≫ 517-3 ≪.

■ Ambulatory Health Care Center

Ambulatory Health Care Center is a building (or part thereof) used to provide simultaneous service (or treatment) to four or more patients, which meet either of the following: (1) Those facilities that provide treatment for outpatients that would render them incapable of self-preservation under emergency conditions without assistance (such as hemodialysis units or freestanding emergency medical units). (2) Those facilities that provide outpatient surgical treatment requiring general anesthesia ≫ 517-3 ≪.

Essential electrical systems for *ambulatory health care centers* must comply with Section 517-45 provisions.

The essential electrical system must supply power for the following: (1) task illumination related to the safety of life and necessary for safe cessation of procedures in progress; (2) all anesthesia and resuscitative equipment used where inhalation anesthetics are administered, including alarm and alerting devices; and (3) all electrical life-support equipment where procedures require such equipment for patient life support ≫ 517-45(b) ≪.

The essential electrical system's alternate power source must be specifically designed for this purpose and must be either a generator, battery system, or self-contained battery integral with the equipment. If critical care areas are present in the facility, the essential electrical system must meet Section 517-30 through 517-35 requirements ≫ 517-45(c)(1) ≪.

The alternate power source must be separate and independent of the normal source, and must be capable of sustaining its connected loads for a minimum of 1½ hours after normal source loss ≫ 517-45(c)(2) ≪.

The essential electrical system must be arranged so that, in the event of a normal power source failure, the alternate source automatically connects to the load within 10 seconds ≫ 517-45(c)(3) ≪.

N O T E

Section 517-50 requirements apply to essential electrical systems for clinics, medical/dental offices, and other health care facilities not covered in Sections 517-30, 517-40, and 517-45. There are only two small differences between Sections 517-45 and 517-50. One is the first fine print note in each section. The second is Section 517-45(c)(1) compared to Section 517-50(c)(1). Section 517-45(c)(1) states "Where critical care areas are present in the facility . . ." and Section 517-50(c)(1) states "Where electrical life-support equipment is required. . . ."

■ Electrical System Terminology

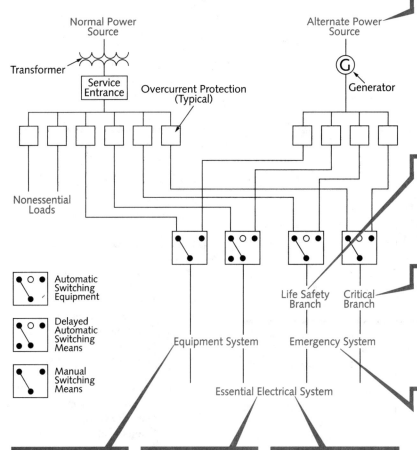

Alternate Power Source: One or more generator sets (or battery systems where permitted) to provide power during normal electrical service interruption, or the public utility electrical service intended to provide power during interruption of service normally provided by on-site generating facilities ≫ 517-3 ≪ .

Life Safety Branch: An Article 700 compliant subsystem of the emergency system consisting of feeders and branch circuits, providing adequate power to ensure patient and personnel safety, and which automatically connects to alternate power sources during interruption of normal power ≫ 517-3 ≪ .

Critical Branch: An emergency system subsystem consisting of feeders and branch circuits supplying energy to task illumination, special power circuits, and selected receptacles serving patient care related areas and functions, and which connects to alternate power sources by transfer switch(es) during interruption of the normal power ≫ 517-3 ≪ .

Emergency System: A system of feeders and branch circuits meeting Article 700 requirements (unless amended by Article 517), and intended to automatically supply alternate power within 10 seconds of interruption, to a limited number of prescribed functions vital to the protection of life and patient safety ≫ 517-3 ≪ .

Equipment System: A system of feeders and branch circuits arranged for delayed, automatic, or manual connection to the alternate power source, primarily serving 3-phase power equipment ≫ 517-3 ≪ .

Essential Electrical System: A system comprised of alternate power sources and all connected distribution systems (and ancillary equipment), designed both to ensure electrical power continuity to designated areas and functions of a health care facility during disruption of normal power sources, and to minimize disruption within the internal wiring system ≫ 517-3 ≪ .

The essential electrical system for these facilities shall comprise a system capable of supplying a limited amount of lighting and power service, considered essential for life safety and orderly cessation of procedures, during normal electrical service interruption. This includes clinics, medical/dental offices, hospitals, and other patient serving health care facilities ≫ 517-25 ≪ .

Electrical Life-Support Equipment: Electrically powered equipment whose continuous operation is necessary to maintain a patient's life ≫ 517-3 ≪ .

Isolated Power System: A system comprising an isolating transformer (or equivalent), a line isolation monitor, and its ungrounded circuit conductors ≫ 517-3 ≪ . (See related definitions, which include: *Hazard Current, Fault Hazard Current, Monitor Hazard Current, Total Hazard Current,* and *Line Isolation Monitor.*)

Isolation Transformer: A multiple-winding type transformer, with physically separated primary and secondary windings, which inductively couples its secondary winding to the grounded feeder systems energizing its primary winding ≫ 517-3 ≪ .

■ Grounding and Bonding

All general care and critical care area receptacles must be listed and identified as "hospital grade," unless an exception is met ≫517-18(b) and 517-19(b)(2) ≪.

Hospital Grade marking

Hospital Grade Receptacle (not isolated ground)

Metal Device Box

Insulated Grounding Conductor

Isolated Ground marking

Metal Raceway

To Panelboard

Receptacles with insulated grounding terminals, as permitted in Section 250-146(d), must be visibly identifiable after installation ≫517-16 ≪. Since the grounding impedance is controlled only by the grounding conductors, and does not benefit functionally from any parallel grounding paths, be cautious when specifying such a system ≫517-16 FPN ≪.

Normal and essential branch-circuit panelboard equipment grounding terminal buses, serving the same individual patient vicinity, must be bonded together with an insulated continuous copper conductor no smaller than No. 10. If more than two panels serve the same location, this conductor must be continuous from panel to panel, but can be broken in order to terminate on each panel's ground bus ≫517-14 ≪.

In a patient care area, all receptacle grounding terminals and all noncurrent-carrying conductive surfaces of fixed electric equipment likely to become energized that are subject to personal contact (operating at over 100 volts) must be grounded by an insulated copper conductor ≫517-13(a) ≪.

The grounding conductor, sized per Table 250-122, must be enclosed in metal raceways together with the branch-circuit conductors supplying these receptacles (or fixed equipment), unless an exception is met ≫517-13(a) ≪.

In addition to Section 517-13(a) requirements, all branch circuits serving patient care areas must have a ground path for fault current by installation in a metal raceway system (or cable assembly). The metal raceway system, or cable armor (or sheath) assembly, must qualify as an equipment grounding return path according to Section 250-118. Type MC cable and Type MI cable must have an outer metal armor/sheath identified as an acceptable grounding return path ≫517-13(b) ≪.

Metal raceways are not required if listed Type MI, MC, or AC cables are used, provided the cable's outer metal armor (or sheath) is identified as an acceptable grounding return path ≫517-13(a) Exception No. 1 ≪.

Metal faceplates can be grounded by means of metal mounting screw(s) securing the faceplate to a grounded outlet box (or grounded wiring device) ≫517-13(a) Exception No. 2 ≪.

Light fixtures more than 7½ feet above the floor, and switches located outside the patient vicinity, do not require grounding by an insulated grounding conductor ≫517-13(a) Exception No. 3 ≪.

Patient Equipment Grounding Point: A jack or terminal bus acting as the collection point for electric appliance redundant grounding in a patient vicinity, or for grounding other items to eliminate electromagnetic interference problems ≫517-3 ≪.

Reference Grounding Point: The ground bus of the panelboard, or isolated power system panel, supplying a patient care area ≫517-3 ≪.

Exposed Conductive Surfaces: Those surfaces capable of carrying electric current and unprotected, unenclosed, or unguarded, which permit personal contact. Paint, anodizing, and similar coatings do not provide suitable insulation, unless listed for such use ≫517-3 ≪.

■ Ground-Fault Protection

If ground-fault protection is provided for service disconnecting means (or feeder disconnecting means) operation as specified by Sections 230-95 or 215-10, an additional ground-fault protection step must be provided in the next level of feeder disconnecting means downstream toward the load. Such protection must consist of overcurrent devices and current transformers (or other equivalent protective equipment) which cause the feeder disconnecting means to open »517-17(a)«.

Ground-Fault Protection

Additional Level of Ground-Fault Protection Required

To Feeder Disconnecting Means

Feeder Conductors

3,000 Amperes
480/277 Volts

Ground-fault protection for the service and feeder disconnecting means operation must be fully selective so that the feeder device, and not the service device, opens for ground faults on the feeder device's load side. A six-cycle minimum separation, between the service and feeder ground-fault tripping bands, must be provided. To achieve 100 percent selectivity, disconnecting device operating time must be considered when selecting the time spread between these two bands »517-17(b)«.

Upon installation of equipment ground-fault protection, each level must be performance tested to ensure compliance with Section 517-17(b) »517-17(c)«.

> **NOTE**
> *Do not install additional levels of ground-fault protection: (1) on the load side of an essential electrical system transfer switch; or (2) between the on-site generating unit(s) described in Section 517-35(b) and the essential electrical system transfer switch(es); or (3) on electrical systems that are not solidly grounded wye systems with greater than 150 volts to ground, but no more than 600 volts phase-to-phase »517-17(a)«.*

■ Wet Locations

Health care facility patient care areas are classified as **general care areas** or **critical care areas**, either of which may be a *wet location* »517-3«.

If an isolated power system is utilized, the equipment must be listed for the purpose and must be compliant with Section 517-160 requirements »517-20(b)«.

> **NOTE**
> *All receptacles and fixed equipment within the wet location area must have ground-fault circuit-interrupter protection for personnel if interruption of power under fault conditions can be tolerated. If such interruption cannot be tolerated, the receptacles must be served by an isolated power system »517-20(a)«.*

Branch circuits (supplying only listed, fixed, therapeutic and diagnostic equipment) can be supplied from a normal grounded service, single- or 3-phase system, provided that: (a) grounded and isolated circuit wiring does not occupy the same raceway; and (b) all conductive equipment surfaces are grounded »517-20(a) *Exception* «.

Wet Locations: Are those patient care areas normally subject to wet conditions while patients are present. These include fluids either standing on the floor or drenching the work area. Routine housekeeping procedures and incidental liquid spillage do not constitute a wet location »517-3«.

PATIENT CARE AREAS

■ Patient Bed Location Branch Circuits (General Care Areas)

Each patient bed location must be supplied by at least two branch circuits: one from the emergency system; one from the normal system. All normal system branch circuits must originate within the same panelboard »517-18(a)«.

Branch circuits serving only special-purpose outlets or receptacles (such as portable X-ray outlets) do not require service from the same distribution panel(s) »517-18(a) Exception No. 1«.

If served from two separate emergency system transfer switches, a general care location does not require circuits from the normal system »517-18(a) Exception No. 3«.

To X-Ray Panel

To Normal Panel

To Emergency Panel

Alternative Method

To Emergency Panel No. 2 (Transfer Switch No. 2)

To Emergency Panel No. 1 (Transfer Switch No. 1)

X-RAY

Hospital Grade Receptacles

General Care Area

Section 517-18(a) requirements do not apply to patient bed locations in clinics, medical/dental offices, and outpatient facilities; psychiatric, substance abuse, and rehabilitation hospitals; nursing home sleeping rooms and limited care facilities where Section 517-10(b)(2) requirements are met »517-18(a) Exception No. 2«.

NOTE

General Care Areas are patient bedrooms, examining rooms, treatment rooms, clinics, and similar areas where the patient contacts ordinary appliances such as a nurse call system, electrical beds, examining lamps, telephone, and entertainment devices. In such areas, patients may also be connected to electro-medical devices (such as heating pads, electrocardiographs, drainage pumps, monitors, otoscopes, ophthalmoscopes, intravenous lines, etc.) »517-3«.

■ Patient Bed Location Receptacles (General Care Areas)

Special purpose receptacles are not counted as required receptacles.

Each patient bed location must have a minimum of four receptacles. They can be either single, duplex, or a combination of both types »518-18(b)«.

NOTE

All receptacles, must be listed and identified as "hospital grade." Each receptacle must be grounded by an insulated copper conductor sized per Table 250-122 »517-18(b)«.

X-RAY

Hospital Grade Receptacles

OR

OR

General Care Area

Section 517-18(b) requirements do not apply to psychiatric, substance abuse, and rehabilitation hospitals meeting Section 517-10(b)(2) requirements »517-18(b) Exception No. 1«.

It is not required that psychiatric security rooms have receptacle outlets »517-18(b) Exception No. 2«.

A total, immediate replacement of existing non-hospital grade receptacles is not necessary. It is intended, however, that non-hospital grade receptacles be replaced with hospital grade receptacles upon modification of use, during renovation, or as receptacles need replacing »517-18(b) FPN«.

■ Pediatric Location Receptacles (General Care Areas)

Tamper Resistant Receptacle

Fifteen- and 20-ampere, 125-volt receptacles which supply patient care areas of pediatric wards, rooms, or areas must either be **listed** tamper resistant, or must employ a **listed** tamper resistant cover ≫517-18(c) ≪.

■ Patient Bed Location Receptacles (Critical Care Areas)

Each patient bed location must have a minimum of six receptacles. At least one of the receptacles must be connected to: (1) the normal system branch circuit as required by Section 517-19(a); or (2) an emergency system branch circuit supplied by a different transfer switch not associated with other receptacles at the same location ≫517-19(b)(1) ≪.

The receptacles can be either single, duplex, or a combination of both types ≫517-19(b)(2) ≪.

The equipment grounding conductor for special-purpose receptacles (such as the operation of mobile X-ray equipment), must extend to the branch circuit reference grounding points for all locations potentially served from such receptacles. Where an isolated ungrounded system serves such a circuit, the grounding conductor does not have to run with the power conductors; however, the special-purpose receptacle's equipment grounding terminal must connect to the reference grounding point ≫517-19(g) ≪.

Hospital Grade Receptacles

Critical Care Area

All receptacles, must be listed and identified as "hospital grade." Each receptacle must be grounded to the reference grounding point by means of an insulated copper equipment grounding conductor ≫517-19(b)(2) ≪.

Although not required, a patient vicinity can have a patient equipment grounding point. Where one is supplied, it can contain one (or more) jacks listed for the purpose. An equipment bonding jumper, no smaller than No. 10, must connect the grounding terminal of all grounding-type receptacles to the patient equipment grounding point. The bonding conductor can be arranged centrically or looped as convenient ≫517-19(c) ≪.

If there is no patient equipment grounding point, it is important that the distance between the reference grounding point and the patient vicinity be as short as possible thereby minimizing potential differences ≫517-19(c) FPN ≪.

Isolated power systems must comply with Section 517-19(e) and (f) provisions.

N O T E

*A grounded electrical distribution system, having metal feeder raceway or Type MC or MI cable, requires that panelboard (or switchboard) grounding be ensured at **each** termination (or junction point) of the raceway or Type MC or MI cable by one of the following means:*

(1) A grounding bushing and a continuous copper bonding jumper, sized according to Section 250-122, with the bonding jumper connected to the junction enclosure or the panel ground bus

(2) Connection of feeder raceways or Type MC or MI cables to threaded hubs or bosses on terminating enclosures

(3) Other approved devices such as bonding-type locknuts or bushings ≫517-19(d) ≪.

■ Patient Bed Location Branch Circuits (Critical Care Areas)

Branch circuits serving only special-purpose receptacles, or equipment in critical care areas, can be served by other panelboards ≫517-19(a) *Exception No. 1* ≪.

Each patient bed location must be supplied by at least two branch circuits: one (or more) from the emergency system; one (or more) from the normal system. At least one emergency system branch circuit must supply outlet(s) at that bed location only ≫517-19(a) ≪.

CAUTION ≪≪≪≪≪≪≪≪≪≪

All normal system branch circuits must originate from a single panelboard ≫517-19(a) ≪.

Emergency system receptacles must be identified, and must also indicate the supplying panelboard and circuit number ≫517-19(a) ≪.

NOTE

Critical Care Areas are those special care units, intensive care units, coronary care units, angiography laboratories, cardiac catheterization laboratories, delivery rooms, operating rooms, and similar areas where patients are subjected to invasive procedures while connected to line-operated, electromedical devices ≫517-3 ≪.

To X-Ray Panel

To Normal Panel

To Emergency Panel

Hospital Grade Receptacles

Alternative Method

To Emergency Panel No. 2 (Transfer Switch No. 2)

To Emergency Panel No. 1 (Transfer Switch No. 1)

Critical care locations served from two separate emergency system transfer switches do not require circuits from the normal system ≫517-19(a) *Exception No. 2* ≪.

Critical Care Area

HOSPITALS

■ Life Safety Branch

The emergency system life safety branch must supply power for lighting, receptacles, and equipment as listed in Section 517-32(a) through(f) ≫517-32 ≪.

Illumination of egress means is required, such as lighting for corridors, passageways, stairways, landings at exit doors, and all paths of approach to exits. Switching arrangement to transfer patient corridor lighting from general illumination circuits is permitted, provided only one of two circuits can be selected, and both circuits cannot be simultaneously extinguished ≫517-32(a) ≪.

Task illumination battery charger for emergency battery-powered lighting unit(s) and selected receptacles at the generator set location ≫517-32(e) ≪.

To Normal Source

To Alternate Source

Life Safety Branch

Overcurrent Protection

Illumination of Means of Egress

Generator Set Location

Exit Signs

Alarm and Alerting Systems

Elevators

Communications Systems

Other

Only functions listed in 517-32(a) through (f) can be connected to the life safety branch ≫517-32 ≪.

Hospital communication systems, used for issuing instructions during emergency conditions ≫517-32(d) ≪.

Elevator cab lighting, control, communications, and signal systems ≫517-32(f) ≪.

Exit and exit directional signs ≫517-32(b) ≪.

Alarm and alerting systems, including: (1) fire alarms and (2) alarms required by systems used for piping nonflammable medical gasses ≫517-32(c) ≪.

■ Electrical Systems

A facility having a maximum essential electrical system demand of 150 kVA can have a single transfer switch serving one or more branches (critical and life safety) or systems (emergency and equipment) »517-30(b)(4)«.

Loads, served by generating equipment not specifically named in Sections 517-32, 517- 33, and 517-34, must be served by their own transfer switches so that these loads (a) are not transferred if the transfer will overload the generating equipment; and (b) are automatically shed upon generating equipment overloading »517-30(b)(5)«.

Hospital power sources and alternate power sources can serve the essential electrical systems of contiguous, or same site, facilities »517-30(b)(6)«.

The number of transfer switches used is based on reliability, design, and load considerations. Each essential electrical system (emergency and equipment) branch must be served by transfer switch(es) »517-30(b)(4)«.

The hospital's emergency system wiring must be either mechanically protected by installation in nonflexible metal raceways, or must be wired with Type MI cable. Where installed as patient care area branch circuit conductors, the installation must comply with Section 517-13 requirements, unless an exception is met »517-30(c)(3)«.

If isolated power systems are installed in any of the areas in Sections 517-33(a)(1) and (a)(2), each system must be supplied by an individual circuit serving no other load »517-30(c)(2)«.

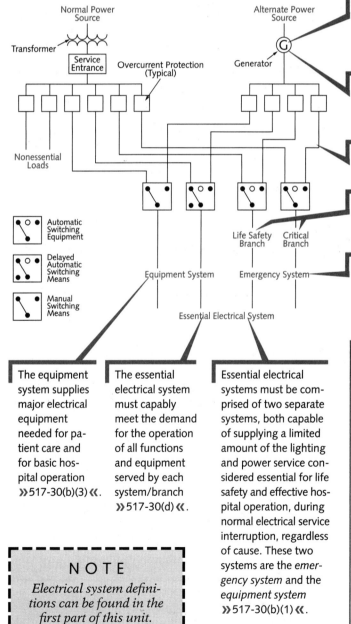

Demand calculations for generator set(s) sizing must be based on the following: (1) prudent demand factors and historical data; (2) connected load; (3) Article 220 feeder calculation procedures; or (4) any combination of the above »517-30(d)«.

The generator set(s) must have sufficient capacity and be properly rated to meet the demand produced by the essential electrical system(s) load at any given time »517-30(d)«.

Feeders must be sized per Articles 215 and 220 »517-30(d)«.

Limit the emergency system to only those circuits essential to life safety and critical patient care. These carry *life safety branch* and the *critical branch* designations »517- 30(b)(2) and 517-31«.

Emergency system branches must be installed and connected to the alternate power source so that all functions specified herein are automatically restored to operation within ten seconds after normal source interruption »517-31«.

The equipment system supplies major electrical equipment needed for patient care and for basic hospital operation »517-30(b)(3)«.

The essential electrical system must capably meet the demand for the operation of all functions and equipment served by each system/branch »517-30(d)«.

Essential electrical systems must be comprised of two separate systems, both capable of supplying a limited amount of the lighting and power service considered essential for life safety and effective hospital operation, during normal electrical service interruption, regardless of cause. These two systems are the *emergency system* and the *equipment system* »517-30(b)(1)«.

NOTE
Electrical system definitions can be found in the first part of this unit.

WARNING

The emergency system's life safety branch and critical branch must remain entirely independent of all other wiring (and equipment), and must not enter the same raceways, boxes, or cabinets with one another or any other wiring. Life safety branch and critical branch wiring can occupy the same raceways, boxes, or cabinets of other circuits (not part of the branch) where such wiring is:
(1) In transfer equipment enclosures, or
(2) In exit/emergency lighting fixtures supplied from two sources, or
(3) In a common junction box attached to exit/emergency lighting fixtures supplied from two sources, or
(4) For multiple emergency circuits supplied from the same branch.
The equipment system wiring can occupy the same raceways, boxes, or cabinets of other circuits that are not part of the life safety branch »517-30(c)(1)«.

■ Critical Branch

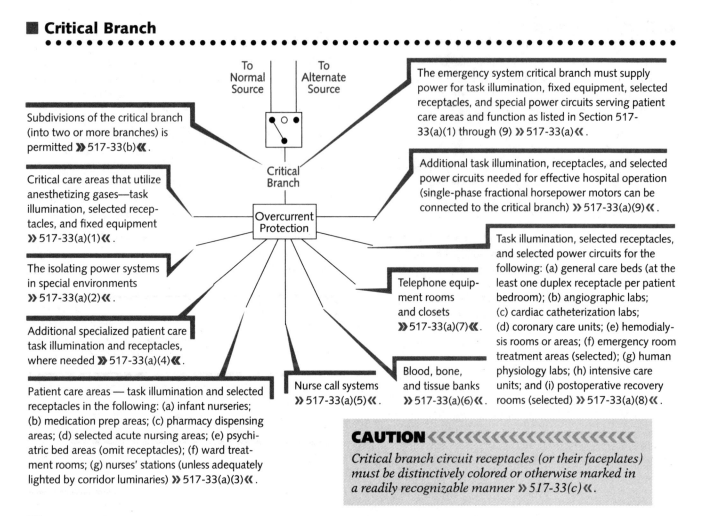

To Normal Source

To Alternate Source

Critical Branch

Overcurrent Protection

Subdivisions of the critical branch (into two or more branches) is permitted ≫ 517-33(b) ≪.

The emergency system critical branch must supply power for task illumination, fixed equipment, selected receptacles, and special power circuits serving patient care areas and function as listed in Section 517-33(a)(1) through (9) ≫ 517-33(a) ≪.

Critical care areas that utilize anesthetizing gases—task illumination, selected receptacles, and fixed equipment ≫ 517-33(a)(1) ≪.

Additional task illumination, receptacles, and selected power circuits needed for effective hospital operation (single-phase fractional horsepower motors can be connected to the critical branch) ≫ 517-33(a)(9) ≪.

The isolating power systems in special environments ≫ 517-33(a)(2) ≪.

Task illumination, selected receptacles, and selected power circuits for the following: (a) general care beds (at the least one duplex receptacle per patient bedroom); (b) angiographic labs; (c) cardiac catheterization labs; (d) coronary care units; (e) hemodialysis rooms or areas; (f) emergency room treatment areas (selected); (g) human physiology labs; (h) intensive care units; and (i) postoperative recovery rooms (selected) ≫ 517-33(a)(8) ≪.

Additional specialized patient care task illumination and receptacles, where needed ≫ 517-33(a)(4) ≪.

Telephone equipment rooms and closets ≫ 517-33(a)(7) ≪.

Patient care areas — task illumination and selected receptacles in the following: (a) infant nurseries; (b) medication prep areas; (c) pharmacy dispensing areas; (d) selected acute nursing areas; (e) psychiatric bed areas (omit receptacles); (f) ward treatment rooms; (g) nurses' stations (unless adequately lighted by corridor luminaries) ≫ 517-33(a)(3) ≪.

Nurse call systems ≫ 517-33(a)(5) ≪.

Blood, bone, and tissue banks ≫ 517-33(a)(6) ≪.

CAUTION ≪≪≪≪≪≪≪≪≪≪≪≪≪≪≪≪≪≪≪≪
Critical branch circuit receptacles (or their faceplates) must be distinctively colored or otherwise marked in a readily recognizable manner ≫ 517-33(c) ≪.

■ Power Sources

Facilities, whose normal power source is supplied by multiple separate central station-fed services, experience greater than normal electrical service reliability than those having only a single feed. Such a dual source of normal power consists of multiple electrical services, fed from separate generator sets, or a utility distribution network having multiple power input sources. These dual services are arranged to provide mechanical and electrical separation when a fault between the facility and the generating sources is unlikely to cause an interruption of more than one of the facility service feeders ≫ 517-35(c) FPN ≪.

N O T E

Carefully consider the location of the essential electrical system components (housing and spaces) to minimize interruptions caused by natural forces common to the area (e.g., storms, floods, earthquakes, or hazards created by adjoining structures or activities). Consideration must also be given to the potential normal electrical service interruptions resulting from similar causes, as well as possible service disruption due to internal wiring and equipment failures ≫ 517-35(c) ≪.

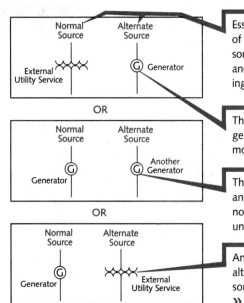

Normal Source

Alternate Source

External Utility Service

Generator

OR

Normal Source

Alternate Source

Generator

Another Generator

OR

Normal Source

Alternate Source

Generator

External Utility Service

Essential electrical systems must have a minimum of two independent power sources: a normal source generally supplying the entire system; and, one or more alternate sources for use during normal source interruption ≫ 517-35(a) ≪.

The alternate power source can be on-site generator(s) driven by some form of prime mover(s) ≫ 517-35(b) ≪.

The alternate power source can also be another generating unit(s) provided the normal source consists of on-site generating unit(s) ≫ 517-35(b) ≪.

An external utility service can act as the alternate power source when the normal source consists of on-site generating unit(s) ≫ 517-35(b) ≪.

■ Equipment Systems

Manual Switching Means

Delayed Automatic Switching Means

Equipment System

The following equipment must have either delayed automatic or manual connection to the alternate power source:

(1) Heating equipment to providing heat for operating, delivery, labor, recovery, intensive care, coronary care, nurseries, infection/isolation rooms, emergency treatment spaces, and general patient rooms. (See exception pertaining to heating of general patient rooms and infection/isolation rooms.)

(2) Elevator(s) providing service to patient, surgical, obstetrical, and ground floors during normal power interruption. (Where interruption of normal power would result in other elevators stopping between floors, throw-over facilities must be provided to allow temporary operation of any elevator for patient/other person release.)

(3) Supply, returns, and exhaust ventilating systems for surgical and obstetrical delivery suites, intensive care, coronary care, nurseries, infection/isolation rooms, emergency treatment spaces, and exhaust fans for laboratory fume hoods, nuclear medicine areas where radioactive material is used, ethylene oxide evacuation, and anesthesia evacuation

(4) Hyperbaric facilities

(5) Hypobric facilities

(6) Automatic doors

(7) Minimal electrically heated autoclaving equipment can be arranged for either automatic or manual connection to the alternate source.

(8) Controls for equipment listed in Section 517-34

(9) Other selected equipment can be served by the equipment system »517-34(b) «.

Equipment system installation and connection to the alternate power source, for equipment described in Section 517-34(a), must automatically restore operation at appropriate time-lag intervals following emergency system activation. The arrangement must also provide for subsequent connection of equipment described in Section 517-34(b) »517-34 «.

Delayed automatic connection to the alternate power source must be provided for the following equipment:

(1) Central suction systems serving medical/surgical functions, including controls. (Such suction systems are permitted on the critical branch.)

(2) Sump pumps and other equipment required for the safe operation of major apparatus, including associated control systems and alarms

(3) Compressed air systems serving medical/surgical functions, including controls. (Such air systems are permitted on the critical branch.)

(4) Smoke control, stair pressurization systems, or both

(5) Kitchen hood supply, exhaust systems, or both, if operation is required during a fire in (or under) the hood »517-34(a) «.

NOTE

Sequential delayed automatic connection to the alternate power source, thereby preventing generator overload, is permitted where indicated by engineering studies »517-34(a) Exception «.

NURSING HOMES AND LIMITED CARE FACILITIES

■ Electrical Systems

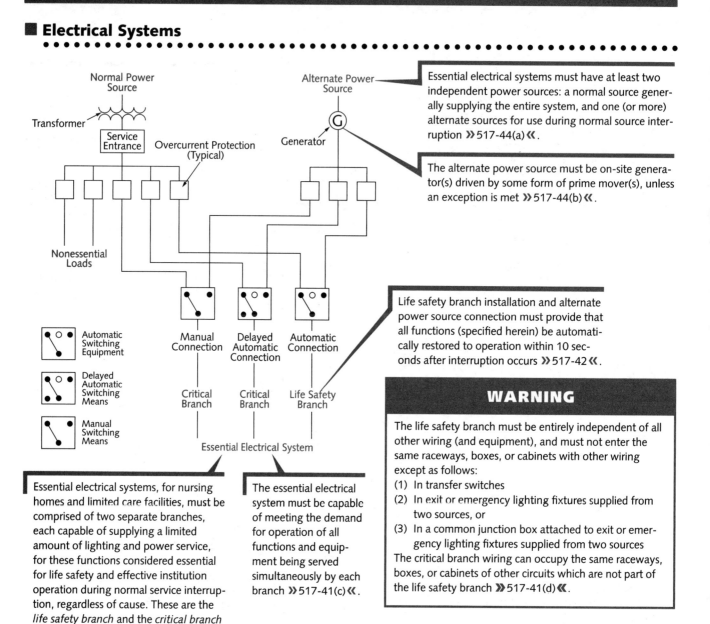

Essential electrical systems must have at least two independent power sources: a normal source generally supplying the entire system, and one (or more) alternate sources for use during normal source interruption **》517-44(a)《**.

The alternate power source must be on-site generator(s) driven by some form of prime mover(s), unless an exception is met **》517-44(b)《**.

Life safety branch installation and alternate power source connection must provide that all functions (specified herein) be automatically restored to operation within 10 seconds after interruption occurs **》517-42《**.

Essential electrical systems, for nursing homes and limited care facilities, must be comprised of two separate branches, each capable of supplying a limited amount of lighting and power service, for these functions considered essential for life safety and effective institution operation during normal service interruption, regardless of cause. These are the *life safety branch* and the *critical branch* **》517-41(a)《**.

The essential electrical system must be capable of meeting the demand for operation of all functions and equipment being served simultaneously by each branch **》517-41(c)《**.

WARNING

The life safety branch must be entirely independent of all other wiring (and equipment), and must not enter the same raceways, boxes, or cabinets with other wiring except as follows:
(1) In transfer switches
(2) In exit or emergency lighting fixtures supplied from two sources, or
(3) In a common junction box attached to exit or emergency lighting fixtures supplied from two sources
The critical branch wiring can occupy the same raceways, boxes, or cabinets of other circuits which are not part of the life safety branch **》517-41(d)《**.

NOTE

Give careful consideration to the essential electrical system components location to minimize interruptions caused by natural forces common to the area (e.g., storms, floods, earthquakes, or hazards created by adjoining structures or activities). Consideration must also be given to possible normal services interruption resulting from similar causes, as well as, potential service disruption due to internal wiring and equipment failures **》517-44(c)《**.

Where the normal source consists of on-site generating units, the alternate source must be either another generator set, or an external utility service **》517-44(b)** *Exception No. 1《*.

Nursing homes or limited care facilities meeting the requirements of Section 517-40(a), *Exception*, can use a battery system or an equipment self-contained, integral battery **》517-44(b)** *Exception No. 2《*.

The number of transfer switches used is based on reliability, design, and load considerations. Each essential electrical system branch (life safety and critical) must be served by one (or more) transfer switches. A facility whose maximum demand is 150 kVA can have a single transfer switch serving one or more branches/systems **》517-41(b)《**.

Life Safety Branch

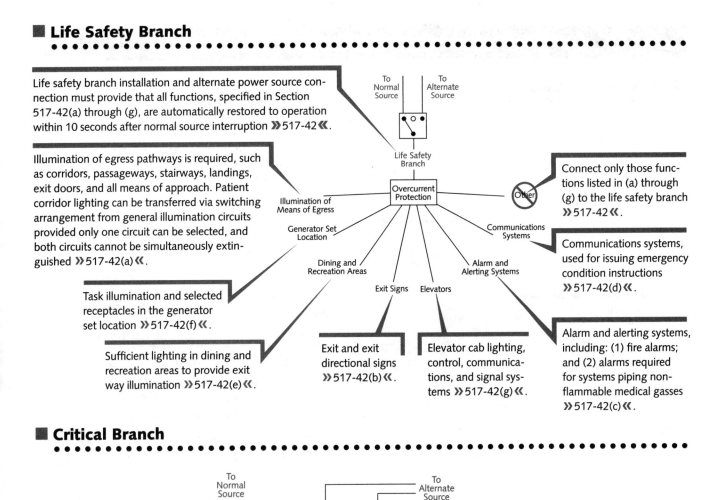

Life safety branch installation and alternate power source connection must provide that all functions, specified in Section 517-42(a) through (g), are automatically restored to operation within 10 seconds after normal source interruption ≫517-42≪.

Illumination of egress pathways is required, such as corridors, passageways, stairways, landings, exit doors, and all means of approach. Patient corridor lighting can be transferred via switching arrangement from general illumination circuits provided only one circuit can be selected, and both circuits cannot be simultaneously extinguished ≫517-42(a)≪.

Task illumination and selected receptacles in the generator set location ≫517-42(f)≪.

Sufficient lighting in dining and recreation areas to provide exit way illumination ≫517-42(e)≪.

Exit and exit directional signs ≫517-42(b)≪.

Elevator cab lighting, control, communications, and signal systems ≫517-42(g)≪.

Connect only those functions listed in (a) through (g) to the life safety branch ≫517-42≪.

Communications systems, used for issuing emergency condition instructions ≫517-42(d)≪.

Alarm and alerting systems, including: (1) fire alarms; and (2) alarms required for systems piping non-flammable medical gasses ≫517-42(c)≪.

Critical Branch

The following equipment connects to the critical branch, and must be arranged for either delayed automatic, or manual connection, to the alternate power source:

(1) Heating equipment providing patient room heating unless an exception is met.
(2) Elevator service where power disruption would result in between-floor stops. Throw-over facilities must be provided to allow temporary operation of any elevator passenger release. See section 517-42(g) for elevator cab lighting, control, and signal system requirements.
(3) Additional illumination, receptacles, and equipment can be connected only to the critical branch ≫517-43(b)≪.

Critical branch installation and alternate power source connection must provide that equipment listed in Section 517-43(a) is automatically restored to operation at appropriate time-lag intervals following the restoration of life safety branch operation. Arrangement must also provide for additional connection of equipment listed in Section 517-43(b), by either delayed automatic or manual operation ≫517-43≪.

The following equipment must be connected to the critical branch; and must be arranged for delayed automatic connection to the alternate power source:

(1) Patient care areas — task illumination and selected receptacles in: (a) medication prep areas; (b) pharmacy dispensing areas; and (c) nurses' stations (unless corridor luminaries provide adequate light)
(2) Sump pumps and other equipment required for major apparatus safe operation, as well as associated control systems/alarms
(3) Smoke control and stair pressurization systems
(4) Kitchen hood supply and/or exhaust systems, if operation is required during a fire in (or under) the hood ≫517-43(a)≪.

INHALATION ANESTHETIZING LOCATIONS

■ Hazardous (Classified) Anesthetizing Locations

Any room (or location) storing flammable anesthetics or volatile flammable disinfecting agents is, from floor to ceiling, a Class I, Division 1 location » 517-60(a)(2)«.

5 Feet

Wherever flammable anesthetics are employed, the entire area is a Class I, Division 1 location, extending from the floor upward to a level 5 feet. The remaining space (from 5 feet to the structural ceiling) is considered to be above a hazardous (classified) location » 517-60(a)(1)«.

Anesthetizing Location: Any health care facility area designated for use in administering any flammable inhalation anesthetic agent (during examination or treatment), including the use of such agents for relative analgesia » 517-3«. (See also the definitions of *Flammable Anesthetics* and *Flammable Anesthetizing Location*.)

Unless otherwise allowed by Section 517-160, each power circuit within (or partially within) a flammable anesthetizing location (Section 517-60) must be isolated from any distribution system by means of an isolated power system » [517-61(a)(1)«.

Any inhalation anesthetizing location, designated for the exclusive use of nonflammable anesthetizing agents, is considered to be other-than-hazardous (classified) location » 517-60(b)«. Installations in other-than-hazardous (classified) locations must comply with Section 517-61(c).

Grounded power systems in anesthetizing locations must meet Section 517-63 provisions.

Low-voltage equipment and instruments must comply with Section 517-64 requirements.

Isolated power system equipment must be listed for the purpose; the system must be designed and installed so that it meets the provisions of Part G » 517-61(a)(2)«.

In hazardous (classified) locations referenced in Section 517-60, all fixed wiring/equipment, and all portable equipment (including lamps and other utilization equipment), operating at more than 10 volts between conductors, must comply with the requirements of Sections 501-1 through 501-15 and Sections 501-16(a) and (b) for Class I, Division 1 locations. All such equipment must be approved for the specific hazardous atmosphere involved » 517-61(a)(3)«.

If a box, fitting, or enclosure is only partially within a hazardous (classified) location(s), the hazardous area extends to include the entire box, fitting, or enclosure » 517-61(a)(4)«.

Receptacles and attachment plugs in a hazardous (classified) location(s) must be listed for use in Class I, Group C hazardous (classified) locations, and must provide a grounding conductor connection » 517-61(a)(5)«.

Flexible cords used in hazardous (classified) locations for connection of portable utilization equipment (including lamps operating at more than 8 volts between conductors), must be of a type approved for extra-hard usage per Table 400-4 and must include an additional conductor for grounding » [517-61(a)(6)«.

The flexible cord must include a storage device which does not subject the cord to less than a 3-inch bending radius » 517-61(a)(7)«.

■ Above Hazardous (Classified) Anesthetizing Locations

Wiring above a hazardous (classified) location (referenced in Section 517-60) must be installed in: rigid metal conduit, electrical metallic tubing, intermediate metal conduit, Type MI cable, or Type MC cable employing a continuous gas/vaportight metal sheath ≫517-61(b)(1)≪.

Installed equipment that may produce arcs, sparks, or particles of hot metal (such as lamps and lampholders for fixed lighting, cutouts, switches, generators, motors, or other equipment having make-and-break or sliding contacts) must be of the totally enclosed type, or by design must prevent the escape of sparks or hot metal particles ≫517-61(b)(2)≪. Wall-mounted receptacles, installed above the hazardous (classified) location in flammable anesthetizing locations, require neither total enclosure nor guarded or screened openings to prevent particle dispersion ≫517-61(b)(2) *Exception*≪.

Approved seals must be provided in conformance with Section 501-5. Section 501-5(a)(4) applies to both horizontal and to vertical boundaries of the defined hazardous (classified) locations ≫517-61(b)(4)≪.

Wiring above a hazardous (classified) location must meet Section 517-61(b) provisions.

Surgical and other lighting fixtures must conform to Section 501-9(b) ≫517-61(b)(3)≪. (See the two exceptions associated with this provision.)

5 Feet

Receptacles and attachment plugs, located above hazardous (classified) anesthetizing locations, must be listed for hospital use for services of prescribed voltage, frequency, rating, and number of conductors with grounding conductor connection provisions. This requirement applies to attachment plugs and receptacles of the 2-pole, 3-wire grounding type for single-phase, 120-volt, nominal, ac service ≫517-61(b)(5)≪.

Plugs and receptacles rated 250-volts, for connection of 50- and 60-amperes ac medical equipment above hazardous (classified) locations, must be arranged so that the 60-ampere receptacle accepts either the 50- or 60-ampere plug. Fifty-ampere receptacles must *not* accept a 60-ampere attachment plug. The plugs must be on the 2-pole, 3-wire design, with a third contact connecting to the insulated (green or green with yellow stripe) electrical system equipment grounding conductor ≫517-61(b)(6)≪.

X-RAY INSTALLATIONS

■ Mobile, Portable, and Transportable X-Ray Equipment

X-Ray Installations (Mobile): X-ray equipment mounted on a permanent base with wheels, casters, or a combination of both, facilitating movement of equipment while fully assembled ≫517-3≪.

X-Ray Installations (Portable): X-ray equipment designed to be hand carried ≫517-3≪.

X-Ray Installations (Transportable): X-ray equipment whose design facilitates vehicular transport (via ready disassembly) and/or vehicular installation ≫517-3≪.

Individual branch circuits are not required for portable, mobile, and transportable medical X-ray equipment requiring a capacity of 60 amperes or less ≫517-71(b)≪.

For portable X-ray equipment connected to a 120-volt branch circuit of 30 amperes (or less), a properly rated grounding-type attachment plug and receptacle can serve as a disconnecting means ≫517-72(c)≪.

■ Fixed and Stationary X-Ray Equipment

Fixed and stationary X-ray equipment must be connected to the power supply by a wiring method which meets the *Code's* general requirements **》517-71(a)《**. Equipment, properly supplied by a branch circuit rated at 30 amperes (or less), can be supplied through a suitable attachment plug and hard-service cable (or cord) **》517-71 *Exception*《**.

A supply circuit disconnecting means of adequate capacity for at least 50 percent of the input (for momentary rating) or 100 percent of the input (for long-time rating) of the X-ray equipment, whichever is greater, must be provided **》517-72(a)《**.

Location and operation of the disconnecting means must be readily accessible from the X-ray control **》517-72(b)《**.

X-Ray Installations (Momentary Rating): a rating based on an operating interval of 5 seconds or less **》517-3《**.

X-Ray Installations (Long-Time Rating): a rating based on an operating interval of 5 minutes or longer **》517-3《**.

Supply branch-circuit conductor ampacity and the current rating of overcurrent protective devices must not be less than 50 percent of the momentary rating or 100 percent of the long-time rating, whichever is greater **》517-73(a)(1)《**.

Supply feeder ampacity and the current rating of overcurrent protective devices, supplying multiple branch circuits feeding X-ray units must not be less than 50 percent of the largest unit's momentary demand rating, plus 25 percent of the momentary demand rating of the next largest unit, plus 10 percent of the momentary demand rating of each additional unit. If the X-ray unit performs simultaneous biplane examinations, the supply conductors and overcurrent protective devices must be 100 percent of each unit's momentary demand rating **》517-73(a)(2)《**. The manufacturer typically specifies minimum distribution transformer and conductor sizes, disconnecting means rating, and overcurrent protection **》517-73(a)(2) FPN《**.

Conductor ampacity and overcurrent protective device ratings must not be less than 100 percent of the current rating of medical X-ray **therapy** equipment **》517-73(b)《**.

Circuits and equipment operating on a supply circuit of over 600 volts must comply with Article 490 **》517-71(c)《**.

Control-circuit conductor installation must comply with 517-74.

X-ray equipment transformers and capacitors are not required to comply with Articles 450 and 460. Capacitors must be mounted within enclosures of insulating material or grounded metal **》517-76《**.

Guarding and grounding must comply with Section 517-78.

WARNING

Do not construe anything in Article 517, Part E, as specifying safeguards against the useful beam or stray X-ray radiation **》Part E《**. (See FPN Nos. 1 and 2 for information pertaining to safety and performance.)

NOTE

Noncurrent-carrying metal parts of X-ray and associated equipment (controls, tables, X-ray tube supports, transformer tanks, shielded cables, X-ray tube heads, etc.) must be grounded per Article 250, as modified by Sections 517-13(a) and (b) **》517-78(c)《**.

Summary

- Article 517 provisions apply to health care facility electrical construction and installation.
- Many commonly used health care facility terms are defined in Section 517-3.
- Article 517's installation criteria and wiring methods serve to minimize electrical hazards.

- Nursing homes and limited care facilities share the same provisions, found in Sections 517-40 through 517-44.
- Each patient bed location, in general care areas, must have at least four receptacles.

- Within critical care areas, each patient bed location requires a minimum of six receptacles.
- The required receptacles, in both general and critical care areas, can be of the single, duplex, or a combination of both types.
- Fifteen- and 20-ampere, 125-volt receptacles supplying pediatric ward patient care areas or rooms must be listed tamper resistant, or must employ a listed tamper resistant cover.
- A hospital essential electrical system is comprised of equipment system(s) and emergency system(s).
- The life safety branch and critical branch make up a hospital's emergency system.
- An alternate power source for a hospital can be: (1) on-site generator(s) driven by some form of prime mover(s); (2) another generating unit(s) where generating unit(s) located on the premises

is the normal source; or (3) an external utility service when the normal source consists of on-site generating unit(s).
- Essential electrical systems for nursing homes and limited care facilities are comprised of two separate branches: the life safety branch and the critical branch.
- The entire area surrounding any location employing flammable anesthetics is a Class I, Division 1 location, including the area extending upward from the floor to a level of 5 feet.
- Wiring and equipment within anesthetizing locations must comply with Section 517-61 requirements.
- X-ray installations must comply with Section 517, Part E requirements.

Unit 16 Competency Test

NEC® Reference   **Answer**

_____ _____   1. X-ray low-voltage cables connecting to oil-filled units that are not completely sealed, such as transformers, condensers, oil coolers, and high-voltage switches, shall have insulation of the ____ type.

_____ _____   2. The branches of the hospital emergency system shall be installed and connected to the alternate power source so that all functions specified herein for the emergency system shall be automatically restored to operation within ____ after interruption of the normal source.

_____ _____   3. The equipment-grounding terminal buses of the normal and essential branch-circuit panelboards serving the same individual patient vicinity shall be bonded together with an insulated continuous copper conductor not smaller than ____.

_____ _____   4. The ____ is a subsystem of the health care emergency system consisting of feeders and branch circuits supplying energy to task illumination, special power circuits, and selected receptacles serving areas and functions related to patient care, and which are connected to alternate power sources by one or more transfer switches during interruption of the normal power source.

_____ _____   5. The receptacles or the faceplates for receptacles supplied by the hospital's critical branch shall have a distinctive ____ so as to be readily recognizable.

_____ _____   6. ____ is a jack or terminal bus that serves as the collection point for redundant grounding of electric appliances serving a patient vicinity or for grounding other items in order to eliminate electromagnetic interference problems.

_____ _____   7. In any anesthetizing area, all metal raceways and metal-sheathed cables, and all noncurrent-carrying conductive portions of fixed electric equipment, shall be ____.

_____ _____   8. A disconnecting means of adequate capacity for at least ____ percent of the input required for the long-time rating or ____ percent of the input required for the momentary rating of the X-ray equipment, whichever is greater, shall be provided in the supply circuit.

NEC® Reference   Answer

—————— —————— 9. Each critical care area patient bed location shall be supplied by at least _____ branch circuit(s).

—————— —————— 10. Typically, in a patient room the patient vicinity encloses a space within the room not less than _____ beyond the perimeter of the bed in its nominal location, and extending vertically not less than _____ feet above the floor.

—————— —————— 11. The line isolation monitor shall not alarm for a fault hazard of less than _____ mA or for a total hazard current of less than _____mA.

—————— —————— 12. Fifteen- and 20-ampere, 125-volt receptacles intended to supply patient care areas of pediatric wards, rooms, or areas shall be _____ resistant or shall employ a _____ resistant cover.

—————— —————— 13. A nursing home's alternate source of power shall be separate and independent of the normal source and shall have a capacity to sustain its connected loads for a minimum of _____ after loss of the normal system.

—————— —————— 14. _____ are patient bedrooms, examining rooms, treatment rooms, clinics, and similar areas in which it is intended that the patient shall come in contact with ordinary appliances such as a nurse call system, electrical beds, examining lamps, telephone, and entertainment devices.

—————— —————— 15. In locations where flammable anesthetics are employed, the entire area shall be considered to be a Class I, Division 1 location that shall extend upward to a level _____ above the floor.

—————— —————— 16. A(n) _____ minimum separation between the service and feeder ground-fault tripping bands shall be provided in health care facilities.

—————— —————— 17. For portable X-ray equipment connected to a 120-volt branch circuit of _____ or less, a grounding-type attachment plug and receptacle of proper rating shall be permitted to serve as a disconnecting means.

—————— —————— 18. The _____ is a subsystem of the health care emergency system consisting of feeders and branch circuits, meeting the requirements of Article 700 and intended to provide adequate power needs to ensure safety to patients and personnel, and which are automatically connected to alternate power sources during interruption of the normal power source.

—————— —————— 19. The ampacity of conductors and the rating of overcurrent protective devices shall not be less than _____ of the current rating of medical X-ray therapy equipment.

—————— —————— 20. As a general rule, the wiring of a hospital's emergency system shall be mechanically protected by installation in _____ metal raceways, or shall be wired with Type _____ cable.

—————— —————— 21. A(n) _____ is any area of the health care facility that has been designated to be used for the administration of any flammable inhalation anesthetic agents in the normal course of examination or treatment.

—————— —————— 22. Each critical care area patient bed location shall be provided with a minimum of _____ receptacle(s).

—————— —————— 23. In areas used for patient care, the grounding terminals of all receptacles and all noncurrent-carrying conductive surfaces of fixed electric equipment likely to become energized that are subject to personal contact, operating at over _____, shall be grounded by an insulated copper conductor.

—————— —————— 24. Wall-mounted remote-control stations for remote-control switches operating at _____ or less shall be permitted to be installed in any anesthetizing location.

_____ _____ 25. A(n) _____ is a system comprised of alternate sources of power and all connected distribution systems and ancillary equipment, designed to ensure continuity of electrical power to designated areas and functions of a health care facility during disruption of normal power sources, and also designed to minimize disruption within the internal wiring system.

_____ _____ 26. In critical care areas, a patient vicinity shall be permitted to have a patient equipment grounding point. An equipment-bonding jumper, not smaller than _____, shall be used to connect the grounding terminal of all grounding-type receptacles to the patient equipment grounding point.

_____ _____ 27. Individual branch circuits shall not be required for portable, mobile, and transportable medical X-ray equipment requiring a capacity of not over _____.

_____ _____ 28. Each general care area patient bed location shall be provided with a minimum of _____ receptacle(s).

_____ _____ 29. _____ areas are those special care units, intensive care units, coronary care units, angiography laboratories, cardiac catheterization laboratories, delivery rooms, operating rooms, and similar areas in which patients are intended to be subjected to invasive procedures and connected to line-operated, electromedical devices.

_____ _____ 30. One transfer switch shall be permitted to serve one or more branches or systems in a limited care facility with a maximum demand on the essential electrical system of _____ kVA.

_____ _____ 31. Size No. 18 or No. 16 fixture wires as specified in Section 725-27 and flexible cords shall be permitted for the control and operating circuits of X-ray and auxiliary equipment where protected by not larger than _____ ampere overcurrent devices.

_____ _____ 32. A flexible cord storage device, within a hazardous (classified) anesthetizing location, shall not subject the cord to bending at a radius of less than _____.

_____ _____ 33. In an area used for patient care, Type MC cable and Type MI cable shall have an outer metal armor or sheath that is _____ as an acceptable grounding return path.

_____ _____ 34. Essential electrical systems for nursing homes and limited care facilities shall be comprised of two separate branches capable of supplying a limited amount of lighting and power service, which is considered essential for the protection of life safety and effective operation of the institution during the time normal electrical service is interrupted for any reason. These two separate branches shall be the _____ branch and the _____ branch.

_____ _____ 35. The ampacity of supply branch-circuit conductors and the current rating of overcurrent protective devices, for X-ray equipment, shall not be less than _____ percent of the long-time rating or _____ percent of the momentary rating, whichever is greater.

_____ _____ 36. In critical care areas, the equipment grounding conductor for special-purpose receptacles, such as the operation of mobile X-ray equipment, shall be extended to the _____ of branch circuits for all locations likely to be served from such receptacles.

_____ _____ 37. In a clinic, the essential electrical system shall be arranged so that, in the event of a failure of the normal power source, the alternate source of power shall be automatically connected to the load within _____.

Industrial Locations

Objectives

After studying this unit, the student will:

✔ be familiar with service entrance provisions, with or without a single main disconnecting means.

✔ know that a raceway (or cable) cannot contain both service conductors and non-service conductors.

✔ have a good understanding of feeder tap rules.

✔ understand the provisions pertaining to conductors supplying a transformer.

✔ be familiar with transformer vault provisions.

✔ know how to correctly bond service raceways and equipment.

✔ be able to accurately connect taps to grounding electrode conductors.

✔ understand the different methods for sizing bonding conductors in paralleled service raceways.

✔ be familiar with the provision requiring a full size equipment-grounding conductor (when used) in each paralleled raceway.

✔ have a thorough understanding of cable tray provisions.

✔ be familiar with, both general and specific motor provisions, i.e., motor and branch circuit overload protection, branch circuit conductors, branch circuit protection, feeder conductors, feeder protection, etc.

✔ understand specific equipment provisions for cranes and hoists, electric welders, electroplating equipment, industrial machinery, and capacitors.

Introduction

Because industrial wiring covers such a broad spectrum of electrical applications, one unit simply cannot contain all of the related provisions. As with other topics, this text's building block approach has laid the foundation for industrial locations, and has steadily built upon that foundation with layers of additional information. For instance, previous units address items pertinent to industrial projects, such as: raceways, services, clearances, busways, transformers, etc. Provisions for services and service equipment can be found in Units 9, 10 and 14, as well as here (as they apply more specifically to industrial applications).

In particular, this unit explains various feeder tap installation provisions, as found in Section 240-21. These include, among others: the 10- and 25-foot tap rules, plus transformer tap rules. While general transformer provisions were introduced in Unit 14, this unit expands that topic by discussing transformer vault requirements. Grounding and bonding provisions make up a part of this unit. Included are grounding electrode conductor taps, bonding service raceways, and equipment grounding conductors in paralleled raceways. Cable tray requirements are also addressed in Unit 17. Cable tray installations, of course, are not limited to industrial environments. They are often part of commercial projects, and are used in many other areas as well.

Article 430 (Motors, Motor Circuits, and Controllers) applies a great number of requirements to the installation of motors, motor branch-circuit and feeder conductors, their protection, motor overload protection, motor control circuits, motor controllers, and motor control centers. The most commonly used of these provisions are illustrated and explained within this unit. Finally, specific equipment, typically found in industrial type locations, are presented. Such specific equipment includes: cranes and hoists, electric welders, electroplating equipment, industrial machinery, and capacitors.

The aim of this unit is to introduce, in a broad manner, the more common components of a very complex area. See Delmar's extensive list of offerings, for a more in-depth study of industrial electrical wiring.

GENERAL

■ Service Entrance: One Main Disconnect

The next higher standard overcurrent device rating (above the ampacity of the conductors being protected) is permitted, provided all the following conditions are met.
1. The protected conductors are not part of a multioutlet branch circuit supplying receptacles for cord- and plug-connected portable loads.
2. The conductors' ampacity does not correspond with a fuse/circuit breaker's standard ampere rating without overload trip adjustments above its rating (but it can have other trip or rating adjustments).
3. The next higher standard rating does not exceed 800 amperes ≫ 240-3(b)≪ .

The number of disconnects on the load side of the service disconnecting means is not limited to six.

Cartridge fuse and fuse-holder overcurrent protection provisions are located in Sections 240-60 and 61.

Service equipment rated no more than 600 volts must be identified by markings as suitable for service equipment ≫ 230-66≪ .

Each ungrounded circuit conductor must have overcurrent protection located at the point where the conductors receive their supply, except as specified in Section 240-21(a) through (g) ≫ 240-21≪ .

All switches, and circuit breakers used as switches, must be located in a readily accessible and operable place. The center of the switch (or circuit breaker) operating handle's grip must not, in its highest position, be more than 79 inches (6' 7") above the floor/working platform ≫ 380-8≪ .

Service Conductors 760 Amperes

Service Equipment

Not Service Equipment

Line Load

800 Amperes

Permitted

Feeder Conductors

Feeder or Branch Circuit Conductors

6'-7" Maximum

Services exceeding 600 volts, nominal, must comply with Section 230, Part H provisions.

If an orderly shutdown is required to minimize the personnel/equipment hazard(s), a system of coordination based on the following two conditions is permitted: (1) Coordinated short-circuit protection; and (2) Overload indication based on monitoring systems or devices ≫ 240-12≪ .

Knife switches rated more than 1,200 amperes at 250 volts or less, and at over 600 amperes at 251 to 600 volts, must be used only as isolating switches and must not be opened under load ≫ 380-13(a)≪ .

Use a circuit breaker or a switch of special design listed for such purpose to interrupt currents over 1,200 amperes at 250 volts, nominal, or less; or, over 600 amperes at 251 to 600 volts, nominal ≫ 380-13(b)≪ .

Comply with the appropriate dimensions from Section 373-6 where insulated conductors are deflected within a wireway, either at the ends or where conduits, fittings, or other raceways or cables enter or leave the wireway; or, where the wireway's direction deflects more than 30 degrees ≫ 362-6≪ .

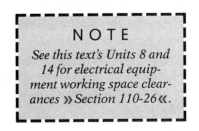

NOTE

See this text's Units 8 and 14 for electrical equipment working space clearances ≫ Section 110-26≪ .

■ Service Entrance: Six Disconnects

The service disconnecting means for each service (permitted by Section 230-2), or for each set of service-entrance conductors (permitted by Section 230-40, Exception Nos. 1 or 3), must consist of no more than six switches/circuit breakers in a single enclosure, in a group of separate enclosures, or in (or on) a switchboard ≫ 230-71(a) ≪.

Two to six disconnects, permitted by Section 230-71, must be grouped. Markings on each disconnect must indicate the load served ≫ 230-72(a) ≪.

Two to six circuit breakers (or sets of fuses) can serve as the overcurrent device providing overload protection. The sum of the circuit breakers/fuses' ratings can exceed the service conductor's ampacity, provided the calculated load (according to Article 220) does not exceed the ampacity of the service conductors ≫ 230-90 *Exception No. 3* ≪.

Each service disconnect must be permanently marked to identify it as a service disconnect ≫ 230-70(b) ≪.

Service Conductors 760 Amperes

Each Disconnect Is Rated 200 Amperes

Service Equipment (Typical)

Supply
Load

Where two to six separately enclosed service disconnecting means (supplying separate loads from one service drop or lateral) are grouped at one location, one set of service-entrance conductors can supply each, or several such, service equipment enclosures ≫ 230-40 *Exception No. 2* ≪.

Wireway

Service Conductors

Feeder or Branch Circuit Conductors

Article 362 contains wireway provisions.

Service-entrance conductors can be spliced (or tapped) by either clamped or bolted connections. Splices must be made in enclosures or, if directly buried, with a listed underground splice kit. Conductor splices must be made in accordance with Sections 110-14, 300-5(e), 300-13, and 300-15 ≫ 230-46 ≪.

Only service conductors can be installed in the same service raceway or service cable ≫ 230-7 ≪.

Means must be provided to disconnect all conductors in a building (or other structure) from the service-entrance conductors ≫ 230-70 ≪.

Install the service disconnecting means at a readily accessible location, either outside of a building/structure, or inside nearest the service conductors' entry point ≫ 230-70(a) ≪.

No more than six disconnects per service can be grouped in any one location. For this section's purpose, a disconnecting means used solely for power monitoring equipment or for a ground-fault protection system's control circuit, installed as part of the listed equipment, is not considered a service disconnecting means ≫ 230-71(a) ≪.

A set of fuses is considered all the fuses required to protect all of the circuit's ungrounded conductors ≫ 230-90(a) ≪.

> ### NOTE
> *Each service disconnect must simultaneously disconnect all ungrounded service conductors under its control from the premises wiring system ≫ 230-74 ≪.*

> ### NOTE
> *Section 230-2 permits one (or more) additional service disconnecting means for fire pumps, legally required standby, or optional standby services. These additional service disconnecting means must be installed at a sufficient distance from the normal (one to six) service disconnecting means in order to minimize the possibility of simultaneous interruption of supply ≫ 230-72(b) ≪.*

Feeder Taps

Because of dissimilar metals characteristics, devices such as pressure terminal/splicing connectors and soldering lugs must be identified for the conductor's material, must be properly installed, and must be properly used. Conductors of dissimilar metals must not be intermixed in a terminal (or splicing) connector where physical contact occurs between dissimilar conductors (such as copper and aluminum, copper and copper-clad aluminum, etc.), unless the device is identified for the purpose and conditions of use »110-14«.

While single-phase system illustrations are used, be aware that the same rules also apply to three-phase systems.

400-Ampere Main Disconnect

225-Ampere Disconnect

225-Ampere Disconnect

All conductor splices, joints, and free ends must either be covered with an insulation equivalent to that of the conductors, or with an insulating device identified for the purpose »110-14(b)«.

Tap conductors can be protected against overcurrent in accordance with Sections 210-19(d), 240-21, 364-11, 364-12, and 430-53(d) »240-3(e)«.

Conductors can be tapped to a feeder, without having overcurrent protection at the tap, as specified in 240-21(b)(1) through(5) »240-21(b)«.

Per Article 240, a **tap conductor** is a conductor other than a service conductor having overcurrent protection ahead of its point of supply, which exceeds the value permitted for similar conductors that are protected as otherwise described in Section 240-3 »240-3(e)«.

Ten Foot Feeder Tap Rule

NOTE
Section 210-19 specifies overcurrent protection locations for branch-circuit tap conductors meeting the Section's requirements »240-21(a)«.

A 60-ampere rated conductor is the minimum size allowed if the overcurrent protection on the tap conductor's line side is 600 amperes ($600 \div 10 = 60$) »240-21(b)(1)(d)«.

The tap conductor's ampacity cannot be less than the rating of the device supplied by the tap conductors, or less than the overcurrent-protective device rating at the tap conductor's termination »240-21(b)(1)(a)(2)«.

Lighting and appliance branch-circuit panelboards must be individually protected (on the supply side) by no more than two main circuit breakers (or two sets of fuses), having a combined rating no greater than the panelboard »384-16(a)«.

60-Ampere Disconnects

The tap conductor's ampacity cannot be less than the combined computed loads on the circuits it supplies »240-21(b)(1)(a)(1)«.

The tap conductors must not extend beyond the switchboard, panelboard, disconnecting means, or control devices they supply »240-21(b)(1)(b)«.

600 Amperes

10 Feet Maximum

Feeder Tap Conductor

Field installations with the tap conductors exiting the enclosure (or vault) where the tap is made, require that the rating of the overcurrent device on the tap conductor's line side not exceed 10 times the tap conductor's ampacity »240-21(b)(1)(d)«.

If the tap conductor's length does not exceed 10 feet, and all of Section 240-21(b)(1)(a) through (d) stipulations are satisfied, overcurrent protection at the tap to the feeder is not required »240-21(b)(1)«.

Except where connected to the feeder, tap conductors must be enclosed in a raceway, extending from the tap to the enclosure of an enclosed switchboard, panelboard, or control device, or to the back of an open switchboard »240-21(b)(1)(c)«.

■ Twenty-Five Foot Feeder Tap Rule

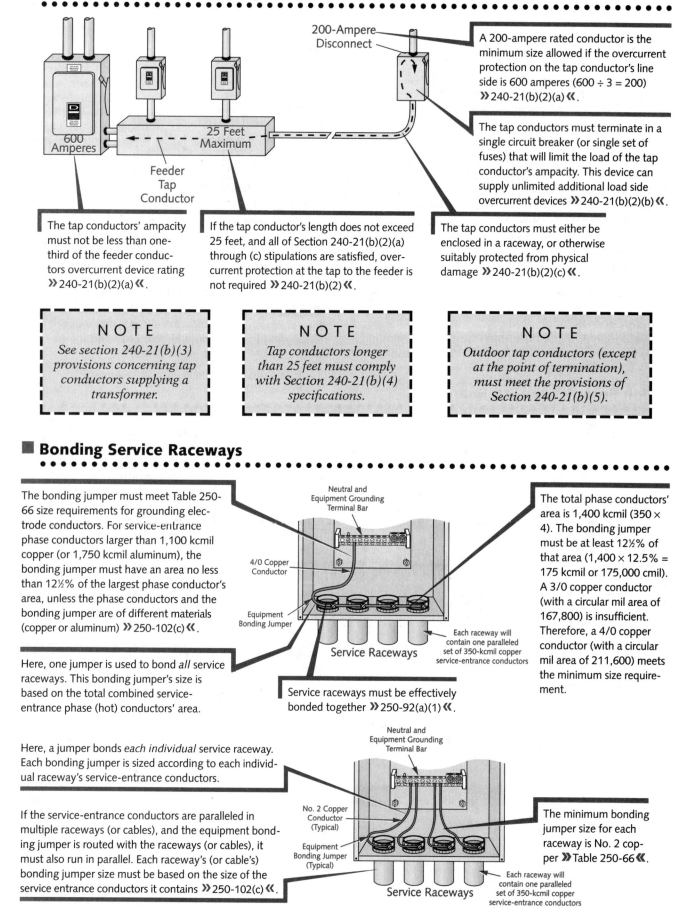

A 200-ampere rated conductor is the minimum size allowed if the overcurrent protection on the tap conductor's line side is 600 amperes (600 ÷ 3 = 200) »240-21(b)(2)(a)«.

The tap conductors must terminate in a single circuit breaker (or single set of fuses) that will limit the load of the tap conductor's ampacity. This device can supply unlimited additional load side overcurrent devices »240-21(b)(2)(b)«.

The tap conductors' ampacity must not be less than one-third of the feeder conductors overcurrent device rating »240-21(b)(2)(a)«.

If the tap conductor's length does not exceed 25 feet, and all of Section 240-21(b)(2)(a) through (c) stipulations are satisfied, overcurrent protection at the tap to the feeder is not required »240-21(b)(2)«.

The tap conductors must either be enclosed in a raceway, or otherwise suitably protected from physical damage »240-21(b)(2)(c)«.

> **N O T E**
> *See section 240-21(b)(3) provisions concerning tap conductors supplying a transformer.*

> **N O T E**
> *Tap conductors longer than 25 feet must comply with Section 240-21(b)(4) specifications.*

> **N O T E**
> *Outdoor tap conductors (except at the point of termination), must meet the provisions of Section 240-21(b)(5).*

■ Bonding Service Raceways

The bonding jumper must meet Table 250-66 size requirements for grounding electrode conductors. For service-entrance phase conductors larger than 1,100 kcmil copper (or 1,750 kcmil aluminum), the bonding jumper must have an area no less than 12½% of the largest phase conductor's area, unless the phase conductors and the bonding jumper are of different materials (copper or aluminum) »250-102(c)«.

Here, one jumper is used to bond *all* service raceways. This bonding jumper's size is based on the total combined service-entrance phase (hot) conductors' area.

The total phase conductors' area is 1,400 kcmil (350 × 4). The bonding jumper must be at least 12½% of that area (1,400 × 12.5% = 175 kcmil or 175,000 cmil). A 3/0 copper conductor (with a circular mil area of 167,800) is insufficient. Therefore, a 4/0 copper conductor (with a circular mil area of 211,600) meets the minimum size requirement.

Service raceways must be effectively bonded together »250-92(a)(1)«.

Here, a jumper bonds *each individual* service raceway. Each bonding jumper is sized according to each individual raceway's service-entrance conductors.

If the service-entrance conductors are paralleled in multiple raceways (or cables), and the equipment bonding jumper is routed with the raceways (or cables), it must also run in parallel. Each raceway's (or cable's) bonding jumper size must be based on the size of the service entrance conductors it contains »250-102(c)«.

The minimum bonding jumper size for each raceway is No. 2 copper »Table 250-66«.

■ Tap Conductors Supplying a Transformer

If the tap conductor's length does not exceed 25 feet, and all of Section 240-21(b)(2)(a) through (c) stipulations are met, overcurrent protection at the tap to the feeder is not required ≫240-21(b)(2)≪.

Primary and secondary conductors must be adequately protected from physical damage ≫240-21(b)(3)(d)≪.

A transformer's primary conductors must have an ampacity at least one-third of the feeder conductors overcurrent device rating ≫240-21(b)(3)(a)≪.

Secondary conductors must terminate in a single circuit breaker or (set of fuses) that will limit the load current to no more than the conductor ampacity permitted by Section 310-15 ≫240-21(b)(3)(e)≪.

A 100-ampere rated conductor is the minimum size allowed if the overcurrent protection on the tap conductor's line side is 300 amperes (300 ÷ 3 = 100) ≫240-21(b)(3)(a)≪.

Conductors supplied by the transformer's secondary must have an ampacity that, when multiplied by the ratio of the secondary-to-primary voltage, is at least one-third of the feeder conductors overcurrent device rating ≫[240-21(b)(3)(b)≪.

The combined length of one primary plus one secondary conductor (excluding any portion of the primary conductor protected at its' ampacity) must not exceed 25 feet ≫240-21(b)(3)(c)≪.

Single Circuit Breaker

Feeder Tap Conductors

Combined Length 25 Feet Maximum

300 Amperes

Conductors, supplied from a set of fuses (or circuit breaker), feeding a transformer are feeder conductors, not feeder tap conductors.

Overcurrent Protection as per Section 450-3

Single Circuit Breaker

Feeder Conductors... **Not** Feeder Tap Conductors

300 Amperes

Transformer secondary conductor provisions are found in Section 240-21(c).

Refer to this text's Unit 14 for transformer provisions (overcurrent protection, grounding, etc.).

Transformer Vaults

Wherever practicable, transformers must be located so that ventilation to outside air does not rely on flues or ducts **»450-41 «**.

Where required by Section 450-9, openings for ventilation must also comply with Section 450-45(a) through (f). That section includes provisions pertaining to location, arrangement, size, covering, dampers, and ducts.

Each doorway from the building interior into the vault must have a tight-fitting door with a minimum fire rating of 3 hours, unless the exception is met. The authority having jurisdiction may require such a door for exterior wall openings, should conditions warrant **»450-43(a) «**.

Doors must be equipped with fully engaged locks, thereby limiting access to qualified persons only **»450-43(c) «**.

Personnel doors must swing out, being equipped with panic bars, pressure plates, or other devices that are normally latched but which open under simple pressure **»450-43(c) «**.

Vaults containing more than 100 kVA transformer capacity must have a drain or other means to carry off any oil or water accumulation, unless local conditions make this impracticable. The vault floor must pitch toward the drain, where provided **»450-46 «**.

Non-electrical pipe or duct systems must neither enter nor pass through a transformer vault. Piping or other facilities providing vault fire protection, or transformer cooling, are allowed **»450-47 «**.

Transformer vaults must not be used for material storage **»450-48 «**.

Vault walls and roofs must be constructed of materials having adequate structural strength for the conditions, and a minimum fire resistance of 3 hours. For this section's purpose, studs and wallboard construction are not acceptable **»450-42 «**.

Transformers protected by automatic sprinkler, water spray, carbon dioxide, or halon systems, and 1-hour rated construction, are permitted **»450-42** *Exception* and 450-43(a) *Exception* **«**.

A door sill (or curb) of a sufficient height (at least 4 inches) to confine oil from the largest transformer within the vault must be provided **»450-43(b) «**.

Curb 4 inch Minimum Height

Vault floors in contact with the earth must be of concrete, no less than 4 inches thick. Where the vault is constructed above a vacant space (or other stories), the floor must adequately support the load imposed thereon; and, must have a minimum fire resistance of 3 hours **»450-42 «**.

NOTE

Entrances to all buildings, rooms, or enclosures containing exposed live parts or exposed conductors operating at over 600 volts, nominal, must be kept locked, unless such entrances are constantly observed by qualified personnel. Where the voltage exceeds 600 volts, nominal, permanent and conspicuous warning signs must be posted, reading as follows: DANGER— HIGH VOLTAGE — KEEP OUT **»***110-34(c)* **«**.

■ Grounding Electrode Conductor Taps

Taps can connect the grounding electrode conductors in a service consisting of more than a single enclosure as permitted in Section 230-40, Exception No. 2. In such a configuration, the tap conductor must extend to the inside of the enclosure. The grounding electrode conductor must be sized according to Section 250-66, while the tap conductors can be sized in accordance with the grounding electrode conductors (also 250-66) for the largest conductor serving their respective enclosures ❯❯250-64(d)❮❮.

Service enclosures containing service conductors (including meter fittings, boxes, or the like) which are interposed in the service raceway (or armor) must be effectively bonded together ❯❯250-92(a)(2)❮❮.

Paribbled Service Raceways

Service Equipment (Typical)

Neutral and Equipment Grounding Terminal Bar (Typical)

Service raceways must be effectively bonded together ❯❯250-92(a)(1)❮❮.

Service Raceways

Bonding Jumper (Typical)

Wireway

Grounding Electrode Terminal Bar

Grounding Electrode Conductor

The grounding electrode conductor must be of one continuous length (without splice or joint), unless spliced by exclusive use of irreversible compression-type connections listed for the purpose, or by exothermic welding ❯❯250-64(c)❮❮.

Any metallic raceway (or armor) which encloses a grounding electrode conductor, as specified in Section 250-64(b), must be effectively bonded together. Bonding applies at each end as well as to all intervening raceways, boxes, and enclosures between the service equipment and the grounding electrode ❯❯250-92(a)(1)❮❮.

The bonding jumper must meet Table 250-66 size requirements for grounding electrode conductors. If the service-entrance phase conductors are larger than 1,100 kcmil copper (or 1,750 kcmil aluminum), the bonding jumper must have an area not less than 12½% of the largest phase conductor's area, unless the phase conductors and the bonding jumper are of different materials (copper or aluminum) ❯❯250-102(c)❮❮.

If the service-entrance conductors are paralleled in multiple raceways (or cables), and the equipment bonding jumper is routed with the raceways (or cables), it must also run in parallel. Base the bonding jumper size for each raceway (or cable) on the size of their corresponding service-entrance conductors ❯❯250-102(c)❮❮.

■ Equipment Grounding Conductors

Since the overcurrent protection has a rating of 1,200 amperes, the minimum size equipment grounding conductor is 3/0 copper (or 250 kcmil aluminum) » Table 250-122 «.

Equipment Grounding Terminal Bar

One 3/0 Copper Conductor in **Each** Raceway

Raceways Will Contain Paralleled Feeder Conductors

A **Full Size** Equipment Grounding Conductor Is Required in **Each** Raceway

Conductors and Equipment Will Be Protected by a 1,200 Ampere Overcurrent Protective Device

Unlike service-equipment bonding jumpers, which can be sized according to each paralleled raceway's service-entrance conductor size, a **full size** equipment grounding conductor (when used) must be installed in **each** raceway.

Each parallel equipment grounding conductor's size is based on the overcurrent device's ampere rating protecting the circuit conductors in the raceway (or cable) per Table 250-122 » 250-122(f)(1) «.

Equipment grounding conductor provisions are not limited to feeders. For example, an alternative drawing could have shown one piece of equipment supplied by a paralleled set of branch circuit conductors. Equipment grounding conductor provisions also apply to branch circuits.

For paralleled conductors in multiple raceways (or cables) as permitted by Section 310-4, any equipment grounding conductors installed must also run in parallel within **each** raceway. Use one of the methods in Section 250-122(f) to ensure protection of the equipment grounding conductors » 250-122(f) «.

> **NOTE**
> *If conductor size is adjusted to compensate for voltage drop, any equipment grounding conductors installed must be proportionately adjusted according to circular mil area » 250-122(b) «.*

> **NOTE**
> *If a single raceway supplies the equipment, only one 3/0 copper (or 250 kcmil aluminum) equipment grounding conductor is required.*

If a single equipment grounding conductor is run with multiple circuits within the same raceway (or cable), it must be sized for the largest overcurrent device protecting the raceway's (or cable's) conductors » 250-122(c) «.

See Article 250, Part F for equipment grounding and equipment grounding conductor provisions.

While copper, aluminum, or copper-clad aluminum equipment grounding conductors of the wire type must not be smaller than shown in Table 250-122, they do not have to be larger than the circuit conductors supplying the equipment » 250-122(a) «.

If a raceway or cable armor (or sheath) serves as the equipment grounding conductor, as provided in Section 250-118 and 250-134(a), it must comply with Section 250-2(d) » 250-122(a) «.

Where ground-fault protection of equipment is installed, each parallel equipment grounding conductor in a multiconductor cable can be sized per Table 250-122 on the basis of the ground-fault protection's trip rating, provided Section 250-122(f)(2)(1) through (3) conditions are met » 250-122(f)(2) «.

Equipment grounding conductor types are listed in Section 250-118.

■ Cable Trays

Cable trays can extend transversely through partitions and walls, or vertically through platforms and floors, in either wet or dry locations where the completed installation meets Section 300-21 requirements ≫318-6(g)≪.

A cable tray system is a unit (or assembly of units or sections) and associated fittings which together form a rigid structural system used to securely fasten/support cables and raceways ≫318-2≪.

Cable trays must be exposed and accessible except as permitted by Section 318-6(g) ≫318-6(h)≪.

Sufficient space must be provided and maintained about cable trays to permit adequate access for cable installation and maintenance ≫318-6(i)≪.

Cable trays must include fittings, or other suitable means, to facilitate changes in direction and elevation ≫318-5(e)≪.

Cable trays must be of adequate strength and rigidity to provide support for all contained wiring ≫318-5(a)≪.

Each cable tray run must be completed before cables are installed ≫318-6(b)≪.

Cable trays must have side rails or similar structural members ≫318-5(d)≪.

Metallic cable trays can serve as equipment grounding conductors where: (1) continuous supervision ensures that only qualified persons will maintain the cable tray system; and (2) the cable tray complies with Section 318-7 provisions ≫318-3(c)≪.

CAUTION ≪≪≪≪≪≪≪≪≪≪≪≪≪
Hazardous (classified) location cable trays must contain only the cable types permitted in Sections 501-4, 502-4, 503-3, and 504-20 ≫318-3(d)≪.

Cable tray installations are not limited to industrial establishments. Section 318-3(a) listed materials can be installed in cable tray systems, under the conditions described in their respective articles and sections ≫318-3≪.

Only in industrial establishments, where maintenance and supervision conditions restrict cable tray system service to qualified personnel, can any of the cables in Section 318-3(b)(1) and (2) be installed in ladder, ventilated troughs, or ventilated channel cable trays ≫318-3(b)≪.

Cable tray systems must be corrosion resistant. If made of ferrous material, the system must be protected from corrosion per Section 300-6 ≫318-5(c)≪.

Cable tray installations must meet Section 318-6 requirements.

N O T E
Cable tray systems must not be used in hoistways, where subject to severe physical damage, or in environmental airspaces, except as permitted in Section 300-22, to support appropriate wiring methods ≫318-4≪.

Steel or aluminum cable tray systems can be used as equipment grounding conductors provided all Section 318-7(b)(1) through (4) conditions are met.

Refer to Section 318-8 for cable installation provisions.

While provisions pertaining to the number of multiconductor cables (rated 2,000 volts or less) can be found in Section 318-9, Section 318-10 contains single conductor cable provisions.

Requirements for cable ampacities (multiconductor and single) are located in Section 318-11.

MOTORS

■ General Motor Provisions

Article 430, Part E specifies protective devices designed to protect feeder conductors supplying motors against overcurrents due to short circuits or grounds »430-61 «.

Sections 430-24 through 26 contain provisions for conductors supplying multiple motors, or a motor(s) and other load(s). These provisions are applicable to more than feeder conductors.

Article 430, Part D lists devices which are used to protect motor branch-circuit conductors, motor control apparatus, and motors against overcurrent due to short circuits or grounds. The rules add to, or amend, Article 240 provisions »430-51 «.

Article 430, Part B specifies conductor sizes capable of carrying the motor(s) current without overheating, under specified conditions »430-21 «.

Article 430, Part C specifies overload devices intended to protect motors, motor-control apparatus, and motor branch-circuit conductors against excessive heating caused by motor overloads and failure to start. Electrical apparatus overload is an operating overcurrent, when of sufficient duration, can result in damage or dangerous overheating of the apparatus. Short circuits or ground faults are not included »431-31 «.

Motor Feeder Overcurrent Protection

Feeder Conductors

Branch-Circuit Overcurrent Protection

Branch-Circuit Conductors

Motor Starter with Motor Overload Protection

General motor provisions are found in Sections 430-1 through 430-18 (Part A).

Motor disconnecting means provisions are found in Sections 430-101 through 430-113 (Part J).

For general motor applications, base current ratings on Section 430-6(a)(1) and (2). (1) Use values given in Tables 430-147 through 430-150 (including notes) to determine conductor ampacity or ampere ratings of switches, branch-circuit short-circuit and ground-fault protection. Do not use the actual current ratings marked on the motor nameplate, unless an exception applies. (2) Separate motor overload protection is based on the motor nameplate current rating.

Part M of Article 430 addresses the grounding of exposed noncurrent-carrying metal parts, of motor and controller frames which are likely to become energized, in order to prevent a voltage above ground should accidental contact between energized parts and frames occur. Insulating, isolating, or guarding are suitable alternatives to motor grounding under certain conditions »430-141 «.

Article 430 covers motors and all of the following as they relate specifically to motors: branch-circuit and feeder conductors and their protection, overload protection, control circuits, controllers, and motor control centers »430-1 «.

Motor and multimotor equipment must be marked according to Section 430-7 requirements.

Motor terminal provisions (markings, conductors, and torque requirements) are located in Section 430-9.

Wires passing through an opening in an enclosure, conduit box, or barrier, require a bushing to protect the conductors from any sharp opening edges »430-13 «.

Motor control circuit requirements are found in Sections 430-71 through 430-74 (Part F).

Motor controller requirements are addressed by Sections 430-81 through 430-91 (Part G).

Refer to Sections 430-92 through 430-98 (Part H) for motor control center provisions.

Requirements for motors operating at over 600 volts, nominal, are located in Sections 430-121 through 430-127 (Part K).

Provisions pertaining to protection of live parts can be found in Sections 430-131 through 430-133 (Part L).

Motor information is located in Part N, Tables 430-147 through 430-152.

■ Motor and Branch Circuit Overload Protection

As an approved method of protecting a motor against overload, Section 430-32(a)(1) lists a separate overload device, responsive to motor current. This device must be selected to trip, or must be rated at no more than the percentage of the motor nameplate full-load current rating shown below:

Motors with a marked service
factor not less than 1.25......................125%
Motors with a marked temperature
rise not over 40°C 125%
All other motors................................115%

Controllers must be marked according to Section 430-8 provisions.

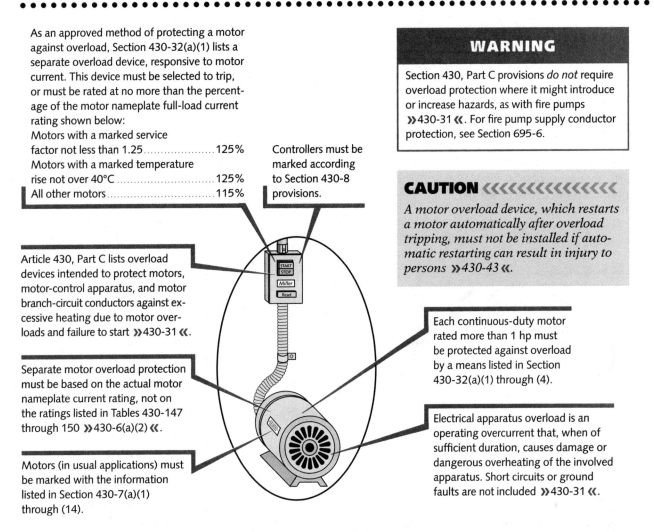

WARNING

Section 430, Part C provisions *do not* require overload protection where it might introduce or increase hazards, as with fire pumps ≫430-31 ≪. For fire pump supply conductor protection, see Section 695-6.

CAUTION ≪≪≪≪≪≪≪≪≪≪≪≪≪

A motor overload device, which restarts a motor automatically after overload tripping, must not be installed if automatic restarting can result in injury to persons ≫*430-43* ≪.

Article 430, Part C lists overload devices intended to protect motors, motor-control apparatus, and motor branch-circuit conductors against excessive heating due to motor overloads and failure to start ≫430-31 ≪.

Separate motor overload protection must be based on the actual motor nameplate current rating, not on the ratings listed in Tables 430-147 through 150 ≫430-6(a)(2) ≪.

Motors (in usual applications) must be marked with the information listed in Section 430-7(a)(1) through (14).

Each continuous-duty motor rated more than 1 hp must be protected against overload by a means listed in Section 430-32(a)(1) through (4).

Electrical apparatus overload is an operating overcurrent that, when of sufficient duration, causes damage or dangerous overheating of the involved apparatus. Short circuits or ground faults are not included ≫430-31 ≪.

Each continuous-duty motor, rated 1 hp or less, *not* permanently installed, nonautomatically started, and within sight from the controller, can be protected against overloads by the branch-circuit short-circuit and ground-fault device. Comply with branch-circuit overcurrent protection size specifications found in Article 430, Part D ≫430-32(b) ≪.

Any automatically started motor, 1 hp or less, must be protected against overload by one of the methods found in Section 430-32(c)(1) through (4).

If the overload relay selected per Section 430-32(a)(1) and (c)(1) is not sufficient to start the motor, or to carry the load, use the next higher size overload, provided the overload relay's trip current does not exceed Section 430-34 percentages.

A motor controller can serve as an overload device if the number of overload units complies with Table 430-37; and, if these units are operative in both the starting and running position (dc motor), or in the running position (ac motor) ≫430-39 ≪.

Part C provisions do not apply to motor circuits rated over 600 volts, nominal ≫430-31 ≪. See Part K.

Overload relays, and other devices for motor and overload protection, incapable of opening short circuits must be protected by fuses (or circuit breakers) rated or set according to Section 430-52, or by a motor short-circuit protector per Section 430-52. [430-40]

Overload protection for motors on general-purpose branch circuits (permitted by Article 210) is required and must meet Section 430-42(a), (b), (c), or (d) specifications.

Appendix D contains a motor application example (Example No. D8).

NOTE

Any motor application must be considered continuous duty unless the driven apparatus, by nature, means the motor cannot, under any conditions, operate continuously with the load ≫*430-33* ≪.

■ Motors on General-Purpose Branch Circuits

Where a motor, without individual overload protection [Section 430-42(a)], is connected to a branch circuit by means of an attachment plug and receptacle the attachment plug and receptacle rating must not exceed 15 amperes at either 125 or 250 volts. Where individual overload protection is required per Section 430-42(b) for a motor or motor-operated appliance attached to a branch circuit through an attachment plug and receptacle, the overload device must be an integral part of the motor/appliance. The rating of the attachment plug and receptacle determines the rating of the circuit to which the motor is connected, as provided in Article 210 ≫430-42(c)≪.

The branch-circuit short-circuit and ground-fault protective device of a circuit to which a motor (or motor-operated appliance) is connected must have a time delay sufficient to start the motor and allow it to accelerate ≫430-42(d)≪.

One or more motors, without individual overload protection, can be connected to a general-purpose branch circuit only if the installation complies with the limiting conditions of Sections 430-32(b) and (c) and Sections 430-53(a)(1) and (a)(2) ≫430-42(a)≪.

1 Horsepower or Less

Over 1 Horsepower

Motors used on general-purpose branch circuits (as permitted in Article 210) require overload protection per Section 430-42(a), (b), (c), or (d) specifications ≫430-42≪.

Motors, of larger ratings than those found in Section 430-53(a), can be connected to general-purpose branch circuits only if each motor has overload protection specifically listed for that motor as specified in Section 430-32. In the case of more than one motor, both the controller and the motor overload device must be approved for group installation, and the short-circuit and ground-fault protective device must be selected according to Section 430-53 ≫430-42(b)≪.

■ Motor Branch Circuit Conductors

Conductors for motors used in short-time, intermittent, periodic, or varying duty applications must have an ampacity of no less than the percentage of the motor nameplate current rating shown in Table 430-22(b), unless the authority having jurisdiction grants special permission for conductors of lower ampacity ≫430-22(b)≪.

Section 430-23 contains wound-rotor secondary conductor provisions.

Conductors supplying several motors, or a motor(s) and other load(s), must comply with Section 430-24.

The ampacity of conductors supplying multimotor and combination-load equipment must meet Section 430-25 requirements.

Where motor circuits contain capacitors, conductors must comply with Section 460-8 and 460-9 ≫430-27≪.

For multispeed motors, base the selection of branch-circuit conductors (between the controller and the motor) on the current rating of the winding(s) energized by the conductors ≫430-22(a)≪.

Branch-circuit conductors, supplying a single continuous-duty motor, must have an ampacity of at least 125% of the motor's full load current rating as determined by Section 430-6(a)(1) ≫430-22(a)≪.

Appendix D contains a motor application example (Example No. D8).

Placement of motors must allow adequate ventilation and facilitate maintenance, such as bearing lubrication and brush replacement ≫430-14(a)≪.

■ Motor Branch-Circuit Short-Circuit and Ground-Fault Protection

Use a protective device whose rating (or setting) does not exceed the value calculated according to Table 430-152 »430-52(c)(1) «.

Should the values for branch-circuit short-circuit and ground-fault protective devices (determined by Table 430-152) not correspond to a standard size or rating of fuses, nonadjustable circuit breakers, thermal protective devices, or possible settings of adjustable circuit breakers, then the next higher standard size, rating or possible setting is permitted »430-52(c)(1) *Exception No.1* «.

The motor branch-circuit short-circuit and ground-fault protective device must comply with 430-52(b) and either (c) or (d), as applicable »430-52(a) «.

Branch-circuit protective devices must comply with Section 240-20 provisions »430-56 «.

CAUTION ‹‹‹‹‹‹‹‹‹‹‹‹

If the maximum branch-circuit short-circuit and ground-fault protective device ratings are shown in the manufacturer's overload relay table (for use with a motor controller), or are otherwise marked on the equipment, they must not be exceeded even if Section 430-52(c)(1) allows higher values »430-52(c)(2) «.

N O T E
The motor branch-circuit short-circuit and ground-fault protective device must be able to carry the motor's starting current »430-52(b) «.

Part D specifies protective devices for motor branch-circuit conductors, motor control apparatus, and motors against overcurrent due to short circuits or grounds »430-51 «.

If the rating specified in Table 430-152, as modified by Exception No. 1, is not sufficient for the motor's starting current, apply Section 430-52(c)(1) Exception No. 2 (a) through (d) provisions »430-52(c)(1) *Exception No. 2* «.

Article 430, Part D provisions do not apply to motor circuits rated over 600 volts, nominal. See Part K »430-51 «.

Suitable fuses are permitted in lieu of devices listed in Table 430-152 for power electronic devices in a solid state motor controller system, provided replacement fuse markings are adjacent to the fuses »430-52(c)(5) «.

A listed self-protected combination controller may be substituted for the devices specified in Table 430-152 »430-52(c)(6) «.

A motor short-circuit protector is permitted in lieu of Table 430-152 listed devices if the motor short-circuit protector is part of a listed combination motor controller which has coordinated motor overload protection and short-circuit and ground-fault protection per Section 430-52(c)(7).

Torque motor branch circuits must be protected at the motor nameplate current rating according to Section 240-3(b) »[430-52(d) «.

Several motors, or loads, on one branch circuit must comply with Section 430-53 provisions.

The branch-circuit short-circuit and ground-fault protective device rating, for multimotor and combination-load equipment, must not exceed the equipment's marked rating in accordance with Section 430-7(d) »430-54 «.

Where fuses are used for motor branch-circuit short-circuit and ground-fault protection, the fuseholders must be of adequate size to accommodate the fuses specified by Table 430-152 »430-57 «. (An exception is provided for fuses having time delay appropriate for the motor's starting characteristics.)

A circuit breaker for motor branch-circuit short-circuit and ground-fault protection must have a current rating as determined by Section 430-52 and 430-110 »430-58 «.

■ Motor Disconnecting Means

Motor branch-circuit and ground-fault protection, and motor overload protection, can be combined into a single device, where the device's rating (or setting) provides the overload protection required by Section 430-32 **》** 430-55**《** .

NOTE
*At least one of the disconnecting means must be **readily** accessible* **》** *430-107***《** .

Motor circuits rated 600 volts, nominal, or less, must have a disconnecting means with an ampere rating of at least 115% of the motor's full-load current rating **》** 430-110(a)**《** .

Combination Motor Starter

Article 430, Part J requires a disconnecting means capable of breaking the connection between the motors/controllers and the circuit **》** 430-101**《** .

The phrase "in sight from" indicates that specified items of equipment are visible and are no more than 50 feet apart **》** Article 100**《** .

According to Article 430, a controller is any switch (or device) normally used to start and stop a motor by making and breaking motor circuit current **》** 430-81(a)**《** .

50 Feet Maximum

An individual, fully-functional disconnecting means must be provided for each controller. The disconnecting means must be located in sight from the controller, unless an exception is met **》** 430-102(a)**《** .

A separate disconnecting means must be located in sight from the motor and driven machinery location **》** 430-102(b)**《** .

The disconnecting means must open all ungrounded supply conductors, and must be designed so that no pole operates independently. The disconnecting means can occupy the same enclosure as the controller **》** 430-103**《** .

The disconnecting means must plainly indicate whether it is in the open (off) or closed (on) position **》** 430-104**《** .

Motor control circuits arrangement must accommodate disconnection from all supply sources when the disconnecting means is in the open position. Should the disconnecting means consist of two or more separate devices, each must comply with Section 430-74(a).

A motor disconnecting means, in addition to the controller disconnecting means, per Section 430-102(a), is not required where the controller's disconnecting means is individually capable of being locked in the open position **》** 430-102(b) *Exception***《** .

Multiple motors used together, or one or more motors used in combination with other loads, must comply with Section 430-110(c) provisions.

A switch (or circuit breaker) can serve as both controller and disconnecting means if it complies with 430-111(a) and is one of a type listed in 430-111(b) **》** 430-111**《** .

Every motor-circuit disconnecting means, between the feeder's point of attachment and the motor's connection point, must comply with the requirements of Sections 430-109 and 430-110 **》** 430-108**《** .

Motor controllers must comply with Article 430, Part G.

The disconnecting means must be a type specified in Section 430-109(a), unless otherwise permitted in (b) through (g), under the conditions given **》** 430-109**《** .

■ Motor Control Centers

● ●

> **NOTE**
> *Motor control units within a motor control center must comply with Section 430-8* ≫ *430-98(b)* ≪.

Motor control centers must have overcurrent protection based on the rating of the common power bus, in accordance with Article 240. This protection must be provided by: (1) an overcurrent protective device located ahead of the motor control center; or (2) a main overcurrent protective device in the motor control center ≫ 430-94 ≪.

> **NOTE**
> *Motor control circuit installations must comply with Article 430, Part F (430-71 through 74).*

Horizontal travel through vertical sections is acceptable where a barrier isolates the conductors from the busbars ≫ 430-97(a) *Exception* ≪.

Article 430, Part H covers motor control centers installed for the control of motors, lighting, and power circuits ≫ 430-92 ≪.

Multisection motor control centers must be bonded together with an equipment grounding conductor, or an equivalent grounding bus, sized according to Table 250-122. Equipment grounding conductors must either terminate on the grounding bus, or to a grounding termination point within a single-section motor control center ≫ 430-96 ≪.

Secure busbars firmly in place, and provide adequate protection from physical damage. Other than those required for interconnections and control wiring, only the conductors that terminate can be located within that vertical section ≫ 430-97(a) ≪.

At motor control center terminals, minimum wire bending space as well as minimum gutter space must comply with article 373 ≫ 430-97(c) ≪.

Any motor control center used as service equipment must comply with Section 430-95.

Spacing between motor control center bus terminals and other bare metal parts must meet Table 430-97 specifications ≫ 430-97(d) ≪.

Barriers are required in all service-entrance motor control centers to isolate service busbars and terminals from the remainder of the motor control center ≫ 430-97(e) ≪.

Motor control centers must be marked according to Section 110-21, so that the markings remain plainly visible after installation. Common power bus current rating and motor control center short-circuit rating must be included in the marking ≫ 430-98(a) ≪.

■ Feeder Conductors

Where one or more of the group's motors are used for short-time, intermittent, periodic, or varying duty, the ampere rating of such motors (used in the calculation) is determined by Section 430-22(b). For the highest rated motor, the greater of: (1) the ampere rating from 430-22(b); or (2) the largest continuous duty motor full-load current multiplied by 1.25, is used ≫430-24 *Exception No. 1*≪.

The ampacity of conductors supplying motor-operated fixed electric space-heating equipment must conform to Section 424-3(b) specifications ≫430-24 *Exception No. 2*≪.

Appendix D contains a motor application example (Example No. D8).

Where interlocked circuitry prevents operation of selected motors (or other loads) at the same time, the conductor ampacity can be based on the highest possible total of motor and other load currents to be operated at the same time ≫430-24 *Exception No. 3*≪.

Feeder tap conductors must comply with Section 430-28 provisions.

> **NOTE**
>
> *To ensure compliance with Sections 430-24, 430-53(b), and 430-53(c), the highest rated, or smallest rated, motor must be based on the rated full-load current selected from Tables 430-147, 430-148, 430-149, and 430-150* ≫430-17≪.

Conductors supplying several motors, or a motor(s) and other load(s), must have an ampacity at least equal to the sum of the full-load current rating, per Section 460-6(a)(1), of all the motors plus 25% of the group's highest rated motor, plus the ampere rating of other loads determined by Article 220 and other applicable articles ≫430-24≪. In other words, multiply the largest motor full-load current (FLC) rating by 125% and add the FLC ratings of all other motors (or loads) in the group.

Where reduced heating of the conductors results form duty-cycle operation, intermittent operation, or from all motors not operating at one time, the authority having jurisdiction may allow a lower feeder-conductor ampacity than that specified in Section 430-24, provided the conductors' ampacity is sufficient for the maximum load determined according to the size and number of motors supplied, and the character of their respective loads and duties ≫430-26≪.

Feeder Conductors Supplying Motors

■ Motor Feeder Short-Circuit and Ground-Fault Protection

A feeder consisting of conductor sizes based on Section 430-24 that supplies a specific fixed motor load(s) must have a protective device with a rating (or setting) no greater than the largest rating (or setting) of the branch-circuit short-circuit and ground-fault protective device for any motor supplied by the feeder (based on the maximum permitted value for the specific protective device shown in Table 430-152, or Section 440-22(a) for hermetic refrigerant motor-compressors), plus the sum of the full-load currents of the other motors in the group »430-62(a)«.

Article 430, Part E, specifies devices for protection of feeder conductors supplying motors against overcurrents due to short circuits or grounds »430-61«.

> ### NOTE
> *Where a feeder supplies a motor load as well as a lighting (or a lighting and appliance) load, the feeder protective device can have a rating (or setting) sufficient to carry the lighting (or the lighting and appliance) load as determined by Articles 210 and 220; plus, for a single motor, the rating permitted by Section 430-52, and, for multiple motors, the rating per Section 430-62 »430-63«.*

Motor Feeder Overcurrent Protection

Where instantaneous trip circuit breaker(s), or motor short-circuit protector(s), are used for motor branch-circuit short-circuit and ground-fault protection, per Section 430-52(c), the above procedure for determining the maximum feeder protective device rating applies with the following provision. For calculation purposes, each instantaneous trip circuit breaker or motor short-circuit protector is assumed to have a rating not exceeding the maximum percentage of motor full-load current permitted by Table 430-152 for the type of feeder protective device used »430-62(a) *Exception*«.

Where feeder conductor ampacity is greater than Section 430-24 requires, the rating (or setting) of the feeder overcurrent protective device can be based on the feeder conductor's ampacity »430-62(b)«. Conductor ampacities are found in Tables 310-16 through 19.

If the same rating (or setting) of the branch-circuit short-circuit and ground-fault protective device is used on multiple branch circuits, supplied by the feeder, one of the protective devices is considered the largest for the above calculation »430-62(a)«.

Appendix D contains a motor application example (Example No. D8).

SPECIFIC EQUIPMENT

■ Cranes and Hoists

Article 610 covers electrical equipment and wiring installation for use with cranes, monorail hoist, hoists, and all runways »610-1«.

All exposed noncurrent-carrying metal parts of cranes, monorail hoist, hoists, and accessories (including pendant controls) must be joined metallically into a continuous electrical conductor so that the entire crane or hoist will be grounded according to Article 250. Moving parts, other than removable accessories or attachments, having metal-to-metal bearing surfaces qualify as electrically connected (via bearing surfaces) for grounding purposes. The frames (trolley and bridge) are electrically grounded through the wheels and respective tracks, unless conditions, such as paint or other insulating material, prevent reliable metal-to-metal contact. In such cases, a separate bonding conductor must be provided »610-61«.

Cranes, hoists, and monorail hoist overload protection must comply with Section 610-43.

Crane, hoist, and monorail hoist motor branch circuits must meet Section 610-42(a) requirements.

> **NOTE**
> *All equipment operating in a hazardous (classified) location must comply with Article 500 »610-2(a)«.*

Conductors must either be enclosed in raceways or be Type AC cable with insulated grounding conductor, Type MC cable, or Type MI cable, unless Section 610-11(a) through (b) otherwise permits »610-11«.

If a crane, hoist, or monorail hoist operates above readily combustible material, the resistors must be located as outlined in Section 610-2(b)(1) and (2) »610-2(b)«.

Conductors must comply with Table 310-13 unless otherwise permitted in Section 610-13(a) through (d) »610-13«.

Conductors exiting raceways/cables must comply with Section 610-12(a) or (b) »610-12«.

Table 610-14(a) dictates the allowable ampacities of conductors »610-14(a)«.

Where the secondary resistor and the controller are separate, calculate the minimum size conductors (between controller and resistor) by multiplying the motor secondary current by the appropriate Table 610-14(b) factor, and select a wire from Table 610-14(a) »610-14(b)«.

Motor and control external conductors must not be smaller than No. 16, unless otherwise permitted in Section 610-14(c)(1) and (2) »610-14(c)«.

Contact conductors must comply with Sections 610-14(d) and 610-21(a) through (h).

Motor loads must be calculated according to Section 610-14(e).

A disconnecting means with a continuous ampere rating not less than that computed in Sections 610-14(e) and (f) must be provided between the runway contact conductors and the power supply. Such a disconnecting means must consist of a motor-circuit switch, circuit breaker, or molded case switch. This switch must meet Section 610-31(1) through (4) provisions »610-31«.

Leads from crane and monorail hoist runway contact conductors (or other power supply) must have a motor-circuit switch (or circuit breaker) which can be locked in the open position »610-32«.

Section 610-32 requires that the continuous ampere rating of the switch or circuit breaker not be less than 50% of the combined short-time ampere rating of the motors, nor less than 75% of the sum of the short-time ampere rating of the motors required for any single motor »610-33«.

Each motor must have an individual controller unless Section 610-51(a) or (b) allows otherwise »610-51«.

Electric Welders

The ampacity of arc welder conductors supplying a group of welders can be less than the sum of the welder supplied currents, as determined by Section 630-11(a). In each case, the conductor rating must be determined by the welder loading, based on each welder's use and the allowance permissible should all welders supplied by the conductors not be used simultaneously. Each welder's load value must take into account both the magnitude and duration of the load while in use **》630-11(b)《**.

Section 630-12(a) and (b) requires arc welder overcurrent protection. Where the resulting values do not correspond with Section 240-6(a) standard ampere ratings; or the rating or setting specified results in unnecessary opening of the overcurrent device, the next higher standard rating or setting is acceptable **》630-12《**.

> ### NOTE
> *Resistance welders must comply with Article 630, Part C (630-31 through 630-34) provisions.*

Supply conductor ampacity, for a single arc welder, must not be less than the $I_{1\,eff}$ value on the rating plate. If the $I_{1\,eff}$ is not given, the supply conductor ampacity must meet Section 630-11 requirements **》630-11(a)《**.

Branch-Circuit Overcurrent Protection

Branch-Circuit Conductors

Feeder Conductors

Conductors supplying one or more arc welders must have an overcurrent protective device rated (or set) not more than 200% of the conductor rating **》630-12(b)《**. (See Section 630-12(b) FPN for calculation explanation.)

Article 630 covers electric arc welding, resistance welding apparatus, and similar welding equipment connected to an electric supply system **》630-1《**.

Per Section 630-11(a), assume arc welder conductor ratings, based on 100% of the current, of the two largest welders, 85% for the third largest, 70% for the fourth largest, and 60% for all remaining welders, to provide an ample safety margin under high-production conditions with respect to the maximum permissible temperatures. Lower percentages are permitted in cases where the work is such that a high-operating duty cycle for individual welders is possible **》630-11(b) FPN《**.

Arc welders must have a rating plate with the information listed in Section 630-14.

Conductors used in the secondary circuit of electric welders must have flame retardant insulation **》630-41《**.

Each arc welder must have overcurrent protection rated (or set) at no more than 200% of $I_{1\,max}$. If the $I_{1\,max}$ is not given, the overcurrent protection must be rated (or set) at no more than 200% of the welder's rated primary current **》630-12(a)《**.

An overcurrent device is not required for an arc welder having supply conductors protected by an overcurrent device rated (or set) at no more than 200% of $I_{1\,max}$, or the welder's rated primary current **》630-12(a)《**.

Supply conductors for an arc welder, protected by an overcurrent device rated (or set) at no more than 200 percent of $I_{1\,max}$ or rated primary welder current, do not require a separate overcurrent device **》630-12(a)《**.

Welding cables can be installed in a dedicated cable tray as indicated by Section 630-42(a), (b), and (c) **》630-42《**.

■ Electroplating

Unless modified by Article 669, electroplating wiring and equipment must comply with the applicable requirements of Chapters 1 through 4 »669-2«.

Equipment used in the electroplating process must be specifically identified for such service »669-3«.

Branch-circuit conductors supplying equipment unit(s) must have an ampacity of at least 125% of the total connected load. Busbar ampacities must meet Section 374-6 requirements »669-5«.

The following stipulations apply to conductors connecting the electrolyte tank equipment to the conversion equipment: (a) Insulated conductors in systems not exceeding 50 volts direct current can be run without insulated support, provided they are protected from physical damage. Bare copper (or aluminum) conductors are permitted where supported on insulators. (b) Insulated conductors in systems exceeding 50 volts direct current can be run on insulated supports, provided they are protected from physical damage. Bare copper (or aluminum) conductors are permitted where (1) supported on insulators and (2) guarded against accidental contact up to the termination point, per Section 110-27 »669-6«.

Where multiple power supplys serve the same dc system, the dc side of each power supply must have a disconnecting means »669-8(a)«.

Removable links/conductors can serve as the disconnecting means »669-8(b)«.

Direct-current conductors must be protected from overcurrent by at least one of the following: (1) fuses or circuit breakers; (2) a current-sensing device that operates a disconnecting means; or (3) other approved means »669-9«.

Article 669 provisions apply to the installation of the electrical components and accessory equipment that supply the power and controls for: electroplating, anodizing, electropolishing, and electrostripping, herein referred to simply as *electroplating* »669-1«.

CAUTION ‹‹‹‹‹‹‹‹‹‹

Warning signs must be posted to indicate the presence of bare conductors »669-7«.

■ Industrial Machinery

Article 670 covers the definition of, the nameplate data for, and the size and overcurrent protection of supply conductors to industrial machinery »670-1«.

Where controls are enclosed, the cabinet door(s) must either open at least 90 degrees, or be removable »670-5«.

Article 110, Part B provisions apply to working space around electrical equipment operating at 600 volts, nominal, or less to ground.

The overcurrent protective device's rating (or setting) for the circuit supplying the machine must comply with Section 670-4(b).

The selected supply conductor size must have an ampacity not less than 125% of the full-load current rating for all resistance heating loads, plus 125% of the highest rated motor's full-load current rating, plus the sum of the full-load current ratings of all other connected motors and apparatus that may be simultaneously operated »670-4(b)«.

Industrial Machinery (Machine) is defined as a power-driven machine (or a group of machines working together in a coordinated manner), not portable by hand during operation, used to process material by cutting; forming; pressure; electrical, thermal, or optical techniques; lamination; or a combination of these processes. It can include associated equipment used to transfer material or tooling (including fixtures), assemble/disassemble, inspect, test, or package. (Associated electrical equipment, including logic controller(s) and associated software (or logic) together with the machine actuators and sensors, are all part of the industrial machine.) »670-2«

A permanent nameplate listing supply voltage, phase, frequency, full-load current, maximum short-circuit and ground-fault protective device ampere rating, largest motor or load ampere rating, short-circuit interrupting capacity of the machine overcurrent-protective device, if furnished, and diagram number must be attached to the control equipment enclosure/machine, remaining plainly visible after installation »670-3(a)«.

> **NOTE**
>
> *Where maintenance and supervision conditions ensure that only qualified persons will service the installation, the working space dimensions in the direction of access to live parts (operating at not over 150 volts line-to-line or line-to-ground) likely to require examination, adjustment, servicing, or maintenance while energized, must be a minimum of 2½ feet* »670-5«.

A machine is considered to be an individual unit, and therefore must have a disconnecting means. Branch circuits protected by either fuses (or circuit breakers) can be the disconnecting means. The disconnecting means is not required to incorporate overcurrent protection. Where overcurrent protection (single circuit breaker or set of fuses) is furnished as part of the machine, Section 670-3 required markings must be used and the supply conductors must be considered either as feeders (or taps), as covered by Section 240-21 »670-4(b)«.

An **Industrial Manufacturing System** is a systematic array of industrial machine(s) not portable by hand and including associated material handling, manipulating, gauging, measuring, or inspection equipment »670-2«.

If the machine has no branch-circuit short-circuit and ground-fault protective device, the overcurrent protective device's rating (or setting) must be based on Sections 430-52 and 430-53, as applicable »670-4(b)«.

Where overcurrent protection is provided, per Section 670-4(b), the machine must be marked "overcurrent protection provided at machine supply terminals" »670-3(b)«.

If opening a machine's enclosure requires a tool, and only troubleshooting/diagnostic testing is involved on live parts, the clearances can be less than 2½ feet »670-5 *Exception*«.

■ Capacitors

Each capacitor must have a nameplate stating the manufacturer's name, rated voltage, frequency, kilovar or amperes, number of phases; and, if filled with a combustible liquid, the amount of liquid in gallons. Any nonflammable liquid filling must be indicated on the nameplate. The nameplate must also identify a capacitor having a discharge device inside the case ≫ 460-12 ≪.

Capacitor residual voltage must be reduced to 50 volts, nominal, or less within 1 minute after the capacitor disconnects from the supply source ≫ 460-6(a) ≪. The discharge circuit shall either be permanently connected to the capacitor/capacitor-bank terminals, or have automatic means of connecting it to the capacitor-bank terminals upon line voltage removal. Manual means of switching or connecting the discharge circuit must not be used ≫ 460-6(b) ≪.

Each capacitor bank's ungrounded conductor must have a disconnecting means, unless connected on the motor controller's load side ≫ 460-8(c)(1) ≪. The disconnecting means must simultaneously open all ungrounded conductors ≫ 460-8(c)(2) ≪. It can disconnect the capacitor from the line as a regular operating procedure ≫ 460-8(c)(3) ≪. The rating of the disconnecting means must be at least 135% of the capacitor's rated current ≫ 460-8(c)(4) ≪.

Capacitors must have a means of discharging stored energy ≫ 460-6 ≪.

Article 460 covers the installation of capacitors on electric circuits ≫ 460-1 ≪.

Capacitor cases must be grounded according to Article 250 stipulations ≫ 460-10 ≪. Do not ground capacitor cases where capacitor units are supported on a structure designed to operate at other than ground potential ≫ 460-10 Exception ≪.

The capacitor circuit conductor's ampacity must be at least 135% of the capacitor's rated current. The ampacity of the conductors that connect a capacitor to a motor's terminals, or to motor circuit conductors, must not be less than one-third of the motor circuit conductor's ampacity; and, in no case less than 135% of the capacitor's rated current ≫ 460-8(a) ≪.

Each capacitor bank's ungrounded conductor must have an overcurrent device, unless connected on the motor overload protective device's load side ≫ 460-8(b)(1) ≪. The overcurrent device's rating (or setting) must be as low as practicable ≫ 460-8(b)(2) ≪.

> **NOTE**
> *Capacitors rated over 600 volts, nominal, must comply with Article 460, Part B (460-24 through 28).*

Surge capacitors, or capacitors included as a component part of other apparatus and conforming with such apparatus' requirements, are excluded from Article 460 requirements ≫ 460-1 ≪.

Article 460 also covers capacitor installation in hazardous (classified) locations as modified by Articles 501 through 503 ≫ 460-1 ≪.

Capacitors containing more than 3 gallons of flammable liquid must be enclosed in vaults or outdoor fenced enclosures in order to comply with Article 110, Part C. This limit applies to any single unit in capacitor installations ≫ 460-2(a) ≪.

If a motor installation includes a capacitor connected on the load side of a motor overload device, the motor overload device's rating/setting must be based on the improved power factor of the motor circuit. Disregard the capacitor effect in determining the motor circuit conductor rating per Section 430-22 ≫ 460-9 ≪.

A capacitor connected on the load side of a motor overload protective device does not require a separate overcurrent device ≫ 460-8(b)(1) Exception ≪.

A separate disconnecting means is not required where a capacitor is connected on a motor controller's load side ≫ 460-8(c)(1) Exception ≪.

> **NOTE**
> *Capacitors, accessible to unauthorized/unqualified persons, must be enclosed, located, or guarded so that neither persons nor conducting materials can come into accidental contact with exposed energized parts, terminals, or associated buses. However, no additional guarding is required for enclosures accessible only to authorized, qualified persons ≫ 460-2(b) ≪.*

Summary

- A service may have up to six disconnecting means in a single enclosure; in a group of separate enclosures; or in (or on) a switchboard.
- A service-entrance conductors' raceway cannot contain non-service conductors.
- Conductors can be tapped, without overcurrent protection at the tap, to a feeder as specified in Section 240-21(b)(1) through (5).
- Conductors can connect to a transformer secondary, without overcurrent protection at the secondary, as specified in Section 240-21(c)(1) through (4).
- Article 450 not only covers transformers, but also includes transformer vault provisions.
- Taps can connect the grounding electrode conductors in a service consisting of more than a single enclosure.
- Cable tray provisions are located in Article 318.
- Article 430 covers motors; motor branch-circuit and feeder conductors (and their protection); motor overload protection; motor control circuits; motor controllers; and motor control centers.
- Article 610 covers electrical equipment and wiring installations used in connection with cranes, monorail hoists, and all runways.
- Electric arc welding, resistance welding apparatus, and similar welding equipment provisions are located in Article 630.
- Article 669 requirements apply to the installation of the electrical components and accessory equipment supplying electroplating, anodizing, electropolishing, and electrostripping power and controls.
- Article 670 covers the definition of, the nameplate data for, and the size and overcurrent protection of supply conductors to industrial machinery.
- Provisions pertaining to the installation of capacitors on electric circuits is contained in Article 460.

Unit 17 Competency Test

NEC® Reference	Answer

1. Terminals of motors and controllers shall be suitably marked or colored where necessary to indicate _____.

2. Capacitors containing more than _____ gallons of flammable liquid shall be enclosed in vaults or outdoor fenced enclosures complying with Article 110, Part C.

3. Cable tray systems shall not be used in _____ or where subject to severe physical damage.

     I. hoistways

     II. hazardous locations

     III. corrosive areas

    a) I only   b) I and II only   c) I and III only   d) I, II, and III

4. Wires that are used as crane (or hoist) contact conductors shall be secured at the ends by means of approved strain insulators and shall be mounted on approved insulators so that the extreme limit of displacement of the wire will not bring the latter within less than _____ from the surface wired over.

5. Multispeed motors shall be marked with the code letter designating the locked-rotor kilovolt-ampere (kVA) per horsepower for the highest speed at which the motor _____.

6. The walls and roofs of transformer vaults shall be constructed of materials that have adequate structural strength for the conditions with a minimum fire resistance of _____.

———————— —————— 7. Where cable trays support individual conductors and where the conductors pass from one cable tray to another, or from a cable tray to raceways or to equipment where the conductors are terminated, the support distance between cable trays or between the cable tray and the equipment shall not exceed _____.

———————— —————— 8. The residual voltage of a capacitor shall be reduced to _____, nominal, or less, within _____ after the capacitor is disconnected from the source of supply.

———————— —————— 9. The disconnecting means for motor circuits rated 600 volts, nominal, or less, shall have an ampacity rating of at least _____ of the full-load current rating of the motor.

———————— —————— 10. Steel cable trays shall not be used as equipment grounding conductors for circuits with ground-fault protection above _____.

———————— —————— 11. Where the conditions of maintenance and supervision ensure that only qualified persons will service the industrial machinery installation, the dimensions of the working space in the direction of access to live parts operating at not over 150 volts line-to-line or line-to-ground that are likely to require examination, adjustment, servicing, or maintenance while energized must be a minimum of _____.

———————— —————— 12. Branch-circuit conductors supplying one or more units of electroplating equipment shall have an ampacity of not less than _____ of the total connected load.

———————— —————— 13. Where practicable, vaults containing more than _____ transformer capacity shall be provided with a drain or other means that will carry off any accumulation of oil or water in the vault unless local conditions make this impracticable.

———————— —————— 14. Conductors that supply one or more arc welders shall be protected by an overcurrent device rated or set at not more than _____ of the conductor rating.

———————— —————— 15. The full-load current of a 25 horsepower, synchronous, 460 volt, three-phase motor at unity power factor is _____.

———————— —————— 16. The ampacity of capacitor circuit conductors shall not be less than _____ of the rated current of the capacitor.

———————— —————— 17. Hoist and monorail hoist and their trolleys that are not used as part of an overhead traveling crane shall not require individual motor overload protection, provided the largest motor does not exceed _____ and all motors are under manual control of the operator.

———————— —————— 18. A door sill or curb that is of sufficient height to confine the oil from the largest transformer within the vault shall be provided, and in no case shall the height be less than _____.

———————— —————— 19. Branch-circuit conductors that supply a single motor used in a continuous duty application shall have an ampacity of not less than _____ of the motor's full-load current rating as determined by Section 430-6(a)(1).

———————— —————— 20. A(n) _____ is a systematic array of one or more industrial machines not portable by hand and that includes any associated material handling, manipulating, gauging, measuring, or inspection equipment.

———————— —————— 21. Cable trays containing welding cables must have a permanent sign attached to the cable tray at intervals not greater than _____.

———————— —————— 22. For raceway terminating at the tray, a listed cable tray clamp or adapter shall be used and no nearby support, such as a support within _____, shall be required.

NEC® Reference	Answer
_____	_____

23. Where the motor control circuit transformer rated primary current is less than 2 amperes, an overcurrent device rated or set at not more than _____ of the rated primary current shall be permitted in the primary circuit.

24. The dimension of the working space in the direction of access to live parts that are likely to require examination, adjustment, servicing, or maintenance while energized shall be a minimum of _____.

25. All transformer vault ventilation openings to the indoors shall be provided with automatic closing fire dampers that operate in response to a vault fire. Such dampers shall possess a standard fire rating of not less than _____.

26. The ampacity of the supply conductors for a resistance welder that may be operated at different times values of primary current or duty cycle shall not be less than _____ of the rated primary current for manually operated non-automatic welders.

27. Conductors supplying several motors, or a motor(s) and other load(s), shall have an ampacity at least equal to the sum of the full-load current rating, as determined by Section 460-6(a)(1) of all the motors, plus _____ of the highest rated motor in the group, plus the ampere rating of other loads determined in accordance with Article 220 and other applicable articles.

28. Cable trays shall be permitted to extend _____ through partitions and walls or _____ through platforms and floors in wet or dry locations where the installations, complete with installed cables, are made in accordance with the requirements of Section 300-21.

29. Conductors for a motor used in a short-time, intermittent, periodic, or varying duty application shall have an ampacity of not less than the percentages of the motor nameplate current ratings shown in _____.

30. Each doorway leading into a transformer vault from the building interior shall be provided with a tight-fitting door that has a minimum fire rating of _____.

31. _____ shall be placed in all service-entrance motor control centers to isolate service busbars and terminals from the remainder of the motor control center.

32. Crane, hoist, and monorail hoist motor branch circuits shall be protected by fuses or inverse-time circuit breakers that have a rating in accordance with _____.

33. For a nonautomatically started motor, the overload protection shall be permitted to be shunted or cut out of the circuit during the starting period of the motor if the device by which the overload protection is shunted or cut out cannot be left in the starting position and if fuses or inverse time circuit breakers rated or set at not over _____ of the full-load current of the motor are located in the circuit so as to be operative during the starting period of the motor.

34. The discharge circuit of a capacitor or capacitor bank shall not be connected in which way(s)?

   I. permanently connected to the terminals

   II. automatic means of connecting it to the terminals

   III. manual means of connecting the discharge circuit

   a) I or II only   b) I or III only   c) III only   d) I, II, or III

35. Each controller shall be capable of starting and stopping the motor it controls and shall be capable of interrupting the _____ of the motor.

Special Occupancies

Objectives

After studying this unit, the student will:

✔ know that portions of occupancies, intended for the assembly more than 99 persons, are referred to as *Places of Assembly*.

✔ be familiar with definitions pertinent to motion picture (and television studio) audience areas, theaters, and similar locations.

✔ not only have a basic understanding of motion picture (and television studio) audience area, theater, and similar location general requirements, but also stage switchboards, stage equipment, and theater dressing room provisions.

✔ know the location of provisions pertaining to carnivals, circuses, exhibitions, fairs, traveling attractions, and similar functions.

✔ have a solid grasp of requirements for motion picture/television studios (and similar locations), including cellulose nitrate film

storage vaults and separately derived systems with 60 volts to ground.

✔ be well acquainted with motion picture projector provisions.

✔ thoroughly understand manufactured building requirements.

✔ be familiar with agricultural building provisions including what constitutes a bonding and equipotential plane.

✔ be appraised of requirements for mobile homes, manufactured homes, and mobile home parks.

✔ know the location of provisions relating to recreational vehicles and recreational vehicle parks.

✔ have a good understanding of floating building requirements.

✔ be familiar with the provisions for marinas and boatyards.

Introduction

Unit 18, "Special Occupancies," covers a variety of mostly unrelated occupancies. This unit addresses most of the occupancy types found in Article 518 through Article 555. Unlike other Chapter 5 articles in the *NEC*®, *Places of Assembly* (Article 518) is not a specific occupancy type. Places of assembly are buildings, or portions of buildings, designed (or intended) for the assembly of 100 or more persons. *Theaters, Audience Areas of Motion Picture and Television Studios, and Similar Locations* (Article 520) covers all buildings, or that part of a building or structure, designed for presentation, dramatic, musical, motion picture projection, or similar purposes; and specific audience seating areas within motion picture or television studios. *Carnivals, Circuses, Fairs, and Similar Events* (Article 525) addresses portable wiring and equipment installations for carnivals, circuses, exhibitions, fairs, traveling attractions, and similar functions, including wiring in/on all structures. Article 530 (*Motion Picture and Television Studios and Similar Locations*) requirements apply to television studios and motion picture studios using either film or electronic cameras, except as provided in Section 520-1, and exchanges, factories, laboratories, stages, or a segment of the building in which film or tape [more than ⅞ inch (22 mm) in width] is exposed, developed, printed, cut, edited, rewound, repaired, or stored. Article 540 (*Motion Picture Projectors*) provisions apply to motion picture projection rooms, motion picture projectors, and associated equipment (both professional and nonprofessional types) using incandescent, carbon arc, xenon, or other light source equipment which is known to develop hazardous gases, dust, or radiation. *Manufactured Buildings* (Article 545) details the requirements for manufactured buildings and building components. Article 547 (*Agricultural Buildings*) provisions apply to certain agricultural buildings, parts of buildings, and adjacent areas. Article 550 (*Mobile Homes, Manufactured Homes, and Mobile Home Parks*) requirements apply to the location types identified by the Article's title. Article 551 (*Recreational Vehicles and Recreational Vehicle Parks*) provisions apply to the electrical conductors and equipment installed within (or on) recreational vehicles, the conductors connecting recreational vehicles to an electrical supply, as well as the installation of equipment and devices related to electrical systems within a recreational vehicle park. *Floating Buildings* (Article 553) covers wiring, services, feeders, and grounding for, of course, floating buildings. Finally, Article 555 (*Marinas and Boatyards*) describes wiring and equipment installations in locations comprising fixed or floating piers, wharfs, docks, and other areas in marinas, boatyards, boat basins, boathouses, and similar occupancies that are, or can be, used for the purpose of repair, berthing, launching, storage, or fueling of small craft and the moorage of floating buildings.

MOTION PICTURE (AND TELEVISION STUDIO) AUDIENCE AREAS, THEATERS, AND SIMILAR LOCATIONS

■ General Requirements

Article 520 covers all buildings/structures, or parts thereof, intended for presentation, dramatic, musical, motion picture projection, or similar purposes as well as specific audience seating areas within motion picture or television studios ≫520-1≪.

The fixed wiring methods employed must be metal raceway, nonmetallic raceway encased in 2 inches or more of concrete, Type MI cable, or Type MC cable ≫520-5(a)≪.

Fixed wiring method stipulations in Article 640 apply to audio signal processing, amplification, and reproduction equipment; Article 800 to communication circuits; Article 725 to Class 2 and Class 3 remote control and signaling circuits; and Article 760 to fire alarm circuits ≫520-5(a) *Exception*≪.

Portable switchboards, stage set lighting, stage effects, and other wiring without a fixed location is permitted with approved flexible cords and cables as provided elsewhere in Article 520. Such wiring cannot be fastened by uninsulated staples or nailing ≫520-5(b)≪.

Nonmetallic-sheathed cable, Type AC cable, electrical nonmetallic tubing, and rigid nonmetallic conduit, can be installed where applicable building code does not require the building (or portions thereof) to be of fire-rated construction ≫520-5(c)≪.

The number of conductors permitted in any metal conduit, rigid nonmetallic conduit, or electrical metallic tubing for border or stage pocket circuits, or for remote-control conductors, must not exceed the percentage fill shown in Table 1 of Chapter 9. Where contained within an auxiliary gutter or wireway, the sum of the cross-sectional area of all internal conductors at any point must not exceed 20% of the interior cross-sectional area of the auxiliary gutter or wireway. The 30-conductor limitation of Sections 362-5 and 374-5 does not apply ≫520-6≪.

CAUTION ≪≪≪≪≪≪≪≪≪≪≪≪≪≪≪≪≪≪≪≪≪≪

*Any size branch circuit supplying receptacle(s) can also supply stage set lighting. The receptacle's voltage rating must not be less than the circuit voltage. Receptacle ampere rating and branch-circuit conductor ampacity must at least equal the branch-circuit overcurrent device ampere rating. **Table 210-21(b)(2) shall not apply*** ≫520-9≪.

Live parts must be enclosed or guarded preventing accidental contact by persons and objects. All switches must be of the externally operable type. Dimmers, including rheostats, must be installed in cases (or cabinets) which effectively surround all live parts ≫520-7≪.

Emergency system controls must comply with Article 700 ≫520-8≪.

Portable stage/studio lighting equipment, and portable power distribution equipment, are permitted for temporary outdoor use, provided the equipment is (1) supervised by qualified personnel while energized and (2) barriered from the general public ≫520-10≪.

■ Definitions

● ●

Border Light: A permanently installed overhead strip light ❯❯ 520-2 ❮❮ .

Portable Equipment: Equipment intended for movement from place to place and fed with portable cords or cables ❯❯ 520-2 ❮❮ .

Breakout assemblies can contain listed, hard usage (junior hard service) cords where all of the following conditions are met:
(1) The cords connect a single multipole connector (containing two or more branch circuits) and multiple two-pole, 3-wire connectors.
(2) No cord in the breakout assembly exceeds 20 feet in length.
(3) The breakout assembly's entire length is protected from physical damage by attachment to a pipe, truss, tower, scaffold, or other substantial support structure.
(4) All branch circuits feeding the breakout assembly are protected by overcurrent devices rated at no more than 20 amperes ❯❯ 520-68(a)(4) ❮❮ .

Flexible conductors, including cable extensions, supplying portable stage equipment must be listed extra-hard usage ❯❯ 520-68(a)(1) ❮❮ .

Breakout Assembly: An adapter used to join a multipole connector (containing two or more branch-circuits) to multiple individual branch circuit connectors ❯❯ 520-2 ❮❮ .

Connector Strip: A metal wireway containing pendant or flush receptacles ❯❯ 520-2 ❮❮ .

Portable Power Distribution Unit: A power distribution box containing receptacles and overcurrent devices ❯❯ 520-2 ❮❮ .

Proscenium: The wall and arch separating the stage from the auditorium (house) ❯❯ 520-2 ❮❮ .

Footlight: A border light installed on (or in) the stage ❯❯ 520-2 ❮❮ .

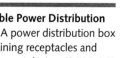

Portable power distribution units must comply with (a) through (e):
(a) The construction must prevent the exposure of current-carrying parts.
(b) Receptacles must comply with Section 520-45 and must have branch-circuit overcurrent protection within the box. Fuses and circuit breakers must be protected from physical damage. Cords/cables supplying pendant receptacles must be listed for extra-hard usage.
(c) Busbars must have an ampacity equal to the sum of the ampere ratings of all the connected circuits. Lugs must be provided for the master cable's connection.
(d) Power accepting flanged surface inlets (recessed plugs) must be rated in amperes.
(e) Cable arrangement must alleviate tension at the termination points, being adequately protected when passing through enclosures ❯❯ 520-62 ❮❮ .

Stand Lamp (Work Light): A portable stand containing a general-purpose lighting fixture, or guarded lamp holder for stage or auditorium general illumination ❯❯ 520-2 ❮❮ .

Stand lamps can be supplied by reinforced cord provided the cord is exempt from severe physical damage and is protected by an overcurrent device rated at not over 20 amperes ❯❯ 520-68(a)(2) ❮❮ .

Strip Light: A lighting fixture with a row of multiple lamps ❯❯ 520-2 ❮❮ .

Two-Fer: An adapter cable containing one male plug and two female cord connectors used to join two loads to one branch circuit ❯❯ 520-2 ❮❮ .

Adapters, two-fers, and other single/multiple circuit outlet devices must comply with all of the following:
(a) Each receptacle and corresponding cable must have the same current and voltage rating as its supply plug. Utilization in a stage circuit having a greater current rating is not allowed.
(b) All conductors must be wired in accordance with Section 520-67.
(c) Adapter and two-fer conductors must be listed either extra-hard usage or hard usage (junior hard service) cord. The overall length of hard usage (junior hard service) cord is restricted to 3.3 feet ❯❯ 520-69 ❮❮ .

■ Fixed Stage Switchboards

Stage switchboards must be of the dead-front type and must comply with Article 384, Part D unless a qualified testing laboratory (via recognized test standards and principles) has approved it as suitable for such use »520-21 «.

A stage-lighting switchboard, to which load circuits are connected, must have a means of branch circuit overcurrent protection. This includes branch circuits supplying stage and auditorium receptacles used for cord- and plug-connected stage equipment. Where the stage switchboard contains nonstage lighting dimmers, the overcurrent protective devices for these branch circuits can be located at the stage switchboard »520-23 «.

A stage switchboard, not completely enclosed dead-front and dead-rear, or recessed into a wall, must have a full length metal hood to protect all equipment on the board from falling objects »520-24 «.

Dimmers installed in ungrounded conductors, must have overcurrent protection no greater than 125% of the dimmer rating, and must be disconnected from all ungrounded conductors when the dimmer's master or individual switch (or circuit breaker) is in the open position »520-25(a) «.

Resistance- or series reactor-type dimmers must comply with Section 520-25(b) requirements.

Autotransformer-type dimmers must meet Section 520-25(c) specifications.

Section 520-25(d) provisions apply to solid-state dimmers.

A stage switchboard must be one (or a combination) of the following types:

(a) Dimmers and switches operate by handles mechanically linked to control devices.

(b) Devices operate electrically from a pilot-type control console or panel. Pilot control panels can be part of the switchboard, or can be located elsewhere.

(c) A stage switchboard having interconnected circuits is a secondary switchboard (patch panel) or panelboard remote to the primary stage switchboard, and must contain overcurrent protection. If the dimmer panel provides the required branch-circuit overcurrent protection, it can be omitted from the intermediate switchboard »520-26 «.

Stage switchboard supply feeders must be one of the following:

(1) A single feeder disconnected by a single disconnect device.

(2) Unlimited feeder quantities are permitted, provided all feeders are part of a single system. Where combined, neutral conductors in a given raceway must have sufficient ampacity to carry the maximum unbalanced current supplied by multiple feeder conductors within the same raceway. It is not necessary that the ampacity be greater than that of the neutral supplying the primary stage switchboard. Parallel neutral conductors must comply with Section 310-4.

(3) Separate feeders of a single primary stage switchboard must each have a disconnecting means. The primary stage switchboard must be permanently and conspicuously labeled with the number and location of disconnecting means. If the disconnecting means are located in multiple distribution switchboards, the primary stage switchboard must have barriers corresponding with these multiple locations »520-27(a) «.

When computing supply capacity to switchboards, it is permissible to consider the maximum intended switchboard load in a given installation, provided that: (1) all switchboard supply feeders are protected by an overcurrent device with a rating no greater than the feeder's ampacity; (2) opening the overcurrent device does not affect proper operation of egress or emergency lighting systems »520-27(c) «.

■ Portable Switchboards on Stage

The enclosure must contain a pilot light connected to the board's supply circuit so that opening the master switch does not cut off the lamp's supply. This lamp must be on an individual branch circuit having overcurrent protection rated (or set) at no more than 15 amperes »520-53(g)«.

Portable switchboards and feeders for on-stages use must comply with Section 520-53(a) through (p) »520-53«.

Only externally operable, enclosed type switches and circuit breakers can be used »520-53(c)«.

Portable switchboards must be housed within substantially constructed enclosures, which remain open during operation. Wooden enclosures must be fully lined with sheet metal of no less than No. 24 MSG. The sheet metal must either be properly galvanized, enameled, or otherwise coated to prevent corrosion, or be of a corrosion-resistant material »520-53(a)«.

Road Show Connection Panel (A Type of Patch Panel): A panel designed to allow for road show connection of portable stage switchboards to fixed lighting outlets via permanently installed supplementary circuits. The panel, supplementary circuits, and outlets must comply with all of the following:
(a) Circuits must terminate in grounding-type polarized inlets whose current and voltage rating match the fixed-load receptacles.
(b) Circuits transferred between fixed and portable switchboards must transfer all circuit conductors simultaneously.
(c) Supplementary circuit supply devices must have branch-circuit overcurrent protection. The individual supplementary circuit, within the road show connection panel and theater, must be protected by branch-circuit overcurrent protective devices of suitable ampacity internal to the road show connection panel.
(d) Panel construction must be in accordance with Article 384 »520-50«.

CAUTION
Unless an exception is met, only qualified personnel can route portable supply conductors, make/break supply connectors (and other supply connections), and energize/deenergize supply services. The portable switchboard must, by means of permanent and conspicuous markings, reflect this requirement »520-53(p)«.

NOTE
Supply conductors can have a maximum of three interconnections (mated connector pairs) where the total length from supply to switchboard does not exceed 100 feet. Should the total length from supply to switchboard exceed 100 feet, one additional supply conductor interconnection is permitted for each additional 100 feet »520-53(j)«.

Circuits from portable switchboards directly supplying equipment containing incandescent lamps of not over 300 watts must be protected by overcurrent protective devices having a rating or setting of not over 20 amperes. Circuits for lampholders over 300 watts are permitted if overcurrent protection complies with Article 210 »520-52«.

With the exception of busbars, all conductors within the switchboard must be stranded »520-53(f)(1)«.

Only listed extra-hard usage cords or cables can supply a portable switchboard. The supply cords (or cable) must terminate within the switchboard enclosure, in an externally operable fused master switch, circuit breaker, or a connector assembly identified for such purpose. The supply cords (or cable), and connection assembly, must (1) have sufficient ampacity to carry the total switchboard-connected load, and (2) be protected by overcurrent devices »520-53(h)(1)«.

Section 520-53(f) covers portable switchboard interior conductors.

Single-conductor portable supply cable sets must comply with Section 520-53(h)(2) requirements.

Supply conductors, not exceeding 10 feet in length between supply and switchboard or supply and a subsequent overcurrent device, can be reduced in size where all Section 520-53(h)(3)(a) through (g) conditions are met »520-53(h)(3)«.

Supply conductors, not exceeding 20 feet in length between supply and switchboard or supply and a subsequent overcurrent protection device, can be reduced in size where all Section 520-53(h)(4)(a) through (g) conditions are met »520-53(h)(4)«.

Single-pole portable cable connectors, where used, must be listed and of the locking type »520-53(k)«. (See the remainder of Section 520-53[k] for more single-pole separable connector provisions.)

■ Stage Equipment Other Than Switchboards

Borders and proscenium sidelights must be: (1) constructed per Section 520-43 specifications, (2) suitably secured and supported; and (3) designed so that the reflector flanges (or other adequate guards) protect the lamps from mechanical damage and from accidental contact with any combustible material ≫520-44(a)≪.

Border Lights

Cords and cables for the border lights' supply must be listed for extra-hard usage and must be suitably supported. Such cords and cables must be employed only where flexible conductors are necessary. Conductor ampacity must satisfy Section 400-5 provisions ≫520-44(b)(1)≪.

If the metal trough construction specified in Section 520-43(a) is not used, footlights must consist of individual outlets whose lampholders are wired with rigid metal conduit, intermediate metal conduit, or flexible metal conduit, Type MC cable, or mineral-insulated, metal-sheathed cable. The circuit conductors must be soldered to the lampholder terminals ≫520-43(b)≪.

The current supply to disappearing footlights must automatically disconnect when the footlights are fully retracted ≫520-43(c)≪.

Footlights

Metal trough footlights, containing circuit conductors, must be made of oxidation-resistant sheet metal no lighter than No. 20 MSG. Lampholder terminals must be spaced at least ½ inch from the metal trough. The circuit conductors must be soldered to the lampholder terminals ≫520-43(a)≪.

Arrange foot lights, border lights, and proscenium side lights so that no supplying branch circuit carries more than a 20 ampere load ≫520-41(a)≪.

Article 210 provisions can be applied to heavy-duty lampholder circuits ≫520-41(b)≪.

The conductors for foot, border, proscenium, or portable strip light fixtures and connector strips must have insulation suitable for the conductor's operating temperature, but in no case less than 125°C (257°F). The ampacity of the 125°C (257°F) conductors must be that of 60°C (140°F) conductors ≫520-42≪.

All drops from connector strips must be 90°C (194°F) wire sized to the ampacity of 60°C (140°F) cords and cables, with no more than 6 inches of conductor extending into the connector strip. Section 310-15(b)(2)(a) does not apply ≫520-42≪.

The ampacity of listed multiconductor extra-hard usage-type cords and cables not in direct contact with equipment containing heat-producing elements can be determined by Table 520-44. If so determined, the maximum load current in any conductor must not exceed the values in Table 520-44 ≫520-44(b)(2)≪.

Receptacles for on-stage electrical equipment (or fixtures) must be rated in amperes. Conductors supplying receptacles must comply with Articles 310 and 400 ≫520-45≪.

Connector strip, drop box, floor pocket, and similar receptacles used to connect portable stage-lighting equipment must be pendant, or mounted in suitable pockets or enclosures, and must satisfy Section 520-45 requirements. Connector strip and drop box supply cables must meet Section 520-44(b) specifications ≫520-46≪.

Bare-bulb lamps in backstage and ancillary areas, where contact with scenery is possible, must be located and guarded to remain free from physical damage. A minimum air space of at least 2 inches between such lamps and any combustible material must be provided ≫520-47≪. Scenery-installed decorative lamps are not be regarded as backstage lamps. [520-47 *Exception*]

The operating circuit of an electrical device releasing stage smoke ventilators must be (1) normally closed and (2) controlled by at least two externally operable switches with one such switch being readily accessible on stage, and the other positioned as dictated by the authority having jurisdiction. The device must be designed for the connected circuit's full voltage, with no resistance inserted. The device must be located in the loft above the scenery having been enclosed in a suitable metal box with a tight, self-closing door ≫520-49≪.

> **NOTE**
> *Portable stage equipment (other than switchboards) must comply with Article 520, Part E (520-61 through 520-69).*

■ Theater Dressing Rooms

NOTE

Fixed wiring in motion picture and television studio dressing rooms must be installed according to Chapter 3 methods »530-31«. (Sections 520-71 through 73 provisions apply only to theater dressing rooms, and not motion picture and television studio dressing rooms.)

In theater dressing rooms, all exposed incandescent lamps less than 8 feet from the floor, must be equipped with open-end guards riveted to the outlet box cover or otherwise appropriately secured »520-72«.

All dressing room lights and receptacles adjacent to the mirror(s) and above the dressing table counter(s) must be controlled by wall switches within the dressing room(s) »520-73«.

Each switch controlling receptacles adjacent to the mirror(s) and above the dressing table counter(s) must have a pilot light located outside the dressing room, adjacent to the door, indicating energized receptacles »520-73«. Note: although the pilot light must be installed outside the dressing room, the switch must be located inside the dressing room.

Pilot Light

Open-end Guards

Receptacles away from dressing table counter(s) and mirror(s) do not require switches »520-73«.

CAUTION «««««««««««
Pendant lampholders are not allowed in dressing rooms »520-71«.

CARNIVALS, CIRCUSES, FAIRS, AND SIMILAR EVENTS

■ Ground-Fault Circuit-Interrupter Protection

NOTE

GFCI protection is not required for receptacles supplying items which are incompatible with ground-fault circuit-interrupter devices (such as cooking and refrigeration equipment) »525-18(b)«.

All personnel utilized 125-volt, single-phase, 15- and 20-ampere receptacle outlets must have listed ground-fault circuit-interrupter protection (GFCI), either as an integral part of the attachment plug or located in the power-supply cord, within 12 inches of the attachment plug. Egress lighting must not be connected to the GFCI receptacle's load side terminals »525-18(a)«.

Receptacle outlets, other than those covered in Section 525-18(a) or (b), can have ground-fault circuit-interrupter protection for personnel. If GFCI protection is not provided, designated on-site personnel must continuously enforced a written procedure which ensures the safety of cord set and receptacle equipment grounding conductors, as described in Section 305-6(b)(2) »525-18(c)«.

General Requirements

Permanent structures must comply with Articles 518 and 520 »525-3(a)«.

Article 250 requirements apply to all system and equipment grounding »525-20«.

Portable distribution or termination boxes must comply with Section 525-15(a) through (d).

The following common-source connected equipment must be bonded: (1) metal raceways and metal sheathed cable; (2) metal enclosures of electric equipment; and (3) metal frames and metal parts of rides, concessions, trailers, trucks, or other equipment that contain or support electrical components »525-21«.

Approved nonconductive mats must cover publically accessible flexible cords (or cable) run on the ground. Arrangement of cables and mats must not present a tripping hazard »525-13(f)«.

Separately derived systems and services must comply with Section 525-10 requirements.

No amusement ride, attraction, tent, or similar structure can support wiring for any other ride or structure, unless specifically designed for that purpose »525-13(e)«.

Flexible cords or cables must be continuous (without splice or tap) between boxes or fittings. Only wet location listed cord connectors can be laid on the ground. Connectors and cable connections must not be placed in publically accessible areas, including traffic paths, unless adequately guarded »525-13(d)«.

All equipment which requires grounding must be grounded by a Section 250-118 recognized equipment grounding conductor (type and size) installed per Article 250. The equipment grounding conductor must be bonded to the system grounded conductor at the service disconnecting means; or, in the case of a separately derived system such as a generator, at the generator or the generator's first disconnecting means. Do not connect the grounded circuit conductor to the equipment grounding conductor on the load side of the service disconnecting means, nor on the load side of a separately derived system disconnecting means »525-22«.

Amusement rides/attractions must be at least 15 feet in any direction away from overhead conductors operating at 600 volts or less, except, of course, for the ride/attraction supply conductors. Amusement rides/attractions must not be situated under, nor within 15 feet horizontally, of conductors operating in excess of 600 volts »525-12(b)«.

Associated electrical equipment and wiring methods of rides, concessions, or similar units must be mechanically protected where subject to physical damage »525-6«.

Conductors must have a vertical clearance to ground per Section 225-18. These clearances apply only to exterior wiring of tents and concessions (not within) »525-12(a)«.

Tent and concession interior electrical wiring for temporary lighting, must be securely installed, and mechanically protected where subject to physical damage. Protect all temporary general illumination lamps from accidental breakage by a suitable fixture or lampholder with a guard »525-13(g)«.

Article 525 covers the installation of portable wiring and equipment for carnivals, circuses, exhibitions, fairs, traveling attractions, and similar functions, including all structure related wiring »525-1«.

Each ride and concession must have a visible fused disconnect switch or circuit breaker within 6 feet of the operator's station. The disconnecting means must be readily accessible to the operator, even during ride operation. If accessible to unqualified persons, the switch or circuit breaker enclosure must be lockable. A shunt trip device, that opens the fused disconnect device or circuit breaker when a switch located in the ride operator's console is closed, is an acceptable method of opening the circuit »525-30«.

PLACES OF ASSEMBLY

■ Comprehensive Provisions

Places of assembly include, but are not limited to: armories, assembly halls, auditoriums (whether or not a part of business/mercantile establishments, schools, or other occupancies), bowling lanes, church chapels, club rooms, courtrooms, dance halls, dining facilities, exhibition halls, gymnasiums, mortuary chapels, multipurpose rooms, museums, transportation access waiting areas, pool rooms, restaurants, skating rinks, etc. »518-2(a)«.

Assemblage of fewer than 100 persons in any room (or space) within a building of other occupancy (where said assembly is incidental to the other occupancy), is part of the other occupancy by classification and is subject to the provisions applicable thereto »518-2(b)«.

Electrical nonmetallic tubing and rigid nonmetallic conduit can be installed within restaurants, hotel/motel conference and meeting rooms, dining facilities and church chapels where:
(1) The electrical nonmetallic tubing or rigid nonmetallic conduit is concealed within walls, floors, and ceilings where the enclosing structure provides a thermal material barrier having at least a 15-minute finish rating as identified in fire-rated assembly listings.
(2) Such tubing or conduit is installed above suspended ceilings where the ceilings act as a thermal material barrier having at least a 15-minute finish rating according to fire-rated assembly listings »518-4(c)«.

Designed for 100 (or more) Persons

Non-Fire-Rated Construction / Fire-Rated Construction / Corridor

Nonmetallic-sheathed cable, Type AC cable, electrical nonmetallic tubing, and rigid nonmetallic conduit can be installed in those buildings (or portions thereof) for which the applicable building code does not require fire-rated construction »518-4(b)«.

Article 518 covers all buildings, portions of buildings, or structures serving as a place of assembly for more than 99 persons »518-1«.

The wiring of any such building/area containing a projection booth, stage platform, or area for the presentation of theatrical/musical productions (fixed or portable) must comply with Article 520. These requirements also apply to associated audience seating, and all equipment used in the production, including portable equipment and associated wiring which is not connected to permanent wiring »518-2(c)«.

Temporary wiring for display booths within exhibition halls (as in trade shows) must be installed in accordance with Article 305. Flexible cables and cords, approved for hard or extra-hard usage, laid on floors must be protected from contact by the general public. Section 305-6 ground-fault circuit-interrupter requirements do not apply »518-3(b)«.

Only metal raceways, flexible metal raceways, nonmetallic raceways (encased in 2 inches or more of concrete), and

Type MI, MC, or AC cable having an insulated equipment grounding conductor (sized per Table 250-122) can be used as fixed wiring methods »518-4(a)«.

Electrical nonmetallic tubing and rigid nonmetallic conduit are not approved for environmental air spaces in accordance with Section 300-22(c) »518-4(c)«.

Portable switchboards and portable power distribution equipment must be supplied only from listed power outlets of sufficient voltage/ampere rating. These power outlets must be protected by overcurrent devices. Neither the overcurrent devices nor the power outlets can be publically accessible. A method for equipment grounding conductor connection must be provided. The neutral of feeders supplying solid-state, 3-phase, 4-wire dimmer systems qualifies as a current-carrying conductor »518-5«.

MOTION PICTURE (AND TELEVISION) STUDIOS, AND SIMILAR LOCATIONS

■ Comprehensive Requirements

Article 530 requirements apply to motion picture/television studios using either film or electronic cameras (except as provided in Section 520-1) as well as exchanges, factories, laboratories, stages, or a building segment in which film or tape wider than ⅞ inch (22 mm) is exposed, developed, printed, cut, edited, rewound, repaired, or stored »530-1 «.

Stage Set: A specific area set up with temporary scenery and properties planned for a particular motion picture or television production scene »530-2 «.

> ┌─────────────────────────────────┐
> │   N O T E   │
> │ *Type MC cable, Type MI cable, metal* │
> │ *raceways, and all noncurrent-carrying* │
> │ *metal parts of appliances, devices,* │
> │ *and equipment must be grounded* │
> │ *as specified in Article 250. Pendant* │
> │ *and portable lamps, stage lighting,* │
> │ *and stage sound equipment operating* │
> │ *at not over 150 volts dc to ground,* │
> │ *are exempt* »530-20 «. │
> └─────────────────────────────────┘

Television Studio or Motion Picture Stage (Sound Stage): All or part of a building, usually insulated from outside noise and natural light, used for motion picture, television, or commercial production purposes »530-2 «.

See Section 530-2 for additional word and phrase definitions relevant to motion picture and television studios, and similar locations.

Portable stage/studio lighting equipment and portable power distribution equipment can be temporary use outdoors, if (1) supervised by qualified personnel while energized and (2) barriered from the general public »530-6 «.

Stage or set permanent wiring must be Type MC cable, Type MI cable, or approved raceways, unless an exception is met »530-11 «.

The wiring for stage set lighting, and other supply wiring not fixed as to location, must be achieved with listed hard usage flexible cords/cables. Where subject to physical damage, such flexible cords/cables must be of the listed extra-hard usage type. Cable splices or taps are allowed, provided the total connected load does not exceed the cable's maximum ampacity »530-12(a) «.

Stage effects and electrical equipment used as stage properties can be wired with single- or multi-conductor listed flexible cords or cables where protected from physical damage and secured to the scenery by approved cable ties or insulated staples. Splices (or taps) can only be made with listed devices in a circuit protected at no more than 20 amperes »530-12(b) «.

Use externally operable type switches for studio stage set lighting and effects whether on the stage, lot, or on location. Where contactors serve as fuse disconnecting means, and individual externally operable switches (such as tumbler switches) control each contactor, these switches must be located no more than 6 feet from the contactor. Remote-control switches must also be provided. A single externally operable switch, capable of simultaneous disconnection of all contactors on any one location board, is permitted where located within 6 feet of that location board »530-13 «. (*Location Board* is defined in Section 530-2, definitions.)

Portable lamps and work lights must have flexible cords, composition or metal-sheathed porcelain sockets, and substantial guards »530-16 «. Portable lamps used as properties in a motion picture or television stage set, on a studio stage or lot, or on location are, for this purpose, not considered portable lamps »530-16 *Exception* «.

Automatic overcurrent protective devices (circuit breakers or fuses) for motion picture studio stage set lighting and associated cables must comply with Section 530-18(a) through (g).

■ Comprehensive Requirements (*continued*)

The maximum ampacity allowed on a given conductor, cable, or cord size is dictated by the applicable tables of Articles 310 and 400 ≫ 530-18 ≪.

It is acceptable to apply Table 530-19(a) demand factors to that portion of the maximum possible connected load for studio or stage set lighting for all permanently installed feeders between substations and stages and to all permanently installed feeders between the main stage switchboard and stage distribution centers (or location boards) ≫ 530-19(a) ≪.

Portable feeders can have a demand of 50% of maximum possible connected load ≫ 530-19(b) ≪.

Plugs and receptacles must be rated in amperes. The voltage rating of plugs and receptacles must be at least equal to the circuit voltage. All ac circuit plug and receptacle ampere ratings must not be less than the feeder (or branch-circuit) overcurrent device ampere rating. Table 210-21(b)(2) does not apply ≫ 530-21(a) ≪.

Any ac single-pole portable cable connectors used must be listed and of the locking type. Sections 400-10 and 410-56 do not apply to listed single-pole separable connections, nor to single-conductor cable assemblies incorporating listed single-pole separable connectors. Paralleled sets of current-carrying single-pole separable connectors acting as input devices must carry prominent warning labels indicating the presence of internal parallel connections. Single-pole separable connectors must comply with at least one provision of Section 530-22(a)(1) through (3) ≫ 530-22(a) ≪.

■ Cellulose Nitrate Film Storage Vaults

CAUTION 〈〈〈〈〈〈〈〈〈〈〈〈〈〈〈〈〈〈〈〈〈〈〈〈〈〈〈

Unless otherwise permitted in Section 530-51, no receptacles, outlets, electric motors, heaters, portable lights, or other portable electric equipment are allowed inside cellulose nitrate film storage vaults ≫ 530-52 ≪.

Cellulose nitrate film storage vault lamps must be installed in glass-enclosed, gasketed-type rigid fixtures ≫ 530-51 ≪.

Glass Enclosed Fixture with Gasket

Lamps must be controlled outside the vault by a switch having a pole in each ungrounded conductor and provided with a pilot light indicating "on" or "off." This switch must disconnect every ungrounded conductor terminating in any outlet within the vault from any supply source ≫ 530-51 ≪.

Switch with Pilot Light

■ Separately Derived Systems with 60 Volts to Ground

All feeders and branch-circuit conductors installed under this section must be identified as to system at all splices and terminations by color, marking, tagging, or equally effective means. The means of identification must appear at each branch-circuit panelboard and at the building's disconnecting means ≫ 530-71(c) ≪.

Permanently wired utilization equipment and receptacles must be grounded via an equipment grounding conductor, run with the circuit conductors, to an equipment grounding bus (prominently marked "Technical Equipment Ground") in the originating branch-circuit panelboard. The grounding bus must be connected to the grounded conductor on the line side of the separately derived system's disconnecting means. The grounding must be sized as specified in Table 250-122 and be run with the feeder conductors. The technical equipment grounding bus need not be bonded to the panelboard enclosure. Other *Code* approved grounding methods are permitted where the grounding return path impedance does not exceed the impedance of equipment grounding conductors sized and installed according to Article 530, Part G ≫ 530-72(b) ≪.

Standard single-phase panelboards and distribution equipment with higher voltage ratings can be used ≫ 530-71(a) ≪.

A separately derived 120-volt, single-phase, 3-wire system with 60 volts on each of two ungrounded conductors to a grounded neutral conductor can be used to reduce objectionable noise in audio/video production (or other similar sensitive electronic equipment locations) provided use is restricted to electronic equipment only and all requirements in Section 530-71 through 73 are met ≫ 530-70 ≪.

Standard Single-Phase Panelboard

Common-Trip Two-Pole Circuit-Breakers

Common-trip, two-pole circuit breakers that are identified for operation at the system voltage must be provided for both ungrounded conductors in all feeders and branch circuits ≫ 530-71(a) ≪.

TECHNICAL POWER 60/120 Volt 3-Wire Single-Phase

The system must be grounded per Section 250-30 as a separately derived single-phase, 3-wire system ≫ 530-72(a) ≪.

The system must be clearly marked on the panel's face or inside the panel's door ≫ 530-71(a) ≪.

All outlet strips, adapters, receptacle covers, and faceplates must be marked as required by Section 530-73(a)(2):

WARNING — TECHNICAL POWER
Do not connect to lighting equipment.
For electronic equipment use only.
60/120 volt 1ø ac
GFCI protected.

WARNING - TECHNICAL POWER
Do not connect to lighting equipment
For electronic equipment use only
60/120 volt, single-phase ac
GFCI protected

Control Room

6 Feet Maximum

GFCI Protected

A 125-volt, single-phase, 15- or 20-ampere rated receptacle, having one of its current-carrying poles connected to a grounded circuit conductor, must be located within 6 feet of all permanently installed 15- or 20-ampere rated 60/120-volt technical power-system receptacles ≫ 530-73(a)(3) ≪.

All 125-volt receptacles used for 60/120-volt technical power must be uniquely configured and identified for use with this class of system. All 125-volt, single-phase, 15- or 20-ampere rated receptacle outlets and attachment plugs, identified for use with grounded circuit conductors, are acceptable in machine rooms, control rooms, equipment rooms, equipment racks, and similar locations that are restricted to use by qualified personnel ≫ 530-73(a)(4) ≪.

All 15- and 20-ampere receptacle outlets must be ground-fault circuit-interrupter protected ≫ 530-73(a)(1) ≪.

Receptacles used for equipment connection must meet Section 530-73(a)(1) through (4) conditions ≫ 530-73(a) ≪.

While isolated ground receptacles are permitted as described in Section 250-146(d), the branch-circuit equipment grounding conductor must be terminated per Section 530-72(b) ≫ 530-73(b) ≪.

MOTION PICTURE PROJECTORS

■ General Provisions

Article 540 provisions apply to motion picture projection rooms, motion picture projectors, and associated equipment, both professional and nonprofessional types, which use incandescent, carbon arc, xenon, or other light source equipment capable of producing hazardous gases, dust, or radiation ≫540-1≪.

Professional Projector: Either of two types of projectors: (1) using 35- or 70-mm film (a minimum of 1⅜ inch wide with 5.4 edge perforations per inch); or (2) a type using carbon arc, xenon, or other light source equipment producing hazardous gases, dust, or radiation ≫540-2≪.

Any professional-type projectors must be located within a permanently constructed projection room approved for the type of building in which it is located. All projection ports, spotlight ports, viewing ports, and similar openings must be completely closed by use of glass or other approved material. Such rooms do not qualify as hazardous (classified) locations as defined in article 500 ≫540-10≪.

Projectors and enclosures for arc, xenon, and incandescent lamps and rectifiers, transformers, rheostats, and similar equipment must be listed ≫540-20≪.

A minimum 30-inch wide working space must be provided on each side and at the rear of each motion picture projector, floodlight, spotlight, or similar equipment ≫540-12≪. Adjacent pieces of equipment can be served by one such space ≫540-12 Exception≪.

Projectors and other equipment must be marked with the manufacturer's name (or trademark) as well as the voltage and current for which they are designed per Section 110-21 ≫540-21≪.

Nonprofessional Projectors: Any type other than described in Section 540-2 ≫540-3≪. These projectors, including miniature types, employing cellulose acetate (safety) film, may be operated without a projection room ≫540-31≪.

Motor generator sets, transformers, rectifiers, rheostats, and similar equipment for the supply (or control) of current to projection (or spotlight equipment) using nitrate film must be located in a separate room. If inside the projection room, they must be located (or guarded) so that arcs/sparks cannot contact film, and the commutator end(s) of motor generator sets must comply with *one* of the following conditions:

(1) Be of the totally enclosed, enclosed fan-cooled, or enclosed pipe-ventilated type

(2) Be enclosed in separate rooms or housings built of noncombustible construction to exclude flyings or lint, and having proper clean air ventilation

(3) Have the motor-generator brush or sliding-contact end enclosed by solid metal covers

(4) Brushes or sliding contacts must be enclosed in substantial, tight metal housings

(5) The upper half of the brush or sliding-contact end of the motor-generator must be enclosed by a wire screen (or perfo-

rated metal) while the lower half must be enclosed by solid metal covers

(6) Have wire screens or perforated metal placed at the commutator of brush ends.

Extraneous equipment (switches, overcurrent devices, etc.), not normally a part of projection, sound reproduction, flood or other special effect lamps, must not be placed in projection rooms, unless an exception is met ≫540-11(b)≪.

Emergency system control must comply with Article 700 ≫540-11(c)≪.

Conductors supplying outlets for arc and xenon projectors (professional-type) must be of sufficient size for the projector, but in no case smaller than No. 8. Conductors for incandescent-type projectors must conform to the normal wiring standards of Section 210-24 ≫540-13≪.

Insulated conductors having an operating temperature rating no less than 200°C (392°F) must be used on all lamps (or other equipment) where the ambient temperature at the installed conductors will exceed 50°C (122°F) ≫540-14≪.

Use flexible cords approved for hard usage, per Table 400-4, on portable equipment ≫540-15≪.

MANUFACTURED BUILDINGS

Manufactured Building Requirements

Article 545 covers manufactured building and building components requirements »545-1«.

Manufactured Building: Any closed construction building made (or assembled) in manufacturing facilities, whether on or off the building site, for installation, or assembly and installation on the building site, excluding manufactured homes, mobile homes, park trailers, or recreational vehicles »545-3«.

NOTE
Service equipment must be installed in accordance with Section 230-70 »545-7«.

All *Code* approved raceways and cable wiring methods, and such other wiring systems specifically intended and listed for use in manufactured buildings, are permitted with both listed fittings and fittings listed/identified for manufactured buildings »545-4(a)«.

Wherever other *Code* article requirements differ from Article 545, the requirements of Article 545 apply »545-2«.

Fittings and connectors to be concealed at the time of on-site assembly (where tested, identified, and listed to applicable standards) are permitted for on-site module/component interconnection. Such fittings and connectors must equal the employed wiring method in insulation, temperature rise, and fault-current withstand. They must also be able to endure the vibration and minor relative motions occurring in the manufactured building components »545-13«.

In closed construction, cables can be secured only at cabinets, boxes, or fittings where using No. 10 or smaller conductors and protection against physical damage is provided »545-4(b)«.

Service-entrance conductors must meet the requirements of Article 230. Service-entrance conductor routing provisions must be made from the service equipment to the service drop or service lateral's point of attachment »545-5«.

Install service-entrance conductors after the building is site-erected, unless the point of attachment location is known prior to manufacture »545-6«.

Exposed conductors and equipment must be protected during manufacturing, packaging, transit, and setup at the building site »545-8«.

Boxes of dimensions other than those required in Table 370-16(a) can be installed where tested, identified, and listed to applicable standards »545-9(a)«.

Any box no larger than 100 cubic inches, intended for closed-construction mounting, must be affixed with anchors (or clamps) to provide a rigid and secure installation »545-9(b)«.

A receptacle (or switch) with integral enclosure and mounting means can be installed where such is tested, identified, and listed to applicable standards »545-10«.

Prewired panels and building components must provide for the bonding, or bonding and grounding, of all exposed metals likely to be energized, per Article 250, Parts E, F, and G »545-11«.

Provisions must be made to route a grounding electrode conductor from the service equipment to the point of attachment to the grounding electrode »545-12«.

Closed Construction: Any building, building component, assembly, or system manufactured in such a manner that concealed parts of manufacture processes cannot be inspected prior to building site installation without disassembly, damage, or destruction »545-3».

Building Component: Any subsystem, subassembly, or other system designed for use in, integral with, or as part of a structure. A building component can be structural, electrical, mechanical, plumbing, fire protection, and other health and safety systems »545-3«.

Building System: Plans, specifications, and documentation for a manufactured building system, or for a type of system building component, such as structural, electrical, mechanical, plumbing, fire protection, and other systems affecting health and safety, including variations thereof specifically permitted by regulation, provided the variations are submitted as part of the building system or as an amendment thereto »545-3«.

AGRICULTURAL BUILDINGS

■ Agricultural Building Provisions

This article's provisions apply to agricultural buildings, part of buildings, or adjacent areas of similar nature as specified in (a) and (b).

(a) Agricultural buildings where excessive dust, and dust in combination with water, may accumulate. Included are all areas of poultry, livestock, and fish confinement systems, where litter dust or feed dust (including mineral feed particles) may collect.

(b) Agricultural buildings with a corrosive atmosphere. Such buildings include areas where:
   (1) poultry and animal excrement can cause corrosive vapors
   (2) corrosive particles may mix with water
   (3) periodic washing for cleaning and sanitizing with water and cleansing agents creates a damp, even wet, environment.
   (4) similar conditions exist ≫ Section 547-1 ≪.

CAUTION ≪≪≪≪≪≪≪≪≪≪≪≪≪≪
Protect all electrical wiring and equipment subject to physical damage ≫ 547-4(e) ≪.

All boxes must be secured within 8 inches of each cabinet, box, or fitting. The ¼-inch airspace required for nonmetallic boxes, fittings, conduit, and cables in Section 300-6(c) does not apply in buildings covered by this article ≫ 547-4(b) ≪.

Switches, circuit breakers, controllers, and fuses (including pushbuttons, relays, and similar devices), must have enclosures as specified in (a) and (b).

(a) Buildings described in Section 547-1(a) (*Excessive Dust and Dust with Water*) must utilize dustproof and weatherproof enclosures.

(b) Buildings described in Section 547-1(b) (*Corrosive Atmosphere*) must utilize enclosures suitable for the encountered conditions ≫ 547-5 ≪.

The wiring method(s) employed must be types UF, NMC, copper SE cables, rigid nonmetallic conduit, liquidtight flexible nonmetallic conduit, or other location-suitable cables/raceways having approved fittings. Article 320 and Article 502 wiring methods are permitted for areas described in Section 547-1(a) ≫ 547-4(a) ≪.

MILLER FARMS

Lighting fixtures must comply with all of the following:

(a) By installation, the entrance of dust, foreign matter, moisture, and corrosive material must be minimized.

(b) Any fixture subject to physical damage must have a suitably protective guard.

(c) A fixture exposed to moisture, whether from condensation, cleansing water, or solution must be watertight ≫ 547-7 ≪.

A **Bonding and Equipotential Plane** is a livestock accessible area where a wire mesh (or other conductive elements) are: (1) embedded in concrete; (2) bonded to all metal structures and fixed nonelectrical metal equipment that may become energized; and (3) connected to the electrical grounding system, thus preventing a voltage difference from developing within the plane. Poultry is not included as livestock for Section 547-9 ≫ 547-9(a) ≪.

The concrete floor of livestock confinement areas must contain conductive elements bonded to the structure's grounding electrode system thereby providing an equipotential plane that may have voltage gradient entrance/exit ramps being traversed daily by livestock. The bonding conductor must be copper (insulated, covered or bare) and no smaller than No. 8. Use pressure connectors or clamps of brass, copper, copper alloy or an equally substantial approved means, to bond wire mesh or conductive elements ≫ 547-9(b) ≪. (See FPN No. 1 through 3.)

Where grounding is required for non-current-carrying metal parts of equipment, raceways, and other enclosures, a copper equipment grounding conductor must be installed between the equipment and the building disconnecting means. If underground, the equipment grounding conductor must be insulated or covered ≫ 547-4(f) ≪.

NOTE

All 125-volt, single-phase, 15- and 20-ampere general purpose receptacles in equipotential plane areas must have personnel ground-fault circuit-interrupter protection ≫ 547-9(c) ≪.

Motors and other rotating electrical machinery must either be totally enclosed or designed to minimize the entrance of dust, moisture, or corrosive particles ≫ 547-6 ≪.

Slatted floors, supported by structures that are part of an equipotential plane, do not require bonding ≫ 547-9(b) Exception No. 2 ≪.

An equipotential plane is not required if there is no electric service to the building, nor any metal equipment accessible to livestock that is

likely to become energized ≫ 547-9(b) Exception No. 1 ≪.

The disconnecting means and the grounding of services and feeders for agricultural building(s) supplied from a distribution point must comply with Section 547-8(a), (b), or (c) ≫ 547-8 ≪. A **Distribution Point** is a centrally located electrical supply structure normally supplying services or feeders to agricultural buildings, as well as to the associated farm dwelling ≫ 547-8 ≪.

MOBILE HOMES, MANUFACTURED HOMES, AND MOBILE-HOME PARKS

■ General Requirements

Article 550 provisions cover the electrical conductors and equipment in/on mobile homes, mobile home electrical supply conductors, and the installation of electrical wiring, fixtures, equipment, and appurtenances related to electrical installations within a mobile-home park up to the mobile-home service-entrance conductors or, if none, the service equipment »550-1 «.

For *Code* purposes (unless otherwise indicated), the term **mobile home** includes manufactured homes »550-2 Manufactured Home «.

This article's provisions apply to mobile homes designed to connect to a wiring system rated 120/240 volts, nominal, 3-wire ac, with grounded neutral »540-4(c) «.

N O T E
Mobile homes, not installed in mobile-home parks, must comply with the provisions of this article »540-4(b) «.

Outdoor or under-chassis line-voltage (120 volts, nominal, or higher) wiring, exposed to moisture/physical damage, must be protected by rigid or intermediated metal conduit. Conductors must be suitable for wet locations »550-10(h)(1) «. Electrical metallic tubing or rigid nonmetallic conduit is permitted if closely routed against frames and equipment enclosures »550-10(h)(1) *Exception* «.

CAUTION «««««««««««««««
All electrical materials, devices, appliances, fittings, and other equipment must be (1) listed or labeled by a qualified testing agency and (2) connected in an approved manner when installed »540-4(d) «.

Mobile Home: A factory-assembled structure transportable in one or more sections, built on a permanent chassis, designed for use as a dwelling having no permanent foundation where connected to the required utilities, and including the plumbing, heating, air-conditioning, and electric systems contained therein »550-2 «.

Manufactured Home: A factory-assembled structure bearing a label identifying it as a manufactured home, transportable in one or more sections, built on a permanent chassis and designed for dwelling use (with or without a permanent foundation) where connected to the required utilities, inclusive of plumbing, heating, air-conditioning, and electrical systems contained therein »550-2 «.

Mobile home receptacle outlets must comply with Section 550-8.

At least one receptacle outlet must be installed outdoors. A receptacle outlet located in an outside accessible compartment qualifies as an outdoor receptacle. These receptacle outlets must be protected as required in Section 550-8(b) »550-8(e) «.

If a pipe heating cable outlet is installed, the outlet must be:

(1) Located within 2 feet of the cold water inlet.

(2) Connected to an interior branch circuit, other than a small appliance branch circuit. A bathroom receptacle circuit can be utilized for this purpose.

(3) On a circuit where all outlets are on the load side of the ground-fault circuit-interrupter protection for personnel.

(4) Mounted on the mobile home's underside, although this is not considered to be the outdoor receptacle outlet required in Section 550-8(e) »550-8(g) «.

Fixtures and appliances must meet Section 550-9 requirements.

Wiring method and material provisions are covered in Section 550-10(a) through (k).

Section 550-13 provides procedures for calculating the supply cord and distribution panelboard load for each mobile home feeder assembly.

Use only approved and listed fixed-type wiring methods to join portions of a circuit that must be electrically joined, and which are located in adjacent mobile-home sections after installation on its support foundation. The circuit's junction must be accessible for disassembly when preparing the home for relocation »550-14 «.

Outdoor outlets, fixtures, air-cooling equipment, etc., are covered in Section 550-15.

■ General Requirements (*continued*)

A mobile home not intended as a dwelling unit is not required to meet this article's provisions pertaining to the number or capacity of circuits required. Included are those equipped for sleeping purposes only, contractor's on-site offices, construction job dormitories, mobile studio dressing rooms, banks, clinics, mobile stores, or intended for the display or demonstration of merchandise or machinery. Such non-dwelling units must, however, meet all other applicable requirements of this article if provided with an electrical installation to be energized from a 120-volt or 120/240-volt ac power supply system. If design or available power supply systems dictate a different voltage, make adjustment in accordance with the appropriate articles/sections for the voltage used ≫ 550-4(a) ≪ .

■ Power Supply

The attachment plug cap must not only be a 3-pole, 4-wire, grounding type, rated 50 amperes, 125/250 volts with a configuration as shown in Figure 550-5(c) but also must be intended for use with a 50-ampere, 125/250-volt receptacle configuration as shown in Figure 550-5(c). It must be listed, individually or as part of a power-supply cord assembly, for such purpose; and molded to (or installed on) the flexible cord at the point where the cord enters the attachment plug cap ≫ 550-5(c) ≪ .

The cord must be a four-conductors listed type. One conductor must be identified by a continuous green color, or a continuous green color with one or more yellow stripes, for use as the grounding conductor ≫ 550-5(b) ≪ .

Cords with adapters and pigtail ends, extension cords, and similar items must not be attached to, or shipped with, a mobile home ≫ 550-5(b) ≪ .

A cord passing through walls or floors, must be protected by means of conduits and bushings (or equivalent). The cord can be installed within the walls, provided a continuous raceway (maximum 1¼ inch in size) is installed from the branch-circuit panelboard to the underside of the mobile home floor ≫ 550-5(g) ≪ .

The power-supply cord's attachment plug cap and any connector cord assembly or receptacle must be permanently protected against corrosion and mechanical damage, if such devices are externally located while the mobile home is in transit ≫ 550-5(h) ≪ .

If the calculated load exceeds 50 amperes, or if a permanent feeder is used, the supply must be by means of:

(1) One mast weatherhead installation, installed per Article 230, containing four continuous, insulated, color-coded feeder conductors; one of which must be an equipment grounding conductor; or

(2) A metal raceway or rigid nonmetallic conduit from the internal disconnecting means to the mobile home's underside, with provisions for attachment to the raceway on the mobile home's underside via a suitable junction box (or fitting) (with or without conductors as in Section 550-5[i][1]) ≫ 550-5(i) ≪ .

In a right-angle cap configuration the grounding member must be the farthest from the cord ≫ 550-5(c) ≪ .

The mobile home power supply must be a feeder assembly consisting of a single listed 50-ampere mobile home power-supply cord having an integrally molded (or securely attached) plug cap, or a permanently installed feeder ≫ 550-5(a) ≪ .

A suitable clamp (or equivalent) must be provided at the distribution panelboard knockout to afford cord strain relief and effectively prevent strain transmission to the terminals when the power-supply cord is handled as intended ≫ 550-5(b) ≪ .

A mobile home power-supply cord, if present, must be permanently attached to the distribution panelboard either directly or via a junction box also permanently connected thereto, with the free end terminating in an attachment plug cap ≫ 550-5(b) ≪ .

The power-supply cord must bear the marking: "FOR USE WITH MOBILE HOMES — 40 AMPERES" or "FOR USE WITH MOBILE HOMES — 50 AMPERES" ≫ 550-5(e) ≪ .

From the end of the power-supply cord (including bared leads), to the face of the attachment plug cap, the cord must not be less than 21 feet nor more than 36½ feet in length. From the face of the attachment plug cap to the mobile home entry point, the cord must be at least 20 feet long ≫ 550-5(d) ≪ .

> **NOTE**
> *The feeder assembly must enter the mobile home through an exterior wall, the floor, or roof* ≫ 550-5(f) ≪ .

Distribution Panelboards and Branch Circuits

Each mobile home must have a single disconnecting means consisting of a circuit breaker, or a switch and fuses, and its accessories installed in a readily accessible location near the supply cord (or conductors) entry point into the mobile home. The main circuit breakers (or fuses) must be plainly marked "Main." This equipment must contain a solderless-type grounding connector (or bar), with sufficient terminals for all grounding conductors. The neutral bar termination of the grounded circuit conductors must be insulated in accordance with Section 550-11(a) ≫ 550-6(a) ≪.

While the distribution panelboard must be in an accessible location, it cannot be located in a bathroom or clothes closet. A clear working space at least 30 inches wide and 30 inches in front of the panelboard must be provided. This space must extend from the floor to the top of the distribution panelboard ≫ 550-6(a) ≪.

The bottom of the distribution equipment, whether circuit breaker or fused type, must be at least 24 inches above the mobile home's floor level ≫ 550-6(a) ≪.

MAIN

24 Inches Minimum

The distribution equipment must be suitably rated for the connected load ≫ 550-6(a) ≪.

Where circuit breakers provide branch-circuit protection, 240-volt circuits must be protected by a 2-pole common or companion trip, or handle-tied paired circuit breakers ≫ 550-6(c) ≪.

Determine the required number of branch circuits in accordance with Section 550-7(a) through (c) ≫ 550-7 ≪.

The branch-circuit equipment can be combined with the disconnecting means as a single assembly. Such a combination can be designated as a distribution panelboard. If a fused distribution panelboard is used, the maximum fuse size for the mains must be plainly marked with ¼-inch (or taller) lettering visible during fuse replacements. If plug fuses and fuseholders are used, they must be tamper-resistant Type S, enclosed in dead-front fuse panelboards. Electrical distribution panelboards containing circuit breakers must also be dead-front type ≫ 550-6 ≪.

A distribution panelboard must be rated no less than 50 amperes and employ a 2-pole circuit breaker rated 40 amperes for a 40-ampere supply cord, or 50 amperes for a 50-ampere supply cord. A distribution panelboard employing a disconnect switch and fuses must be rated 60 amperes and employ a single 2-pole, 60-ampere fuseholder with 40- or 50-ampere main fuses for 40- or 50-ampere supply cords, respectively. The fuse size must be plainly marked on the outside of the distribution panelboard ≫ 550-6(a) ≪.

Each mobile home must have branch-circuit distribution equipment which includes circuit breaker or fuse overcurrent protection for each branch circuit. The branch-circuit overcurrent devices must be rated as follows: (1) not more than the circuit conductors; and (2) not more than 150% of a single appliance rated 13.3 amperes (or more) supplied by an individual branch circuit; but (4) not more than the overcurrent protection size and type marked on the air conditioner (or other motor-operated appliance). A 15-ampere multiple receptacle can be connected to a 20-ampere laundry circuit ≫ 550-6(b) ≪.

A metal nameplate on the outside adjacent to the feeder assembly entrance must read:

THIS CONNECTION FOR 120/240-VOLT, 3-POLE, 4-WIRE, 60 HERTZ, _____ AMPERE SUPPLY

Fill in the blank with the correct ampere rating ≫ 550-6(d) ≪.

Multiply 3 volt-amperes (per square foot) by the outside dimensions of the mobile home (coupler excluded) and divide by 120 volts to determine the number of 15- or 20-ampere lighting circuits. Lighting circuits can serve built-in gas ovens with electric service only for lights, clocks or timers; or listed cord-connected garbage disposal units ≫ 550-7(a) ≪.

Small appliance branch circuits must be installed according to Section 210-52(b) ≫ 550-7(b) ≪.

There must be one or more adequately rated circuits in accordance with the following:

(1) Ampere rating of fixed appliances not over 50% of circuit rating if lighting outlets (receptacles, other than kitchen, dining area, and laundry, considered as lighting outlets) exist on the same circuit.

(2) For fixed appliances on a circuit without lighting outlets, the sum of rated amperes must not exceed the branch-circuit rating. Motor loads or other continuous duty loads must not exceed 80% of the branch-circuit rating.

(3) The rating of a single cord- and plug-connected appliance on a circuit having no other outlets must not exceed 80% of the circuit rating.

(4) The rating of a range branch circuit must be based on the range demand as specified in Section 550-13(b)(5) ≫ 550-7(c) ≪.

■ Grounding

Grounding terminals must be of the solderless type, listed as pressure-terminal connectors, and recognized for the wire size employed »550-11(c)(2)«.

The supply cord or permanent feeder's green-colored insulated grounding wire must be connected to the grounding bus in the distribution panelboard or disconnecting means »550-11(b)(1)«.

Grounding Bus

The bonding conductor must be solid or stranded, insulated or bare, but must be No. 8 copper minimum, or the equivalent. The bonding conductor routing must prevent physical damage exposure »550-11(c)(2)«.

To Chassis

To Service Equipment

A bonding conductor must be connected between the distribution panelboard and an accessible terminal on the chassis »550-11(c)(1)«.

Remove and discard bonding screws, straps, or buses in the distribution panelboard or in appliances »550-11(a)(1)«.

The grounded circuit conductor (neutral) must be insulated from the grounding conductors, from equipment enclosures, and other grounded parts. The grounded (neutral) circuit terminals in distribution panelboards, ranges, clothes dryers, counter-mounted cooking units, and wall-mounted ovens must be insulated from the equipment enclosure »550-11(a)(1)«.

Main Bonding Jumper **Not** Installed

Green-colored Insulated Grounding Conductor

In a mobile home, grounding of metal parts (both electrical and non-electrical) must be accomplished by connection to a distribution panelboard's grounding bus. The grounding bus must be grounded through the green colored insulated conductor in the supply cord (or the feeder wiring) to the service ground in the service-entrance equipment, located adjacent to the mobile home »550-11«.

NOTE

Connections of ranges and clothes dryers with 120/240-volt, 3-wire ratings must be made with (a) 4-conductor cord and 3-pole, 4-wire, grounding-type plugs or (b) by Type AC cable, Type MC cable, or (c) conductors enclosed in flexible metal conduit »550- 11(a)(2)«.

CAUTION «««««««««««««

Neither the mobile home's frame nor any appliance frame shall be connected to the grounded circuit conductor (neutral) in the mobile home »550-11«.

Cord-connected appliances (such as washing machines, clothes dryers, refrigerators, and the electrical system of gas ranges, etc.) must be grounded via a cord having a grounding conductor and grounding-type attachment plug »550-11(b)(3)«.

Metallic pipes (gas, water, and waste) as well as metallic air-circulating ducts are considered bonded if connected to the chassis terminal by clamps, solderless connectors, or by appropriate grounding-type straps »550-11(c)(3)«.

Any metallic exterior covering is considered bonded if (1) the metal panels overlap one another and are securely attached to the wood or metal frame parts by metallic fasteners and (2) the lower panel of the metallic exterior covering is secured at opposite ends by metallic fasteners at a chassis cross member by two metal straps per mobile home unit (or section). The bonding strap must be at least 4 inches wide and of the same material as the skin, or of a material of equal or better electrical conductivity. The straps must be fastened with paint-penetrating fittings such as screws and star-washers or equally appropriate substitute »550-11(c)(4)«.

All potentially energizable exposed noncurrent-carrying metal parts must be effectively bonded to either the grounding terminal or the distribution panelboard's enclosure »550-11(c)(1)«.

In the electrical system, all exposed metal parts, enclosures, frames, lamp fixtures canopies, etc., must be effectively bonded to the grounding terminal or enclosure of the distribution panelboard »550-11(b)(2)«.

■ Services and Feeders

Underground Service

Mobile home service equipment must be rated no less than 100 amperes at 120/240 volts. A mobile home feeder assembly must be connected by a permanent wiring method. Power outlets used as mobile-home service equipment can also contain receptacles rated up to 50 amperes with appropriate overcurrent protection. Fifty-ampere receptacles must conform to Figure 550-5(c)'s configuration 》550-23(c)《.

Additional receptacles are allowed for connection of electrical equipment located outside the mobile home. All such 125-volt, single-phase, 15- and 20-ampere receptacles must be protected by a listed ground-fault circuit-interrupter 》550-23(e)《.

Any mobile home service equipment utilizing a 125/250-volt receptacle must be marked as follows.

TURN DISCONNECTING SWITCH
OR CIRCUIT BREAKER OFF BEFORE INSERTING
OR REMOVING PLUG. PLUG MUST
BE FULLY INSERTED OR REMOVED.

The marking must be located adjacent to the service equipment's receptacle outlet 》550-23(g)《.

The mobile-home park secondary electrical distribution system supplying mobile home lots must be single-phase, 120/240 volts, nominal. For Part C purposes, where the park service exceeds 240 volts, nominal, transformers and secondary distribution panelboards must be treated as services 》551-21《.

The **manufactured home** service equipment can be installed in, or on, a manufactured home, provided all of the following conditions are met:

(1) The manufactured home is secured to a permanent foundation which meets applicable building codes.

(2) The service equipment is installed in a manner acceptable to the authority having jurisdiction.

(3) The installation of the service equipment complies with Article 230.

(4) Means are provided for connecting a grounding electrode conductor to the service equipment, and routing it outside the structure 》550-23(b)《.

Overhead Service

2 Feet Minimum

The **mobile home** service equipment must be located adjacent to, not mounted in or on, the mobile home. It must be located in sight from and within 30 feet from the exterior wall of the mobile home it serves. The service equipment can be located elsewhere on the premises, provided a disconnecting means suitable for service equipment is located as stated above for service equipment. Grounding at the disconnecting means must be compliant with Section 250-32 》550-23(a)《.

Mobile home service equipment must provide a means for connecting (by a fixed wiring method) an accessory building, structure, or additional electrical equipment located outside the mobile home 》550-23(d)《.

Outdoor mobile-home disconnecting means must be installed so that the bottom of the disconnecting means enclosure is at least 2 feet above finished grade (or working platform). Installation must ensure that the center of the operating handle's grip, in the highest position, is no more than 6 feet 7 inches above the finished grade (or working platform) 》550-23(f)《.

CAUTION ≪≪≪≪≪≪≪≪≪≪≪≪≪≪≪≪≪≪≪≪

Mobile home lot feeder circuit conductors must have adequate capacity for the loads supplied and must be rated no less than 100 amperes at 120/240 volts 》550-24(b)《.

Park electrical wiring systems must be calculated (at 120/240 volts) on the larger of (1) 16,000 volt-amperes for each mobile home lot, or (2) the load calculated per Section 550-13 for the largest typical mobile home that each lot accepts. The feeder or service load can be computed in accordance with Table 550-22. No demand factor is allowed for any other load, except as provided in this *Code*. Mobile home service and feeder conductors, in compliance with Section 310-15(b)(6), are permitted 》550-22《.

Mobile home feeder conductors must consist of either a listed cord, factory-installed in accordance with Section 550-5(b), or a permanently installed feeder consisting of four, insulated, color-coded conductors, each identified by the factory or field marking, in compliance with Section 310-12. Stripping the insulation is not an acceptable method of identifying equipment grounding conductors 》550-24(a)《.

If a mobile home feeder is installed between service equipment and a mobile home disconnecting means, as covered in Section 550-23(a), the equipment grounding conductor can be omitted where the grounded circuit conductor is grounded at the disconnecting means as required in Section 250-32(b) 》550-24(a) *Exception*《.

─────────────────────────────

NOTE

For the purpose of this Code *and unless otherwise indicated, the term* mobile home *includes manufactured homes 》550-2 Manufactured Home《.*

─────────────────────────────

RECREATIONAL VEHICLES AND RECREATIONAL-VEHICLE PARKS

■ Recreational Vehicles

Article 551 provisions cover the electrical conductors and equipment installed within (or on) recreational vehicles, the conductors that connect recreational vehicles to an electrical supply, and the installation of electrical related equipment and devices within a recreational-vehicle park »551-1 «.

Recreational vehicle electrical equipment and material, intended for connection to a wiring system rated 120 volts, nominal, 2-wire with ground, or a wiring system rated 120/240 volts, nominal, 3-wire with ground, must be listed and installed in accordance with the requirements of Parts A, C, D, E, and F of Article 551 »551-40(a) «.

Recreational Vehicle: A vehicular-type unit primarily designed as temporary living quarters for recreational, camping, or travel use, which either has its own motive power or is mounted on (or drawn) by another vehicle. Basically this includes: travel trailers, camping trailers, truck campers, and motor homes »551-2 «. (See Section 551-2 for other word and phrase definitions relating to recreational vehicles and recreational vehicle parks.)

NOTE
A recreational vehicle, not used for the purpose defined in Section 551-2, is not required to comply with Part A provisions relating to the number (or capacity) of circuits. It must, however, meet all other applicable requirements of this article if the recreational vehicle has an electrical installation which can be energized from a 120- or 120/240-volt, nominal, ac power-supply system »551-4(a) «.

Every appliance must be accessible for inspection, service, repair, and replacement without removal of permanent construction. Means must be provided to securely fasten appliances in place during travel »551-57 «.

Motor Home: A vehicular unit designed to provide temporary living quarters for recreational, camping, or travel use built on, or permanently attached to, a self-propelled motor vehicle chassis, or on a chassis cab/van that is an integral part of the total vehicle »551-2 «.

Recreational Vehicle Park: A plot of land on which multiple recreational vehicle sites are located, established, or maintained for general public recreational vehicle occupancy as temporary living quarters for recreation or vacation purposes »551-2 «.

Recreational Vehicle Site: A section of ground within a recreational vehicle park intended for the temporary accommodation of either a recreational vehicle, tent, or other individual camping unit »551-2 «.

Recreational Vehicle Stand: The area of a recreational vehicle site intended for the placement of a recreational vehicle »551-2 «.

Low-voltage circuits furnished and installed by the recreational vehicle manufacturer, other than automotive vehicle circuits or extensions thereof, are subject to this *Code*. Circuits supplying lights subject to federal or state regulations must comply with applicable government regulations in addition to this *Code* »551-10 «.

Recreational vehicle low-voltage systems are covered in Article 551, Part B (Section 551- 10).

Voltage converters (120-volt alternating current to low-voltage direct current) must comply with Section 551-20(b). Voltage converters must be bonded per Section 551-20(c).

Generator installations must comply with Section 551-30.

Section 551-41 covers required receptacle outlet provisions.

Recreational vehicle branch circuit requirements are located in Section 551-42.

Branch-circuit overcurrent protection must comply with Section 551-43.

Recreational vehicle main power-supply assemblies must meet Section 551-44 requirements.

Recreational vehicle distribution panelboard provisions are found in Section 551-45.

continued

■ Recreational Vehicles (*continued*)

The power-supply assembly (or assemblies) must be factory supplied/installed and be of a type specified in (1) or (2) »Section 551-46(a)«.

The main power-supply cord must comply with Section 551-46(b) specifications.

Recreational vehicle power-supply attachment plugs must meet Section 551-46(c) requirements.

Each recreational vehicle must be labeled at the electrical entrance per Section 551-46(d) specifications.

Section 551-46(e) covers location requirements for the electrical power-supply entrance.

Recreational vehicle wiring methods must comply with Section 551-47(a) through (r).

Power-supply grounding, distribution panelboard grounding, and insulated neutral conductor requirements are located in Section 551-54.

While Section 551-55(a) through (f) covers interior equipment grounding provisions, bonding of noncurrent-carrying metal parts are found in Section 551-56(a) through (f).

Each recreational vehicle is subject to the tests specified in Section 551-60(a) and (b).

■ Recreational-Vehicle Parks

The recreational-vehicle park secondary electrical distribution system to 50-ampere recreational vehicle sites must be derived from a single-phase, 120/240-volt, 3-wire system. Other recreational vehicle sites with 125-volt, 20- and 30-ampere receptacles may be derived from any grounded distribution system supplying 120-volt single-phase power. The neutral conductors can be reduced in size below the minimum required ungrounded conductor size for 240-volt, line-to-line, permanently connected loads only »551-72«.

Every electrical equipped recreational vehicle site must have at least one 20-ampere, 125-volt receptacle »551-71«.

At least 70% of all electrically supplied recreational vehicle sites must be equipped with a 30-ampere, 125-volt receptacle conforming to Figure 551-46(c). Additional receptacle configurations conforming to Section 551-81 can be included. Dedicated tent sites having a 15- or 20-ampere electrical supply can be excluded when determining 30- or 50-ampere receptacles »551-71«.

Five percent (or more) of all electrically supplied recreational vehicle sites must be equipped with a 50-ampere, 125/250-volt receptacle conforming to the configuration identified in Article 551, Figure 551-46(c). These electrical supplies can include additional receptacles configured per Section 551-81 »551-71«.

> **NOTE**
> *In all areas subject to vehicular movement, open conductors of not over 600 volts, nominal, must have a vertical clearance of at least 18 feet and a minimum 3-foot horizontal clearance. In all other areas, clearances must conform to Sections 225-18 and 225-19* »551-79«.

CAUTION «««««««

All 125-volt, single-phase, 15- and 20-ampere receptacles must have listed ground-fault circuit-interrupter protection for personnel »551-71«.

■ Recreational-Vehicle Parks (*continued*)

Recreational vehicle site supply equipment grounding must comply with Section 551-76(a) through (d).

Where park service exceeds 240 volts, transformers and secondary distribution panelboards must be treated as services »551-73(b)«.

Recreational site feeder-circuit conductors must have adequate ampacity for the loads supplied and must be rated no less than 30 amperes. The grounded and ungrounded conductors must have the same ampacity »551-73(d)«. (Also see this section's FPN pertaining to voltage drop.)

No grounding electrode connection can be made to the neutral conductor on the load side of the service disconnecting means, or to the transformer distribution panelboard »551-76(d)«.

If provided, the recreational vehicle site electrical equipment must be located on the parked vehicle's left (road) side, on a line that is 9 feet, ± 1 foot, from the stand's longitudinal centerline and shall be located at any point along the line from the rear of the stand to 15 feet forward of the rear of the stand »551-77(a)«.

For pull-through sites, the electrical supply equipment can be located at any point along the line from 16 to 32 feet forward of the rear of the stand »551-77(a)«.

A disconnecting switch (or circuit breaker) must be provided in the site supply equipment for disconnecting the recreational vehicle power supply »551-77(b)«.

All site supply equipment must be accessible by an unobstructed entrance or passageway at least 2 feet wide and 6½ feet high »551-77(c)«.

Sufficient space must be constantly available around all electrical equipment for ready and safe operation, in accordance with Section 110-26 »551-77(e)«.

Where the site supply equipment contains a 125/250-volt receptacle, the equipment must be marked as follows: "Turn disconnecting switch or circuit breaker off before inserting or removing plug. Plug must be fully inserted or removed." The marking must be located on the equipment adjacent to the receptacle outlet »551-77(f)«.

Site supply equipment must be located no less than 2 feet, nor more than 6½ feet, above the ground »551-77(d)«.

Electrical service and feeders must be calculated on the basis of no less than 9,600 volt-amperes per site equipped with 50-ampere, 120/240-volt supply facilities; 3,600 volt-amperes per site equipped with both 20-ampere and 30-ampere supply facilities; 2,400 volt-amperes per site equipped with only 20-ampere supply facilities; and 600 volt-amperes per dedicated tent site. Table 551-73 demand factors are the minimum allowable demand factors that can be used in calculating load for service and feeders. If the recreational vehicle site's electrical supply has more than one receptacle, the calculated load is computed for the highest rated receptacle only »551-73(a)«. (Also see Section 551-73(c) and FPN.)

Underground service, feeder, branch-circuit, and recreational vehicle site feeder-circuit conductors must be installed according to Section 551-80(a) and (b).

FLOATING BUILDINGS

Floating Building Provisions

NOTE

All electrical system enclosures and exposed metal parts must be bonded to the grounding bus »553-10(a)«.

Article 553 covers wiring, services, feeders, and grounding for floating buildings »553-1«.

Floating Building: A building unit as defined in Article 100 that floats on water, is moored in a permanent location, and has a premises wiring system served through connection by permanent wiring to a remote electrical supply system »553-2«.

The service equipment for a floating building must be located adjacent to, but not in (or on), the building »553-4«.

One set of service conductors can serve multiple service equipment sets »553-5«.

Each floating building must be supplied by a single set of feeder conductors from its service equipment »553-6«. Where the floating building has multiple occupancy, each occupant can be supplied by a single set of feeder conductors extended from the occupant's service equipment to the corresponding panelboard »553-6 *Exception*«.

Wiring system flexibility must be maintained between the floating buildings and the supply conductors. All wiring must be installed so that water surface motion, and changes in water level, do not result in unsafe conditions »553-7(a)«.

Liquidtight flexible metal conduit, or liquidtight flexible non-metallic conduit with approved fittings, are allowed for feeders and wherever flexible connections are required. Extra-hard usage portable power cable, listed for both wet locations and sunlight resistance, are permitted for a feeder to a floating building where flexibility is required »553-7(b)«.

In a floating building, electrical and nonelectrical parts grounding must be provided via connection to a grounding bus in the building panelboard. The grounding bus must be grounded through a green-colored insulated equipment grounding conductor run with the feeder conductors and connected to a service equipment grounding terminal. The

CAUTION ««««««««««««««««««««««
All metal parts in contact with the water, all metal piping, and all noncurrent-carrying metal parts which may become energized, must be bonded to the panelboard grounding bus »553-11«.

grounding terminal in the service equipment must be grounded by connection through an insulated grounding electrode conductor to an on-shore grounding electrode »553-8«.

The grounded circuit conductor (neutral) must be an insulated conductor identified in conformance with Section 200-6. The neutral conductor must be connected to the equipment grounding terminal in the service equipment, and, except for that connection, it must be insulated from the equipment grounding conductors, equipment enclosures, and all other grounded parts. The neutral circuit terminals in the panelboard and in ranges, clothes dryers, counter-mounted cooking units, and the like, must be insulated from the enclosures »553-9«.

Where cord-connected appliances require grounding, it must be accomplished by means of an equipment grounding conductor in the cord and a grounding-type attachment plug »553-10(b)«.

Wiring for floating buildings must comply with the applicable provisions of other articles of this *Code*, except as modified by this article »553-3«.

MARINAS AND BOATYARDS

■ Marina and Boatyard Requirements

> ### CAUTION ‹‹‹‹‹‹‹‹‹‹‹‹‹‹‹‹‹‹‹‹
>
> *Each single receptacle supplying shore power to boats must be supplied from a power outlet or panelboard by an individual branch circuit of the voltage class and rating corresponding to the receptacle's rating* »555-5‹‹.

Marina and boatyard wiring and equipment must comply with this article and also with the applicable provisions of other *Code* articles »555-2‹‹.

A wet location approved wiring method must be used. Extra-hard usage portable power cable, listed for both wet locations and sunlight resistance, can be used for a feeder requiring flexibility »555-7‹‹.

The following items must be connected to an equipment grounding conductor run with the circuit conductors inside a raceway or cable:

(1) Boxes, cabinets, and all other metal enclosures

(2) Metal frames of utilization equipment

(3) Grounding terminals of grounding-type receptacles »555-8(a)‹‹.

The equipment grounding conductor must be an insulated copper conductor with a continuous outer finish, either green or green with yellow stripe(s) »555-8(b)‹‹. (See exception pertaining to Type MI cable.)

The insulated copper equipment grounding conductor must be sized per Section 250-122, but in no case smaller than No. 12 »555-8(c)‹‹.

Branch-circuit insulated equipment grounding conductors must terminate either at a grounding terminal in a remote panelboard, or at the main service equipment grounding terminal »555-8(d)‹‹.

A feeder suppling a remote panelboard must have an insulated equipment grounding conductor extending from a service equipment grounding terminal to a grounding terminal in the remote panelboard »555-8(e)‹‹.

Wiring above and below navigable water is subject to approval by the authority having jurisdiction »555-9‹‹.

Floating dock or marina service equipment must be located adjacent to, rather than on or in, the floating structure.

The load calculation specifications for each feeder and/or service circuit supplying receptacles providing shore power for boats are located in Section 555-6.

Article 555 covers wiring and equipment installation in the areas comprising fixed (or floating) piers, wharfs, docks, and other areas in marinas, boatyards, boat basins, boathouses, and similar occupancies designed (or planned) for the purpose of repair, berthing, launching, storage, or fueling of small craft and the moorage of floating buildings.

A readily accessible disconnecting means must be provided by which each boat can be isolated from its supply circuit. The disconnecting means (1) must consist of a circuit breaker, switch, or both; (2) must be located within sight of the shore power connection; and, (3) be intended to constitute the means of supply cutoff to the boat »555-4‹‹.

> ┌─────────────────────────────┐
> ### NOTE
> *Fifteen- and 20-ampere, single-phase, 125-volt receptacles (other than those supplying shore power to boats located at piers, wharfs, and other locations) must be protected by ground-fault circuit-interrupters* »555-3‹‹.
> └─────────────────────────────┘

Where shore power is supplied to accommodations for boats longer than 20 feet, shore-power receptacles of a locking and grounding type rated 30 amperes or more must be provided »555-3‹‹.

Shore Power for Boats Longer than 20 Feet in Lenght

30 Ampere Locking and Grounding Type Receptacle

Shore Power for Boats 20 Feet or Less in Lenght

20 Ampere Locking and Grounding Type Receptacle

Where shore power is supplied, those accommodations for boats a maximum of 20 feet in length must be equipped with shore-power receptacles of a locking and grounding type rated no less than 20 amperes »555-3‹‹.

Summary

- Article 518 covers all buildings, portions of buildings, or structures intentionally designed for the assembly of 100 or more persons.
- Article 530 provisions apply to motion picture (and television) studio audience areas, theaters, and similar locations.
- While Article 520, Part B applies to fixed stage switchboards, Part D applies to portable on-stage switchboards.
- Stage equipment, other than switchboards (fixed and portable), are covered in Article 520, Parts C and E, respectively.
- Theater dressing room provisions are located in Article 520, Part F (520-71 through 73).
- Article 525 covers the installation of portable wiring and equipment for carnivals, circuses, exhibitions, fairs, traveling attractions, and similar functions, including wiring in (or on) all structures.
- Motion picture/television studios, and similar locations are located in Article 530.
- Article 530, Part G (530-70 through 73) covers separately derived systems with 60 volts to ground.
- The provisions of Article 540 apply to motion picture projector rooms, motion picture projectors, and associated equipment.

- Article 545 covers requirements for manufactured building and their components.
- The requirements of Article 547 apply to agricultural building (all or in part), or adjacent areas of like nature as specified in Section 547-1(a) and (b).
- Article 550 (mobile homes, manufactured homes, and mobile-home parks) are divided into three parts: Part A (general), Part B, (mobile homes), and Part C (services and feeders).
- The provisions of Article 551 cover the electrical conductors and equipment installed within (or on) recreational vehicles, the conductors connecting recreational vehicles to an electrical supply, and the installation of electrical equipment and devices within a recreational vehicle park.
- Article 553 covers floating building wiring, services, feeders, and grounding.
- Article 555 covers wiring and equipment installation in the areas comprising fixed (or floating) piers, wharfs, docks, and other areas in marinas, boatyards, boat basins, boathouses, and similar occupancies designed (or planned) for repair, berthing, launching, storage, or fueling of small craft and the moorage of floating buildings.

Unit 18 Competency Test

NEC® Reference   **Answer**

_____ _____

1. A(n) _____ is any building that is of closed construction and is made or assembled in manufacturing facilities on or off the building site for installation, or assembly and installation on the building site, other than manufactured homes, mobile homes, park trailers, or recreational vehicles.

2. Every recreational vehicle site with electrical supply shall be equipped with at least _____ 20-ampere, 125-volt receptacle. A minimum of _____ percent of all recreational vehicle sites with electrical supply shall each be equipped with a 30-ampere, 125-volt receptacle.

3. Outdoor mobile home disconnecting means shall be installed so the bottom of the enclosure containing the disconnecting means is not less than _____ feet above finished grade or working platform.

4. A(n) _____ is an adapter cable containing one male and two female cord connectors used to connect two loads to one branch circuit.

5. Agricultural building lighting fixtures shall be installed to minimize the entrance of _____.

6. The voltage drop on any separately derived 120-volt, single-phase, 3-wire system (with 60 volts to ground) branch circuit shall not exceed _____ %. The combined voltage drop of feeder and branch-circuit conductors shall not exceed _____ %.

NEC® Reference   Answer

7. All exposed incandescent lamps in theater dressing rooms, where less than _____ feet from the floor, shall be equipped with open-end guards riveted to the outlet box cover or otherwise sealed or locked in place.

8. In exhibition halls used for display booths, as in trade shows, the temporary wiring shall be installed in accordance with Article _____.

9. Recreational vehicle site feeder-circuit conductors shall have adequate ampacity for the loads supplied and shall be rated at not less than _____.

10. A(n) _____ is a factory-assembled structure or structures transportable in one or more sections that is built on a permanent chassis and designed to be used as a dwelling without a permanent foundation where connected to the required utilities, and includes the plumbing, heating, air-conditioning, and electric systems contained therein.

11. The service equipment for floating docks or marinas shall be located _____ the floating structure.

12. Stage cables for motion picture studio stage set lighting shall be protected by means of overcurrent devices set at not more than _____ % of the ampacity given in the applicable tables of Articles 310 and 400.

13. A theater stage switchboard that is not completely enclosed dead-front and dead-rear or recessed into a wall shall be provided with a(n) _____ extending the full length of the board to protect all equipment on the board from falling objects.

14. Each carnival ride and concession shall be provided with a fused disconnecting switch or circuit breaker located within sight and within _____ feet of the operator's station.

15. Each motion picture projector, floodlight, spotlight or similar equipment shall have clear working space not less than _____ wide on each side and at the rear thereof.

16. The overall length of a mobile home power-supply cord, measured from the end of the cord, including bared leads, to the face of the attachment plug cap shall not be less than _____ feet and shall not exceed _____ feet.

17. A travel trailer is a vehicular unit, mounted on wheels, designed to provide temporary living quarters for recreational, camping, or travel use, of such size or weight as not to require special highway movement permits when towed by a motorized vehicle, and of gross trailer area less than _____ square feet.

18. A 15-ampere duplex receptacle, installed in a television audience seating area, can be supplied by a _____ ampere branch circuit.

      I. 15

      II. 20

      III. 30

   a) II only   b) I only   c) either I or II   d) I, II, or III

19. Where shore power is supplied to accommodations for boats longer than 20 feet in length, shore-power receptacles of a(n) _____ type rated at _____ amperes or more shall be provided.

20. Occupancy of any room or space for assembly purposes by less than _____ persons in a building of other occupancy, and incidental to such other occupancy, shall be classified as part of the other occupancy and subject to the provisions applicable thereto.

NEC® Reference   **Answer**

_____   _____ 21. Open conductors of not over 600 volts, nominal, shall have a horizontal clearance of not less than _____ feet and a vertical clearance of not less than _____ feet in all areas subject to recreational vehicle movement.

_____   _____ 22. A 125-volt, single-phase, 15- or 20-ampere rated receptacle having one of its current-carrying poles connected to a grounded circuit conductor shall be located within _____ of all permanently installed 15- or 20-ampere rated 60/120-volt technical power-system receptacles.

_____   _____ 23. Manufactured building service-entrance conductors must be installed after erection at the building site unless the _____ is known prior to manufacturer.

_____   _____ 24. Circuits from portable switchboards (within theaters) directly supplying equipment containing incandescent lamps of not over _____ watts shall be protected by overcurrent protective devices having a rating or setting of not over 20 amperes.

_____   _____ 25. A _____ is a centrally located electrical structure from which services or feeders to agricultural buildings and other building, including the associated farm dwellings, are normally supplied.

_____   _____ 26. The service equipment shall be located in sight from and not more than _____ feet from the exterior wall of the mobile home it serves.

_____   _____ 27. Conductors supplying outlets of arc and xenon projectors of the professional type shall not be smaller than _____ and shall be of sufficient size for the projector employed.

_____   _____ 28. Each recreational vehicle site service shall be calculated on the basis of not less than _____ volt-amperes per site equipped with 50-ampere, 120/240-volt supply facilities; and _____ volt-amperes per site equipped with only 20-ampere supply facilities that are not dedicated to tent sites.

_____   _____ 29. Fixed stage switchboards (within theaters) shall be of the _____ type and shall comply with Part D of Article 384 unless approved based on suitability as a stage switchboard as determined by a qualified testing laboratory and recognized test standards and principles.

_____   _____ 30. In locating marina receptacles, consideration should be given to the _____.

_____   _____ 31. A _____ is a building unit as defined in Article 100 that floats on water, is moored in a permanent location, and has a premises wiring system served through connection by permanent wiring to an electricity supply system not located on the premises.

_____   _____ 32. The power supply to the mobile home shall be a feeder assembly consisting of not more than one listed _____ ampere mobile power-supply cord with an integrally molded or securely attached plug cap, or a permanently installed feeder.

_____   _____ 33. Amusement rides and amusement attractions shall be maintained not less than _____ feet in any direction from overhead conductors operating at 600 volts or less, except for the conductors supplying the amusement ride or attraction.

_____   _____ 34. Lamps in cellulose nitrate film storage vaults shall be installed in rigid fixtures of the _____ type.

_____   _____ 35. A professional projector is a type of projector using _____ film that has a minimum width of _____ inch and has on each edge _____ perforations per inch, or a type using carbon arc, xenon, or other light source equipment that develops hazardous gases, dust, or radiation.

NEC® Reference   Answer

_____   _____   36. In closed construction of manufactured buildings, cables shall be permitted to be secured only at cabinets, boxes, or fittings where _____ or smaller conductors are used and protection against physical damage is provided.

_____   _____   37. All cables installed in agricultural buildings shall be secured within ____ of each cabinet, box, or fitting.

_____   _____   38. Where shore power is supplied, those accommodations for boats 20 feet or less in length shall be equipped with shore-power receptacles of a(n) _____ type rated at not less than _____ amperes.

_____   _____   39. A minimum of ____ % of all recreational vehicle sites, with electrical supply, shall each be equipped with a 50-ampere, 125/250-volt receptacle.

_____   _____   40. Footlights, border lights, and proscenium side lights shall be arranged so that no branch circuit supplying such equipment will carry a load exceeding ____.

Specific Equipment

Objectives

• •

After studying this unit, the student will:

✔ have a good understanding of electric sign and outline lighting system requirements, including neon and skeleton tubing.

✔ understand manufactured wiring system provisions.

✔ be able to properly install office partition wiring (power, communication, etc.)

✔ be familiar with electric vehicle charging system stipulations.

✔ thoroughly understand information technology (data processing) equipment requirements.

✔ be able to locate provisions relating to fire pumps.

✔ have a basic understanding of elevator, dumbwaiter, escalator, moving walk, wheelchair lift, and stairway chair lift requirements.

✔ be familiar with definitions pertinent to swimming pools, fountains, and similar installations.

✔ understand what constitutes a common bonding grid (for permanently installed pools) and what items must be bonded thereto.

✔ have a solid grasp of equipment, panelboard, receptacle, and lighting (including underwater lighting) provisions pertaining to permanently installed swimming pools.

✔ be able to accurately define storable pool and know the applicable requirements.

✔ be familiar with spa and hot tub provisions.

✔ know where to find requirements for therapeutic pools and tubs.

✔ understand the fundamental hydromassage bathtub provisions.

Introduction

• •

Bringing this text to a close, this unit addresses a wide variety of electrical equipment requirements. *National Electrical Code*® Chapter 6 is the basis for numerous equipment illustrations throughout this unit. While most of the articles of Chapter 6 are covered here, four additional articles are described under *Industrial Locations* (Unit 17). Those articles are: 610 *(Cranes and Hoists)*, 630 *(Electric Welders)*, 669 *(Electroplating)*, and 670 *(Industrial Machinery)*. Some of the less commonly used articles are not referenced.

Beginning with illustrations for both *Electric Signs and Outline Lighting* and *Neon and Skeleton Tubing*, this unit recounts some of the many Article 600 requirements. Subsequently described in illustrative detail (with corresponding articles numbers in parentheses) are: *Manufactured Wiring Systems* (604); *Office Furnishings* (605); *Electric Vehicle Charging Systems* (625); *Audio Signal Processing, Amplification, and Reproduction Equipment* (640); *Information Technology Equipment* (645); *Pipe Organs* (650); and *Fire Pumps* (695). Next, this unit devotes several illustrations to Article 620's topics: *Elevators, Dumbwaiters, Escalators, Moving Walks, Wheelchair Lifts, and Stairway Chair Lifts*. Finally, the expansive subject of *Swimming Pools, Fountains, and Similar Locations* (Article 680) is introduced. Because Article 680 covers a broad range of equipment and installation applications, a significant number of illustrations have been dedicated to these requirements. It is imperative that the reader have a thorough understanding of the words and phrases relating specifically to swimming pools, fountains, and similar locations, found in

Section 680-4. One particularly notable definition is that of **storable pool**. A misconception held by many is that a storable pool is, in simple terms, any above-ground swimming pool. Observe, however, that a storable pool has a maximum depth of only 42 inches. While some of the smaller above-ground pools fall within this parameter; many, if not most, have a depth of at least 48 inches. It must be concluded, therefore, that any above ground pool having a maximum depth greater than 42 inches qualifies as a permanently installed pool, and must be compliant with Article 680, Parts A and B. There are seven subdivisions of Article 680: **A**—*General*, **B**—*Permanently Installed Pools*, **C**—*Storable Pools*, **D**—*Spas and Hot Tubs*, **E**—*Fountains*, **F**—*Pools and Tubs for Therapeutic Use*, and **G**—*Hydromassage Bathtubs*. While space constraints prohibit an all-encompassing discussion of these provisions, the more predominant features are illustrated and discussed. Because of the potentially fatal nature of mistakes, study this unit carefully together with Article 680 when installing/wiring a pool, spa, hydromassage bathtub, or similar item.

As throughout this text, state and/or local jurisdictions may alter electrical requirements. Obtain a copy of any additional rules and regulations for your specific area.

EQUIPMENT

■ Pipe Organs

Electronic organs must comply with the appropriate Article 640 provisions ≫650-2≪.

The power source must be a transformer-type rectifier, having a dc potential of 30 volts or less ≫650-3≪.

The rectifier must be grounded according to Section 205-112(b) provisions ≫650-4≪.

Conductors must comply with Section 650-5(a) through (d) ≫650-5≪.

Use a minimum of No. 28 conductors for electronic signal circuits, and at least No. 26 for electromagnetic valve supply and the like. A main common-return conductor in the electromagnetic supply must be No. 14 or larger ≫650-5(a)≪.

Conductors must have thermoplastic or thermosetting insulation ≫650-5(b)≪.

Except for the common-return conductor and conductors inside the organ proper, the organ sections and the console conductors must be cabled. An additional covering can enclose both cable and common-return conductors. Alternately, a separate common-return conductor is allowed to contact the cable ≫650-5(c)≪.

Each cable must have an outer covering, either overall or on individual subassemblies of grouped conductors. Tape is an acceptable covering. If not housed in metal raceway, the covering must be flame retardant or the cable/cable subassembly must be wrapped by closely wound fireproof tape ≫650-5(d)≪.

Article 650 covers electrically operated pipe organ circuits and components employed for sounding apparatus and keyboard control ≫650-1≪.

Securely fastened cables can attach directly to the organ structure without insulating supports. Cables must not be placed in contact with other conductors (except for the common return conductor) ≫650-6≪.

Arrange circuits so that all conductors are protected from overcurrent by an overcurrent device rated at no more than 6 amperes ≫650-7≪. This provision does not apply to the main supply and common-return conductors ≫650-7 *Exception*≪.

■ Electric Signs and Outline Lighting

Each sign and outline lighting system, or its supplying feeder/branch, must be controlled by an externally operable switch (or circuit breaker) that opens all ungrounded conductors ≫ 600-6 ≪. The disconnecting means must be within sight of the sign or outline lighting system that it controls. If out of the line of sight from any energizable section, it must be possible to lock the disconnecting means in the open position ≫ 600-6(a) ≪.

Lockable Disconnect Located Inside

Article 600 covers conductor and equipment installation for electric signs and outline lighting as defined in Article 100 ≫ 600-1 ≪. By definition, electric signs and outline lighting includes all products and installations utilizing neon tubing, such as signs, decorative elements, skeleton tubing, or art forms ≫ 600-1 FPN ≪.

14 Feet

Sports Store

Signs and outline lighting systems must be marked with the manufacturer's name, trademark, or other identification, and input voltage and current rating ≫ 600-4(a) ≪.

Signs and outline lighting system equipment in wet locations, unless of the listed watertight type, must be weatherproof, and if necessary, have drain holes in accordance with Section 600-9(d)(1) through (3) ≫ 600-9(d) ≪.

Unless protected from physical damage, sign or outline lighting system equipment must be at least 14 feet above areas accessible to vehicles ≫ 600-9(a) ≪.

Signs and outline lighting system metal equipment must be grounded ≫ 600-7 ≪. (See the additional grounding provisions in Section 600-7.)

NOTE

Each commercial building and occupancy open to pedestrians must have at least one accessible outlet at every tenant space entrance for sign or outline lighting system use. The outlet(s) must be supplied by an exclusively dedicated branch circuit rated 20 amperes (or more). Service halls or corridors are not considered pedestrian accessible ≫ 600-5(a) ≪.

Supplying branch circuits of signs and outline lighting systems containing incandescent and fluorescent forms of illumination must not exceed a 20-ampere rating ≫ 600-5(b)(1) ≪.

The wiring method supplying signs and outline lighting systems must terminate within: a sign, an outline lighting system enclosure, a suitable box, or a conduit body ≫ 600-5(c)(1) ≪.

Sign-supporting metal poles can enclose supply conductors, provided the stipulations of Section 410-15(b) are met ≫ 600-5(c)(3) ≪.

Other than lamps and neon tubing, all live parts must be enclosed ≫ 600-8 ≪.

Signs and outline lighting systems having incandescent lamp holders must be marked with the maximum allowable lamp wattage. The permanently installed markings must be in letters at least ¼ inch high, and must be visible during relamping ≫ 600-4(b) ≪.

Portable or mobile signs must meet Section 600-10 requirements.

If operated by external electronic or electromechanical controllers, sign or outline lighting systems can have a disconnecting means located within sight of the controller or within the controller enclosure. The disconnecting means must disconnect both the sign (or outline lighting system) and the controller from all ungrounded supply conductors ≫ 600-6(a) ≪.

Ballasts, transformers, and electronic power supplies must meet the requirements of Sections 600-21 through 600-23.

■ Neon and Skeleton Tubing

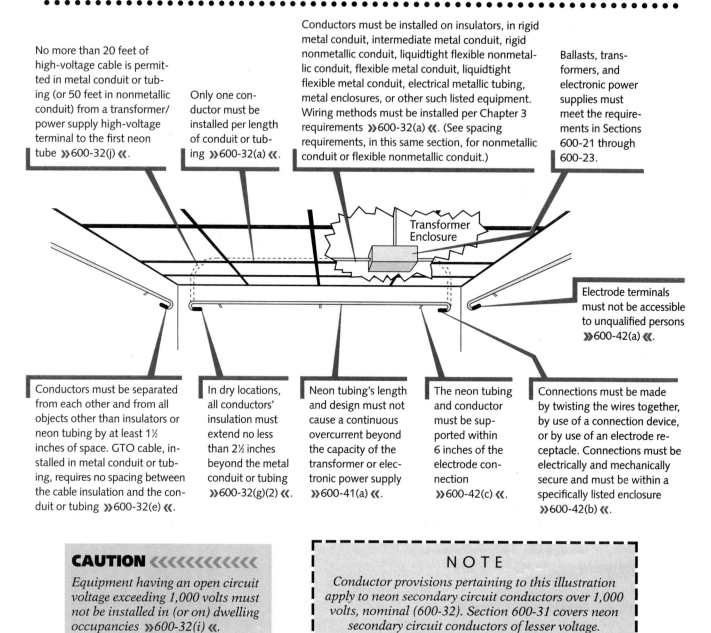

No more than 20 feet of high-voltage cable is permitted in metal conduit or tubing (or 50 feet in nonmetallic conduit) from a transformer/power supply high-voltage terminal to the first neon tube »600-32(j) «.

Only one conductor must be installed per length of conduit or tubing »600-32(a) «.

Conductors must be installed on insulators, in rigid metal conduit, intermediate metal conduit, rigid nonmetallic conduit, liquidtight flexible nonmetallic conduit, flexible metal conduit, liquidtight flexible metal conduit, electrical metallic tubing, metal enclosures, or other such listed equipment. Wiring methods must be installed per Chapter 3 requirements »600-32(a) «. (See spacing requirements, in this same section, for nonmetallic conduit or flexible nonmetallic conduit.)

Ballasts, transformers, and electronic power supplies must meet the requirements in Sections 600-21 through 600-23.

Transformer Enclosure

Electrode terminals must not be accessible to unqualified persons »600-42(a) «.

Conductors must be separated from each other and from all objects other than insulators or neon tubing by at least 1½ inches of space. GTO cable, installed in metal conduit or tubing, requires no spacing between the cable insulation and the conduit or tubing »600-32(e) «.

In dry locations, all conductors' insulation must extend no less than 2½ inches beyond the metal conduit or tubing »600-32(g)(2) «.

Neon tubing's length and design must not cause a continuous overcurrent beyond the capacity of the transformer or electronic power supply »600-41(a) «.

The neon tubing and conductor must be supported within 6 inches of the electrode connection »600-42(c) «.

Connections must be made by twisting the wires together, by use of a connection device, or by use of an electrode receptacle. Connections must be electrically and mechanically secure and must be within a specifically listed enclosure »600-42(b) «.

CAUTION ‹‹‹‹‹‹‹‹‹‹‹‹

Equipment having an open circuit voltage exceeding 1,000 volts must not be installed in (or on) dwelling occupancies »600-32(i) «.

NOTE

Conductor provisions pertaining to this illustration apply to neon secondary circuit conductors over 1,000 volts, nominal (600-32). Section 600-31 covers neon secondary circuit conductors of lesser voltage.

Article 600, Part B provisions apply only to field-installed skeleton tubing, and are in addition to Part A requirements »600-30 «.

Neon Tubing: Electric-discharge tubing produced in shapes forming letters (or parts thereof), skeleton tubing, outline lighting, other decorative elements or art forms, and filled with various inert gases »600-2 «.

Skeleton Tubing: Neon tubing that is itself the sign or outline lighting and not attached to an enclosure or sign body »600-2 «.

Conductors must be insulated, listed for the purpose, rated for the voltage, not smaller than No. 18, and have a minimum temperature rating of 105°C »600-32(b) «.

The supply branch circuits of neon tubing installations must not be rated in excess of 30 amperes »600-5(b)(2) «.

Sharp bends in insulated conductors must be avoided »600-32(d) «.

Pedestrian accessible neon tubing, other than dry-location portable signs, must be protected from physical damage »600-9(b) «.

Installed conductors must not be subject to physical damage »600-32(c) «.

In damp or wet locations, all conductors' insulation must extend no less than 4 inches beyond the metal conduit or tubing »600-32(g)(1). «

■ Manufactured Wiring Systems

Article 604 provisions apply to field-installed wiring using off-site manufactured subassemblies for branch circuits, remote-control circuits, signaling circuits, and communications circuits in accessible areas »604-1 «.

Manufactured Wiring System: A system containing components assembled in the process of manufacture which after assembly cannot be inspected without damage or destruction »604-2 «.

Cable must be listed armored cable or metal-clad cable, containing nominal 600-volt No. 10 or 12 copper-insulated conductors, having a bare or insulated copper equipment grounding conductor equal in size to the ungrounded conductor »604-6(a)(1) «. (See *Exceptions No. 1 through 3*.)

The type of cable or conduit must be clearly marked on each section »604-6(a)(3) «.

Receptacles and connectors must be (1) of the locking type, uniquely polarized and identified for the purpose, and (2) part of a listed assembly appropriate for the system »604-6(b) «.

Armored cable must meet the support provisions found in Article 333, while metal-clad cable support requirements are found in Article 334.

Ceiling grids cannot be used to support cables (and raceways). Secure support must be provided via wires and associated fittings *and* must be installed in addition to the ceiling grid support wires »300-11(a) «. (This illustration does not show cable supports.)

Manufactured wiring systems are permitted in accessible and dry locations, and in plenums and spaces used for environmental air, where listed for this application and installed per Section 300-22 »604-4 «. In concealed spaces, one end of tapped cable can extend into hollow walls for direct switch/outlet termination »604-4 *Exception No. 1* «.

Except as modified by this article's requirements, all other relevant *NEC®* articles apply »604-3 «.

Where listed for the purpose, manufactured wiring systems can be used in outdoor locations »604-4 *Exception No. 2* «.

Conduit must be listed flexible metal conduit, containing nominal 600-volt No. 10 or 12 copper-insulated conductors, having a copper equipment grounding conductor (bare or insulated) equal in size to the ungrounded conductor »604-6(a)(2) «. (See *Exceptions No. 1 through 3*.)

While a fixture tap (no longer than 6 feet) connecting to a **single fixture** can contain conductors smaller than No. 12, they cannot be smaller than No. 18 »604-6(a) *Exception No. 1 to (1) and (2)* «.

Conductors smaller than No. 12 are permitted for remote-control, signaling, or communications circuits. The assembly must be listed for the purpose »604-6(a) *Exception No. 2 to (1) and (2)* «.

Hard-usage flexible cord (minimum No. 12 conductors) can be part of a listed factory-made assembly (not exceeding 6 feet in length) as a transition method between manufactured wiring system components and utilization equipment, not permanently secured to the building structure. The cord must be visible along its entire length and must not be subject to strain or physical damage »604-6(a) *Exception No. 3 to (1) and (2)* «.

CAUTION ‹‹‹‹‹

All unused outlets must be capped to effectively close connector openings »604-7 «.

NOTE

Manufactured wiring systems cannot be used where conductors or cables are limited by the provisions of Articles 333 and 334 »604-5 «.

Office Furnishings

• •

> ### N O T E
> *Fixed wired partitions (secured to building surfaces) must be permanently connected to the building electrical system by one of Chapter 3's wiring methods* ≫605-6≪.

Lighting equipment listed and identified for wired partition use must comply with all of the following:
(a) A secure attachment or support means is provided.
(b) If cord and plug connection is provided, the cord length, while suitable for the intended application, must not exceed 9 feet in length. The cord must be No. 18 or larger, must contain an equipment grounding conductor, and must be of the hard usage type. Any other means of connection must be identified as suitable for the existing conditions.
(c) Convenience receptacles are not permitted in lighting fixtures ≫605-5≪.

Article 605 covers electrical equipment, lighting accessories, and wiring systems which connect to, are contained within, or are installed on relocatable wired partitions ≫605-1≪.

Wiring systems must be identified as suitable power providers for lighting accessories and appliances in wired partitions. The partitions must not extend from floor to ceiling, unless the exception is met ≫605-2≪.

The electrical connection between partitions must be either a flexible assembly (identified for wired partition usage), or be of flexible cord, provided *all* of the following conditions are met: (1) the cord is extra-hard usage type; (2) the partitions are mechanically contiguous; (3) the cord, no longer than necessary for maximum partition positioning, is in no case to exceed 2 feet; and (4) the cord terminates at an attachment plug and cord connector with strain relief ≫605-4≪.

All conductors and connections must be contained within wiring channels (metal or other material) identified as suitable for the conditions of use. These channels must be free of projections or other conditions that may damage conductor insulation ≫605-3≪.

Except as modified by Article 605 requirements, all other *NEC*® articles apply ≫605-2(b)≪.

Freestanding partitions (not fixed) can be permanently connected to the building's electrical system by one of Chapter 3's approved wiring methods ≫605-7≪.

If used in hazardous (classified) locations, these assemblies must also conform with Article 500 through 517 ≫605-2(c)≪.

Individual freestanding partitions or groups of individual partitions that are electrically connected, and mechanically contiguous, not exceeding 30 feet when assembled, can connect to the building electrical system by means of a single flexible cord and plug, provided all of the following are met:

(a) The flexible power-supply cord must be of the extra-hard usage type, with No. 12 or larger conductors, an insulated equipment grounding conductor, and not more than 2 feet in length.

(b) The receptacle(s) supplying power must be on a separate circuit serving only panels (no other loads), and must be located within 12 inches of the partition to which it is connected.

(c) Individual partitions, or groups of interconnected individual partitions, must contain no more than thirteen 15-ampere, 125-volt receptacle outlets.

(d) Individual partitions, or interconnected groups must not contain multiwire circuits ≫605-8≪.

Electric Vehicle Charging Systems

Unless specifically listed for the purpose and location, the electric vehicle supply equipment's coupling means must be situated no less than 1½ feet and no more than 4 feet above floor level »625-29(b)«.

Article 625 provisions cover the electrical conductors and equipment external to an electric vehicle which connect it to an electrical supply by conductive or inductive means, as well as the installation of equipment and devices related to electric vehicle charging »625-1«.

Electric Vehicle: A highway worthy automotive-type vehicle (such as passenger automobiles, buses, trucks, vans, and the like) primarily powered by an electric motor drawing current from a rechargeable storage battery, fuel cell, photovoltaic array, or other electric current source. Article 625 excludes electric motorcycles and similar type vehicles as well as off-road self-propelled electric vehicles (such as industrial trucks, hoists, lifts, transports, golf carts, airline ground support equipment, tractors, boats, etc.) »625-2«.

NOTE

All electrical materials, devices, fittings, and associated equipment must be listed or labeled »625-5«.

Electric Vehicle Connector: A device (part of the electric coupler) that, by insertion into an electric vehicle inlet, establishes an electrical connection for the purpose of charging and information exchange »625-2«.

Electric Vehicle Coupler: A mating electric vehicle inlet and electric vehicle connector set »625-2«.

Electric Vehicle Inlet: The electric vehicle device into which the electric vehicle connector is inserted for charging and information exchange. For *NEC®* purposes, the electrical vehicle inlet is considered part of the vehicle rather than part of the electric vehicle supply equipment »625-2«.

Electric Vehicle Supply Equipment: Includes the conductors (ungrounded, grounded, and equipment grounding conductors), electric vehicle connectors, attachment plugs, and all other fittings, devices, power outlets, or apparatuses purposely installed to transfer energy from the premises wiring to the electric vehicle »625-2«.

Electric Vehicle Nonvented Storage Battery: A hermetically-sealed battery comprised of rechargeable electrochemical cell(s) without provisions for: excessive gas pressure release, the addition of water (or electrolyte), or external electrolyte specific gravity measurements »625-2«.

The electric vehicle coupler must comply with Section 625-9(a) through (f) »625-9«.

Personnel Protection System: A system of personnel protection devices and structural features working together to provide personnel protection against electric shock »625-2«.

Article 625, Part C (625-13 through 19) contains equipment construction requirements.

Feeders and branch circuits supplying electric vehicle supply equipment requires overcurrent protection, sized for continuous duty, and having a rating of no less than 125% of the maximum electric vehicle equipment load. Where the same feeder or branch circuit supplies noncontinuous loads, the overcurrent device must have a rating at least equal to the sum of the noncontinuous loads plus 125% of the continuous loads »625-21«.

Mechanical ventilation is not required where electric vehicle nonvented storage batteries are used, or where the electric vehicle supply equipment is listed/labeled and marked as suitable for charging electric vehicles indoors without ventilation per Section 625-15(b) »625-29(c)«.

Outdoor site requirements are located in Section 625-30.

WARNING

Mechanical ventilation (such as a fan) must be provided where the electric vehicle supply equipment is listed/labeled and marked as suitable for charging electric vehicles that require ventilation for indoor charging, in accordance with Section 625-15(c). Both supply and exhaust ventilation equipment must be permanently installed to intake from, and vent directly to, the outdoors. Positive pressure ventilation systems are only permitted in locations that have been specifically designed and approved for that application. Mechanical ventilation requirements must be determined using one of the three methods found in Section 625-29(d) »625-29(d)«.

■ Audio Signal Processing, Amplification, and Reproduction Equipment

Article 640 covers equipment and wiring for audio signal generation, recording, processing, amplification and reproduction; distribution of sound; public address; speech input systems; temporary audio system installations; and electronic musical instruments (including organs). This encompasses audio systems subject to Article 517, Part F, and Articles 518, 520, 525, and 530 ≫ 640-1 ≪.

Audio Signal Processing Equipment: Electrically operated equipment that produces and/or processes electronic signals which, when appropriately amplified and reproduced via loudspeaker, output an acoustic signal within the range of normal human hearing. Within Article 640, the terms **equipment** and **audio equipment** are equivalent to audio signal processing equipment ≫ 640-2 ≪.

N O T E
Equipment and cabling must be installed in a neat, workmanlike manner ≫ 640-6 ≪.

Loudspeaker: Equipment that converts an ac electric signal into an acoustic signal. The term **speaker** is commonly used to mean loudspeaker ≫ 640-2 ≪.

Mixer: Equipment used to combine and level-match a multiplicity of electronic signals, as from microphones, electronic instruments, and recorded audio ≫ 640-2 ≪.

Audio System: The totality of equipment and interconnected wiring used to fabricate a fully functional audio signal processing, amplification, and reproduction system ≫ 640-2 ≪.

Equipment Rack: A framework for the equipment support and/or enclosure, either portable or stationary ≫ 640-2 ≪.

Section 640-2 provides a comprehensive list of the most commonly used words and phrases relevant to Article 640.

Grounding of separately derived systems with **60 volts to ground** must meet Section 530-72 requirements ≫ 640-7(b) ≪.

Equipment access must not be denied by accumulated wires and cables that prevent panel removal, including suspended ceiling panels ≫ 640-5 ≪.

Wireways and auxiliary gutters must be grounded and bonded per Article 250. If the wireway (or auxiliary gutter) contains no power-supply wires, a No. 14 copper equipment grounding conductor (or equivalent) is sufficient. If the wireway (or auxiliary gutter) contains power-supply wires, the equipment grounding conductor must meet Section 250-122 size specifications ≫ 640-7(a) ≪.

Isolated grounding-type receptacles are permitted as described in Section 250-146(d), as well as for the implementation of other Article 250 compliant technical power systems. For separately derived systems with **60 volts to ground**, the branch-circuit equipment grounding conductor must be terminated per Section 530-72(b) ≫ 640-7(c) ≪.

Portable Equipment: Equipment fed with portable cords or cables intended for movement from one place to another ≫ 640-2 ≪.

Temporary Equipment: Portable wiring and equipment used for events of a transient or temporary nature where equipment removal is presumed at the event's conclusion ≫ 640-2 ≪.

Technical Power System (often referred to as **tech power**): an electrical distribution system with Section 250-146(d) approved grounding, where the equipment grounding conductor is isolated from the premises grounded conductor except at a single grounded termination point within a branch circuit panelboard, the origination (main breaker) branch-circuit panelboard, or at the premises grounding electrode ≫ 640-2 ≪.

Permanent audio system installations must comply with Article 640, Part B (640-21 through 25).

Portable and temporary audio system installations must meet the provisions found in Article 640, Part C (640-41 through 46).

Grouped (or bundled) insulated conductors of different systems, in close physical contact within the same raceway (or other enclosure), or in portable cords (or cables), must comply with Section 300-3(c)(1) ≫ 640-8 ≪.

Information Technology Equipment

A single power disconnect means must be provided for all electronic equipment in the information technology equipment room. Likewise, a means to disconnect the power to all of the room's dedicated HVAC systems serving the room, closing all required fire/smoke dampers, must be provided. These disconnecting means' controls must be grouped, identified, and readily accessible at the principal exit doors. A single means to control both the electronic equipment and HVAC systems is acceptable 》645-10《.

Article 645 covers equipment, power-supply wiring, equipment interconnected wiring, as well as grounding of information technology equipment (and systems), including information technology equipment room terminal units 》645-1《.

Article 645 applies, provided that listed information technology equipment is installed 》645-2(c)《.

The data processing system can be connected to a branch circuit by any of the following listed means: (1) computer/data processing cable and attachment plug cap; (2) flexible cord and attachment plug cap; or (3) cord set assembly (protected from physical damage where run on the floor's surface) 》645-5(b)《.

Branch circuit conductors supplying data processing system unit(s) must have an ampacity of at least 125% of the total connected load 》645-5(a)《.

Cables (power, communications, connecting, and interconnecting), and receptacles associated with the information technology equipment are permitted under a raised floor, provided Section 645-5(d)(1) through (5) requirements are met 》645-5(d)《.

It is not necessary to secure in place cables (power, communications, connecting, and interconnecting) and associated boxes, connectors, plugs, and receptacles that are listed as part of, or for, information technology equipment 》645-5(e)《.

Unless (a) or (b) permits otherwise, **uninterruptible power supply** (UPS) systems installed within the information technology room, together with their supply and output circuits, must comply with Section 645-10. The battery must also be disconnected from its load by the disconnecting means.
(a) Installations qualifying under Article 685 provisions.
(b) A Section 645-10 compliant disconnecting means is not required for power sources capable of supplying 750 volt-amperes (or less), derived either from UPS equipment or from battery circuits integral to electronic equipment, provided all other Section 645-11 requirements are met 》645-11《.

UPS

Dedicated HVAC Unit

Article 645 does apply when a dedicated heating/ventilating/air-conditioning (HVAC) system (separate from other occupancy areas) exclusively serves the information technology equipment. Any HVAC system, serving other occupancies, can also serve the information technology equipment room if fire/smoke dampers exist at room boundary penetration points. Such dampers must operate both upon activation of smoke detectors and by disconnecting means operation as required by Section 645-10 》645-2(b)《.

All exposed noncurrent-carrying metal parts of an information technology system must either be grounded (per Article 250) or be double insulated. Power systems (derived within listed information technology equipment) supplying information technology systems through receptacles (or cable assemblies), which are part of this equipment, are not considered separately derived for the purpose of applying Section 250-20(d). Signal reference structures, where installed, must be bonded to the equipment grounding system provided for the information technology equipment 》645-15《.

Article 645 applies, provided the following conditions are met:
(1) Disconnecting means complying with Section 645-10 are provided 》645-2(a)《.
(2) The information technology equipment room is occupied only by necessary maintenance and operations personnel 》645-2(d)《.
(3) Fire-resistant-rated walls and floors, and ceilings with protected openings separate the room from other occupancies 》645-2(e)《.
(4) The building construction, rooms, areas, and occupancy comply with applicable building codes 》645-2(f)《.

> **N O T E**
> *Article 645 applies, provided **all** of the conditions listed in 645-2(a) through (f) are met* 》645-2《.

■ Fire Pumps

All wiring from the controllers to the pump motors must be in rigid metal conduit, intermediate metal conduit, liquidtight flexible metal conduit, or liquidtight flexible nonmetallic conduit Type LFNC-B, or Type MI cable ≫695-6(e)≪.

Conductors supplying fire pump motor(s), pressure maintenance pumps, and associated accessory equipment must have a minimum rating of 125% of the sum of the fire pump motor(s) and pressure maintenance motor(s) full-load current(s), plus 100% of the associated fire pump accessory equipment ≫695-6(c)(1)≪. Conductors supplying only fire pump motor(s) must have a minimum rating of 125% of the fire pump motor(s) full-load current(s) ≫695-6(c)(2)≪.

Article 695 covers the installation of: (1) electric power sources and interconnecting circuits; and (2) switching and control equipment dedicated to fire pump drivers ≫695-1(a)≪.

NOTE
Taps used to supply fire pump equipment, if provided with service equipment and installed per service-entrance conductor requirements, can be connected to the supply side of the service disconnecting means ≫230-82(4)≪. (See Section 230-94 Exception No. 4.)

NOTE
Article 695 does not cover: (1) performance, maintenance, and acceptance testing of the fire pump system, and the system components' internal wiring; and (2) pressure maintenance (jockey or makeup) pumps ≫695-1(b)≪.

Except where altered by Article 695, the installation of fire pump electrical conductors and equipment must comply with all applicable requirements in Chapters 1 through 4 ≫695-2≪.

Electric motor-driven fire pumps must have a reliable power source ≫695-3≪. See Section 695-3(a) for individual source requirements and Section 695-3(b) for multiple source requirements.

A fire pump can be supplied by a separate service ≫230-2(a) and 695-3(a)(1)≪.

Electric motor-driven fire pump supply circuits must be guarded against inadvertent disconnection per Section 695-4(a) or (b) ≫695-4≪.

If service or system voltage differs from the fire pump motor's utilization voltage, one or more transformers (protected by disconnecting means and overcurrent protective devices) can be installed between the system supply and the fire pump controller in accordance with Section 695-5(a), (b), or (c). Only transformers covered in (c) can supply loads unrelated to the fire pump system ≫695-5≪.

Power circuits and wiring methods must comply with the requirements in (a) through (g), and, as permitted in Section 230-90(a), *Exception No. 4;* Section 230-94, *Exception No. 4;* Section 230-95, *Exception No. 2;* Section 240-13(3); Section 230-208; Section 240-3(a); and Section 430-31 ≫695-6≪.

The location of electric motor-driven fire pump controllers and power transfer switches must be within sight, and as close as practicable to, the motors they control ≫695-12(a)≪. Engine-driven fire pump controllers must be located as close as practical to, and within sight of, the engines they control ≫695-12(b)≪.

Fire pump controllers and power transfer switches, by location and/or protection, must remain undamaged by water escaping from pumps or pump connections ≫695-12(e)≪.

The disconnecting means must be marked "Fire Pump Disconnecting Means." The minimum 1-inch-high letters must be visible without opening enclosure doors or covers ≫695-4(b)(3)≪.

Fire pump supply conductors on the final disconnecting means load side, and overcurrent device(s) permitted by Section 695-4(b), must be entirely independent of other wiring. They must supply only those loads directly associated with the fire pump system, and must be amply protected from potential damage by fire, structural failure, or operational accident. These conductors can be routed through a building(s) encased in 2 inches of concrete or within enclosed construction dedicated to the fire pump circuit(s) having a minimum of 1-hour fire resistance. Installation must comply with any/all restrictions outlined in the listing of the electrical circuit protective system used ≫695-6(b)≪. (See the exception pertaining to this requirement.)

Automatic protection against overloads is not allowed on power circuits. Except as noted in Section 695-5(c)(2), branch-circuit and feeder conductors must be protected against short circuits only ≫695-6(d)≪.

All engine controller/battery wiring must be protected against mechanical injury, being installed in accordance with the controller and engine manufacturer's instructions ≫695-6(g)≪.

ELEVATORS, DUMBWAITERS, ESCALATORS, MOVING WALKS, WHEELCHAIR LIFTS, AND STAIRWAY CHAIR LIFTS

■ General

Working space must be provided about controllers, disconnecting means, and other electrical equipment. The minimum working space must meet Section 110-26(a) specifications. Where maintenance and supervision conditions ensure that only qualified persons will examine, adjust, service, and maintain the equipment, Section 110-26(a) clearance requirements can be waived per Section 620-5(a) through (d) »620-5«.

Circuit voltage of heating/air-conditioning equipment branch circuits, located on the elevator car, must not exceed 600 volts »620-3(c)«.

Lighting circuits must comply with Article 410 requirements »620-3(b)«.

The minimum conductor size for traveling cables which feed lighting circuits is No. 14 copper. No. 20 (or larger) copper conductors are permitted in parallel, provided the ampacity equals or surpasses that of No. 14 copper. For traveling cable circuits, other than lighting, No. 20 copper is the minimum size »620-12(a)«.

The motor controller rating must comply with Section 430-83. The rating can be less than the nominal rating of the elevator motor, when the controller inherently limits the available power to the motor, being appropriately marked as power limited »620-15«.

Article 620 covers the installation of electric equipment and wiring relating to elevators, dumbwaiters, escalators, moving walks, wheelchair lifts, and stairway chair lifts »620-1«.

All live parts of electrical apparatus in (and around): elevator and dumbwaiter hoistways, landings, and cars; escalator and moving walk wellways or landings; and machinery space runways of wheelchair lifts and stairway chair lifts, must be enclosed to guard against accidental contact »620-4«.

Control System: Overall system governing starting, stopping, direction of motion, acceleration, speed, and retardation of the moving member »620-2«.

Motion Controller: Electrical device(s) for that portion of the control system which governs acceleration, speed, retardation, and stoppage of the moving member »620-2«.

Motor Controller: The control system's operative units, i.e., the starter device(s) and power conversion equipment which drive an electric motor, or the pumping unit used to power hydraulic control equipment »620-2«.

Operation Controller: Electric device(s) for that portion of the control system which initiates starting, stopping, and direction of motion in response to an operating device's signal »620-2«.

Supply voltage between conductors must not exceed 300 volts unless Section 620-3(a) through (c) permits otherwise »620-3«.

Door operator controller and door motor branch circuits, as well as feeders to motor controllers, driving machine motors, machine brakes, and motor-generator sets must have a circuit voltage of 600 volts or less »620-3(a)«.

Operating Device: Devices used to activate the operation controller, such as the car switch, push buttons, key or toggle switch(s) »620-2«.

Conductors and optical fibers within hoistways, escalator and moving walk wellways, wheelchair lifts, stairway chair lift runways, and machinery spaces, in/on cars, and machine/control rooms (exclusive of traveling cables connecting the car or counterweight and hoistway wiring), must be installed in rigid metal conduit, intermediate metal conduit, electrical metallic tubing, rigid nonmetallic conduit, or wireways; or must be Type MC, MI, or AC cable unless otherwise permitted in Section 620-21(a) through (c) »620-21«.

Conductor insulation must comply with Section 620-11(a) through (d).

The minimum conductor size, other than traveling cables and conductors that form an integral part of control equipment, is No. 24 copper »620-12«.

The motor controller rating must comply with Section 430-83. The rating can be less than the elevator motor's nominal rating, when the controller inherently limits motor power, and is marked as power limited »620-15«.

Feeder and branch circuit conductor ampacity must comply with Section 620-13(a) through (d).

■ Elevator Wiring

Flexible metal conduit, liquidtight flexible metal conduit, or liquidtight flexible nonmetallic conduit of ⅜ inch nominal trade size or larger, no longer than 6 feet, can be installed between control panels and machine motors, machine brakes, motor-generator sets, disconnecting means; and pumping unit motors and valves ≫ 620-21(a)(3)(a) ≪. Liquidtight flexible nonmetallic conduit, as defined in Section 351-22(2), can be installed in lengths exceeding 6 feet ≫ 620-21(a)(3)(a) *Exception* ≪.

Flexible metal conduit, liquidtight flexible metal conduit, or liquidtight flexible nonmetal conduit of ⅜ inch nominal trade size (or larger), not exceeding 6 feet in length, allowed for installation on cars, must be securely fastened in an oil free location ≫ 620-21(a)(2)(a) ≪. Liquidtight flexible nonmetallic conduit of ⅜ inch nominal trade size or larger, as defined by Section 351-22(2), is permitted in lengths greater than 6 feet ≫ 620-21(a)(2)(a) *Exception* ≪.

Hard-service cords and junior hard-service cords, conforming to Article 400 (Table 400-4) requirements, are permitted as flexible connections between the car's fixed wiring and the car door/gate devices. Only hard-service cords are permitted as flexible connections for top-of-car operating devices or car-top work lights. Devices (or fixtures) must be grounded via an equipment grounding conductor run with the circuit conductors. Cables having smaller conductors as well as other insulation (jacket) types and thicknesses can serve as flexible connections between the car's fixed wiring and the car door/gate devices, if specifically listed for this use ≫ 620-21(a)(2)(b) ≪.

Flexible metal conduit, liquidtight flexible metal conduit, or liquidtight flexible nonmetallic cable can be installed within hoistways between risers and limit switch interlocks, operating buttons, and similar devices ≫ 620-21(a)(1)(a) ≪.

Class 2 power-limited circuit cables installed in hoistways between risers, signal equipment, and operating devices must be of a jacketed flame-retardant type, securely supported and protected from physical damage ≫ 620-21(a)(1)(b) ≪.

Flexible metal conduit, liquidtight flexible metal conduit, liquidtight flexible nonmetallic conduit or flexible cords and cables, or conductors grouped together (taped or corded) that are part of listed equipment, a driving machine (or driving machine brake) are permitted on the counter-weight assembly, in lengths not to exceed 6 feet without being installed in a raceway if of a jacketed flame-retardant type, and protected from physical damage ≫ 620-21(a)(4) ≪.

Flexible metal conduit, liquidtight flexible metal conduit, liquidtight flexible nonmetallic conduit or flexible cords and cables, or conductors grouped together (taped or corded) which are part of listed equipment, a driving machine (or driving machine brake) can be installed on the car assembly, in lengths not to exceed 6 feet without being installed in a raceway if of a jacketed flame-retardant type, and by location protected from physical damage ≫ 620-21(a)(2)(d) ≪.

Flexible cords and cables that are components of listed equipment, used in circuits operating at 30 volts rms (or less), or 42 volts dc (or less), are permitted in lengths not exceeding 6 feet provided the cords and cables are (1) supported and protected from physical damage and (2) of a jacketed and flame-retardant type ≫ 620-21(a)(2)(c) ≪.

Motor-generators, machine motors, or pumping unit motors and valves, located adjacent to, or underneath, control equipment and provided with extra-length terminal leads (no longer than 6 feet), can be connected directly to controller terminal studs without regard to Articles 430 and 445 carrying-capacity requirements. Auxiliary gutters are permitted in machine/control rooms between controllers, starters, and similar apparatus ≫ 620-21(a)(3)(b) ≪.

Existing or listed equipment conductors can be grouped together (taped or corded) without being installed in a raceway. Such cable groups should by location be protected from physical damage, being supported at intervals of 3 feet or less ≫ 620-21(a)(3)(d) ≪.

■ Escalator and Moving Walkway Wiring

> **NOTE**
> *Wheelchair lift and stairway chair lift raceways must be installed according to Section 620-21(c) provisions.*

> **NOTE**
> *Escalators, moving walks, wheelchair lifts, and stairway chair lifts must comply with Article 250* »620-84«.

Flexible metal conduit, liquidtight flexible metal conduit, or liquidtight flexible nonmetallic conduit are acceptable in escalator and moving walk wellways. Flexible metal conduit or liquidtight flexible conduit of ⅜ inch nominal trade size is permitted in lengths of 6 feet or less »620-21(b)(1)«. Liquidtight flexible nonmetallic conduit of ⅜ inch nominal trade size or larger, as defined in Section 351-22(2), can be installed in lengths greater than 6 feet »620-21(b)(1) *Exception*«.

Class 2 power-limited circuit cables can be installed within escalators and moving walkways provided the cables are (1) supported and protected from physical damage and (2) of a jacketed and flame-retardant type »620-21(b)(2)«.

Hard-service cords, conforming to Article 400 (Table 400-4) requirements, are permitted as flexible connections on escalator and moving walk control panels and disconnecting means, where the entire control panel and disconnecting means can be removed from machine spaces per Section 620-5 »620-21(b)(3)«.

■ Conductor Installation

Auxiliary gutters are not subject to the length restrictions of Section 374-2, nor to the number of conductors per Section 374-5 »620-35«.

Only electric wiring, raceways, and cables used in direct connection with the elevator or dumbwaiter are permitted inside the hoistway and machine room. Wiring for signals; for communication with the car; for lighting, heating, air conditioning; and for ventilating the hoistway are allowed »620-37(a)«.

Cables or raceway supports in a hoistway, escalator/moving walk wellway, or wheelchair lift and stairway chair lift runway, must be securely fastened to the guide rail; escalator/moving walk truss; or to the hoistway, wellway, or runway construction »620-34«.

Optical fiber cables and conductors for operating devices, operation and motion control, power, signaling, fire, alarm, lighting, heating, and air-conditioning circuits of 600 volts (or less) can run in the same traveling cable (or raceway) system if (1) all conductors are insulated for the maximum voltage applied to any conductor within the cable (or raceway) system; and (2) all live parts of the equipment are insulated from ground for the same maximum voltage. Such a traveling cable (or raceway) can also include shielded conductors and/or coaxial cable(s), if insulated for the maximum voltage applied to any conductor within the cable (or raceway) system. Conductors can be suitably shielded for telephone, audio, video, or higher frequency communications circuits »620-36«.

> **CAUTION** «««««««««««««««
> *Electrical equipment and wiring used for elevators, dumbwaiters, escalators, moving walks, and wheelchair lifts and stairway lifts in garages must comply with Article 511 requirements* »620-38«.

Bonding of elevator rails (car and/or counterweight) to a lightning protection system grounding-down conductor(s) is permitted. The lightning protection system must be located outside the hoistway. Elevator rails or other hoistway equipment cannot be used as the grounding-down conductor for lightning protection systems »620-37(b)«.

Main feeders supplying elevator and dumbwaiter power must be installed outside the hoistway unless: (1) by special permission, elevator feeders are permitted within an existing hoistway containing no spliced conductors; or (2) feeders are permitted inside the hoistway for elevators with driving machine motors located in the hoistway, or on the car, or on the counterweight »620-37(c)«.

Wireway

The cross-sectional area's sum of the individual conductors in a wireway must be no more than 50% of the wireway's interior cross-sectional area. Vertically run wireways must be securely supported at intervals of 15 feet or less, and must have a maximum of one joint between supports. Adjoining wireway sections must be securely fastened together to form a rigid joint »620-32«.

■ Branch Circuits and Traveling Cables

• •

Each machine room and/or machinery space requires at least one 125-volt, single-phase, duplex receptacle ≫620-23(c)≪.

A separate branch circuit must supply the machine room/machinery space lighting and receptacle(s). Required lighting must not be connected to a GFCI's load side ≫620-23(a)≪. Locate the machine room lighting switch at the machine room/machinery space point of entry ≫620-23(b)≪.

A separate branch circuit must supply each elevator car's lights, receptacle(s), auxiliary lighting power source, and ventilation. The branch circuit's overcurrent protective device must be located in the elevator machine room/machinery space ≫620-22(a)≪. A dedicated branch circuit is required for each elevator car's air-conditioning/heating unit. Locate this branch-circuit's overcurrent protective device in the elevator machine room/machinery space ≫620-22(b)≪.

Metal raceways and cables of Types MC, MI, or AC attached to an elevator car must be bonded to the car's grounded metal parts ≫620-81≪.

Traveling cables must be suspended at both the car and hoistway ends (or counterweight end if applicable) thereby reducing the strain on individual copper conductors to a minimum. Support traveling cables by any one of the following means:
(1) steel supporting member(s)
(2) looping the cables around supports for unsupported lengths less than 100 feet
(3) suspending from the supports by a means which automatically tightens around the cable with increased tension (for unsupported lengths up to 200 feet) ≫620-41≪.

Traveling cable can be run without the use of a raceway for a distance of 6 feet or less (measured from the first support point on the elevator car, hoistway wall, or counterweight (where applicable), provided the conductors are grouped and taped (or corded) together, or are still in the original sheath. Traveling cables can continue, as fixed wiring, to elevator controller enclosures as well as elevator car and machine room connections, if suitable support and physical damage protection is provided ≫620-44≪.

Careful location of traveling cable supports can reduce to a minimum the possibility of damage due to cables contacting hoistway construction/equipment. Provide suitable guards where necessary to protect the cables against damage ≫620-43≪.

Hoistway pit lighting and receptacle(s) must be supplied by a separate branch circuit. In no case should required lighting be connected to the load side of a ground-fault circuit-interrupter ≫620-24(a)≪. The lighting switch's location must be readily accessible from the pit access door ≫620-24(b)≪.

Each hoistway pit must have at least one 125-volt, single-phase, duplex receptacle ≫620-24(c)≪.

NOTE

For electric elevators, all motor frames, elevator machines, controllers, and metal electrical-equipment enclosures in (or on) the car, or in the hoistway, must be grounded per Article 250 ≫620-82≪.

CAUTION ≪≪≪≪≪≪≪≪≪≪≪≪≪≪

All 125-volt, single-phase, 15- and 20-ampere receptacles installed in machine rooms, machinery spaces, pits, on elevator car tops, and in escalator/moving walk wellways must be of the ground-fault circuit-interruption type. A single receptacle supplying a permanently installed sump pump does not require GFCI protection ≫620-85≪.

■ Protection and Control

Each unit must have a single means for disconnecting all ungrounded main power supply conductors which is designed so that no pole operates independently. Where multiple driving machines are connected to a single elevator, escalator, moving walk, or pumping unit, a single means must be provided to disconnect the motor(s) and control valve operating magnets. The main power supply disconnecting means must not disconnect the branch circuit required by Sections 620-22, 23, and 24 »620-51«.

The disconnecting means' location must be readily accessible to qualified persons »620-51(c)«. Depending on the application, the disconnecting means' location must comply with 1, 2, 3, or 4 in Section 620-51(c).

Where there are multiple driving machines in a machine room, the disconnecting means must be numbered to correspond with the identifying number of the driving machine being controlled. The disconnecting means must have a sign identifying the location of the supply side overcurrent protective device »620-51(d)«.

Equipment receiving electrical power from multiple sources must satisfy Section 620-52 provisions.

Each elevator car must have a single means for disconnecting all ungrounded car light, receptacle(s), and ventilation power-supply conductors. The disconnecting means, capable of locking in the open position, must be located in that elevator car's machine room/machinery space. In a machine room housing equipment for multiple elevator cars, each disconnecting means must be numbered to correspond with the identifying number of the car whose light source it controls. The disconnecting means must have a sign identifying the location of the supply side overcurrent protective device »620-53«.

Each elevator car must have a single means for disconnecting all ungrounded heating and air-conditioning power-supply conductors »620-54«. (The remainder of this provision is almost identical to Section 620-53.)

Operating devices and control/signaling circuits must be protected against overcurrent according to Sections 725-23 and 725-24. Class 2 power-limited circuits must be protected against overcurrent in accordance with the requirements of Chapter 9, Notes to Tables 11(a) and 11(b) »620-61(a)«.

Duty on elevator and dumbwaiter driving machine motors and driving motors of motor-generators (used with generator field control) is rated as **intermittent**. Such motors must be protected against overload per Section 430-33 »620-61(b)(1)«.

Duty on escalator and moving walk driving machine motors is rated as **continuous**. Protect such motors against overload in accordance with Section 430-32 »620-61(b)(2)«. Escalator and moving walk driving machine motors, and driving motors of motor-generator sets, must be protected against running overload as provided in Table 430-37 »620-61(b)(3)«.

Wheelchair lift and stairway chair lift driving machine motor duty is rated as **intermittent**. Such motors must be protected against overload according to Section 430-33 »620-61(b)(4)«.

CAUTION ⟨⟨⟨⟨⟨⟨⟨⟨⟨⟨⟨⟨⟨⟨⟨⟨⟨⟨⟨⟨⟨

Make no provision to open (or close) the disconnecting means from any remote part of the premises. If sprinklers are installed in hoistways, machine rooms, or machinery spaces, the disconnecting means can automatically open the power supply to the affected elevator(s) prior to water release. Automatic closure of this disconnecting means is prohibited. Power must be restored only by manual intervention »620-51(b)«.

Elevator, dumbwaiter, escalator, and moving walk driving machines; motor-generator sets; motor controllers; and disconnecting means must be installed in a room or enclosure dedicated for that purpose, unless otherwise permitted in Section 620-71(a) or (b). The room, or enclosure, must be secured against unauthorized access »620-71«.

N O T E

Elevator emergency and standby power systems must comply with Section 620-91.

SWIMMING POOLS, FOUNTAINS, AND SIMILAR INSTALLATIONS

■ General

Article 680 provisions apply to the electrical wiring construction (and installation) for equipment in, or adjacent to, all swimming, wading, therapeutic, and decorative pools, fountains, hot tubs, spas, and hydromassage bathtubs, whether permanently installed or storable, as well as to similar metallic auxiliary equipment (such as pumps, filters, etc.) ≫ 680-1 ≪ .

Permanently installed pool underwater lighting fixture requirements are located in Section 680-20.

Permanently Installed Swimming, Wading, and Therapeutic Pools: Those that are constructed totally or partially in the ground, all others with depth capacities greater than 42 inches, and all pools installed inside of a building, regardless of water depth, whether or not served by electrical circuits of any nature ≫ 680-4 ≪ .

Permanently installed pools must comply with Article 680, Part B.

Fountain: Specific to this article, fountain includes fountains, ornamental pools, display pools, and reflection pools. Drinking fountains are not included ≫ 680-4 ≪ .

Part E provisions apply to all permanently installed fountains as defined in Section 680-4. Self-contained, portable fountains (no larger than 5 feet in any dimension) are not covered here. Fountains sharing water with a pool must comply with this article's pool requirements ≫ 680-50 ≪ .

Permanently Installed Decorative Fountains and Reflection Pools: Those that are constructed in, or on, the ground, or in a building so that the fountain cannot be readily disassembled, with or without electrical circuits of any nature. These units are constructed primarily for their aesthetic value, and are not intended for swimming or wading ≫ 680-4 ≪ .

Spa or Hot Tub: A hydromassage pool, or tub for recreational or therapeutic use, not located in health care facilities, designed for immersion of users, and usually having a filter, heater, and motor-driven blower. It may be installed either indoors or out, on, or in, the ground or supporting structure. Generally, a spa or hot tub is not designed to be drained after each use ≫ 680-4 ≪ .

Spa and hot tub requirements can be found in Article 680, Part D (680-38 through 42).

Bonding provisions, for permanently installed pools, are located in Section 680-22.

Pool: Includes swimming, wading, and permanently installed therapeutic pools ≫ 680-4 ≪ .

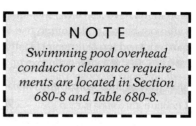

- - - - - - - - - - - - - - - - - -
| N O T E |
Swimming pool overhead conductor clearance requirements are located in Section 680-8 and Table 680-8.
- - - - - - - - - - - - - - - - - -

All electrical equipment installed in the water, walls, or decks of pools, fountains, and similar applications must comply with the provisions of this article ≫ 680-2 ≪ .

Except as modified by this section, wiring and equipment in, or adjacent to, pools and fountains, must comply with the applicable requirements of Chapters 1 through 4 ≫ 680-3(a) ≪ .

Audio equipment installation and wiring, adjacent to pools and fountains, must comply with applicable Article 640 requirements. Underwater loudspeakers must be installed according to Section 680-23 ≫ 680-3(b) ≪ .

Section 680-25 contains requirements for grounding of underwater lighting fixtures, junction boxes, metal transformer enclosures, panelboards, motors, together with other electrical enclosures and equipment.

Storable pool provisions are located in Article 680, Part C (680-30 through 32).

The provisions of Article 680, Part F shall apply to therapeutic pools and tubs in health care facilities, gymnasiums, athletic training rooms, and similar areas ≫ 680-60 ≪ .

Hydromassage bathtub requirements are located in Article 680, Part G.

■ Definitions

Self-Contained Spa or Hot Tub:
Factory-fabricated unit consisting of a spa or hot tub vessel having integrated water-circulating, heating, and control equipment. Equipment may include pumps, air blowers, heaters, lights, controls, sanitizer generators, etc. ≫680-4 ≪.

Storable Swimming or Wading Pool: Those constructed on, or above, the ground having a maximum water depth capacity of 42 inches, or a pool with nonmetallic, molded polymeric walls (or inflatable fabric walls) regardless of dimension ≫680-4 ≪.

Pool Cover, Electrically Operated:
Motor-driven equipment designed to cover and uncover the pool's water surface by means of a flexible sheet or rigid frame ≫680-4 ≪.

Packaged Spa or Hot Tub Equipment Assembly: A factory-fabricated unit consisting of water-circulating, heating, and control equipment mounted on a common base, which operates as a spa or hot tub. Equipment may include pumps, air blowers, heaters, lights, controls, sanitizer generators, etc. ≫680-4 ≪.

Hydromassage Bathtub: A permanently installed bathtub with recirculation piping, pump, and associated equipment. It is designed to accept, circulate, and discharge water upon each use ≫680-4 ≪.

Cord- and Plug-Connected Lighting Assembly: A lighting assembly consisting of a lighting fixture intended for installation in the wall of a spa, hot tub, or storable pool, having a cord- and plug-connected transformer ≫680-4 ≪.

Self-Contained Therapeutic Tubs or Hydrotherapeutic Tanks: A factory-fabricated unit consisting of a therapeutic tub or hydrotherapeutic tank with integrated water-circulating, heating, and control equipment. Equipment may include pumps, air blowers, heaters, lights, controls, sanitizer generators, etc. ≫680-4 ≪.

Wet-Niche Lighting Fixture:
A lighting fixture intended for installation in a pool or fountain structure's forming shell, where completely surrounded by water ≫680-4 ≪.

Forming Shell: A support structure designed for a wet-niche lighting fixture assembly and intended for pool or fountain structure mounting ≫680-4 ≪.

No-Niche Lighting Fixture: A lighting fixture intended for above or below water installation without a niche ≫680-4 ≪. Note: Not to be mistaken for a dry-niche lighting fixture.

Dry-Niche Lighting Fixture: A lighting fixture intended for installation in the wall of a pool or fountain in a niche that is sealed against water entry ≫680-4 ≪.

■ Common Bonding Grid

Structural reinforcing steel, or the walls of bolted or welded metal pool structures, can serve as a common bonding grid for nonelectrical parts where connections are made in accordance with Section 250-8 »680-22(b)«. Grounding conductors and bonding jumpers must be connected by exothermic welding, listed pressure connectors, listed clamps, or other listed means »250-8«.

Metal Walls (bolted together)

A common bonding grid can be any of the following:
(1) The structural reinforcing steel of a concrete pool where the rods are bonded together by the usual steel tie wires, or an equivalent.
(2) The wall of a bolted or welded metal pool.
(3) A solid copper conductor (insulated, covered, or bare), no smaller than No. 8.
(4) Brass rigid metal conduit or intermediate metal conduit, or other appropriately identified corrosion-resistant metal conduit »680-22(b)«.

No. 8 Solid Copper

Corrosion-Resistant Metal Conduit

Reinforcing Steel

CAUTION ««««««««««««««««««««««

*The parts specified in Section 680-22(a) must be connected to a common bonding grid with a No. 8 or larger solid copper conductor (insulated, covered, or bare). Connections must be made by exothermic welding, or by pressure connectors/clamps **labeled** as suitable for the purpose and made of the following material: stainless steel, brass, copper, or copper alloy »680-22(b)«.*

■ Receptacles

CAUTION ««««««««

A receptacle installed in a wet location, unattended while the plugged-in product is in use, must have an enclosure that is weatherproof at all times, i.e., attachment plug cap inserted or removed »410-57(b)(1)«.

Receptacle(s) providing power for a permanently installed pool or fountain's water-pump motor(s), or other loads directly related to the circulation and sanitation system (permitted in Section 680-7), can be located between 5 feet and 10 feet from the inside walls of the pool or fountain. If so located, the receptacle(s) must be **single**, of the **locking and grounding type**, protected by ground-fault circuit interrupter(s) »680-6(a)(1)«.

A dwelling unit's permanently installed pool must have at least one 125-volt 15- or 20- ampere receptacle on a general-purpose branch circuit which is located a minimum of 10 feet, but no more than 20 feet, from the pool's inside wall. This receptacle must not be more than 6 feet 6 inches above the floor, platform, or grade level serving the pool »680-6(a)(2)«.

Dwelling Unit

N O T E

All 125-volt receptacles located within 20 feet of the pool or fountain's inside walls must be protected by a ground-fault circuit-interrupter »680-6(a)(3)«. This distance is determined by measuring the shortest path of the appliance supply cord (connected to the receptacle) would follow without piercing a floor, wall, ceiling, hinged or sliding panel doorway, window opening, or other effective permanent barrier »680-6(a)(3) FPN«.

5 Feet · 5 to 10 Feet · 10 to 20 Feet

Required

Single, Locking and Grounding Type. GFCI Protected

GFCI Protected

Fixed or stationary equipment rated 20 amperes or less, other than an underwater lighting fixture for a permanently installed pool, can be connected with a flexible cord to facilitate removal (or disconnection) during maintenance/ repair. For other than storable pools, the flexible cord must be no longer than 3 feet, and must have a copper equipment grounding conductor (not smaller than No. 12) with a grounding-type attachment plug »680-7«.

■ Bonding

Bond all metal-sheathed cables/race-ways, metal piping, and all fixed metal parts that are not separated from the pool by a permanent barrier within 5 feet horizontally of the pool's inside wall, and within 12 feet above the pool's maximum water level ≫680-22(a)(5)≪. Examples of fixed metal parts include: fences/fence posts, gutter down-spouts, door/window frames, raceways, etc.

All parts listed in Section 680-22(a)(1) through (5) must be bonded together ≫680-22(a)≪. Those parts must also be connected to a common bonding grid ≫680-22(b)≪.

Bond all metal fittings within, or attached to, the pool structure ≫680-22(a)(3)≪. Note: Each ladder/hand rail anchor must be bonded unless not over 4 inches in any dimension.

Bond all metal parts of equipment associated with pool covers, including electric motors ≫680-22(a)(4)≪.

Bond all forming shells and mounting brackets of no-niche fixtures, unless a listed low-voltage lighting system is used which does not require bonding ≫680-22(a)(2)≪.

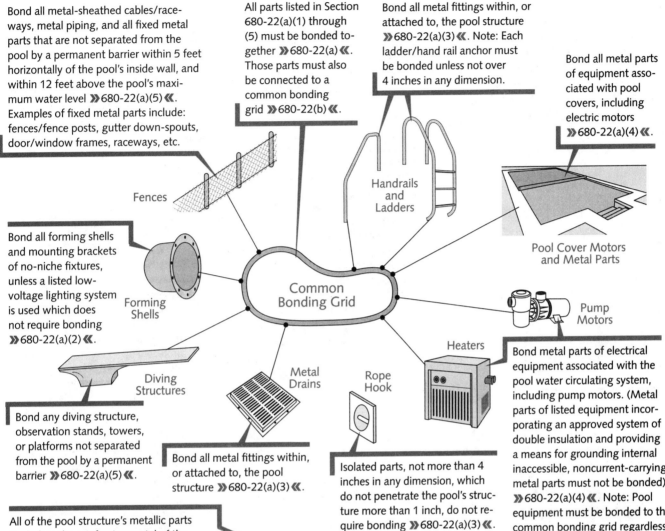

Fences

Handrails and Ladders

Pool Cover Motors and Metal Parts

Forming Shells

Common Bonding Grid

Pump Motors

Diving Structures

Metal Drains

Rope Hook

Heaters

Steel Tie Wires

Reinforcing Steel

Bond any diving structure, observation stands, towers, or platforms not separated from the pool by a permanent barrier ≫680-22(a)(5)≪.

Bond all metal fittings within, or attached to, the pool structure ≫680-22(a)(3)≪.

Isolated parts, not more than 4 inches in any dimension, which do not penetrate the pool's structure more than 1 inch, do not require bonding ≫680-22(a)(3)≪.

Bond metal parts of electrical equipment associated with the pool water circulating system, including pump motors. (Metal parts of listed equipment incorporating an approved system of double insulation and providing a means for grounding internal inaccessible, noncurrent-carrying metal parts must not be bonded) ≫680-22(a)(4)≪. Note: Pool equipment must be bonded to the common bonding grid regardless of the intervening distance and/or location to the pool.

All of the pool structure's metallic parts (including the reinforcing metal of the pool shell, coping stones, and deck) must be bonded. The usual steel tie wires are suitable for bonding the reinforcing steel together; no welding or special clamping is required. These tie wires must be made tight ≫680-22(a)(1)≪. Note: Steel tie wires are permitted to bond only reinforcing steel *together*. Steel tie wires must not be used to connect the No. 8 copper bonding conductor to the reinforcing steel.

If reinforcing steel is effectively insulated at the time of manufacture by a listed encapsulating nonconductive compound, which is maintained during installation, bonding is not required ≫680-22(a)(1)≪.

In the case of pool water heaters (rated more than 50 amperes) having specific bonding and grounding instructions, only those parts designated to be bonded shall be bonded, and only those parts designated to be grounded shall be grounded ≫680-22(c)≪.

Underwater audio equipment must comply with Section 680-23 provisions.

In addition to bonding, some of these items must be grounded according to Section 680-25 requirements.

NOTE

This section's intention is not to require that No. 8 or larger solid copper bonding conductor be extended (or attached) to any remote panelboard, service equipment, or any electrode, but rather that is must be employed to eliminate voltage gradients in the pool area as prescribed ≫680-22≪.

CAUTION ≪≪≪≪≪≪≪≪≪≪≪≪≪≪≪≪

The parts specified in Section 680-22(a) must be connected to a common bonding grid via a solid copper conductor (insulated, covered, or bare) no smaller than No. 8. Connections must be made by exothermic welding, or by pressure connectors/clamps **labeled** *as suitable for that purpose and made of the following material: stainless steel, brass, copper, or copper alloy ≫680-22(b)≪.*

■ Lighting Fixtures, Lighting Outlets, and Ceiling-Suspended (Paddle) Fans

In outdoor pool areas, lighting fixtures, lighting outlets, and ceiling-suspended (paddle) fans must not be installed over the pool, or over the area extending 5 feet horizontally from the pool's inside walls, unless no part of the lighting fixture or ceiling-suspended (paddle) fan is less than 12 feet above the maximum water level »680-6(b)(1)«.

Section 680-6(b)(1) limitations do not apply to indoor pool areas if all of the following conditions are satisfied: (a) fixtures are of a totally enclosed type; (b) the branch circuit supplying the fixtures and/or ceiling-suspended (paddle) fans are ground-fault circuit-interrupter protected; and (c) the bottom of the fixture or ceiling-suspended (paddle) fan is at least 7 feet, 6 inches above the maximum water level »680-6(b)(3)«.

Cord-connected lighting fixtures must meet the same Section 680-7 specifications as other cord- and plug-connected equipment, where installed within 16 feet of any point along the water's surface, measured radially »[680-6(b)(5)«.

NOTE
Locate all switching devices on the property at least 5 feet horizontally from a pool's inside walls, unless separated from the pool by a solid fence, wall, or other permanent barrier »680-6(c)«.

Existing lighting fixtures and lighting outlets located less than 5 feet (measured horizontally from the pool's inside walls) must be: (1) at least 5 feet above the maximum water level surface; (2) rigidly attached to the existing structure; and (3) protected by a ground-fault circuit-interrupter »680-6(b)(2)«.

Lighting fixtures and lighting outlets, installed in the area extending between 5 feet and 10 feet horizontally from a pool's inside walls, must be protected by a ground-fault circuit-interrupter, unless 5 feet above the maximum water level and rigidly attached to the pool's adjacent (or enclosing) structure »680-6(b)(4)«.

■ Hydromassage Bathtubs

CAUTION ‹‹‹‹‹‹‹‹‹‹‹‹‹

A receptacle must not be installed within a bathtub or shower space »410-57(c)«. The bathtub space (zone) measures 3 feet horizontally, and 8 feet vertically, from the top of the bathtub rim or shower stall threshold »410-4(d)«.

All bathroom receptacles must have ground-fault circuit-interrupter protection for personnel »210-8(a)(1) and (b)(1)«.

All 125-volt, single-phase receptacles within 5 feet (measured horizontally) of hydromassage tub's inside walls must be protected by ground-fault circuit interrupter(s) »680-70«.

NOTE
Unrelated lighting fixtures, switches, receptacles, and other electrical equipment located in the same room, must be installed in accordance with the requirements of Chapters 1 through 4 in the NEC® covering the installation of that equipment in bathrooms »680-71«.

Hydromassage bathtubs, and associated electrical components, must be protected by a ground-fault circuit-interrupter »680-70«.

Hydromassage bathtub electrical equipment must be accessible without damage to the building structure or finish »680-72«.

All metal piping systems, electrical equipment metal parts, and pump motors associated with the hydromassage tub must be bonded together using a No. 8 (or larger) solid copper bonding jumper (insulated, covered, or bare). Metal parts of listed equipment incorporating an approved double insulation system and providing a grounding means for internal nonaccessible, noncurrent-carrying metal parts must not be bonded »680-73«.

■ Underwater Lighting Fixtures (Permanently Installed Pools)

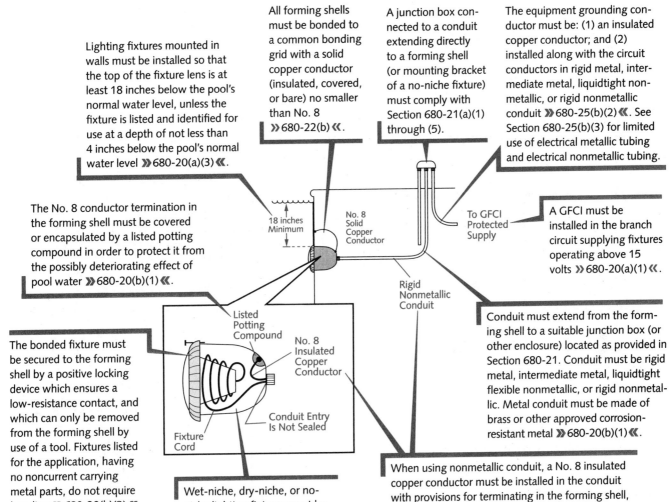

Lighting fixtures mounted in walls must be installed so that the top of the fixture lens is at least 18 inches below the pool's normal water level, unless the fixture is listed and identified for use at a depth of not less than 4 inches below the pool's normal water level ≫680-20(a)(3)≪.

All forming shells must be bonded to a common bonding grid with a solid copper conductor (insulated, covered, or bare) no smaller than No. 8 ≫680-22(b)≪.

A junction box connected to a conduit extending directly to a forming shell (or mounting bracket of a no-niche fixture) must comply with Section 680-21(a)(1) through (5).

The equipment grounding conductor must be: (1) an insulated copper conductor; and (2) installed along with the circuit conductors in rigid metal, intermediate metal, liquidtight nonmetallic, or rigid nonmetallic conduit ≫680-25(b)(2)≪. See Section 680-25(b)(3) for limited use of electrical metallic tubing and electrical nonmetallic tubing.

The No. 8 conductor termination in the forming shell must be covered or encapsulated by a listed potting compound in order to protect it from the possibly deteriorating effect of pool water ≫680-20(b)(1)≪.

A GFCI must be installed in the branch circuit supplying fixtures operating above 15 volts ≫680-20(a)(1)≪.

The bonded fixture must be secured to the forming shell by a positive locking device which ensures a low-resistance contact, and which can only be removed from the forming shell by use of a tool. Fixtures listed for the application, having no noncurrent carrying metal parts, do not require bonding ≫680-20(b)(3)≪.

Conduit must extend from the forming shell to a suitable junction box (or other enclosure) located as provided in Section 680-21. Conduit must be rigid metal, intermediate metal, liquidtight flexible nonmetallic, or rigid nonmetallic. Metal conduit must be made of brass or other approved corrosion-resistant metal ≫680-20(b)(1)≪.

Wet-niche, dry-niche, or no-niche lighting fixtures must be connected to a No. 12 or larger equipment grounding conductor sized per Table 250-122.

When using nonmetallic conduit, a No. 8 insulated copper conductor must be installed in the conduit with provisions for terminating in the forming shell, junction box/transformer enclosure, or ground-fault circuit-interrupter enclosure; unless using a listed low-voltage lighting system which does not require grounding ≫680-20(b)(1)≪.

Dry-niche lighting fixtures must comply with Section 680-20(c) provisions.

No-niche fixtures must comply with Section 680-20(d) requirements.

The junction box or transformer/other enclosure in the supply circuit to a wet-niche (or no-niche) lighting fixture and the field-wiring chamber of a dry-niche lighting fixture, must be grounded to the panelboard's equipment grounding terminal. This terminal must be directly connected to the panelboard enclosure. The equipment grounding conductor must be installed without joint or splice, except as permitted in Section 680-25(b)(4)(a) and (b) ≫680-25(b)(4)≪.

Underwater audio equipment provisions are located in Section 680-23.

> ### NOTE
> *Conductors on the load side of a ground-fault circuit interrupter or transformer, used to comply with Section 680-20(a)(1) provisions, must not occupy raceways, boxes, or enclosures containing other conductors, unless the other conductors are either protected by ground-fault circuit interrupters or are grounding conductors. Feed-through type ground-fault circuit-interrupter supply conductors can occupy the same enclosure. Ground-fault circuit interrupters are permitted in a panelboard containing circuits protected by means other than ground-fault circuit-interrupters ≫680-5(c)≪.*

Junction Boxes for Underwater Lighting

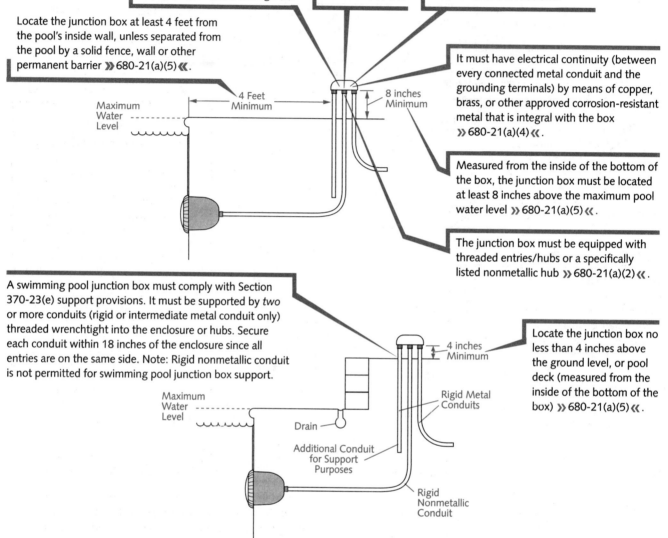

A junction box connected to a conduit extending directly to a forming shell (or mounting bracket of a no-niche fixture) must comply with Section 680-21(a)(1) through (5).

It must be listed and labeled for the purpose ≫ 680-21(a)(1) ≪ .

The junction box must be made of copper, brass, suitable plastic, or other approved corrosion-resistant material ≫ 680-21(a)(3) ≪ .

Locate the junction box at least 4 feet from the pool's inside wall, unless separated from the pool by a solid fence, wall or other permanent barrier ≫ 680-21(a)(5) ≪ .

It must have electrical continuity (between every connected metal conduit and the grounding terminals) by means of copper, brass, or other approved corrosion-resistant metal that is integral with the box ≫ 680-21(a)(4) ≪ .

Measured from the inside of the bottom of the box, the junction box must be located at least 8 inches above the maximum pool water level ≫ 680-21(a)(5) ≪ .

The junction box must be equipped with threaded entries/hubs or a specifically listed nonmetallic hub ≫ 680-21(a)(2) ≪ .

Maximum Water Level

4 Feet Minimum

8 inches Minimum

A swimming pool junction box must comply with Section 370-23(e) support provisions. It must be supported by *two* or more conduits (rigid or intermediate metal conduit only) threaded wrenchtight into the enclosure or hubs. Secure each conduit within 18 inches of the enclosure since all entries are on the same side. Note: Rigid nonmetallic conduit is not permitted for swimming pool junction box support.

Locate the junction box no less than 4 inches above the ground level, or pool deck (measured from the inside of the bottom of the box) ≫ 680-21(a)(5) ≪ .

Maximum Water Level

Drain

Additional Conduit for Support Purposes

4 inches Minimum

Rigid Metal Conduits

Rigid Nonmetallic Conduit

Any enclosure for a transformer, ground-fault circuit-interrupter, or a similar device connected to a conduit extending directly to a forming shell (or mounting bracket of a no-niche fixture) must comply with Section 680-21(b) requirements.

Junction boxes and enclosures, mounted above the grade of the finished walkway around the pool, must not be located in the walkway unless afforded additional protection, such as by location under diving boards, adjacent to fixed structures, and the like. [680-21(c)]

Flush deck box provisions are located in Section 680-21(a)(5).

Junction boxes, transformer enclosures, and ground-fault circuit-interrupter enclosures connected to a conduit extending directly to a forming shell (or mounting bracket of a no-niche fixture) must have a number of grounding terminals which exceeds the number of conduit entries by at least one ≫ 680-21(d) ≪ .

Strain relief must be provided for the termination of an underwater lighting fixture's flexible cord within a junction box, transformer enclosure, ground-fault circuit-interrupter or other enclosure ≫ 680-21(e) ≪ .

■ Equipment and Panelboards

Connect pool-associated motor(s) to a No. 12 or larger equipment grounding conductor sized according to Table 250-122. It must be an insulated copper conductor and must be installed together with the circuit conductors in rigid metal conduit, intermediate metal conduit, rigid nonmetallic conduit, or type MC cable listed for the application. If installed on (or within) buildings, electrical metallic tubing can be used to protect the conductors ≫ 680-25(c) ≪.

Where connections must be flexible at (or adjacent to) the motor, liquidtight flexible metal or nonmetallic conduit with approved fittings is permitted ≫ 680-25(c) ≪.

An accessible disconnecting means must be provided within sight from all pools, spas, and hot tub equipment. It must be located 5 feet or more from the inside walls of the pool, spa, or hot tub ≫ 680-12 ≪.

WARNING

Wiring which supplies pool pump motors rated 15 and 20 amperes, 125 volt or 240 volt, single phase, whether by receptacle or direct connection, must have ground-fault circuit-interrupter protection for personnel, unless located in dwelling units ≫ 680-6(d) ≪.

NOTE
The only underground wiring permitted under the pool, or within the area extending 5 feet horizontally from the pool's inside wall, is the wiring necessary to supply pool equipment (permitted by this article). Where space limitations prevent wiring outside the 5-foot restricted area, such wiring is permitted if installed in rigid metal conduit, intermediate metal conduit, or a nonmetallic raceway system. All metal conduit must be corrosion resistant and suitable for the location ≫ 680-10 ≪. Minimum burial depths are also provided in this section.

A panelboard and a disconnecting means (where installed), which are not part of the service equipment or source of a separately derived system, must have an equipment grounding conductor between its grounding terminal and the grounding terminal of the applicable service equipment or separately derived system source. This conductor, sized according to Table 250-122, must not be smaller than No. 12. This insulated conductor must be installed along with the feeder conductors in rigid metal, intermediate metal, liquidtight flexible nonmetallic, or rigid nonmetallic conduit. Electrical metallic tubing can protect conductors where installed on, or within, the building per Article 348. Electrical nonmetallic tubing can enclose conductors installed within the building in accordance with Article 331. The equipment grounding conductor must be connected to a panelboard's equipment grounding terminal and, where installed, to the disconnecting means enclosure ≫ 680-25(d) ≪.

The following equipment must be grounded:
(1) Wet-niche and no-niche underwater lighting fixtures, other than those low-voltage systems listed for the application without a grounding conductor
(2) Dry-niche underwater lighting fixtures
(3) All electrical equipment located within 5 feet of the pool's inside wall
(4) All electrical equipment associated with the pool's recirculating system
(5) Junction boxes
(6) Transformer enclosures
(7) Ground-fault circuit-interrupters
(8) Panelboards (not part of the service equipment) which supply any pool associated electrical equipment ≫ 680-24 ≪.

A separate building's panelboard can supply swimming pool equipment if the feeder meets Section 250-32 grounding requirements. If an equipment grounding conductor is installed, it must be an insulated conductor ≫ 680-25(d)(2) ≪.

In the interior of a one-family dwelling, or in another associated building/structure, any of Chapter 3's wiring methods, that contain an insulated (or outer sheath covered) copper equipment grounding conductor (no smaller than No. 12),

can be used to connect pool-associated motors. Flexible cord is permitted in accordance with Section 680-7 ≫ 680-25(c) ≪.

The heating elements of electric pool water heaters must be subdivided into loads not exceeding 48 amperes, and protected at no more than 60 amperes. The branch-circuit conductor ampacity and the overcurrent protective devices rating/setting must be at least 125% of the nameplate rating's total load ≫ 680-9 ≪.

Electric equipment must not be installed in rooms (or pits) without adequate drainage to prevent water accumulation during normal operation or filter maintenance ≫ 680-11 ≪.

Electrically operated pool cover requirements are in Section 680-26.

Deck-area heater provisions are located in Section 680-27.

A permanently installed pool can have listed cord- and plug-connected pool pumps which incorporate an approved double insulation system and provides a grounding means for the pump's internal and inaccessible, noncurrent-carrying metal parts ≫ 680-28 ≪.

■ Storable Pools

Weatherproof
While-in-Use
Cover

No Maximum Dimension

42 inches
Maximum

Storable Swimming or Wading Pool: Those constructed on or above ground with a maximum water depth capacity of 42 inches, or a pool with nonmetallic, molded polymeric walls, or inflatable fabric walls, of unlimited dimension 》680-4 《.

GFCI
Protected
Receptacle

No Bonding
Required

A wet location receptacle, into which is plugged a pool pump requires a continuously weatherproof enclosure, i.e., whether the attachment plug is inserted or removed 》410-57(b)(1) 《.

A cord-connected pool filter pump must incorporate an approved double insulation system, or its equivalent, and must have a means for grounding the appliance's internal and nonaccessible noncurrent-carrying metal parts only. The grounding means must consist of an equipment grounding conductor (run with the power-supply conductors) in the flexible cord which properly terminates in a grounding-type attachment plug with a fixed grounding contact member 》680-30 《.

A lighting fixture installed in, or on, a storable pool's wall must be part of a cord- and plug-connected lighting assembly. This assembly must have:

(1) no exposed parts;

(2) a fixture lamp operating at no more than 15 volts;

(3) an impact-resistant polymeric lens, fixture body, and transformer enclosure;

(4) a transformer meeting Section 680-5(a) requirements with a primary rating not over 150 volts; and

(5) been listed as an assembly for the intended purpose 》680-32(a) 《.

CAUTION 《《《《《《

Only ground-fault circuit-interrupter protected electrical equipment (including power-supply cords) can be used with storable pools 》*680-31* 《. *If flexible cords are used, see Section 400-4* 》*680-31 FPN* 《.

A lighting assembly (without a transformer) whose fixture lamp(s) operate at 150 volts or less, can be cord- and plug-connected where the assembly complies with all of the following:

(1) it has no exposed metal parts;

(2) it has an impact-resistant polymeric lens and fixture body;

(3) a ground-fault circuit-interrupter with open neutral protection is provided as an integral part of the assembly;

(4) the fixture lamp is permanently connected to the ground-fault circuit-interrupter with open-neutral protection;

(5) it complies with Section 680-20(a) requirements; and

(6) it is listed as an assembly for this specific purpose 》680-32(b) 《.

■ Spas and Hot Tubs

• •

Lighting fixtures, lighting outlets, and ceiling-suspended (paddle) fans located over the spa/hot tub, less than 5 feet from the spa or hot tub's inside walls must not only be a minimum of 7 feet, 6 inches, above the maximum water level but must also be ground-fault circuit-interrupter protected »680-41(b)(1) «.

Lighting fixtures, meeting either (a) or (b) requirements, *and* protected by a ground-fault circuit-interrupter, can be installed less than 7 feet, 6 inches, over a spa or hot tub:

(a) Recessed fixtures having a glass or plastic lens and nonmetallic or electrically isolated metal trim, suitable for damp location use

(b) Surface-mounted fixtures, with a glass or plastic globe, and a nonmetallic body or a metallic body isolated from contact. Such fixtures must be suitable for use in damp locations »680-41(b)(2) «.

Spa or Hot Tub Installed Indoors

Lighting fixtures, lighting outlets, and ceiling-suspended (paddle) fans, located 12 feet or more above the maximum water level, do not require ground-fault circuit-interrupter protection »680-41(b)(1) «.

Switches must be located at least 5 feet, measured horizontally, from the spa or hot tub's inside walls »680-41(c) «.

A readily accessible, clearly labeled emergency shutoff or control switch, for stopping the motor(s) that provide recirculating and jet system power, must be installed at least 5 feet away, adjacent to, and within sight of the spa or hot tub. Note: This requirement does not apply to single-family dwellings »680-38 «.

Receptacles of 125 volts, within 10 feet of the spa or hot tub's inside walls, must be protected by a ground-fault circuit-interrupter »680-41(a)(2) «. Determine this distance by measuring the shortest path the supply cord of an appliance connected to the receptacle would follow without piercing a building's floor, wall, ceiling or other effective permanent barrier »680-41(a)(2) FPN «.

Locate property receptacles at least 5 feet (measured horizontally) from the spa or hot tub's inside walls »680-41(a)(1) «.

At least one 125-volt, 15- or 20-ampere general-purpose branch-circuit receptacle must be located a minimum of 5 feet, and no more than 10 feet, from the inside wall of an indoor spa or hot tub »680-41(a) «.

NOTE

*A spa or hot tub installed **outdoors** must comply with Article 680, Part A and B provisions unless otherwise permitted in the following: (a) listed packaged units utilizing a factory-installed remote panelboard can be connected with 6 feet or less of liquidtight flexible conduit, or can be cord and plug connected with a cord no longer than 15 feet, if protected by a ground-fault circuit-interrupter. (b) bonding by metal-to-metal mounting on a common frame or base is permitted. The metal bands or hoops used to secure wooden staves do not require bonding as stipulated in Section 680-22 »680-40 «.*

Receptacles providing spa or hot tub power must be ground-fault circuit-interrupter protected »680-41(a)(3) «.

An indoor spa or hot tub must comply with Article 680 provisions (Parts A and B) unless modified by this section, and must be connected by a Chapter 3 approved wiring method. Listed spa and hot tub packaged units, rated 20 amperes or less, can be cord and plug connected to facilitate the removal or disconnection of the unit for maintenance and repair »680-41 «.

Indoor spas and hot tubs must be bonded in accordance with Sections 680-41(d) and (e).

All of the following must be grounded: (1) electric equipment located within 5 feet of the spa or hot tub's inside wall; and (2) electric equipment associated with the spa or hot tub's circulating system »680-41(f) «.

All electrical equipment must be grounded in accordance with Article 250 and be connected by one of the wiring methods of Chapter 3 »680-41(g)(1) «. If the equipment is connected with a flexible cord, the equipment grounding conductor must be connected to a fixed metal part of the assembly »680-41(g)(2) «.

Use only listed electric spa or hot tub water heaters whose elements are subdivided into 48 ampere (and under) loads, and are protected at not more than 60 amperes. The branch-circuit conductor ampacity, and the overcurrent protective device rating/setting, must not be less than 125% of the nameplate rating's total load »680-41(h) «.

■ Pools and Tubs for Therapeutic Use

Article 680, Part F provisions apply to therapeutically used pools and tubs in health care facilities, gymnasiums, athletic training rooms, and similar areas. See Section 517-3 for **health care facilities** definition. Portable therapeutic appliances must comply with Article 422 ≫ 680-60 ≪.

All tub associated metal parts must be bonded by any of the following methods: (1) interconnection of threaded metal piping and fittings; (2) metal-to-metal mounting on a common frame or base; (3) suitable metal clamp connections; or (4) the provision of a solid copper bonding jumper (insulated, covered, or bare) no smaller than No. 8 ≫ 680-62(d) ≪.

Outlet(s) supplying the following must be protected by a ground-fault circuit-interrupter: (1) self-contained therapeutic tub or hydrotherapeutic tank; or (2) packaged therapeutic tub or hydrotherapeutic tank; or (3) field-assembled therapeutic tub or hydrotherapeutic tank with a 50-ampere or less heater load ≫ 680-62(a) ≪.

A listed self-contained unit, or listed packaged equipment assembly, marked to indicate that integral ground-fault circuit-interrupter protection is provided for all electrical parts within the unit or assembly (pumps, air-blowers, heaters, lights, controls, sanitizer generators, wiring, etc.), does not require that the outlet supply be ground-fault circuit-interrupter protected ≫ 680-62(a) ≪.

A field-assembled therapeutic tub or hydrotherapeutic tank, rated greater than 250 volts or rated 3 phase, does not require supply protection by means of a ground-fault circuit interrupter ≫ 680-62(a) ≪.

All electrical equipment must be grounded in accordance with Article 250 and connected by Chapter 3 wiring methods ≫ 680-62(f)(1) ≪.

Equipment connected with a flexible cord, requires that the equipment grounding conductor be connected to a metal part of the assembly ≫ 680-62(f)(2) ≪.

All receptacles within 5 feet of a therapeutic tub must be protected by a ground-fault circuit-interrupter ≫ 680-62(g) ≪.

All of the following must be **grounded**: (1) Electrical equipment located within 5 feet of the tub's inside wall. (2) Electrical equipment associated with the tub's circulating system ≫ 680-62(e) ≪.

The following parts must be bonded together:
(1) All metal fittings within, or attached to, the tub structure.
(2) Metal parts of electrical equipment associated with the tub's water circulating system, including pump motors.
(3) Metal-sheathed cables and raceways and metal piping within 5 feet of the tub's inside walls and not separated from the tub by a permanent barrier.
(4) All metal surfaces within 5 feet of the tub's inside walls not separated by a permanent barrier.
(5) Electrical devices and controls unrelated to the therapeutic tubs must either be located at least 5 feet away from such units, or be bonded to the therapeutic tub system ≫ 680-62(c) ≪.

Tubs used for the therapeutic submersion and treatment of patients, which during normal use are not easily moved from place to place, or which are fastened or secured at a given location (including associated piping systems) must conform to Section 680-62(a) through (e) ≫ 680-62 ≪.

NOTE
A ground-fault circuit-interrupter must protect all therapeutic equipment ≫ *680-62(b)* ≪. *Portable therapeutic appliances must comply with Section 250-114* ≫ *680-62(b) Exception* ≪.

CAUTION ≪≪≪≪≪≪≪
All lighting fixtures used in therapeutic tub areas must be of the totally enclosed type ≫ *680-62(h)* ≪.

Therapeutic pools constructed in, or on, the ground, or within a building, in such a manner that the pool cannot be readily disassembled, must comply with Article 680, Parts A and B ≫ 680-61 ≪. The limitations of Sections 680-(b)(1) and (2) do not apply if all lighting fixtures are of the totally enclosed type ≫ 680-61 *Exception* ≪.

Summary

- Article 600 covers the conductor and equipment installation for electric signs and outline lighting (including neon and skeleton tubing).
- Manufactured wiring system provisions are located in Article 604.
- Article 605 (Office Furnishings) covers electrical equipment, lighting accessories, and wiring systems used to connect, or contained within, wired partitions.
- Electric vehicle charging system provisions are located in Article 625.
- Audio signal processing, amplification, and reproduction equipment requirements can be found in Article 640.
- Article 645 covers information technology (data processing) equipment.
- Electrically operated pipe organ circuits and components employed for sounding apparatus and keyboard control are located in Article 650.
- Article 695 (Fire Pumps) covers the installation of electric power sources and interconnecting circuits as well as switching and control equipment dedicated to fire pump drivers.
- Elevator, dumbwaiter, escalator, moving walk, wheelchair lift, and stairway chair lift provisions are contained in Article 620.

- Article 680 provisions apply to the construction and installation of electrical wiring and equipment in, or near, all swimming, wading, therapeutic, and decorative pools, fountains, hot tubs, spas, and hydromassage bathtubs, whether permanently installed or storable, as well as to metallic auxiliary equipment (such as pumps, filters, etc.).
- Parts specified in Section 680-22(a) must be connected to a permanently installed pool's common bonding grid with a No. 8 or larger solid conductor (insulated, covered, or bare).
- Very specific and detailed receptacle and lighting (including underwater) requirements are stipulated in Article 680.
- While Article 680, Part B covers permanently installed pools, storable pools requirements are found in Part C.
- Indoor spa and hot tub installations are covered in Section 680-41(a) through (i).
- Article 680, Part F provisions apply to pools and tubs for therapeutic use in health care facilities, gymnasiums, athletic training rooms, and similar areas.
- Hydromassage bathtubs (and associated electrical components) are covered in Article 680, Part G.

Unit 19 Competency Test

NEC® Reference	Answer	
_____	_____	1. Storable swimming pools are capable of holding water to a maximum depth of _____.
_____	_____	2. Overcurrent protection for feeders and branch circuits supplying electric vehicle supply equipment shall be sized for _____ duty and shall have a rating of not less than _____ % of the maximum load of the electric vehicle supply equipment.
_____	_____	3. A clearly labeled emergency shutoff or control switch for the purpose of stopping the motor(s) that provide power to the recirculation system and jet system shall be installed readily accessible to the users and at least _____ away, adjacent to, and within sight of the spa or hot tub.
_____	_____	4. Transformers used for the supply of underwater fixtures shall be an isolated winding type that has a _____ between the primary and secondary windings.
_____	_____	5. Duty on elevator driving machine motors shall be rated as _____.
_____	_____	6. Not more than _____ of high-voltage cable shall be permitted in nonmetallic conduit from a high-voltage terminal of a neon transformer (over 1,000 volts, nominal) supply to the first neon tube.

——————— ——————— 7. All fixed metal parts that are within _____ horizontally of the inside walls of a permanently installed swimming pool, and within _____ above the pool's maximum water level must be bonded together.

——————— ——————— 8. Electrically operated pipe organ circuits shall be arranged so that all conductors shall be protected from overcurrent by an overcurrent device rated at not more than _____.

——————— ——————— 9. All electric hot tub water heaters shall be listed and shall have the heating elements subdivided into loads not exceeding _____ amperes and protected at not more than _____ amperes.

——————— ——————— 10. All live parts of electrical apparatus in the hoistways, at the landings, in or on the cars of elevators and dumbwaiters, in the wellways or the landings of escalators or moving walks, or in the runways and machinery spaces of wheelchair lifts and stairway lifts shall be _____.

——————— ——————— 11. Unless specifically listed for the purpose and location, the coupling means of electric vehicle supply equipment shall be stored or located at a height of not less than _____ feet and not more than _____ inches above an outdoor parking surface.

——————— ——————— 12. A(n) _____ is a structure designed to support a wet-niche lighting fixture assembly and intended for mounting in a pool or fountain structure.

——————— ——————— 13. A disconnecting means shall be _____, located within sight from all pools, spas, and hot tub equipment, and shall be located at least 5 feet from the inside walls of the pool, spa, or hot tub.

——————— ——————— 14. Switches, flashers, and similar devices controlling transformers and electronic power supplies shall be rated for controlling inductive loads or have a current rating not less than _____ of the current rating of the transformer.

       a) 80%   b) 100%   c) 125%   d) 200%

——————— ——————— 15. Audio system equipment supplied by branch-circuit power shall not be placed laterally within _____ of the inside wall of a pool, spa, hot tub, or fountain, nor within _____ of the prevailing or tidal water mark.

——————— ——————— 16. Where a permanently installed pool is installed at a dwelling unit(s), at least one 125-volt 15- or 20-ampere receptacle on a general-purpose branch circuit shall be located a minimum of _____ from and not more than _____ from the inside wall of the pool.

——————— ——————— 17. The voltage at a fire pump controller's line terminals shall not drop more than _____ below normal (controller-rated voltage) under motor starting conditions.

——————— ——————— 18. Lighting fixtures located over a spa or within 5 feet from the inside walls of the spa shall be a minimum of _____ above the maximum water level and shall be protected by a ground-fault circuit interrupter.

——————— ——————— 19. Traveling cable shall be permitted to be run without the use of a raceway for a distance not exceeding _____ in length as measured from the first point of support on the elevator car or hoistway wall, or counterweight where applicable, provided the conductors are grouped together and taped or corded, or in the original sheath.

——————— ——————— 20. Underground wiring shall not be permitted under the pool or within the area extending _____ horizontally from the inside wall of the pool unless this wiring is necessary to supply pool equipment permitted by this article.

NEC® Reference	Answer	
_____	_____	21. Hydromassage bathtub electrical equipment shall be _____ without damaging the building structure or building finish.
_____	_____	22. All 125-volt receptacles located within _____ of the inside wall of the pool or fountain shall be protected by a ground-fault circuit-interrupter.
_____	_____	23. Manufactured wiring system cable shall be listed armored cable or metal-clad cable containing nominal 600-volt No. _____ conductors with a bare or insulated copper equipment grounding conductor equivalent in size to the ungrounded conductor.
_____	_____	24. Lighting fixtures mounted in permanently installed swimming pool walls shall be installed with the top of the fixture lens at least _____ below the normal water level of the pool.
_____	_____	25. Sign or outline lighting system equipment shall be at least _____ above areas accessible to vehicles unless protected from physical damage.
_____	_____	26. Existing lighting fixtures and lighting outlets located less than _____ measured horizontally from the inside walls of a pool shall be at least _____ above the surface of the maximum water level, shall be rigidly attached to the existing structure, and shall be protected by a ground-fault circuit-interrupter.
_____	_____	27. Any sign installed inside a fountain shall be at least _____ inside the fountain measured from the outside edges of the fountain.
_____	_____	28. The conductors to the hoistway door interlocks from the hoistway riser shall be flame retardant and suitable for a temperature of not less than _____.
_____	_____	29. In outdoor pool areas, ceiling suspended (paddle) fans shall not be installed over the pool or over the area extending _____ horizontally from the inside walls of the pool unless no part of the fan is less than _____ above the maximum water level.
_____	_____	30. The branch-circuit conductors supplying one or more units of a data processing system shall have an ampacity not less than _____ of the total connected load.
_____	_____	31. A junction boxes connected to a conduit that extends directly to a forming shell shall be provided with a number of grounding terminals that shall be _____.
_____	_____	32. Electric vehicle supply equipment shall be provided with a(n) _____ that de-energizes the electric vehicle connector and its cable whenever the electric connector is uncoupled from the electric vehicle.
_____	_____	33. Fixed or stationary equipment rated _____, other than an underwater lighting fixture for a permanently installed pool, shall be permitted to be connected with a flexible cord to facilitate the removal or disconnection for maintenance or repair.
_____	_____	34. Listed spa (or hot tub) packaged units utilizing a factory-installed remote panelboard shall be permitted to be connected with not more than _____ of liquidtight flexible conduit or be cord and plug connected with a cord not longer than _____ if protected by a ground-fault circuit-interrupter.
_____	_____	35. Branch circuits that supply neon tubing installations shall not be rated in excess of _____.

—————   —————   36. The termination of the No. 8 conductor in a permanently installed swimming pool forming shell shall be covered with, or encapsulated in, a(n) _____ to protect such connection from the possible deteriorating effect of pool water.

—————   —————   37. Conductors supplying a fire pump motor(s), pressure maintenance pumps, and associated fire pump accessory equipment shall have a minimum rating of _____ of the sum of the fire pump motor(s) and pressure maintenance motor(s) full-load current(s), and _____ of the associated fire pump accessory equipment.

—————   —————   38. Switching devices on the property shall be located at least _____ horizontally from the inside walls of a pool unless separated from the pool by a solid fence, wall, or other permanent barrier.

—————   —————   39. Article _____ covers electrical equipment, lighting accessories, and wiring systems used to connect, or contained within, or installed on relocatable wired partitions.

—————   —————   40. A junction box connected to a conduit that extends directly to a forming shell shall be located not less than _____, measured from the inside of the bottom of the box, above the maximum pool water level, or not less than _____ above the ground level, or pool deck.

Index